Theoretical Neuroscience

Computational Neuroscience

Terrence J. Sejnowski and Tomaso Poggio, editors

Theoretical Neuroscience

Computational and Mathematical Modeling of Neural Systems

Peter Dayan and L.F. Abbott

The MIT Press
Cambridge, Massachusetts
London, England

Typeset in Palatino by the authors using LATEX 2_ε.
Printed and bound in the United States of America.

Library of Congress Cataloging-in-Publication Data

Dayan, Peter.
Theoretical neuroscience : computational and mathematical modeling of neural systems / Peter Dayan and L.F. Abbott.
    p. cm. — (Computational neuroscience)
Includes bibliographical references.
ISBN-13: 978-0-262-04199-7 (hc. : alk. paper)—978-0-262-54185-5 (pbk. : alk. paper)
ISBN-10: 0-262-04199-5 (hc. : alk. paper)—0-262-54185-8 (pbk. : alk. paper)
1. Neural networks (Neurobiology) – Computer simulation. 2. Human information processing – Computer simulation.   3. Computational neuroscience. I. Abbott, L.F.   II. Title.   III. Series

QP363.3 .D39   2001
573.8'01'13—dc21

2001044005

10   9

To our families

Contents

Exercises   `http://mitpress.mit.edu/dayan-abbott`

Series Foreword

Computational neuroscience is an approach to understanding the information content of neural signals by modeling the nervous system at many different structural scales, including the biophysical, the circuit, and the systems levels. Computer simulations of neurons and neural networks are complementary to traditional techniques in neuroscience. This book series welcomes contributions that link theoretical studies with experimental approaches to understanding information processing in the nervous system. Areas and topics of particular interest include biophysical mechanisms for computation in neurons, computer simulations of neural circuits, models of learning, representation of sensory information in neural networks, systems models of sensory-motor integration, and computational analysis of problems in biological sensing, motor control, and perception.

Terrence J. Sejnowski
Tomaso Poggio

Preface

Theoretical analysis and computational modeling are important tools for characterizing what nervous systems do, determining how they function, and understanding why they operate in particular ways. Neuroscience encompasses approaches ranging from molecular and cellular studies to human psychophysics and psychology. Theoretical neuroscience encourages crosstalk among these subdisciplines by constructing compact representations of what has been learned, building bridges between different levels of description, and identifying unifying concepts and principles. In this book, we present the basic methods used for these purposes and discuss examples in which theoretical approaches have yielded insight into nervous system function.

The questions what, how, and why are addressed by descriptive, mechanistic, and interpretive models, each of which we discuss in the following chapters. Descriptive models summarize large amounts of experimental *descriptive models* data compactly yet accurately, thereby characterizing what neurons and neural circuits do. These models may be based loosely on biophysical, anatomical, and physiological findings, but their primary purpose is to describe phenomena, not to explain them. Mechanistic models, on the *mechanistic models* other hand, address the question of how nervous systems operate on the basis of known anatomy, physiology, and circuitry. Such models often form a bridge between descriptive models couched at different levels. Interpretive models use computational and information-theoretic principles *interpretive models* to explore the behavioral and cognitive significance of various aspects of nervous system function, addressing the question of why nervous systems operate as they do.

It is often difficult to identify the appropriate level of modeling for a particular problem. A frequent mistake is to assume that a more detailed model is necessarily superior. Because models act as bridges between levels of understanding, they must be detailed enough to make contact with the lower level yet simple enough to provide clear results at the higher level.

Organization and Approach

This book is organized into three parts on the basis of general themes. Part I, Neural Encoding and Decoding, (chapters 1–4) is devoted to the coding of information by action potentials and the representation of information by populations of neurons with selective responses. Modeling of neurons and neural circuits on the basis of cellular and synaptic biophysics is presented in part II, Neurons and Neural Circuits (chapters 5–7). The role of plasticity in development and learning is discussed in part III, Adaptation and Learning (chapters 8-10). With the exception of chapters 5 and 6, which jointly cover neuronal modeling, the chapters are largely independent and can be selected and ordered in a variety of ways for a one- or two-semester course at either the undergraduate or the graduate level.

background   Although we provide some background material, readers without previous exposure to neuroscience should refer to a neuroscience textbook such as Kandel, Schwartz, & Jessell (2000); Nicholls, Martin, & Wallace (1992); Bear, Connors, & Paradiso (1996); Shepherd (1997); Zigmond et al. (1998); or Purves et al. (2000).

Theoretical neuroscience is based on the belief that methods of mathematics, physics, and computer science can provide important insights into nervous system function. Unfortunately, mathematics can sometimes seem more of an obstacle than an aid to understanding. We have not hesitated to employ the level of analysis needed to be precise and rigorous. At times, this may stretch the tolerance of some of our readers. We encourage such readers to consult the Mathematical Appendix, which provides a brief review of most of the mathematical methods used in the text, but also to persevere and attempt to understand the implications and consequences of a difficult derivation even if its steps are unclear.

exercises   Theoretical neuroscience, like any skill, can be mastered only with practice. Exercises are provided for this purpose on the web site for this book, http://mitpress.mit.edu/dayan-abbott. We urge the reader to do them. In addition, it will be highly instructive for the reader to construct the models discussed in the text and explore their properties beyond what we have been able to do in the available space.

Referencing

In order to maintain the flow of the text, we have kept citations within the chapters to a minimum. Each chapter ends with an annotated bibliography containing suggestions for further reading (which are denoted by a bold font), information about works cited within the chapter, and references to related studies. We concentrate on introducing the basic tools of computational neuroscience and discussing applications that we think best help the reader to understand and appreciate them. This means that a number of systems where computational approaches have been applied with significant success are not discussed. References given in the annotated bibliographies lead the reader toward such applications. Many people have contributed significantly to the areas we cover. The books and review articles in the annotated bibliographies provide more comprehensive references to work that we have failed to cite.

Acknowledgments

We are extremely grateful to a large number of students at Brandeis, the Gatsby Computational Neuroscience Unit, and MIT, and colleagues at many institutions who have painstakingly read, commented on, and criticized numerous versions of all the chapters. We particularly thank Bard Ermentrout, Mark Goldman, John Hertz, Mark Kvale, Zhaoping Li, Eve Marder, and Read Montague for providing extensive discussion and advice on the entire book. A number of people read significant portions of the text and provided valuable comments, criticism, and insight: Bill Bialek, Pat Churchland, Nathaniel Daw, Dawei Dong, Peter Földiák, Fabrizio Gabbiani, Zoubin Ghahramani, Geoff Goodhill, David Heeger, Geoff Hinton, Ken Miller, Phil Nelson, Sacha Nelson, Bruno Olshausen, Mark Plumbley, Alex Pouget, Fred Rieke, John Rinzel, Emilio Salinas, Sebastian Seung, Mike Shadlen, Satinder Singh, Rich Sutton, Nick Swindale, Carl van Vreeswijk, Chris Williams, David Willshaw, Charlie Wilson, Angela Yu, and Rich Zemel.

We received significant additional assistance and advice from Greg DeAngelis, Andy Barto, Matt Beal, Sue Becker, Tony Bell, Paul Bressloff, Emery Brown, Matteo Carandini, Frances Chance, Yang Dan, Kenji Doya, Ed Erwin, John Fitzpatrick, David Foster, Marcus Frean, Ralph Freeman, Enrique Garibay, Frederico Girosi, Charlie Gross, Andreas Herz, Mike Jordan, Sham Kakade, Szabolcs Káli, Christof Koch, Simon Laughlin, John Lisman, Shawn Lockery, Guy Mayraz, Josh McDermott, Markus Meister, Earl Miller, Quaid Morris, Tony Movshon, Yuko Munakata, Randy O'Reilly, Simon Osindero, Tomaso Poggio, Clay Reid, Max Riesenhuber, Dario Ringach, Horacio Rotstein, Sam Roweis, Lana Rutherford, Ken Sugino, Alexei Samsonovich, Bob Shapley, Wolfram Schultz, Idan Segev, Terry Sejnowski, Jesper Sjöström, Haim Sompolinsky, Fiona Stevens, David Tank, Emo Todorov, Alessandro Treves, Gina Turrigiano, David Van Essen, Martin Wainwright, Xiao-Jing Wang, Chris Watkins, Max Welling, Jenny Whiting, Matt Wilson, Laurenz Wiskott, Danny Young, and Kechen Zhang. Thanks also to Quentin Huys, Philip Jonkers, Alexander Lerchner, Shih-Chii Liu, Máté Lengyel, Alex Loebel, Hadi Murr, Iain Murray, Jihwan Myung, John van Opstal, David Simon, Ed Snelson, and Rafael Yuste for pointing out errors in the text.

We thank Maneesh Sahani for advice and for indexing a substantial part of the text, Heidi Cartwright for creating the cover art, and Michael Rutter for his patience and consistent commitment. P.D. acknowledges the support of the Gatsby Charitable Foundation. Karen Abbott provided valuable help with the figures and with proofreading. Finally, we apologize to anyone we have inadvertently omitted from these lists.

I Neural Encoding and Decoding

1 Neural Encoding I: Firing Rates and Spike Statistics

1.1 Introduction

Neurons are remarkable among the cells of the body in their ability to propagate signals rapidly over large distances. They do this by generating characteristic electrical pulses called action potentials or, more simply, spikes that can travel down nerve fibers. Neurons represent and transmit information by firing sequences of spikes in various temporal patterns. The study of neural coding, which is the subject of the first four chapters of this book, involves measuring and characterizing how stimulus attributes, such as light or sound intensity, or motor actions, such as the direction of an arm movement, are represented by action potentials.

The link between stimulus and response can be studied from two opposite points of view. Neural encoding, the subject of chapters 1 and 2, refers to the map from stimulus to response. For example, we can catalog how neurons respond to a wide variety of stimuli, and then construct models that attempt to predict responses to other stimuli. Neural decoding refers to the reverse map, from response to stimulus, and the challenge is to reconstruct a stimulus, or certain aspects of that stimulus, from the spike sequences it evokes. Neural decoding is discussed in chapter 3. In chapter 4, we consider how the amount of information encoded by sequences of action potentials can be quantified and maximized. Before embarking on this tour of neural coding, we briefly review how neurons generate their responses and discuss how neural activity is recorded. The biophysical mechanisms underlying neural responses and action potential generation are treated in greater detail in chapters 5 and 6.

Properties of Neurons

Neurons are highly specialized for generating electrical signals in response to chemical and other inputs, and transmitting them to other cells. Some important morphological specializations, seen in figure 1.1, are the dendrites that receive inputs from other neurons and the axon that carries the neuronal output to other cells. The elaborate branching structure of

axons and dendrites

the dendritic tree allows a neuron to receive inputs from many other neurons through synaptic connections. The cortical pyramidal neuron of figure 1.1A and the cortical interneuron of figure 1.1C each receive thousands of synaptic inputs, and for the cerebellar Purkinje cell of figure 1.1B the number is over 100,000. Figure 1.1 does not show the full extent of the axons of these neurons. Axons from single neurons can traverse large fractions of the brain or, in some cases, of the entire body. In the mouse brain, it has been estimated that cortical neurons typically send out a total of about 40 mm of axon and have approximately 4 mm of total dendritic cable in their branched dendritic trees. The axon makes an average of 180 synaptic connections with other neurons per mm of length and the dendritic tree receives, on average, 2 synaptic inputs per μm. The cell body or soma of a typical cortical neuron ranges in diameter from about 10 to 50 μm.

ion channels

Along with these morphological features, neurons have physiological specializations. Most prominent among these are a wide variety of membrane-spanning ion channels that allow ions, predominantly sodium (Na^+), potassium (K^+), calcium (Ca^{2+}), and chloride (Cl^-), to move into and out of the cell. Ion channels control the flow of ions across the cell membrane by opening and closing in response to voltage changes and to both internal and external signals.

membrane potential

The electrical signal of relevance to the nervous system is the difference in electrical potential between the interior of a neuron and the surrounding extracellular medium. Under resting conditions, the potential inside the cell membrane of a neuron is about -70 mV relative to that of the surrounding bath (which is conventionally defined to be 0 mV), and the cell is said to be polarized. Ion pumps located in the cell membrane maintain concentration gradients that support this membrane potential difference. For example, Na^+ is much more concentrated outside a neuron than inside it, and the concentration of K^+ is significantly higher inside the neuron than in the extracellular medium. Ions thus flow into and out of a cell due to both voltage and concentration gradients. Current in the form of positively charged ions flowing out of the cell (or negatively charged ions flowing into the cell) through open channels makes the membrane potential more negative, a process called hyperpolarization. Current flowing into the cell changes the membrane potential to less negative or even positive values. This is called depolarization.

hyperpolarization and depolarization

action potential

If a neuron is depolarized sufficiently to raise the membrane potential above a threshold level, a positive feedback process is initiated, and the neuron generates an action potential. An action potential is a roughly 100 mV fluctuation in the electrical potential across the cell membrane that lasts for about 1 ms (figure 1.2A). Action potential generation also depends on the recent history of cell firing. For a few milliseconds just after an action potential has been fired, it may be virtually impossible to initiate another spike. This is called the absolute refractory period. For a longer interval known as the relative refractory period, lasting up to tens of milliseconds after a spike, it is more difficult to evoke an action potential.

refractory period

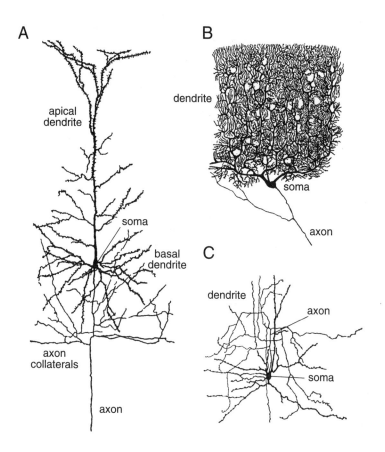

Figure 1.1 Diagrams of three neurons. (A) A cortical pyramidal cell. These are the primary excitatory neurons of the cerebral cortex. Pyramidal cell axons branch locally, sending axon collaterals to synapse with nearby neurons, and also project more distally to conduct signals to other parts of the brain and nervous system. (B) A Purkinje cell of the cerebellum. Purkinje cell axons transmit the output of the cerebellar cortex. (C) A stellate cell of the cerebral cortex. Stellate cells are one of a large class of interneurons that provide inhibitory input to the neurons of the cerebral cortex. These figures are magnified about 150-fold. (Drawings from Cajal, 1911; figure from Dowling, 1992.)

Action potentials are of great importance because they are the only form of membrane potential fluctuation that can propagate over large distances. Subthreshold potential fluctuations are severely attenuated over distances of 1 mm or less. Action potentials, on the other hand, are regenerated actively along axon processes and can travel rapidly over large distances without attenuation.

Axons terminate at synapses where the voltage transient of the action potential opens ion channels, producing an influx of Ca^{2+} that leads to the release of a neurotransmitter (figure 1.2B). The neurotransmitter binds to receptors at the signal-receiving or postsynaptic side of the synapse,

synapse

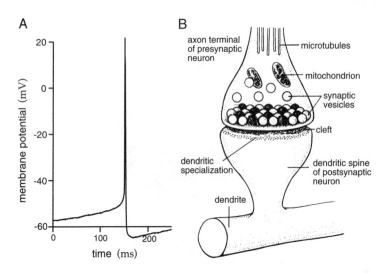

Figure 1.2 (A) An action potential recorded intracellularly from a cultured rat neocortical pyramidal cell. (B) Diagram of a synapse. The axon terminal or bouton is at the end of the axonal branch seen entering from the top of the figure. It is filled with synaptic vesicles containing the neurotransmitter that is released when an action potential arrives from the presynaptic neuron. Transmitter crosses the synaptic cleft and binds to receptors on the dendritic spine, a process roughly 1 μm long that extends from the dendrite of the postsynaptic neuron. Excitatory synapses onto cortical pyramidal cells form on dendritic spines as shown here. Other synapses form directly on the dendrites, axon, or soma of the postsynaptic neuron. (A recorded by L. Rutherford in the laboratory of G. Turrigiano; B adapted from Kandel et al., 1991.)

causing ion-conducting channels to open. Depending on the nature of the ion flow, the synapses can have either an excitatory, depolarizing, or an inhibitory, typically hyperpolarizing, effect on the postsynaptic neuron.

Recording Neuronal Responses

Figure 1.3 illustrates intracellular and extracellular methods for recording neuronal responses electrically (they can also be recorded optically). Membrane potentials are measured intracellularly by connecting a hollow glass electrode filled with a conducting electrolyte to a neuron, and comparing the potential it records with that of a reference electrode placed in the extracellular medium. Intracellular recordings are made either with *sharp and patch* sharp electrodes inserted through the membrane into the cell, or patch *electrodes* electrodes that have broader tips and are sealed tightly to the surface of the membrane. After the patch electrode seals, the membrane beneath its tip is either broken or perforated, providing electrical contact with the interior of the cell. The top trace in figure 1.3 is a schematic of an intracellular recording from the soma of a neuron firing a sequence of action potentials. The recording shows rapid spikes riding on top of a more slowly varying subthreshold potential. The bottom trace is a schematic of an intracellular recording made some distance out on the axon of the neuron. These traces

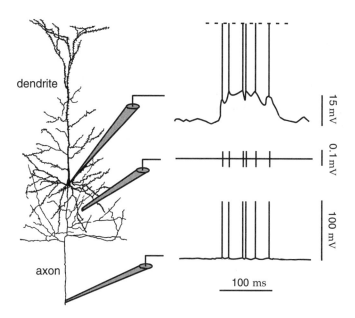

Figure 1.3 Three simulated recordings from a neuron. The top trace represents a recording from an intracellular electrode connected to the soma of the neuron. The height of the action potentials has been clipped to show the subthreshold membrane potential more clearly. The time scale is such that the action potential trajectory cannot be resolved. The bottom trace represents a recording from an intracellular electrode connected to the axon some distance away from the soma. The full height of the action potentials is indicated in this trace. The middle trace is a simulated extracellular recording. Action potentials appear as roughly equal positive and negative potential fluctuations with an amplitude of around 0.1 mV. This is roughly 1000 times smaller than the approximately 0.1 V amplitude of an intracellularly recorded action potential. (Neuron drawing is the same as figure 1.1A.)

are drawings, not real recordings; such intracellular axon recordings, although possible in some types of cells, are difficult and rare. Intracellular recordings from the soma are the norm, but intracellular dendritic recordings are increasingly being made as well. The subthreshold membrane potential waveform, apparent in the soma recording, is completely absent on the axon due to attenuation, but the action potential sequence in the two recordings is the same. This illustrates the important point that spikes, but not subthreshold potentials, propagate regeneratively down axons.

The middle trace in figure 1.3 illustrates an idealized, noise-free extracellular recording. Here an electrode is placed near a neuron but it does not penetrate the cell membrane. Such recordings can reveal the action potentials fired by a neuron, but not its subthreshold membrane potentials. Extracellular recordings are typically used for in vivo experiments, especially those involving behaving animals. Intracellular recordings are sometimes made in vivo, but this is difficult to do. Intracellular recording is more commonly used for in vitro preparations, such as slices of neural tissue. The responses studied in this chapter are action potential sequences that can be recorded either intra- or extracellularly.

extracellular electrodes

From Stimulus to Response

Characterizing the relationship between stimulus and response is difficult because neuronal responses are complex and variable. Neurons typically respond by producing complex spike sequences that reflect both the intrinsic dynamics of the neuron and the temporal characteristics of the stimulus. Isolating features of the response that encode changes in the stimulus can be difficult, especially if the time scale for these changes is of the same order as the average interval between spikes. Neural responses can vary from trial to trial even when the same stimulus is presented repeatedly. There are many potential sources of this variability, including variable levels of arousal and attention, randomness associated with various biophysical processes that affect neuronal firing, and the effects of other cognitive processes taking place during a trial. The complexity and trial-to-trial variability of action potential sequences make it unlikely that we can describe and predict the timing of each spike deterministically. Instead, we seek a model that can account for the probabilities that different spike sequences are evoked by a specific stimulus.

Typically, many neurons respond to a given stimulus, and stimulus features are therefore encoded by the activities of large neural populations. In studying population coding, we must examine not only the firing patterns of individual neurons but also the relationships of these firing patterns to each other across the population of responding cells.

In this chapter, we introduce the firing rate and spike-train correlation functions, which are basic measures of spiking probability and statistics. We also discuss spike-triggered averaging, a method for relating action potentials to the stimulus that evoked them. Finally, we present basic stochastic descriptions of spike generation, the homogeneous and inhomogeneous Poisson models, and discuss a simple model of neural responses to which they lead. In chapter 2, we continue our discussion of neural encoding by showing how reverse-correlation methods are used to construct estimates of firing rates in response to time-varying stimuli. These methods have been applied extensively to neural responses in the retina, lateral geniculate nucleus (LGN) of the thalamus, and primary visual cortex, and we review the resulting models.

1.2   Spike Trains and Firing Rates

Action potentials convey information through their timing. Although action potentials can vary somewhat in duration, amplitude, and shape, they are typically treated as identical stereotyped events in neural encoding studies. If we ignore the brief duration of an action potential (about 1 ms), an action potential sequence can be characterized simply by a list of the times when spikes occurred. For n spikes, we denote these times by t_i with $i = 1, 2, \ldots, n$. The trial during which the spikes are recorded is taken to start at time 0 and end at time T, so $0 \leq t_i \leq T$ for all i. The spike sequence can also be represented as a sum of infinitesimally narrow, idealized spikes in the form of Dirac δ functions (see the Mathematical Appendix),

$$\rho(t) = \sum_{i=1}^{n} \delta(t - t_i) \,. \tag{1.1}$$

We call $\rho(t)$ the neural response function and use it to re-express sums over spikes as integrals over time. For example, for any well-behaved function $h(t)$, we can write

neural response function $\rho(t)$

$$\sum_{i=1}^{n} h(t - t_i) = \int_{-\infty}^{\infty} d\tau \, h(\tau)\rho(t - \tau) \,, \tag{1.2}$$

where the integral is over the duration of the trial. The equality follows from the basic defining equation for a δ function,

δ function

$$\int d\tau \, \delta(t - \tau)h(\tau) = h(t) \,, \tag{1.3}$$

provided that the limits of the integral surround the point t (if they do not, the integral is 0).

Because the sequence of action potentials generated by a given stimulus varies from trial to trial, neuronal responses are typically treated statistically or probabilistically. For example, they may be characterized by firing rates, rather than as specific spike sequences. Unfortunately, the term "firing rate" is applied conventionally to a number of different quantities. The simplest of these is what we call the spike-count rate, which is obtained by counting the number of action potentials that appear during a trial and dividing by the duration of the trial. We denote the spike-count rate by r, where

spike-count rate r

$$r = \frac{n}{T} = \frac{1}{T} \int_0^T d\tau \, \rho(\tau) \,. \tag{1.4}$$

The second equality follows from the fact that $\int d\tau \, \rho(\tau) = n$ and indicates that the spike-count rate is the time average of the neural response function over the duration of the trial.

The spike-count rate can be determined from a single trial, but at the expense of losing all temporal resolution about variations in the neural response during the course of the trial. A time-dependent firing rate can be defined by counting spikes over short time intervals, but this can no longer be computed from a single trial. For example, we can define the firing rate at time t during a trial by counting all the spikes that occurred between times t and $t + \Delta t$, for some small interval Δt, and dividing this count by Δt. However, for small Δt, which allows for high temporal resolution, the result of the spike count on any given trial is apt to be either 0 or 1, giving only two possible firing-rate values. The solution to this problem is to average over multiple trials. Thus, we define the time-dependent firing rate as the average number of spikes (averaged over trials) appearing during a short interval between times t and $t + \Delta t$, divided by the duration of the interval.

trial average ⟨ ⟩

The number of spikes occurring between times t and $t + \Delta t$ on a single trial is the integral of the neural response function over that time interval. The average number of spikes during this interval is the integral of the trial-averaged neural response function. We use angle brackets, ⟨ ⟩, to denote averages over trials that use the same stimulus, so that ⟨z⟩ for any quantity z is the sum of the values of z obtained from many trials involving the same stimulus, divided by the number of trials. The trial-averaged neural response function is denoted by ⟨$\rho(t)$⟩, and the time-dependent firing rate

firing rate r(t)

is given by

$$r(t) = \frac{1}{\Delta t} \int_t^{t+\Delta t} d\tau \, \langle \rho(\tau) \rangle \,. \tag{1.5}$$

We use the notation $r(t)$ for this important quantity (as opposed to r for the spike-count rate), and when we use the term "firing rate" without any modifiers, we mean $r(t)$. Formally, the limit $\Delta t \to 0$ should be taken on the right side of this expression, but, in extracting a time-dependent firing rate from data, the value of Δt must be large enough so there are sufficient numbers of spikes within the interval defining $r(t)$ to obtain a reliable estimate of the average.

For sufficiently small Δt, $r(t)\Delta t$ is the average number of spikes occurring between times t and $t + \Delta t$ over multiple trials. The average number of spikes over a longer time interval is given by the integral of $r(t)$ over that interval. If Δt is small, there will never be more than one spike within the interval between t and $t + \Delta t$ on any given trial. This means that $r(t)\Delta t$ is also the fraction of trials on which a spike occurred between those times. Equivalently, $r(t)\Delta t$ is the probability that a spike occurs during this time

spiking probability

interval. This probabilistic interpretation provides a formal definition of the time-dependent firing rate; $r(t)\Delta t$ is the probability of a spike occurring during a short interval of duration Δt around the time t.

In any integral expression such as equation 1.2, the neural response function generates a contribution whenever a spike occurs. If we use the trial-average response function instead, as in equation 1.5, this generates contributions proportional to the fraction of trials on which a spike occurred. Because of the relationship between this fraction and the firing rate, we can replace the trial-averaged neural response function with the firing rate $r(t)$ within any well-behaved integral, for example,

$$\int d\tau \, h(\tau) \, \langle \rho(t - \tau) \rangle = \int d\tau \, h(\tau) r(t - \tau) \tag{1.6}$$

for any function h. This establishes an important relationship between the average neural response function and the firing rate; the two are equivalent when used inside integrals. It also provides another interpretation of $r(t)$ as the trial-averaged density of spikes along the time axis.

In the same way that the response function $\rho(t)$ can be averaged across trials to give the firing rate r(t), the spike-count firing rate can be averaged over trials, yielding a quantity that we refer to as the average firing rate. This is denoted by $\langle r \rangle$ and is given by

$$\langle r \rangle = \frac{\langle n \rangle}{T} = \frac{1}{T} \int_0^T d\tau \, \langle \rho(\tau) \rangle = \frac{1}{T} \int_0^T dt \, \text{r}(t) \,. \qquad (1.7)$$

The first equality indicates that $\langle r \rangle$ is just the average number of spikes per trial divided by the trial duration. The third equality follows from the equivalence of the firing rate and the trial-averaged neural response function within integrals (equation 1.6). The average firing rate is equal to both the time average of r(t) and the trial average of the spike-count rate r. Of course, a spike-count rate and average firing rate can be defined by counting spikes over any time period, not necessarily the entire duration of a trial.

The term "firing rate" is commonly used for all three quantities, r(t), r, and $\langle r \rangle$. Whenever possible, we use the terms "firing rate", "spike-count rate", and "average firing rate" for r(t), r, and $\langle r \rangle$, respectively, but when this becomes too cumbersome, the different mathematical notations serve to distinguish them. In particular, we distinguish the spike-count rate r from the time-dependent firing rate r(t) by using a different font and by including the time argument in the latter expression (unless r(t) is independent of time). The difference between the fonts is rather subtle, but the context should make it clear which rate is being used.

Measuring Firing Rates

The firing rate r(t) cannot be determined exactly from the limited data available from a finite number of trials. In addition, there is no unique way to approximate r(t). A discussion of the different methods allows us to introduce the concept of a linear filter and kernel that will be used extensively in the following chapters. We illustrate these methods by extracting firing rates from a single trial, but more accurate results could be obtained by averaging over multiple trials.

Figure 1.4 compares a number of ways of approximating r(t) from a spike sequence. Figure 1.4A shows 3 s of the response of a neuron in the inferotemporal cortex recorded while a monkey watched a video. Neurons in the region of cortex where this recording was made are selective for complex visual images, including faces. A simple way of extracting an estimate of the firing rate from a spike train like this is to divide time into discrete bins of duration Δt, count the number of spikes within each bin, and divide by Δt. Figure 1.4B shows the approximate firing rate computed using this procedure with a bin size of 100 ms. Note that with this procedure, the quantity being computed is really the spike-count firing rate over the duration of the bin, and that the firing rate r(t) within a given bin is approximated by this spike-count rate.

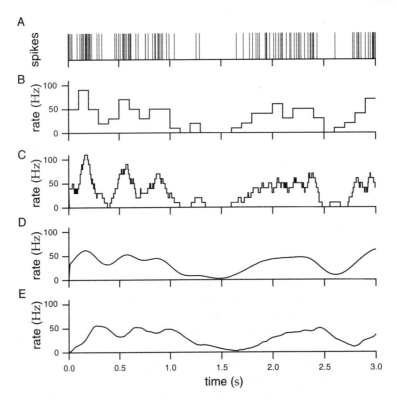

Figure 1.4 Firing rates approximated by different procedures. (A) A spike train from a neuron in the inferotemporal cortex of a monkey recorded while that animal watched a video on a monitor under free viewing conditions. (B) Discrete-time firing rate obtained by binning time and counting spikes with $\Delta t = 100$ ms. (C) Approximate firing rate determined by sliding a rectangular window function along the spike train with $\Delta t = 100$ ms. (D) Approximate firing rate computed using a Gaussian window function with $\sigma_t = 100$ ms. (E) Approximate firing rate using the window function of equation 1.12 with $1/\alpha = 100$ ms. (Data from Baddeley et al., 1997.)

The binning and counting procedure illustrated in figure 1.4B generates an estimate of the firing rate that is a piecewise constant function of time, resembling a histogram. Because spike counts can take only integer values, the rates computed by this method will always be integer multiples of $1/\Delta t$, and thus they take discrete values. Decreasing the value of Δt increases temporal resolution by providing an estimate of the firing rate at more finely spaced intervals of time, but at the expense of decreasing the resolution for distinguishing different rates. One way to avoid quantized firing rates is to vary the bin size so that a fixed number of spikes appears in each bin. The firing rate is then approximated as that fixed number of spikes divided by the variable bin width.

Counting spikes in preassigned bins produces a firing-rate estimate that depends not only on the size of the time bins but also on their placement. To avoid the arbitrariness in the placement of bins, we can instead take a single bin or window of duration Δt and slide it along the spike train, counting the number of spikes within the window at each location. The jagged curve in figure 1.4C shows the result of sliding a 100 ms wide window along the spike train. The firing rate approximated in this way can be expressed as the sum of a window function over the times t_i for $i = 1, 2, \ldots, n$ when the n spikes in a particular sequence occurred,

$$r_{\text{approx}}(t) = \sum_{i=1}^{n} w(t - t_i),$$
(1.8)

where the window function is

$$w(t) = \begin{cases} 1/\Delta t & \text{if } -\Delta t/2 \leq t < \Delta t/2 \\ 0 & \text{otherwise}. \end{cases}$$
(1.9)

Use of a sliding window avoids the arbitrariness of bin placement and produces a rate that might appear to have a better temporal resolution. However, it must be remembered that the rates obtained at times separated by less than one bin width are correlated because they involve some of the same spikes.

The sum in equation 1.8 can also be written as the integral of the window function times the neural response function (see equation 1.2):

$$r_{\text{approx}}(t) = \int_{-\infty}^{\infty} d\tau \, w(\tau) \rho(t - \tau).$$
(1.10)

The integral in equation 1.10 is called a linear filter, and the window function w, also called the filter kernel, specifies how the neural response function evaluated at time $t - \tau$ contributes to the firing rate approximated at time t.

linear filter and kernel

The jagged appearance of the curve in figure 1.4C is caused by the discontinuous shape of the window function used. An approximate firing rate can be computed using virtually any window function $w(\tau)$ that goes to 0 outside a region near $\tau = 0$, provided that its time integral is equal to 1. For example, instead of the rectangular window function used in figure 1.4C, $w(\tau)$ can be a Gaussian:

$$w(\tau) = \frac{1}{\sqrt{2\pi}\sigma_w} \exp\left(-\frac{\tau^2}{2\sigma_w^2}\right).$$
(1.11)

In this case, σ_w controls the temporal resolution of the resulting rate, playing a role analogous to Δt. A continuous window function like the Gaussian used in equation 1.8 generates a firing-rate estimate that is a smooth function of time (figure 1.4D).

Both the rectangular and the Gaussian window functions approximate the firing rate at any time, using spikes fired both before and after that time. A postsynaptic neuron monitoring the spike train of a presynaptic cell has access only to spikes that have previously occurred. An approximation of the firing rate at time t that depends only on spikes fired before t can be calculated using a window function that vanishes when its argument is negative. Such a window function or kernel is called causal. One commonly used form is the α function

$$w(\tau) = [\alpha^2 \tau \exp(-\alpha\tau)]_+ \qquad (1.12)$$

half-wave rectification [ ]$_+$

where $1/\alpha$ determines the temporal resolution of the resulting firing-rate estimate. The notation $[z]_+$ for any quantity z stands for the half-wave rectification operation,

$$[z]_+ = \begin{cases} z & \text{if } z \geq 0 \\ 0 & \text{otherwise}. \end{cases} \qquad (1.13)$$

Figure 1.4E shows the firing rate approximated by such a causal scheme. Note that this rate tends to peak later than the rate computed in figure 1.4D using a temporally symmetric window function.

Tuning Curves

stimulus s

Neuronal responses typically depend on many different properties of a stimulus. In this chapter, we characterize responses of neurons as functions of just one of the stimulus attributes to which they may be sensitive. The value of this single attribute is denoted by s. In chapter 2, we consider more complete stimulus characterizations.

response tuning curve $f(s)$

A simple way of characterizing the response of a neuron is to count the number of action potentials fired during the presentation of a stimulus. This approach is most appropriate if the parameter s characterizing the stimulus is held constant over the trial. If we average the number of action potentials fired over (in theory, an infinite number of) trials and divide by the trial duration, we obtain the average firing rate, $\langle r \rangle$, defined in equation 1.7. The average firing rate written as a function of s, $\langle r \rangle = f(s)$, is called the neural response tuning curve. The functional form of a tuning curve depends on the parameter s used to describe the stimulus. The precise choice of parameters used as arguments of tuning curve functions is partially a matter of convention. Because tuning curves correspond to firing rates, they are measured in units of spikes per second or Hz.

primary visual cortex V1

Figure 1.5A shows extracellular recordings of a neuron in the primary visual cortex (V1) of a monkey. While these recordings were being made, a bar of light was moved at different angles across the region of the visual field where the cell responded to light. This region is called the receptive field of the neuron. Note that the number of action potentials fired depends on the angle of orientation of the bar. The same effect is shown in figure 1.5B in the form of a response tuning curve, which indicates how the average firing rate depends on the orientation of the light bar stimulus.

Gaussian tuning curve

The data have been fitted by a response tuning curve of the form

$$f(s) = r_{max} \exp\left(-\frac{1}{2}\left(\frac{s - s_{max}}{\sigma_f}\right)^2\right), \qquad (1.14)$$

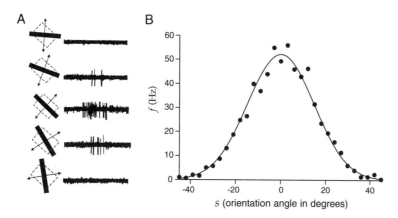

Figure 1.5 (A) Recordings from a neuron in the primary visual cortex of a monkey. A bar of light was moved across the receptive field of the cell at different angles. The diagrams to the left of each trace show the receptive field as a dashed square and the light source as a black bar. The bidirectional motion of the light bar is indicated by the arrows. The angle of the bar indicates the orientation of the light bar for the corresponding trace. (B) Average firing rate of a cat V1 neuron plotted as a function of the orientation angle of the light bar stimulus. The curve is a fit using the function 1.14 with parameters $r_{max} = 52.14$ Hz, $s_{max} = 0°$, and $\sigma_f = 14.73°$. (A adapted from Wandell, 1995, based on an original figure from Hubel and Wiesel, 1968; B data points from Henry et al., 1974).)

where s is the orientation angle of the light bar, s_{max} is the orientation angle evoking the maximum average response rate r_{max} (with $s - s_{max}$ taken to lie in the range between $-90°$ and $+90°$), and σ_f determines the width of the tuning curve. The neuron responds most vigorously when a stimulus having $s = s_{max}$ is presented, so we call s_{max} the preferred orientation angle of the neuron.

Response tuning curves can be used to characterize the selectivities of neurons in visual and other sensory areas to a variety of stimulus parameters. Tuning curves can also be measured for neurons in motor areas, in which case the average firing rate is expressed as a function of one or more parameters describing a motor action. Figure 1.6A shows an example of extracellular recordings from a neuron in primary motor cortex in a monkey that has been trained to reach in different directions. The stacked traces for each direction are rasters showing the results of five different trials. The horizontal axis in these traces represents time, and each mark indicates an action potential. The firing pattern of the cell, in particular the rate at which spikes are generated, is correlated with the direction of arm movement, and thus encodes information about this aspect of the motor action.

primary motor cortex M1

Figure 1.6B shows the response tuning curve of an M1 neuron plotted as a function of the direction of arm movement. Here the data points have been fitted by a tuning curve of the form

cosine tuning curve

$$f(s) = r_0 + (r_{max} - r_0)\cos(s - s_{max}),\qquad(1.15)$$

where s is the reaching angle of the arm, s_{max} is the reaching angle associ-

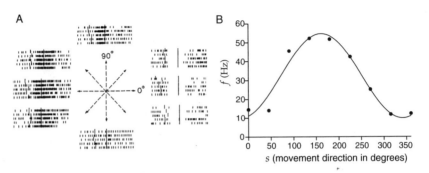

Figure 1.6 (A) Recordings from the primary motor cortex of a monkey performing an arm-reaching task. The hand of the monkey started from a central resting
location, and reaching movements were made in the directions indicated by the
arrows. The rasters for each direction show action potentials fired on five trials.
(B) Average firing rate plotted as a function of the direction in which the monkey moved its arm. The curve is a fit using the function 1.15 with parameters
$r_{max} = 54.69$ Hz, $r_0 = 32.34$ Hz, and $s_{max} = 161.25°$. (A adapted from Georgopoulos et al., 1982, which is also the source of the data points in B.)

ated with the maximum response r_{max}, and r_0 is an offset or background
firing rate that shifts the tuning curve up from the zero axis. The minimum
firing rate predicted by equation 1.15 is $2r_0 - r_{max}$. For the neuron of figure 1.6B, this is a positive quantity, but for some M1 neurons $2r_0 - r_{max} < 0$,
and the function 1.15 is negative over some range of angles. Because firing rates cannot be negative, the cosine tuning curve must be half-wave
rectified in these cases (see equation 1.13),

$$f(s) = [r_0 + (r_{max} - r_0)\cos(s - s_{max})]_+ .$$ (1.16)

Figure 1.7B shows how the average firing rate of a V1 neuron depends on
retinal disparity and illustrates another important type of tuning curve.
Retinal disparity is a difference in the retinal location of an image between
the two eyes (figure 1.7A). Some neurons in area V1 are sensitive to disparity, representing an early stage in the representation of viewing distance.
sigmoidal In figure 1.7B, the data points have been fitted with a tuning curve called
tuning curve a logistic or sigmoidal function,

$$f(s) = \frac{r_{max}}{1 + \exp\left((s_{1/2} - s)/\Delta_s\right)} .$$ (1.17)

In this case, s is the retinal disparity, the parameter $s_{1/2}$ is the disparity
that produces a firing rate half as big as the maximum value r_{max}, and Δ_s
controls how quickly the firing rate increases as a function of s. If Δ_s is
negative, the firing rate is a monotonically decreasing function of s rather
than a monotonically increasing function as in figure 1.7B.

Spike-Count Variability

Tuning curves allow us to predict the average firing rate, but they do not
describe how the spike-count firing rate r varies about its mean value

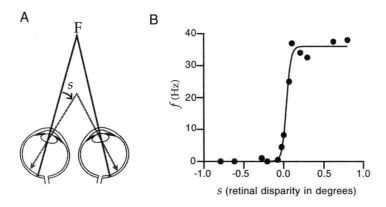

Figure 1.7 (A) Definition of retinal disparity. The gray lines with arrows show the location on each retina of an object located nearer than the fixation point F. The image from the fixation point falls at the fovea in each eye, the small pit where the black lines meet the retina. The image from a nearer object falls to the left of the fovea in the left eye and to the right of the fovea in the right eye. For objects farther away than the fixation point, this would be reversed. The disparity angle s is indicated in the figure. (B) Average firing rate of a cat V1 neuron responding to separate bars of light illuminating each eye, plotted as a function of the disparity. Because this neuron fires for positive s values, it is called a far-tuned cell. The curve is a fit using the function 1.17 with parameters $r_{max} = 36.03$ Hz, $s_{1/2} = 0.036°$, and $\Delta_s = 0.029°$. (A adapted from Wandell, 1995; B data points from Poggio and Talbot, 1981.)

$\langle r \rangle = f(s)$ from trial to trial. While the map from stimulus to average response may be described deterministically, it is likely that single-trial responses such as spike-count rates can be modeled only in a probabilistic manner. For example, r values can be generated from a probability distribution with mean $f(s)$. The trial-to-trial deviation of r from $f(s)$ is considered to be noise, and such models are often called noise models. The standard deviation for the noise distribution either can be independent of $f(s)$, in which case the variability is called additive noise, or it can depend on $f(s)$. Multiplicative noise corresponds to having the standard deviation proportional to $f(s)$.

Response variability extends beyond the level of spike counts to the entire temporal pattern of action potentials. Later in this chapter, we discuss a model of the neuronal response that uses a stochastic spike generator to produce response variability. This approach takes a deterministic estimate of the firing rate, $r_{est}(t)$, and produces a stochastic spiking pattern from it. The spike generator produces variable numbers and patterns of action potentials, even if the same estimated firing rate is used on each trial.

1.3 What Makes a Neuron Fire?

Response tuning curves characterize the average response of a neuron to a given stimulus. We now consider the complementary procedure of averaging the stimuli that produce a given response. To average stimuli in this way, we need to specify what fixed response we will use to "trigger" the average. The most obvious choice is the firing of an action potential. Thus, we ask, "What, on average, did the stimulus do before an action potential was fired?" The resulting quantity, called the spike-triggered average stimulus, provides a useful way of characterizing neuronal selectivity. Spike-triggered averages are computed using stimuli characterized by a parameter $s(t)$ that varies over time. Before beginning our discussion of spike triggering, we describe some features of such stimuli.

Describing the Stimulus

Neurons responding to sensory stimuli face the difficult task of encoding parameters that can vary over an enormous dynamic range. For example, photoreceptors in the retina can respond to single photons or can operate in bright light with an influx of millions of photons per second. To deal with such wide-ranging stimuli, sensory neurons often respond most strongly to rapid changes in stimulus properties and are relatively insensitive to steady-state levels. Steady-state responses are highly compressed functions of stimulus intensity, typically with logarithmic or weak power-law dependences. This compression has an interesting psychophysical correlate. Weber measured how different the intensity of two stimuli had to be for them to be reliably discriminated, the "just noticeable" difference Δs. He found that, for a given stimulus, Δs is proportional to the magnitude of the stimulus s, so that $\Delta s/s$ is constant. This relationship is called

Weber's law
Weber's law. Fechner suggested that noticeable differences set the scale for perceived stimulus intensities. Integrating Weber's law, this means that the perceived intensity of a stimulus of absolute intensity s varies as

Fechner's law
$\log s$. This is known as Fechner's law.

Sensory systems make numerous adaptations, using a variety of mechanisms, to adjust to the average level of stimulus intensity. When a stimulus generates such adaptation, the relationship between stimulus and response is often studied in a potentially simpler regime by describing responses to fluctuations about a mean stimulus level. In this case,

$\int_0^T dt\, s(t)/T = 0$
$s(t)$ is defined so that its time average over the duration of a trial is 0, $\int_0^T dt\, s(t)/T = 0$. We frequently impose this condition.

Our analysis of neural encoding involves two different types of averages: averages over repeated trials that employ the same stimulus, which we denote by angle brackets, and averages over different stimuli. We could introduce a second notation for averages over stimuli, but this can be avoided when using time-dependent stimuli. Instead of presenting a number of different stimuli and averaging over them, we can string together all of the stimuli we wish to consider into a single time-dependent stimulus

stimulus and
sequence and average over time. Thus, stimulus averages are replaced by
time averages
time averages.

Although a response recorded over a trial depends only on the values taken by $s(t)$ during that trial, some of the mathematical analyses presented in this chapter and in chapter 2 are simplified if we define the stimulus at other times as well. It is convenient if integrals involving the stimulus are time-translationally invariant so that for any function h and time interval τ

$$\int_0^T dt\, h(s(t+\tau)) = \int_\tau^{T+\tau} dt\, h(s(t)) = \int_0^T dt\, h(s(t)). \qquad (1.18)$$

To assure the last equality, we define the stimulus outside the time limits of the trial by the relation $s(T+\tau) = s(\tau)$ for any τ, thereby making the stimulus periodic.   *periodic stimulus*

The Spike-Triggered Average

The spike-triggered average stimulus, $C(\tau)$, is the average value of the stimulus a time interval τ before a spike is fired. In other words, for a spike occurring at time t_i, we determine $s(t_i - \tau)$, and then we sum over all n spikes in a trial, $i = 1, 2, \ldots, n$, and divide the total by n. In addition, we average over trials. Thus,   *spike-triggered average $C(\tau)$*

$$C(\tau) = \left\langle \frac{1}{n} \sum_{i=1}^{n} s(t_i - \tau) \right\rangle \approx \frac{1}{\langle n \rangle} \left\langle \sum_{i=1}^{n} s(t_i - \tau) \right\rangle. \qquad (1.19)$$

The approximate equality of the last expression follows from the fact that if n is large, the total number of spikes on each trial is well approximated by the average number of spikes per trial, $n \approx \langle n \rangle$. We make use of this approximation because it allows us to relate the spike-triggered average to other quantities commonly used to characterize the relationship between stimulus and response (see below). Figure 1.8 provides a schematic description of the computation of the spike-triggered average. Each time a spike appears, the stimulus in a time window preceding the spike is recorded. Although the range of τ values in equation 1.19 is unlimited, the response is typically affected only by the stimulus in a window a few hundred milliseconds wide immediately preceding a spike. More precisely, we expect $C(\tau)$ to approach 0 for positive τ values larger than the correlation time between the stimulus and the response. If the stimulus has no temporal correlations with itself, we also expect $C(\tau)$ to be 0 for $\tau < 0$, because the response of a neuron cannot depend on future stimuli. In practice, the stimulus is recorded only over a finite time period, as indicated by the shaded areas in figure 1.8. The recorded stimuli for all spikes are then summed and the procedure is repeated over multiple trials.

The spike-triggered average stimulus can be expressed as an integral of the stimulus times the neural response function of equation 1.1. If we replace the sum over spikes with an integral, as in equation 1.2, and use the approximate expression for $C(\tau)$ in equation 1.19, we find

$$C(\tau) = \frac{1}{\langle n \rangle} \int_0^T dt\, \langle \rho(t) \rangle\, s(t-\tau) = \frac{1}{\langle n \rangle} \int_0^T dt\, r(t) s(t-\tau). \qquad (1.20)$$

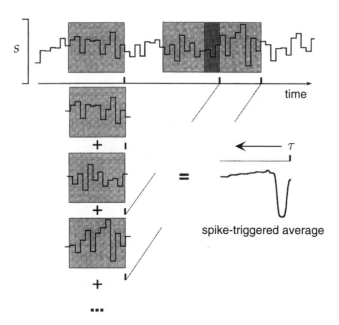

Figure 1.8 Schematic of the procedure for computing the spike-triggered average stimulus. Each gray rectangle contains the stimulus prior to one of the spikes shown along the time axis. These are averaged to produce the waveform shown at the lower right, which is the average stimulus before a spike. The stimulus in this example is a piecewise constant function of time. (Adapted from Rieke et al., 1997.)

The second equality is due to the equivalence of $\langle \rho(t) \rangle$ and $r(t)$ within integrals. Equation 1.20 allows us to relate the spike-triggered average to the correlation function of the firing rate and the stimulus.

Correlation functions are a useful way of determining how two quantities that vary over time are related to one another. The two quantities being *firing-rate stimulus* related are evaluated at different times, one at time t and the other at time *correlation function* $t + \tau$. The correlation function is then obtained by averaging their product Q_{rs} over all t values, and it is a function of τ. The correlation function of the firing rate and the stimulus is

$$Q_{rs}(\tau) = \frac{1}{T} \int_0^T dt\, r(t) s(t + \tau). \qquad (1.21)$$

By comparing equations 1.20 and 1.21, we find that

$$C(\tau) = \frac{1}{\langle r \rangle} Q_{rs}(-\tau), \qquad (1.22)$$

where $\langle r \rangle = \langle n \rangle / T$ is the average firing rate over the set of trials. Because the argument of the correlation function in equation 1.22 is $-\tau$, the spike-triggered average stimulus is often called the reverse correlation function. *reverse correlation* It is proportional to the correlation of the firing rate with the stimulus at *function* preceding times.

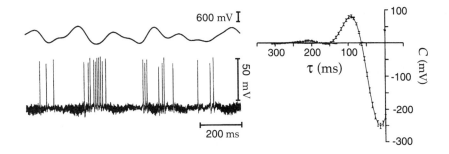

Figure 1.9 The spike-triggered average stimulus for a neuron of the electrosensory lateral-line lobe of the weakly electric fish *Eigenmannia*. The upper left trace is the potential used to generate the electric field to which this neuron is sensitive. The evoked spike train is plotted below the stimulus potential. The plot on the right is the spike-triggered average stimulus. (Adapted from Gabbiani et al., 1996.)

The spike-triggered average stimulus is widely used to study and characterize neural responses. Because $C(\tau)$ is the average value of the stimulus at a time τ before a spike, larger values of τ represent times farther in the past relative to the time of the triggering spike. For this reason, we plot spike-triggered averages with the time axis going backward compared to the normal convention. This allows the average spike-triggering stimulus to be read off from the plots in the usual left-to-right order.

Figure 1.9 shows the spike-triggered average stimulus for a neuron in the electrosensory lateral-line lobe of the weakly electric fish *Eigenmannia*. Weakly electric fish generate oscillating electric fields from an internal electric organ. Distortions in the electric field produced by nearby objects are detected by sensors spread over the skin of the fish. The lateral-line lobe acts as a relay station along the processing pathway for electrosensory signals. Fluctuating electrical potentials, such as that shown in the upper left trace of figure 1.9, elicit responses from electrosensory lateral-line lobe neurons, as seen in the lower left trace. The spike-triggered average stimulus, plotted at the right, indicates that, on average, the electric potential made a positive upswing followed by a large negative deviation prior to a spike being fired by this neuron.

The results obtained by spike-triggered averaging depend on the particular set of stimuli used during an experiment. How should this set be chosen? In chapter 2, we show that there are certain advantages to using a stimulus that is uncorrelated from one time to the next, a white-noise stimulus. A heuristic argument supporting the use of such stimuli is that in asking what makes a neuron fire, we may want to sample its responses to stimulus fluctuations at all frequencies with equal weight (i.e., equal power), and this is one of the properties of white-noise stimuli. In practice, white-noise stimuli can be generated with equal power only up to a finite frequency cutoff, but neurons respond to stimulus fluctuations only within a limited frequency range anyway. Figure 1.9 is based on such an approximate white-noise stimulus. The power in a signal as a function of its frequency is called the power spectrum or power spectral density. White noise has a flat power spectrum.

White-Noise Stimuli

The defining characteristic of a white-noise stimulus is that its value at any one time is uncorrelated with its value at any other time. This condition can be expressed using the stimulus-stimulus correlation function, also called the stimulus autocorrelation, which is defined by analogy with equation 1.21 as

stimulus autocorrelation function Q_{ss}

$$Q_{ss}(\tau) = \frac{1}{T} \int_0^T dt\, s(t)s(t+\tau)\,. \tag{1.23}$$

Just as a correlation function provides information about the temporal relationship between two quantities, so an autocorrelation function tells us about how a quantity at one time is related to itself evaluated at another time. For white noise, the stimulus autocorrelation function is 0 in the range $-T/2 < \tau < T/2$ except when $\tau = 0$, and over this range

$$Q_{ss}(\tau) = \sigma_s^2 \delta(\tau)\,. \tag{1.24}$$

The constant σ_s, which has the units of the stimulus times the square root of the unit of time, reflects the magnitude of the variability of the white noise. In appendix A, we show that equation 1.24 is equivalent to the statement that white noise has equal power at all frequencies.

No physical system can generate noise that is white to arbitrarily high frequencies. Approximations of white noise that are missing high-frequency components can be used, provided the missing frequencies are well above the sensitivity of the neuron under investigation. To approximate white noise, we consider times that are integer multiples of a basic unit of duration Δt, that is, times $t = m\Delta t$ for $m = 1, 2, \ldots, M$ where $M\Delta t = T$. The function $s(t)$ is then constructed as a discrete sequence of stimulus values. This produces a steplike stimulus waveform, like the one that appears in figure 1.8, with a constant stimulus value s_m presented during time bin m. In terms of the discrete-time values s_m, the condition that the stimulus is uncorrelated is

$$\frac{1}{M} \sum_{m=1}^{M} s_m s_{m+p} = \begin{cases} \sigma_s^2/\Delta t & \text{if } p = 0 \\ 0 & \text{otherwise}\,. \end{cases} \tag{1.25}$$

The factor of $1/\Delta t$ on the right side of this equation reproduces the δ function of equation 1.24 in the limit $\Delta t \to 0$. For approximate white noise, the autocorrelation function is 0 except for a region around $\tau = 0$ with width of order Δt. Similarly, the binning of time into discrete intervals of size Δt means that the noise generated has a flat power spectrum only up to frequencies of order $1/(2\Delta t)$.

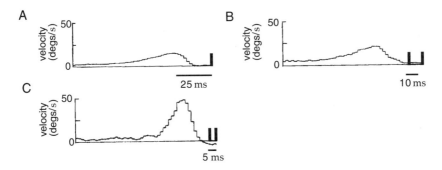

Figure 1.10 Single- and multiple-spike-triggered average stimuli for a blowfly H1 neuron responding to a moving visual image. (A) The average stimulus velocity triggered on a single spike. (B) The average stimulus velocity before two spikes with a separation of 10 ± 1 ms. (C) The average stimulus before two spikes with a separation of 5 ± 1 ms. (Data from de Ruyter van Steveninck and Bialek, 1988; figure adapted from Rieke et al., 1997.)

An approximation to white noise can be generated by choosing each s_m independently from a probability distribution with mean 0 and variance $\sigma_s^2 / \Delta t$. Any reasonable probability function satisfying these two conditions can be used to generate the stimulus values within each time bin. A special class of white-noise stimuli, Gaussian white noise, results when the probability distribution used to generate the s_m values is a Gaussian function. The factor of $1/\Delta t$ in the variance indicates that the variability must be increased as the time bins get smaller. A number of other schemes for efficiently generating approximations of white-noise stimuli are discussed in the references at the end of this chapter.

Multiple-Spike-Triggered Averages and Spike-Triggered Correlations

In addition to triggering on single spikes, stimulus averages can be computed by triggering on various combinations of spikes. Figure 1.10 shows some examples of two-spike triggers. These results come from a study of the H1 movement-sensitive visual neuron of the blowfly. The H1 neuron detects the motion of visual images during flight in order to generate and guide stabilizing motor corrections. It responds to motion of the visual scene. In the experiments, the fly is held fixed while a visual image with a time-varying velocity $s(t)$ is presented. Figure 1.10A, showing the spike-triggered average stimulus, indicates that this neuron responds to positive angular velocities after a latency of about 15 ms. Figure 1.10B is the average stimulus prior to the appearance of two spikes separated by 10 ± 1 ms. In this case, the two-spike average is similar to the sum of two single-spike-triggered average stimuli displaced from one another by 10 ms. Thus, for 10 ms separations, two spikes occurring together tell us no more as a two-spike unit than they would individually. This result changes when shorter separations are considered. Figure 1.10C shows the average stimulus triggered on two spikes separated by 5 ± 1 ms. The average stimulus triggered on a pair of spikes separated by 5 ms is not the same as the sum of the average stimuli for each spike separately.

Spike-triggered averages of other stimulus-dependent quantities can provide additional insight into neural encoding, for example, spike-triggered average autocorrelation functions. Obviously, spike-triggered averages of higher-order stimulus combinations can be considered as well.

1.4   Spike-Train Statistics

A complete description of the stochastic relationship between a stimulus and a response would require us to know the probabilities corresponding to every sequence of spikes that can be evoked by the stimulus. Spike times are continuous variables, and, as a result, the probability for a spike to occur at any precisely specified time is actually zero. To get a nonzero value, we must ask for the probability that a spike occurs within a specified interval, for example, the interval between times t and $t + \Delta t$. For small Δt, the probability of a spike falling in this interval is proportional to the size of the interval, Δt. A similar relation holds for any continuous stochastic variable z. The probability that z takes a value between z and $z + \Delta z$, for small Δz (strictly speaking, as $\Delta z \to 0$), is equal to $p[z]\Delta z$, where $p[z]$ is called a probability density.

Throughout this book, we use the notation $P[\ ]$ to denote probabilities and $p[\ ]$ to denote probability densities. We use the bracket notation $P[\ ]$ generically for the probability of something occurring and also to denote a specific probability function. In the latter case, the notation $P(\ )$ would be more appropriate, but switching between square brackets and parentheses is confusing, so the reader will have to use the context to distinguish between these cases.

The probability of a spike sequence appearing is proportional to the probability density of spike times, $p[t_1, t_2, \ldots, t_n]$. In other words, the probability $P[t_1, t_2, \ldots, t_n]$ that a sequence of n spikes occurs with spike i falling between times t_i and $t_i + \Delta t$ for $i = 1, 2, \ldots, n$ is given in terms of this density by the relation $P[t_1, t_2, \ldots, t_n] = p[t_1, t_2, \ldots, t_n](\Delta t)^n$.

Unfortunately, the number of possible spike sequences is typically so large that determining or even roughly estimating all of their probabilities of occurrence is impossible. Instead, we must rely on some statistical model that allows us to estimate the probability of an arbitrary spike sequence occurring, given our knowledge of the responses actually recorded. The firing rate $r(t)$ determines the probability of firing a single spike in a small interval around the time t, but $r(t)$ is not, in general, sufficient information to predict the probabilities of spike sequences. For example, the probability of two spikes occurring together in a sequence is not necessarily equal to the product of the probabilities that they occur individually, because the presence of one spike may effect the occurrence of the other. If, however, the probability of generating an action potential is independent of the presence or timing of other spikes (i.e., if the spikes are statistically independent) the firing rate is all that is needed to compute the probabilities for all possible action potential sequences.

A stochastic process that generates a sequence of events, such as action potentials, is called a point process. In general, the probability of an event occurring at any given time could depend on the entire history of preceding events. If this dependence extends only to the immediately preceding event, so that the intervals between successive events are independent, the point process is called a renewal process. If there is no dependence at all on preceding events, so that the events themselves are statistically independent, we have a Poisson process. The Poisson process provides an extremely useful approximation of stochastic neuronal firing. To make the presentation easier to follow, we separate two cases, the homogeneous Poisson process, for which the firing rate is constant over time, and the inhomogeneous Poisson process, which involves a time-dependent firing rate.

point process

renewal process

Poisson process

The Homogeneous Poisson Process

We denote the firing rate for a homogeneous Poisson process by $r(t) = r$ because it is independent of time. When the firing rate is constant, the Poisson process generates every sequence of n spikes over a fixed time interval with equal probability. As a result, the probability $P[t_1, t_2, \ldots, t_n]$ can be expressed in terms of another probability function $P_T[n]$, which is the probability that an arbitrary sequence of exactly n spikes occurs within a trial of duration T. Assuming that the spike times are ordered so that $0 \le t_1 \le t_2 \le \ldots \le t_n \le T$, the relationship is

$$P[t_1, t_2, \ldots, t_n] = n! P_T[n] \left(\frac{\Delta t}{T} \right)^n . \qquad (1.26)$$

This relationship is a special case of equation 1.37 derived below.

To compute $P_T[n]$, we divide the time T into M bins of size $\Delta t = T/M$. We can assume that Δt is small enough so that we never get two spikes within any one bin because, at the end of the calculation, we take the limit $\Delta t \to 0$. $P_T[n]$ is the product of three factors: the probability of generating n spikes within a specified set of the M bins, the probability of not generating spikes in the remaining $M - n$ bins, and a combinatorial factor equal to the number of ways of putting n spikes into M bins. The probability of a spike occurring in one specific bin is $r\Delta t$, and the probability of n spikes appearing in n specific bins is $(r\Delta t)^n$. Similarly, the probability of not having a spike in a given bin is $(1 - r\Delta t)$, so the probability of having the remaining $M - n$ bins without any spikes in them is $(1 - r\Delta t)^{M-n}$. Finally, the number of ways of putting n spikes into M bins is given by the

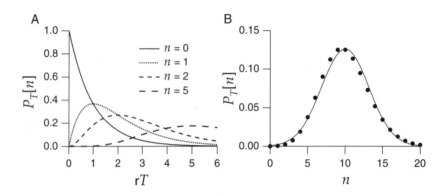

Figure 1.11 (A) The probability that a homogeneous Poisson process generates n spikes in a time period of duration T plotted for $n = 0, 1, 2,$ and 5. The probability is plotted as function of the rate times the duration of the interval, rT, to make the plot applicable for any rate. (B) The probability of finding n spikes during a time period for which $rT = 10$ (dots) compared with a Gaussian distribution with mean and variance equal to 10 (line).

binomial coefficient $M!/(M - n)!n!$. Putting all these factors together, we find

$$P_T[n] = \lim_{\Delta t \to 0} \frac{M!}{(M - n)!n!}(r\Delta t)^n(1 - r\Delta t)^{M-n}. \tag{1.27}$$

To take the limit, we note that as $\Delta t \to 0$, M grows without bound because $M\Delta t = T$. Because n is fixed, we can write $M - n \approx M = T/\Delta t$. Using this approximation and defining $\epsilon = -r\Delta t$, we find that

$$\lim_{\Delta t \to 0}(1 - r\Delta t)^{M-n} = \lim_{\epsilon \to 0}\left((1 + \epsilon)^{1/\epsilon}\right)^{-rT} = e^{-rT} = \exp(-rT) \tag{1.28}$$

because $\lim_{\epsilon \to 0}(1 + \epsilon)^{1/\epsilon}$ is, by definition, $e = \exp(1)$. For large M, $M!/(M - n)! \approx M^n = (T/\Delta t)^n$, so

$$P_T[n] = \frac{(rT)^n}{n!}\exp(-rT). \tag{1.29}$$

Poisson distribution This is called the Poisson distribution. The probabilities $P_T[n]$, for a few n values, are plotted as a function of rT in figure 1.11A. Note that as n increases, the probability reaches its maximum at larger T values and that large n values are more likely than small ones for large T. Figure 1.11B shows the probabilities of various numbers of spikes occurring when the average number of spikes is 10. For large rT, which corresponds to a large expected number of spikes, the Poisson distribution approaches a Gaussian distribution with mean and variance equal to rT. Figure 1.11B shows that this approximation is already quite good for $rT = 10$.

We can compute the variance of spike counts produced by a Poisson process from the probabilities in equation 1.29. For spikes counted over an interval of duration T, the variance of the spike count (derived in appendix B) is

$$\sigma_n^2 = \langle n^2 \rangle - \langle n \rangle^2 = rT. \tag{1.30}$$

Thus the variance and mean of the spike count are equal. The ratio of these
two quantities, $\sigma_n^2/\langle n \rangle$, is called the Fano factor and takes the value 1 for a   *Fano factor*
homogeneous Poisson process, independent of the time interval T.

The probability density of time intervals between adjacent spikes is called
the interspike interval distribution, and it is a useful statistic for character-   *interspike interval*
izing spiking patterns. Suppose that a spike occurs at a time t_i for some   *distribution*
value of i. The probability of a homogeneous Poisson process generating
the next spike somewhere in the interval $t_i + \tau \le t_{i+1} < t_i + \tau + \Delta t$, for
small Δt, is the probability that no spike is fired for a time τ, times the
probability, $r\Delta t$, of generating a spike within the following small interval
Δt. From equation 1.29, with $n = 0$, the probability of not firing a spike
for period τ is $\exp(-r\tau)$, so the probability of an interspike interval falling
between τ and $\tau + \Delta t$ is

$$P[\tau \le t_{i+1} - t_i < \tau + \Delta t] = r\Delta t \exp(-r\tau). \qquad (1.31)$$

The probability density of interspike intervals is, by definition, this probability with the factor Δt removed. Thus, the interspike interval distribution for a homogeneous Poisson spike train is an exponential. The most
likely interspike intervals are short ones, and long intervals have a probability that falls exponentially as a function of their duration.

From the interspike interval distribution of a homogeneous Poisson spike
train, we can compute the mean interspike interval,

$$\langle \tau \rangle = \int_0^\infty d\tau \, \tau r \exp(-r\tau) = \frac{1}{r}, \qquad (1.32)$$

and the variance of the interspike intervals,

$$\sigma_\tau^2 = \int_0^\infty d\tau \, \tau^2 r \exp(-r\tau) - \langle \tau \rangle^2 = \frac{1}{r^2}. \qquad (1.33)$$

The ratio of the standard deviation to the mean is called the coefficient of   *coefficient of*
variation,   *variation C_V*

$$C_V = \frac{\sigma_\tau}{\langle \tau \rangle}, \qquad (1.34)$$

and it takes the value 1 for a homogeneous Poisson process. This is a
necessary, though not sufficient, condition to identify a Poisson spike train.
Recall that the Fano factor for a Poisson process is also 1. For any renewal
process, the Fano factor evaluated over long time intervals approaches the
value C_V^2.

The Spike-Train Autocorrelation Function

The spike interval distribution measures the distribution of times between
successive action potentials in a train. It is useful to generalize this concept and determine the distribution of times between any two spikes in

a train. This is called the spike-train autocorrelation function, and it is particularly useful for detecting patterns in spike trains, most notably oscillations. The spike-train autocorrelation function is the autocorrelation of the neural response function of equation 1.1 with its average over time and trials subtracted out. The time average of the neural response function, from equation 1.4, is the spike-count rate r, and the trial average of this quantity is $\langle r \rangle = \langle n \rangle / T$. Thus, the spike-train autocorrelation function is

spike-train autocorrelation function $Q_{\rho\rho}$

$$Q_{\rho\rho}(\tau) = \frac{1}{T} \int_0^T dt \, \langle (\rho(t) - \langle r \rangle)(\rho(t+\tau) - \langle r \rangle) \rangle. \tag{1.35}$$

Because the average is subtracted from the neural response function in this expression, $Q_{\rho\rho}$ should really be called an autocovariance, not an autocorrelation, but in practice it isn't.

The spike-train autocorrelation function is constructed from data in the form of a histogram by dividing time into bins. The value of the histogram for a bin labeled with a positive or negative integer m is computed by determining the number of the times that any two spikes in the train are separated by a time interval lying between $(m-1/2)\Delta t$ and $(m+1/2)\Delta t$ with Δt the bin size. This includes all pairings, even between a spike and itself. We call this number N_m. If the intervals between the n^2 spike pairs in the train were uniformly distributed over the range from 0 to T, there would be $n^2 \Delta t / T$ intervals in each bin. This uniform term is removed from the autocorrelation histogram by subtracting $n^2 \Delta t / T$ from N_m for all m. The spike-train autocorrelation histogram is then defined by dividing the resulting numbers by T, so the value of the histogram in bin m is $H_m = N_m / T - n^2 \Delta t / T^2$. For small bin sizes, the $m = 0$ term in the histogram counts the average number of spikes, that is $N_0 = \langle n \rangle$ and in the limit $\Delta t \to 0$, $H_0 = \langle n \rangle / T$ is the average firing rate $\langle r \rangle$. Because other bins have H_m of order Δt, the large $m = 0$ term is often removed from histogram plots. The spike-train autocorrelation function is defined as $H_m / \Delta t$ in the limit $\Delta t \to 0$, and it has the units of a firing rate squared. In this limit, the $m = 0$ bin becomes a δ function, $H_0 / \Delta t \to \langle r \rangle \delta(\tau)$.

As we have seen, the distribution of interspike intervals for adjacent spikes in a homogeneous Poisson spike train is exponential (equation 1.31). By contrast, the intervals between any two spikes (not necessarily adjacent) in such a train are uniformly distributed. As a result, the subtraction procedure outlined above gives $H_m = 0$ for all bins except for the $m = 0$ bin that contains the contribution of the zero intervals between spikes and themselves. The autocorrelation function for a Poisson spike train generated at a constant rate $\langle r \rangle = r$ is thus

$$Q_{\rho\rho}(\tau) = r\delta(\tau). \tag{1.36}$$

cross-correlation function

A cross-correlation function between spike trains from two different neurons can be defined by analogy with the autocorrelation function by determining the distribution of intervals between pairs of spikes, one taken

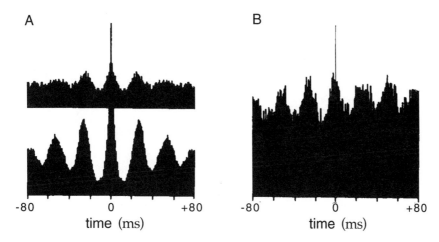

Figure 1.12 Autocorrelation and cross-correlation histograms for neurons in the primary visual cortex of a cat. (A) Autocorrelation histograms for neurons recorded in the right (upper) and left (lower) hemispheres show a periodic pattern indicating oscillations at about 40 Hz. The lower diagram indicates stronger oscillations in the left hemisphere. (B) The cross-correlation histogram for these two neurons shows that their oscillations are synchronized with little time delay. (Adapted from Engel et al., 1991.)

from each train. The spike-train autocorrelation function is an even function of τ, $Q_{\rho\rho}(\tau) = Q_{\rho\rho}(-\tau)$, but the cross-correlation function is not necessarily even. A peak at zero interval in a cross-correlation function signifies that the two neurons are firing synchronously. Asymmetric shifts in this peak away from 0 result from fixed delays between the firing of the two neurons, and they indicate nonsynchronous but phase-locked firing. Periodic structure in either an autocorrelation or a cross-correlation function or histogram indicates that the firing probability oscillates. Such periodic structure is seen in the histograms of figure 1.12, showing 40 Hz oscillations in neurons of cat primary visual cortex that are roughly synchronized between the two cerebral hemispheres.

The Inhomogeneous Poisson Process

When the firing rate depends on time, different sequences of n spikes occur with different probabilities, and $p[t_1, t_2, \ldots, t_n]$ depends on the spike times. Because spikes are still generated independently by an inhomogeneous Poisson process, their times enter into $p[t_1, t_2, \ldots, t_n]$ only through the time-dependent firing rate $r(t)$. Assuming, as before, that the spike times are ordered $0 \leq t_1 \leq t_2 \leq \ldots \leq t_n \leq T$, the probability density for n spike times (derived in appendix C) is

$$p[t_1, t_2, \ldots, t_n] = \exp\left(-\int_0^T dt\, r(t)\right) \prod_{i=1}^n r(t_i). \qquad (1.37)$$

This result applies if the spike times have been written in temporal order. If the spike times are not ordered, so that, for example, we are interested in the probability density for any spike occurring at the time t_1, not necessarily the first spike, this expression should be divided by a factor of $n!$ to account for the number of different possible orderings of spike times.

The Poisson Spike Generator

Spike sequences can be simulated by using some estimate of the firing rate, $r_{est}(t)$, predicted from knowledge of the stimulus, to drive a Poisson process. A simple procedure for generating spikes in a computer program is based on the fact that the estimated probability of firing a spike during a short interval of duration Δt is $r_{est}(t)\Delta t$. The program progresses through time in small steps of size Δt and generates, at each time step, a random number x_{rand} chosen uniformly in the range between 0 and 1. If $r_{est}(t)\Delta t > x_{rand}$ at that time step, a spike is fired; otherwise it is not.

For a constant firing rate, it is faster to compute spike times t_i for $i = 1, 2, \ldots n$ iteratively by generating interspike intervals from an exponential probability density (equation 1.31). If x_{rand} is uniformly distributed over the range between 0 and 1, the negative of its logarithm is exponentially distributed. Thus, we can generate spike times iteratively from the formula $t_{i+1} = t_i - \ln(x_{rand})/r$. Unlike the algorithm discussed in the previous paragraph, this method works only for constant firing rates. However, it can be extended to time-dependent rates by using a procedure called rejection sampling or spike thinning. The thinning technique requires a bound r_{max} on the estimated firing rate such that $r_{est}(t) \leq r_{max}$ at all times. We first generate a spike sequence corresponding to the constant rate r_{max} by iterating the rule $t_{i+1} = t_i - \ln(x_{rand})/r_{max}$. The spikes are then thinned by generating another x_{rand} for each i and removing the spike at time t_i from the train if $r_{est}(t_i)/r_{max} < x_{rand}$. If $r_{est}(t_i)/r_{max} \geq x_{rand}$, spike i is retained. Thinning corrects for the difference between the estimated time-dependent rate and the maximum rate.

Figure 1.13 shows an example of a model of an orientation-selective V1 neuron constructed in this way. In this model, the estimated firing rate is determined from the response tuning curve of figure 1.5B,

$$r_{est}(t) = f(s(t)) = r_{max}\exp\left(-\frac{1}{2}\left(\frac{s(t)-s_{max}}{\sigma_f}\right)^2\right). \qquad (1.38)$$

This is an extremely simplified model of response dynamics, because the firing rate at any given time depends only on the value of the stimulus at that instant of time and not on its recent history. Models that allow for a dependence of firing rate on stimulus history are discussed in chapter 2. In figure 1.13, the orientation angle increases in a sequence of steps. The firing rate follows these changes, and the Poisson process generates an irregular firing pattern that reflects the underlying rate but varies from trial to trial.

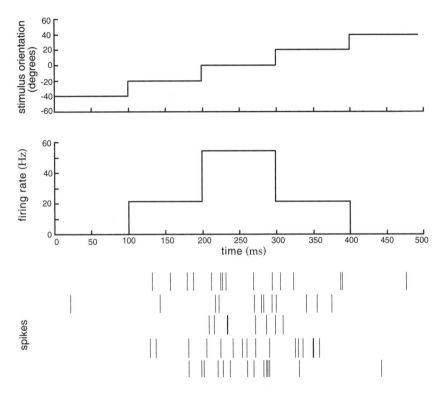

Figure 1.13 Model of an orientation-selective neuron. The orientation angle (top panel) was increased from an initial value of -40° by 20° every 100 ms. The firing rate (middle panel) was used to generate spikes (bottom panel) using a Poisson spike generator. The bottom panel shows spike sequences generated on five different trials.

Certain features of neuronal firing violate the independence assumption that forms the basis of the Poisson model, at least if a constant firing rate is used. We have already mentioned the absolute and relative refractory periods, which are periods of time following the generation of an action potential when the probability of a spike occurring is greatly or somewhat reduced. Frequently, these are most prominent features of real neuronal spike trains that are not captured by a Poisson model. Refractory effects can be incorporated into a Poisson model of spike generation by setting the firing rate to 0 immediately after a spike is fired, and then letting it return to its predicted value according to some dynamic rule such as an exponential recovery.

Comparison with Data

The Poisson process is simple and useful, but does it match data on neural response variability? To address this question, we examine Fano factors, interspike interval distributions, and coefficients of variation.

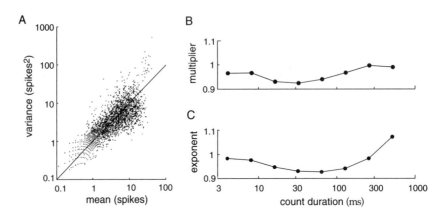

Figure 1.14 Variability of MT neurons in alert macaque monkeys responding to moving visual images. (A) Variance of the spike counts for a 256 ms counting period plotted against the mean spike count. The straight line is the prediction of the Poisson model. Data are from 94 cells recorded under a variety of stimulus conditions. (B) The multiplier A in the relationship between spike-count variance and mean as a function of the duration of the counting interval. (C) The exponent B in this relation as a function of the duration of the counting interval. (Adapted from O'Keefe et al., 1997.)

The Fano factor describes the relationship between the mean spike count over a given interval and the spike-count variance. Mean spike counts $\langle n \rangle$ and variances σ_n^2 from a wide variety of neuronal recordings have been fitted to the equation $\sigma_n^2 = A \langle n \rangle^B$, and the multiplier A and exponent B have been determined. The values of both A and B typically lie between 1.0 and 1.5. Because the Poisson model predicts $A = B = 1$, this indicates that the data show a higher degree of variability than the Poisson model would predict. However, many of these experiments involve anesthetized animals, and it is known that response variability is higher in anesthetized than in alert animals.

area MT

Figure 1.14 shows data for spike-count means and variances extracted from recordings of MT neurons in alert macaque monkeys using a number of different stimuli. The MT (medial temporal) area is a visual region of the primate cortex where many neurons are sensitive to image motion. The individual means and variances are scattered in figure 1.14A, but they cluster around the diagonal which is the Poisson prediction. Similarly, the results show A and B values close to 1, the Poisson values (figure 1.14B). Of course, many neural responses cannot be described by Poisson statistics, but it is reassuring to see a case where the Poisson model seems a reasonable approximation. As mentioned previously, when spike trains are not described very accurately by a Poisson model, refractory effects are often the primary reason.

Interspike interval distributions are extracted from data as interspike interval histograms by counting the number of intervals falling in discrete time bins. Figure 1.15A presents an example from the responses of a non-bursting cell in area MT of a monkey in response to images consisting of

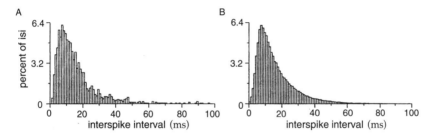

Figure 1.15 (A) Interspike interval distribution from an MT neuron responding to a moving, random-dot image. The probability of interspike intervals falling into the different bins, expressed as a percentage, is plotted against interspike interval. (B) Interspike interval histogram generated from a Poisson model with a stochastic refractory period. (Adapted from Bair et al., 1994.)

randomly moving dots with a variable amount of coherence imposed on their motion (see chapter 3 for a more detailed description). For interspike intervals longer than about 10 ms, the shape of this histogram is exponential, in agreement with equation 1.31. However, for shorter intervals there is a discrepancy. While the homogeneous Poisson distribution of equation 1.31 rises for short interspike intervals, the experimental results show a rapid decrease. This is the result of refractoriness making short interspike intervals less likely than the Poisson model would predict. Data on interspike intervals can be fitted more accurately by a gamma distribution,

gamma distribution

$$p[\tau] = \frac{r(r\tau)^k \exp(-r\tau)}{k!} \tag{1.39}$$

with $k > 0$, than by the exponential distribution of the Poisson model, which has $k = 0$.

Figure 1.15B shows a theoretical histogram obtained by adding a refractory period of variable duration to the Poisson model. Spiking was prohibited during the refractory period, and then was described once again by a homogeneous Poisson process. The refractory period was randomly chosen from a Gaussian distribution with a mean of 5 ms and a standard deviation of 2 ms (only random draws that generated positive refractory periods were included). The resulting interspike interval distribution of figure 1.15B agrees quite well with the data.

C_V values extracted from the spike trains of neurons recorded in monkeys from area MT and primary visual cortex (V1) are shown in figure 1.16. The data have been divided into groups based on the mean interspike interval, and the coefficient of variation is plotted as a function of this mean interval, equivalent to $1/\langle r \rangle$. Except for short mean interspike intervals, the values are near 1, although they tend to cluster slightly lower than 1, the Poisson value. The small C_V values for short interspike intervals are due to the refractory period. The solid curve is the prediction of a Poisson model with refractoriness.

The Poisson model with refractoriness provides a reasonably good description of a significant amount of data, especially considering its sim-

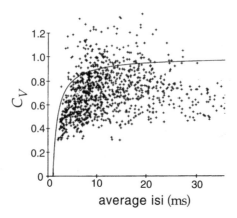

Figure 1.16 Coefficients of variation for a large number of V1 and MT neurons plotted as a function of mean interspike interval. The solid curve is the result of a Poisson model with a refractory period. (Adapted from Softky and Koch, 1992.)

plicity. However, there are cases in which the accuracy in the timing and numbers of spikes fired by a neuron is considerably higher than would be implied by Poisson statistics. Furthermore, even when it successfully describes data, the Poisson model does not provide a mechanistic explanation of neuronal response variability. Spike generation, by itself, is highly reliable in real neurons. Figure 1.17 compares the response of V1 cells to constant current injection in vivo and in vitro. The in vitro response is a regular and reproducible spike train (left panel). The same current injection paradigm applied in vivo produces a highly irregular pattern of firing (center panel) similar to the response to a moving bar stimulus (right panel). Although some of the basic statistical properties of firing variability may be captured by the Poisson model of spike generation, the spike-generating mechanism itself in real neurons is clearly not responsible for the variability. We explore ideas about possible sources of spike-train variability in chapter 5.

Some neurons fire action potentials in clusters or bursts of spikes that cannot be described by a Poisson process with a fixed rate. Bursting can be included in a Poisson model by allowing the firing rate to fluctuate in order to describe the high rate of firing during a burst. Sometimes the distribution of bursts themselves can be described by a Poisson process (such a doubly stochastic process is called a Cox process).

1.5   The Neural Code

The nature of the neural code is a topic of intense debate within the neuroscience community. Much of the discussion has focused on whether neurons use rate coding or temporal coding, often without a clear definition of what these terms mean. We feel that the central issue in neural coding is whether individual action potentials and individual neurons encode inde-

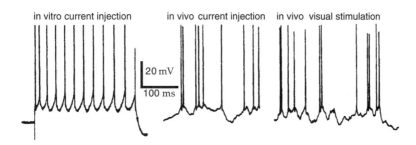

in vitro current injection   in vivo current injection   in vivo visual stimulation

20 mV
100 ms

Figure 1.17 Intracellular recordings from cat V1 neurons. The left panel is the response of a neuron in an in vitro slice preparation to constant current injection. The center and right panels show recordings from neurons in vivo responding to either injected current (center) or a moving visual image (right). (Adapted from Holt et al., 1996.)

pendently of each other, or whether correlations between different spikes and different neurons carry significant amounts of information. We therefore contrast independent-spike and independent-neuron codes with correlation codes before addressing the issue of temporal coding.

Independent-Spike, Independent-Neuron, and Correlation Codes

The neural response, and its relation to the stimulus, are completely characterized by the probability distribution of spike times as a function of the stimulus. If spike generation can be described as an inhomogeneous Poisson process, this probability distribution can be computed from the time-dependent firing rate $r(t)$, using equation 1.37. In this case, $r(t)$ contains all the information about the stimulus that can be extracted from the spike train, and the neural code could reasonably be called a rate code. Unfortunately, this definition does not agree with common usage. Instead, we will call a code based solely on the time-dependent firing rate an independent-spike code. This refers to the fact that the generation of each spike is independent of all the other spikes in the train. If individual spikes do not encode independently of each other, we call the code a correlation code, because correlations between spike times may carry additional information. In reality, information is likely to be carried both by individual spikes and through correlations, and some arbitrary dividing line must be established to characterize the code. Identifying a correlation code should require that a significant amount of information be carried by correlations, for example, as much as is carried by the individual spikes. *independent-spike code*

correlation code

A simple example of a correlation code would occur if significant amounts of information about a stimulus were carried by interspike intervals. In this case, if we considered spike times individually, independently of each other, we would miss the information carried by the intervals between them. This is just one example of a correlation code. Information could be carried by more complex relationships between spikes.

Independent-spike codes are much simpler to analyze than correlation codes, and most work on neural coding assumes spike independence. When careful studies have been done, it has been found that some information is carried by correlations between two or more spikes, but this information is rarely larger than 10% of the information carried by spikes considered independently. Of course, it is possible that, due to our ignorance of the "real" neural code, we have not yet uncovered or examined the types of correlations that are most significant for neural coding. Although this is not impossible, we view it as unlikely and feel that the evidence for independent-spike coding, at least as a fairly accurate approximation, is quite convincing.

The discussion to this point has focused on information carried by single neurons, but information is typically encoded by neuronal populations. When we study population coding, we must consider whether individual neurons act independently, or whether correlations between different neurons carry additional information. The analysis of population coding is easiest if the response of each neuron is considered statistically independent, and such independent-neuron coding is typically assumed in the analysis of population codes (chapter 3). The independent-neuron hypothesis does not mean that the spike trains of different neurons are not combined into an ensemble code. Rather, it means that they can be combined without taking correlations into account. To test the validity of this assumption, we must ask whether correlations between the spiking of different neurons provide additional information about a stimulus that cannot be obtained by considering all of their firing patterns individually.

independent-neuron code

Synchronous firing of two or more neurons is one mechanism for conveying information in a population correlation code. Rhythmic oscillations of population activity provide another possible mechanism, as discussed below. Both synchronous firing and oscillations are common features of the activity of neuronal populations. However, the existence of these features is not sufficient for establishing a correlation code, because it is essential to show that a significant amount of information is carried by the resulting correlations. The assumption of independent-neuron coding is a useful simplification that is not in gross contradiction with experimental data, but it is less well established and more likely to be challenged in the future than the independent-spike hypothesis.

synchrony and oscillations

Place-cell coding of spatial location in the rat hippocampus is an example in which at least some additional information appears to be carried by correlations between the firing patterns of neurons in a population. The hippocampus is a structure located deep inside the temporal lobe that plays an important role in memory formation and is involved in a variety of spatial tasks. The firing rates of many hippocampal neurons, recorded when a rat is moving around a familiar environment, depend on the location of the animal and are restricted to spatially localized areas called the place fields of the cells. In addition, when a rat explores an environment, hippocampal neurons fire collectively in a rhythmic pattern with a frequency in the theta range, 7-12 Hz. The spiking time of an individual

hippocampal place cells

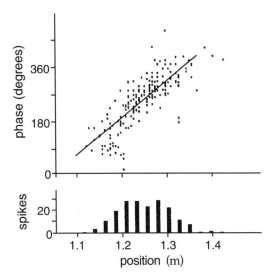

Figure 1.18 Position versus phase for a hippocampal place cell. Each dot in the upper figure shows the phase of the theta rhythm plotted against the position of the animal at the time when a spike was fired. The linear relation shows that information about position is contained in the relative phase of firing. The lower plot is a conventional place field tuning curve of spike count versus position. (Adapted from O'Keefe and Recce, 1993.)

place cell relative to the phase of the population theta rhythm gives additional information about the location of the rat not provided by place cells considered individually. The relationship between location and phase of place-cell firing shown in figure 1.18 means, for example, that we can distinguish two locations on opposite sides of the peak of a single neuron's tuning curve that correspond to the same firing rate, by knowing when the spikes occurred relative to the theta rhythm. However, the amount of additional information carried by correlations between place-field firing and the theta rhythm has not been fully quantified.

Temporal Codes

The concept of temporal coding arises when we consider how precisely we must measure spike times to extract most of the information from a neuronal response. This precision determines the temporal resolution of the neural code. A number of studies have found that this temporal resolution is on a millisecond time scale, indicating that precise spike timing is a significant element in neural encoding. Similarly, we can ask whether high-frequency firing-rate fluctuations carry significant information about a stimulus. When precise spike timing or high-frequency firing-rate fluctuations are found to carry information, the neural code is often identified as a temporal code.

The temporal structure of a spike train or firing rate evoked by a stimulus is determined both by the dynamics of the stimulus and by the nature of the neural encoding process. Stimuli that change rapidly tend to generate precisely timed spikes and rapidly changing firing rates no matter what neural coding strategy is being used. Temporal coding refers to (or should refer to) temporal precision in the response that does not arise solely from the dynamics of the stimulus, but that nevertheless relates to properties of the stimulus. The interplay between stimulus and encoding dynamics makes the identification of a temporal code difficult.

The issue of temporal coding is distinct and independent from the issue of independent-spike coding discussed above. If the independent-spike hypothesis is valid, the temporal character of the neural code is determined by the behavior of $r(t)$. If $r(t)$ varies slowly with time, the code is typically called a rate code, and if it varies rapidly, the code is called temporal. Figure 1.19 provides an example of different firing-rate behaviors for a neuron in area MT of a monkey recorded over multiple trials with three different stimuli (consisting of moving random dots). The activity in the top panel would typically be regarded as reflecting rate coding, and the activity in the bottom panel as reflecting temporal coding. However, the identification of rate and temporal coding in this way is ambiguous because it is not obvious what criterion should be used to characterize the changes in $r(t)$ as slow or rapid.

One possibility is to use the spikes to distinguish slow from rapid, so that a temporal code is identified when peaks in the firing rate occur with roughly the same frequency as the spikes themselves. In this case, each peak corresponds to the firing of only one, or at most a few action potentials. While this definition makes intuitive sense, it is problematic to extend it to the case of population coding. When many neurons are involved, any single neuron may fire only a few spikes before its firing rate changes, but collectively the population may produce a large number of spikes over the same time period. Thus, by this definition, a neuron that appears to employ a temporal code may be part of a population that does not.

Another proposal is to use the stimulus, rather than the response, to establish what makes a temporal code. In this case, a temporal code is defined as one in which information is carried by details of spike timing on a scale shorter than the fastest time characterizing variations of the stimulus. This requires that information about the stimulus be carried by Fourier components of $r(t)$ at frequencies higher than those present in the stimulus. Many of the cases where a temporal code has been reported using spikes to define the nature of the code would be called rate codes if the stimulus were used instead.

The debate between rate and temporal coding dominates discussions about the nature of the neural code. Determining the temporal resolution of the neural code is clearly important, but much of this debate seems uninformative. We feel that the central challenge is to identify relationships

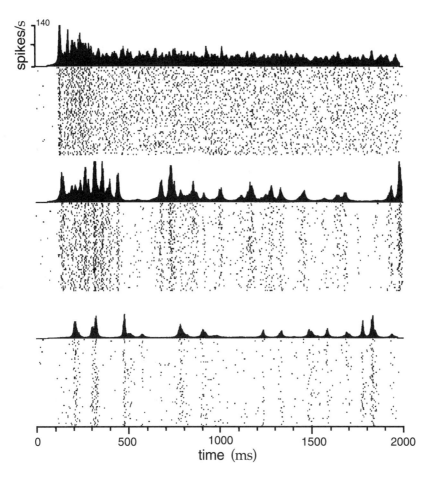

Figure 1.19 Time-dependent firing rates for different stimulus parameters. The rasters show multiple trials during which an MT neuron responded to the same moving, random-dot stimulus. Firing rates, shown above the raster plots, were constructed from the multiple trials by counting spikes within discrete time bins and averaging over trials. The three different results are from the same neuron but using different stimuli. The stimuli were always patterns of moving random dots, but the coherence of the motion was varied (see chapter 3 for more information about this stimulus). (Adapted from Bair and Koch, 1996.)

between the firing patterns of different neurons in a responding population and to understand their significance for neural coding.

1.6   Chapter Summary

With this chapter, we have begun our study of the way that neurons encode information using spikes. We used a sequence of δ functions, the neural response function, to represent a spike train and defined three types of firing rates: the time-dependent firing rate $r(t)$, the spike-count rate r, and the average firing rate $\langle r \rangle$. In the discussion of how the firing rate $r(t)$ could be extracted from data, we introduced the important concepts of a linear filter and a kernel acting as a sliding window function. The average firing rate expressed as a function of a static stimulus parameter is called the response tuning curve, and we presented examples of Gaussian, cosine, and sigmoidal tuning curves. Spike-triggered averages of stimuli, or reverse correlation functions, were introduced to characterize the selectivity of neurons to dynamic stimuli. The homogeneous and inhomogeneous Poisson processes were presented as models of stochastic spike sequences. We defined correlation functions, auto- and cross-correlations, and power spectra, and used the Fano factor, interspike-interval histogram, and coefficient of variation to characterize the stochastic properties of spiking. We concluded with a discussion of independent-spike and independent-neuron codes versus correlation codes, and of the temporal precision of spike timing as addressed in discussions of temporal coding.

1.7 Appendices

A: The Power Spectrum of White Noise

The Fourier transform of the stimulus autocorrelation function (see the Mathematical Appendix),

$$\tilde{Q}_{ss}(\omega) = \frac{1}{T} \int_{-T/2}^{T/2} d\tau \, Q_{ss}(\tau) \exp(i\omega\tau) \,, \tag{1.40}$$

power spectrum   is called the power spectrum. Because we have defined the stimulus as periodic outside the range of the trial T, we have used a finite-time Fourier transform and ω should be restricted to values that are integer multiples of $2\pi/T$. We can compute the power spectrum for a white-noise stimulus using the fact that $Q_{ss}(\tau) = \sigma_s^2 \delta(\tau)$ for white noise,

$$\tilde{Q}_{ss}(\omega) = \frac{\sigma_s^2}{T} \int_{-T/2}^{T/2} d\tau \, \delta(\tau) \exp(i\omega\tau) = \frac{\sigma_s^2}{T} \,. \tag{1.41}$$

This is the defining characteristic of white noise; its power spectrum is independent of frequency.

Using the definition of the stimulus autocorrelation function, we can also write

$$
\begin{aligned}
\tilde{Q}_{ss}(\omega) &= \frac{1}{T} \int_0^T dt\, s(t) \frac{1}{T} \int_{-T/2}^{T/2} d\tau\, s(t+\tau) \exp(i\omega\tau) \\
&= \frac{1}{T} \int_0^T dt\, s(t) \exp(-i\omega t) \frac{1}{T} \int_{-T/2}^{T/2} d\tau\, s(t+\tau) \exp(i\omega(t+\tau)) \,.
\end{aligned} \tag{1.42}
$$

The first integral on the right side of the second equality is the complex conjugate of the Fourier transform of the stimulus,

$$\tilde{s}(\omega) = \frac{1}{T} \int_0^T d\tau \, s(\tau) \exp(i\omega\tau). \tag{1.43}$$

The second integral, because of the periodicity of the integrand (when ω is an integer multiple of $2\pi/T$) is equal to $\tilde{s}(\omega)$. Therefore,

$$\tilde{Q}_{ss}(\omega) = |\tilde{s}(\omega)|^2, \tag{1.44}$$

which provides another definition of the stimulus power spectrum. It is the absolute square of the Fourier transform of the stimulus.

Although equations 1.40 and 1.44 are both sound, they do not provide a statistically efficient method of estimating the power spectrum of discrete approximations to white-noise sequences generated by the methods described in this chapter. That is, the apparently natural procedure of taking a white-noise sequence $s(m\Delta t)$ for $m = 1, 2, \ldots, T/\Delta t$, and computing the square amplitude of its Fourier transform at frequency ω,

$$\frac{\Delta T}{T} \left| \sum_{m=1}^{T/\Delta t} s(m\Delta t) \exp(-i\omega m \Delta t) \right|^2,$$

is a biased and extremely noisy way of estimating $\tilde{Q}_{ss}(\omega)$. This estimator is called the periodogram. The statistical problems with the periodogram, and some of the many suggested solutions, are discussed in almost any textbook on spectral analysis (see, e.g., Percival and Waldron, 1993).

periodogram

B: Moments of the Poisson Distribution

The average number of spikes generated by a Poisson process with constant rate r over a time T is

$$\langle n \rangle = \sum_{n=0}^{\infty} n P_T[n] = \sum_{n=0}^{\infty} \frac{n(rT)^n}{n!} \exp(-rT), \tag{1.45}$$

and the variance in the spike count is

$$\sigma_n^2(T) = \sum_{n=0}^{\infty} n^2 P_T[n] - \langle n \rangle^2 = \sum_{n=0}^{\infty} \frac{n^2(rT)^n}{n!} \exp(-rT) - \langle n \rangle^2. \tag{1.46}$$

To compute these quantities, we need to calculate the two sums appearing in these equations. A good way to do this is to compute the moment-generating function

moment-generating function

$$g(\alpha) = \sum_{n=0}^{\infty} \frac{(rT)^n \exp(\alpha n)}{n!} \exp(-rT). \tag{1.47}$$

The kth derivative of g with respect to α, evaluated at the point $\alpha = 0$, is

$$\left.\frac{d^k g}{d\alpha^k}\right|_{\alpha=0} = \sum_{n=0}^{\infty} \frac{n^k (rT)^n}{n!} \exp(-rT), \tag{1.48}$$

so once we have computed g, we need to calculate only its first and second derivatives to determine the sums we need. Rearranging the terms a bit, and recalling that $\exp(z) = \sum z^n/n!$, we find

$$g(\alpha) = \exp(-rT) \sum_{n=0}^{\infty} \frac{\left(rT \exp(\alpha)\right)^n}{n!} = \exp(-rT) \exp\left(rTe^\alpha\right). \tag{1.49}$$

The derivatives are then

$$\frac{dg}{d\alpha} = rTe^\alpha \exp(-rT) \exp(rTe^\alpha) \tag{1.50}$$

and

$$\frac{d^2 g}{d\alpha^2} = (rTe^\alpha)^2 \exp(-rT) \exp(rTe^\alpha) + rTe^\alpha \exp(-rT) \exp(rTe^\alpha). \tag{1.51}$$

Evaluating these at $\alpha = 0$ and putting the results into equations 1.45 and 1.46 gives the results $\langle n \rangle = rT$ and $\sigma_n^2(T) = (rT)^2 + rT - (rT)^2 = rT$.

C: Inhomogeneous Poisson Statistics

The probability density for a particular spike sequence with spike times t_i for $i = 1, 2, \ldots, n$ is obtained from the corresponding probability distribution by multiplying the probability that the spikes occur when they do by the probability that no other spikes occur. We begin by computing the probability that no spikes are generated during the time interval from t_i to t_{i+1} between two adjacent spikes. We determine this by dividing the interval into M bins of size Δt and setting $M\Delta t = t_{i+1} - t_i$. We will ultimately take the limit $\Delta t \to 0$. The firing rate during bin m within this interval is $r(t_i + m\Delta t)$. Because the probability of firing a spike in this bin is $r(t_i + m\Delta t)\Delta t$, the probability of not firing a spike is $1 - r(t_i + m\Delta t)\Delta t$. To have no spikes during the entire interval, we must string together M such bins, and the probability of this occurring is the product of the individual probabilities,

$$P[\text{no spikes}] = \prod_{m=1}^{M} (1 - r(t_i + m\Delta t)\Delta t). \tag{1.52}$$

We evaluate this expression by taking its logarithm,

$$\ln P[\text{no spikes}] = \sum_{m=1}^{M} \ln (1 - r(t_i + m\Delta t)\Delta t), \tag{1.53}$$

using the fact that the logarithm of a product is the sum of the logarithms of the multiplied terms. Using the approximation $\ln(1 - r(t_i + m\Delta t)\Delta t) \approx -r(t_i + m\Delta t)\Delta t$, valid for small Δt, we can simplify this to

$$\ln P[\text{no spikes}] = -\sum_{m=1}^{M} r(t_i + m\Delta t)\Delta t. \tag{1.54}$$

In the limit $\Delta t \to 0$, the approximation becomes exact and this sum becomes the integral of $r(t)$ from t_i to t_{i+1},

$$\ln P[\text{no spikes}] = -\int_{t_i}^{t_{i+1}} dt\, r(t). \tag{1.55}$$

Exponentiating this equation gives the result we need,

$$P[\text{no spikes}] = \exp\left(-\int_{t_i}^{t_{i+1}} dt\, r(t)\right). \tag{1.56}$$

The probability density $p[t_1, t_2, \ldots, t_n]$ is the product of the densities for the individual spikes and the probabilities of not generating spikes during the interspike intervals, between time 0 and the first spike, and between the time of the last spike and the end of the trial period:

$$
\begin{aligned}
p[t_1, t_2, \ldots, t_n] &= \exp\left(-\int_0^{t_1} dt\, r(t)\right)\exp\left(-\int_{t_n}^{T} dt\, r(t)\right) \times \\
&\quad r(t_n)\prod_{i=1}^{n-1} r(t_i)\exp\left(-\int_{t_i}^{t_{i+1}} dt\, r(t)\right).
\end{aligned} \tag{1.57}
$$

The exponentials in this expression all combine because the product of exponentials is the exponential of the sum, so the different integrals in this sum add up to form a single integral:

$$
\begin{aligned}
&\exp\left(-\int_0^{t_1} dt\, r(t)\right)\exp\left(-\int_{t_n}^{T} dt\, r(t)\right)\prod_{i=1}^{n-1}\exp\left(-\int_{t_i}^{t_{i+1}} dt\, r(t)\right) \\
&= \exp\left(-\left(\int_0^{t_1} dt\, r(t) + \sum_{i=1}^{n-1}\int_{t_i}^{t_{i+1}} dt\, r(t) + \int_{t_n}^{T} dt\, r(t)\right)\right) \\
&= \exp\left(-\int_0^{T} dt\, r(t)\right).
\end{aligned} \tag{1.58}
$$

Substituting this into 1.57 gives the result in equation 1.37.

1.8 Annotated Bibliography

Braitenberg & Schuz (1991) provides some of the quantitative measures of neuroanatomical properties of cortex that we quote. **Rieke et al. (1997)**

describes the analysis of spikes and the relationships between neural responses and stimuli, and is a general reference for material we present in chapters 1–4. **Gabbiani & Koch (1998)** provides another account of some of this material. The mathematics underlying point processes, the natural statistical model for spike sequences, is found in **Cox (1962)** and **Cox & Isham (1980)**, including the relationship between the Fano factor and the coefficient of variation. A general analysis of histogram representations appears in **Scott (1992)**, and white-noise and filtering techniques (our analysis of which continues in chapter 2) are described in de Boer & Kuyper (1968), **Marmarelis & Marmarelis (1978)**, and **Wiener (1958)**. Berry & Meister (1998) discuss the effects of refractoriness on patterns of spiking.

In chapters 1 and 3, we discuss two systems associated with studies of spike encoding; the H1 neuron in the visual system of flies, reviewed by **Rieke et al. (1997)**, and area MT of monkeys, discussed by Parker & Newsome (1998). **Wandell (1995)** introduces orientation and disparity tuning, relevant to examples presented in this chapter.

2 Neural Encoding II: Reverse Correlation and Visual Receptive Fields

2.1 Introduction

The spike-triggered average stimulus introduced in chapter 1 is a standard way of characterizing the selectivity of a neuron. In this chapter, we show how spike-triggered averages and reverse-correlation techniques can be used to construct estimates of firing rates evoked by arbitrary time-dependent stimuli. Firing rates calculated directly from reverse-correlation functions provide only a linear estimate of the response of a neuron, but in this chapter we also present various methods for including nonlinear effects such as firing thresholds.

Spike-triggered averages and reverse-correlation techniques have been used extensively to study properties of visually responsive neurons in the retina (retinal ganglion cells), lateral geniculate nucleus (LGN), and primary visual cortex (V1 or area 17). At these early stages of visual processing, the responses of some neurons (simple cells in primary visual cortex, for example) can be described quite accurately using this approach. Other neurons (complex cells in primary visual cortex, for example) can be described by extending the formalism. Reverse-correlation techniques have also been applied to responses of neurons in visual areas V2, area 18, and MT, but they generally fail to capture the more complex and nonlinear features typical of responses at later stages of the visual system. Descriptions of visual responses based on reverse correlation are approximate, and they do not explain how visual responses arise from the synaptic, cellular, and network properties of retinal, LGN, and cortical circuits. Nevertheless, they provide an important framework for characterizing response selectivities, a reference point for identifying and characterizing novel effects, and a basis for building mechanistic models, some of which are discussed at the end of this chapter and in chapter 7.

retina
LGN
V1, area 17

2.2 Estimating Firing Rates

In chapter 1, we discussed a simple model in which firing rates were estimated as instantaneous functions of the stimulus, using response tuning curves. The activity of a neuron at time t typically depends on the behavior of the stimulus over a period of time starting a few hundred milliseconds prior to t and ending perhaps tens of milliseconds before t. Reverse-correlation methods can be used to construct a more accurate model that includes the effects of the stimulus over such an extended period of time. The basic problem is to construct an estimate $r_{est}(t)$ of the firing rate $r(t)$ evoked by a stimulus $s(t)$. The simplest way to construct an estimate is to assume that the firing rate at any given time can be expressed as a weighted sum of the values taken by the stimulus at earlier times. Because time is a continuous variable, this "sum" actually takes the form of an integral, and we write

firing rate estimate $r_{est}(t)$

$$r_{est}(t) = r_0 + \int_0^\infty d\tau \, D(\tau) s(t - \tau) \,. \tag{2.1}$$

The term r_0 accounts for any background firing that may occur when $s = 0$. $D(\tau)$ is a weighting factor that determines how strongly, and with what sign, the value of the stimulus at time $t - \tau$ affects the firing rate at time t. Note that the integral in equation 2.1 is a linear filter of the same form as the expressions used to compute $r_{approx}(t)$ in chapter 1.

As discussed in chapter 1, sensory systems tend to adapt to the absolute intensity of a stimulus. It is easier to account for the responses to fluctuations of a stimulus around some mean background level than it is to account for adaptation processes. We therefore assume throughout this chapter that the stimulus parameter $s(t)$ has been defined with its mean value subtracted out. This means that the time integral of $s(t)$ over the duration of a trial is 0.

We have provided a heuristic justification for the terms in equation 2.1 but, more formally, they correspond to the first two terms in a systematic expansion of the response in powers of the stimulus. Such an expansion, *Volterra expansion* called a Volterra expansion, is the functional equivalent of the Taylor series expansion used to generate power series approximations of functions. For the case we are considering, it takes the form

$$r_{est}(t) = r_0 + \int d\tau \, D(\tau) s(t - \tau) + \int d\tau_1 d\tau_2 \, D_2(\tau_1, \tau_2) s(t - \tau_1) s(t - \tau_2) +$$

$$\int d\tau_1 d\tau_2 d\tau_3 \, D_3(\tau_1, \tau_2, \tau_3) s(t - \tau_1) s(t - \tau_2) s(t - \tau_3) + \dots \,. \tag{2.2}$$

This series was rearranged by Wiener to make the terms easier to compute. *Wiener expansion* The first two terms of the Volterra and Wiener expansions take the same mathematical forms and are given by the two expressions on the right side *Wiener kernel* of equation 2.1. For this reason, D is called the first Wiener kernel, the linear kernel, or, when higher-order terms (terms involving more than one factor of the stimulus) are not being considered, simply the kernel.

To construct an estimate of the firing rate based on equation 2.1, we choose the kernel D to minimize the squared difference between the estimated response to a stimulus and the actual measured response averaged over the duration of the trial, T,

$$E = \frac{1}{T} \int_0^T dt \, (r_{\text{est}}(t) - r(t))^2 \,. \tag{2.3}$$

This expression can be minimized by setting its derivative with respect to the function D to 0 (see appendix A). The result is that D satisfies an equation involving two quantities introduced in chapter 1, the firing rate-stimulus correlation function, $Q_{rs}(\tau) = \int dt \, r(t) \, s(t+\tau)/T$, and the stimulus autocorrelation function, $Q_{ss}(\tau) = \int dt \, s(t) \, s(t+\tau)/T$:

optimal kernel

$$\int_0^\infty d\tau' \, Q_{ss}(\tau - \tau') D(\tau') = Q_{rs}(-\tau) \,. \tag{2.4}$$

The method we are describing is called reverse correlation because the firing rate–stimulus correlation function is evaluated at $-\tau$ in this equation.

Equation 2.4 can be solved most easily if the stimulus is white noise, although it can be solved in the general case as well (see appendix A). For a white-noise stimulus $Q_{ss}(\tau) = \sigma_s^2 \delta(\tau)$ (see chapter 1), so the left side of equation 2.4 is

$$\sigma_s^2 \int_0^\infty d\tau' \, \delta(\tau - \tau') D(\tau') = \sigma_s^2 D(\tau) \,. \tag{2.5}$$

As a result, the kernel that provides the best linear estimate of the firing rate is

white-noise kernel

$$D(\tau) = \frac{Q_{rs}(-\tau)}{\sigma_s^2} = \frac{\langle r \rangle C(\tau)}{\sigma_s^2} \,, \tag{2.6}$$

where $C(\tau)$ is the spike-triggered average stimulus and $\langle r \rangle$ is the average firing rate of the neuron. For the second equality, we have used the relation $Q_{rs}(-\tau) = \langle r \rangle C(\tau)$ from chapter 1. Based on this result, the standard method used to determine the optimal kernel is to measure the spike-triggered average stimulus in response to a white-noise stimulus.

In chapter 1, we introduced the H1 neuron of the fly visual system, which responds to moving images. Figure 2.1 shows a prediction of the firing rate of this neuron obtained from a linear filter. The velocity of the moving image is plotted in 2.1A, and two typical responses are shown in 2.1B. The firing rate predicted from a linear estimator, as discussed above, and the firing rate computed from the data by binning and counting spikes are compared in figure 2.1C. The agreement is good in regions where the measured rate varies slowly, but the estimate fails to capture high-frequency fluctuations of the firing rate, presumably because of nonlinear effects not captured by the linear kernel. Some such effects can be described by a static nonlinear function, as discussed below. Others may require including higher-order terms in a Volterra or Wiener expansion.

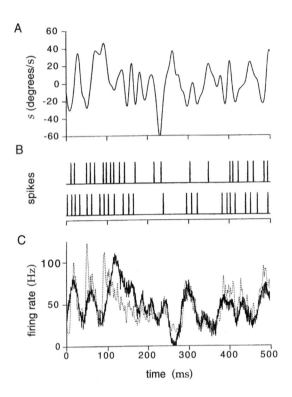

Figure 2.1 Prediction of the firing rate for an H1 neuron responding to a moving visual image. (A) The velocity of the image used to stimulate the neuron. (B) Two of the 100 spike sequences used in this experiment. (C) Comparison of the measured and computed firing rates. The dashed line is the firing rate extracted directly from the spike trains. The solid line is an estimate of the firing rate constructed by linearly filtering the stimulus with an optimal kernel. (Adapted from Rieke et al., 1997.)

The Most Effective Stimulus

Neuronal selectivity is often characterized by describing stimuli that evoke maximal responses. The reverse-correlation approach provides a basis for this procedure by relating the optimal kernel for firing-rate estimation to the stimulus predicted to evoke the maximum firing rate, subject to a constraint. A constraint is essential because the linear estimate in equation 2.1 is unbounded. The constraint we use is that the time integral of the square of the stimulus over the duration of the trial is held fixed. We call this integral the stimulus energy. The stimulus for which equation 2.1 predicts the maximum response at some fixed time subject to this constraint, is computed in appendix B. The result is that the stimulus producing the maximum response is proportional to the optimal linear kernel or, equivalently, to the white-noise spike-triggered average stimulus. This is an important result because in cases where a white-noise analysis has not been done, we may still have some idea what stimulus produces the maximum response.

The maximum stimulus analysis provides an intuitive interpretation of the linear estimate of equation 2.1. At fixed stimulus energy, the integral in 2.1 measures the overlap between the actual stimulus and the most effective stimulus. In other words, it indicates how well the actual stimulus matches the most effective stimulus. Mismatches between these two reduce the value of the integral and result in lower predictions for the firing rate.

Static Nonlinearities

The optimal kernel produces an estimate of the firing rate that is a linear function of the stimulus. Neurons and nervous systems are nonlinear, so a linear estimate is only an approximation, albeit a useful one. The linear prediction has two obvious problems: there is nothing to prevent the predicted firing rate from becoming negative, and the predicted rate does not saturate, but instead increases without bound as the magnitude of the stimulus increases. One way to deal with these and some of the other deficiencies of a linear prediction is to write the firing rate as a background rate plus a nonlinear function of the linearly filtered stimulus. We use L to represent the linear term we have been discussing thus far:

$$L(t) = \int_0^\infty d\tau D(\tau)s(t-\tau). \qquad (2.7)$$

The modification is to replace the linear prediction $r_{est}(t) = r_0 + L(t)$ with the generalization

$$r_{est}(t) = r_0 + F(L(t)), \qquad (2.8)$$

$r_{est}(t)$ *with static nonlinearity*

where F is an arbitrary function. F is called a static nonlinearity to stress that it is a function of the linear filter value evaluated instantaneously at the time of the rate estimation. If F is appropriately bounded from above and below, the estimated firing rate will never be negative or unrealistically large.

F can be extracted from data by means of the graphical procedure illustrated in figure 2.2A. First, a linear estimate of the firing rate is computed using the optimal kernel defined by equation 2.4. Next a plot is made of the pairs of points $(L(t), r(t))$ at various times and for various stimuli, where $r(t)$ is the actual rate extracted from the data. There will be a certain amount of scatter in this plot due to the inaccuracy of the estimation. If the scatter is not too large, however, the points should fall along a curve, and this curve is a plot of the function $F(L)$. It can be extracted by fitting a function to the points on the scatter plot. The function F typically contains constants used to set the firing rate to realistic values. These give us the freedom to normalize $D(\tau)$ in some convenient way, correcting for the arbitrary normalization by adjusting the parameters within F.

Static nonlinearities are used to introduce both firing thresholds and saturation into estimates of neural responses. Thresholds can be described by

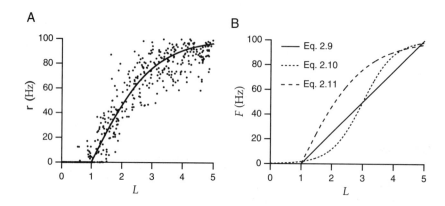

Figure 2.2 (A) A graphical procedure for determining static nonlinearities. Linear estimates L and actual firing rates r are plotted (solid points) and fitted by the function $F(L)$ (solid line). (B) Different static nonlinearities used in estimating neural responses. L is dimensionless, and equations 2.9, 2.10, and 2.11 have been used with $G = 25$ Hz, $L_0 = 1$, $L_{1/2} = 3$, $r_{max} = 100$ Hz, $g_1 = 2$, and $g_2 = 1/2$.

writing

$$F(L) = G[L - L_0]_+ , \tag{2.9}$$

threshold function

rectification

sigmoidal function

where L_0 is the threshold value that L must attain before firing begins. Above the threshold, the firing rate is a linear function of L, with G acting as the constant of proportionality. Half-wave rectification is a special case of this with $L_0 = 0$. That this function does not saturate is not a problem if large stimulus values are avoided. If needed, a saturating nonlinearity can be included in F, and a sigmoidal function is often used for this purpose,

$$F(L) = \frac{r_{max}}{1 + \exp\left(g_1(L_{1/2} - L)\right)} . \tag{2.10}$$

Here r_{max} is the maximum possible firing rate, $L_{1/2}$ is the value of L for which F achieves half of this maximal value, and g_1 determines how rapidly the firing rate increases as a function of L. Another choice that combines a hard threshold with saturation uses a rectified hyperbolic tangent function,

$$F(L) = r_{max}[\tanh\left(g_2(L - L_0)\right)]_+ , \tag{2.11}$$

where r_{max} and g_2 play similar roles as in equation 2.10, and L_0 is the threshold. Figure 2.2B shows the different nonlinear functions that we have discussed.

Although the static nonlinearity can be any function, the estimate of equation 2.8 is still restrictive because it allows for no dependence on weighted autocorrelations of the stimulus or other higher-order terms in the Volterra series. Furthermore, once the static nonlinearity is introduced, the linear kernel derived from equation 2.4 is no longer optimal because it was chosen to minimize the squared error of the linear estimate $r_{est}(t) = r_0 + L(t)$,

stimulus

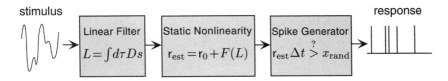

response

Figure 2.3 Simulating spiking responses to stimuli. The integral of the stimulus s times the optimal kernel D is first computed. The estimated firing rate is the background rate r_0 plus a nonlinear function of the output of the linear filter calculation. Finally, the estimated firing rate is used to drive a Poisson process that generates spikes.

not the estimate with the static nonlinearity $r_{est}(t) = r_0 + F(L(t))$. A theorem due to Bussgang (see appendix C) suggests that equation 2.6 will provide a reasonable kernel, even in the presence of a static nonlinearity, if the white noise stimulus used is Gaussian.

In some cases, the linear term of the Volterra series fails to predict the response even when static nonlinearities are included. Systematic improvements can be attempted by including more terms in the Volterra or Wiener series, but in practice it is quite difficult to go beyond the first few terms. The accuracy with which the first term, or first few terms, in a Volterra series can predict the responses of a neuron can sometimes be improved by replacing the parameter s in equation 2.7 with an appropriately chosen function of s, so that

$$L(t) = \int_0^\infty d\tau D(\tau) f(s(t - \tau)).$$ (2.12)

A reasonable choice for this function is the response tuning curve. With this choice, the linear prediction is equal to the response tuning curve, $L = f(s)$, for static stimuli, provided that the integral of the kernel D is equal to 1. For time-dependent stimuli, we can think of equation 2.12 as a dynamic extension of the response tuning curve.

A model of the spike trains evoked by a stimulus can be constructed by using the firing-rate estimate of equation 2.8 to drive a Poisson spike generator (see chapter 1). Figure 2.3 shows the structure of such a model with a linear filter, a static nonlinearity, and a stochastic spike generator. In the figure, spikes are shown being generated by comparing the spiking probability $r(t)\Delta t$ to a random number, although the other methods discussed in chapter 1 could be used instead. Also, the linear filter acts directly on the stimulus s in figure 2.3, but it could act instead on some function $f(s)$, such as the response tuning curve.

2.3 Introduction to the Early Visual System

Before discussing how reverse-correlation methods are applied to visually responsive neurons, we review the basic anatomy and physiology of the

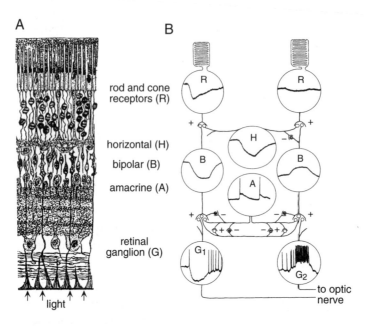

A

B

rod and cone
receptors (R)

horizontal (H)
bipolar (B)
amacrine (A)

retinal
ganglion (G)

light

to optic
nerve

Figure 2.4 (A) An anatomical diagram of the circuitry of the retina of a dog. Cell types are identified at right. In the intact eye, counterintuitively, light enters through the side opposite from the photoreceptors. (B) Intracellular recordings from retinal neurons of the mud puppy responding to a flash of light lasting for 1 s. In the column of cells on the left side of the diagram, the resulting hyperpolarizations are about 4 mV in the rod and retinal ganglion cells, and 8 mV in the bipolar cell. Pluses and minuses represent excitatory and inhibitory synapses, respectively. (A adapted from Nicholls et al., 1992; drawing from Cajal, 1911. B data from Werblin and Dowling 1969; figure adapted from Dowling, 1992.)

early stages of the visual system. The conversion of a light stimulus into an electrical signal and ultimately an action potential sequence occurs in the retina. Figure 2.4A is an anatomical diagram showing the five principal cell types of the retina, and figure 2.4B is a rough circuit diagram. In the retina, light is first converted into an electrical signal by a phototransduction cascade within rod and cone photoreceptor cells. Figure 2.4B shows intracellular recordings made in neurons of the retina of a mud puppy (an amphibian). The stimulus used for these recordings was a flash of light falling primarily in the region of the photoreceptor at the left of figure 2.4B. The rod cells, especially the one on the left side of figure 2.4B, are hyperpolarized by the light flash. This electrical signal is passed along to bipolar and horizontal cells through synaptic connections. Note that in one of the bipolar cells, the signal has been inverted, leading to depolarization. These smoothly changing membrane potentials provide a graded representation of the light intensity during the flash. This form of coding is adequate for signaling within the retina, where distances are small. However, it is inadequate for the task of conveying information from the retina to the brain.

*retinal ganglion
cells*

The output neurons of the retina are the retinal ganglion cells, whose axons form the optic nerve. As seen in figure 2.4B, the subthreshold potentials

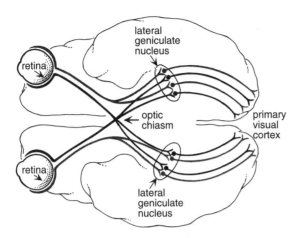

Figure 2.5 Pathway from the retina through the lateral geniculate nucleus (LGN) of the thalamus to the primary visual cortex in the human brain. (Adapted from Nicholls et al., 1992.)

of the two retinal ganglion cells shown are similar to those of the bipolar cells immediately above them in the figure, but now with superimposed action potentials. The two retinal ganglion cells shown in the figure have different responses and transmit different sequences of action potentials. G_2 fires while the light is on, and G_1 fires when it turns off. These are called ON and OFF responses, respectively. The optic nerve conducts the output spike trains of retinal ganglion cells to the lateral geniculate nucleus of the thalamus, which acts as a relay station between the retina and primary visual cortex (figure 2.5). Prior to arriving at the LGN, some retinal ganglion cell axons cross the midline at the optic chiasm. This allows the left and right sides of the visual fields from both eyes to be represented on the right and left sides of the brain, respectively (figure 2.5).

ON and OFF responses

Neurons in the retina, LGN, and primary visual cortex respond to light stimuli in restricted regions of the visual field called their receptive fields. Patterns of illumination outside the receptive field of a given neuron cannot generate a response directly, although they can significantly affect responses to stimuli within the receptive field. We do not consider such effects, although they are of considerable experimental and theoretical interest. In the monkey, cortical receptive fields range in size from around a tenth of a degree near the fovea to several degrees in the periphery. Within the receptive fields, there are regions where illumination higher than the background light intensity enhances firing, and other regions where lower illumination enhances firing. The spatial arrangement of these regions determines the selectivity of the neuron to different inputs. The term "receptive field" is often generalized to refer not only to the overall region where light affects neuronal firing, but also to the spatial and temporal structure within this region.

receptive fields

Visually responsive neurons in the retina, LGN, and primary visual cortex are divided into two classes, depending on whether or not the contribu-

*simple and
complex cells*

tions from different locations within the visual field sum linearly. X cells in the cat retina and LGN, P cells in the monkey retina and LGN, and simple cells in primary visual cortex appear to satisfy this assumption. Other neurons, such as Y cells in the cat retina and LGN, M cells in the monkey retina and LGN, and complex cells in primary visual cortex, do not show linear summation across the spatial receptive field, and nonlinearities must be included in descriptions of their responses. We do this for complex cells later in this chapter.

A first step in studying the selectivity of any neuron is to identify the types of stimuli that evoke strong responses. Retinal ganglion cells and LGN neurons have similar selectivities and respond best to circular spots of light surrounded by darkness or dark spots surrounded by light. In primary visual cortex, many neurons respond best to elongated light or dark bars or to boundaries between light and dark regions. Gratings with alternating light and dark bands are effective and frequently used stimuli for these neurons.

Many visually responsive neurons react strongly to sudden transitions in the level of image illumination, a temporal analogue of their responsiveness to light-dark spatial boundaries. Static images are not very effective at evoking visual responses. In awake animals, images are constantly kept in motion across the retina by eye movements. In experiments in which the eyes are fixed, moving light bars and gratings, or gratings undergoing periodic light-dark reversals (called counterphase gratings) are more effective stimuli than static images. Some neurons in primary visual cortex are directionally selective; they respond more strongly to stimuli moving in one direction than in the other.

To streamline the discussion in this chapter, we consider only gray-scale images, although the methods presented can be extended to include color. We also restrict the discussion to two-dimensional visual images, ignoring how visual responses depend on viewing distance and encode depth. In discussing the response properties of retinal, LGN, and V1 neurons, we do not follow the path of the visual signal, nor the historical order of experimentation, but instead begin with primary visual cortex and then move back to the LGN and retina. In this chapter, the emphasis is on properties of individual neurons; we discuss encoding by populations of visually responsive neurons in chapter 10.

The Retinotopic Map

A striking feature of most visual areas in the brain, including primary visual cortex, is that the visual world is mapped onto the cortical surface in a topographic manner. This means that neighboring points in a visual image evoke activity in neighboring regions of visual cortex. The retinotopic map refers to the transformation from the coordinates of the visual world to the corresponding locations on the cortical surface.

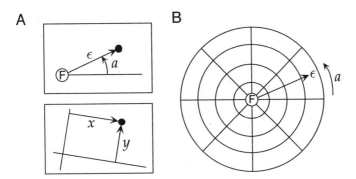

Figure 2.6 (A) Two coordinate systems used to parameterize image location. Each rectangle represents a tangent screen, and the filled circle is the location of a particular image point on the screen. The upper panel shows polar coordinates. The origin of the coordinate system is the fixation point F, the eccentricity ϵ is proportional to the radial distance from the fixation point to the image point, and a is the angle between the radial line from F to the image point and the horizontal axis. The lower panel shows Cartesian coordinates. The location of the origin for these coordinates and the orientation of the axes are arbitrary. They are usual chosen to center and align the coordinate system with respect to a particular receptive field being studied. (B) A bull's-eye pattern of radial lines of constant azimuth, and circles of constant eccentricity. The center of this pattern at zero eccentricity is the fixation point F. Such a pattern was used to generate the image in figure 2.7A.

Objects located a fixed distance from one eye lie on a sphere. Locations on this sphere can be represented using the same longitude and latitude angles used for the surface of the earth. Typically, the "north pole" for this spherical coordinate system is located at the fixation point, the image point that focuses onto the fovea or center of the retina. In this system of coordinates (figure 2.6), the latitude coordinate is called the eccentricity, ϵ, and the longitude coordinate, measured from the horizontal meridian, is called the azimuth, a. In primary visual cortex, the visual world is split in half, with the region $-90° \leq a \leq +90°$ for ϵ from $0°$ to about $70°$ (for both eyes) represented on the left side of the brain, and the reflection of this region about the vertical meridian represented on the right side of the brain.

eccentricity ϵ
azimuth a

In most experiments, images are displayed on a flat screen (called a tangent screen) that does not coincide exactly with the sphere discussed in the previous paragraph. However, if the screen is not too large, the difference is negligible, and the eccentricity and azimuth angles approximately coincide with polar coordinates on the screen (figure 2.6A). Ordinary Cartesian coordinates can also be used to identify points on the screen (figure 2.6A). The eccentricity ϵ and the x and y coordinates of the Cartesian system are based on measuring distances on the screen. However, it is customary to divide these measured distances by the distance from the eye to the screen and to multiply the result by $180°/\pi$ so that these coordinates are ultimately expressed in units of degrees. This makes sense because it is the angular, not the absolute, size and location of an image that is typically relevant for studies of the visual system.

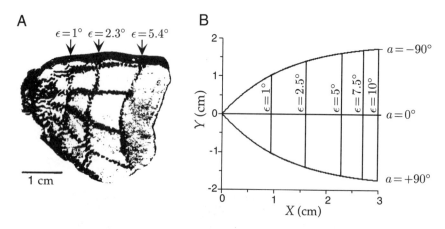

Figure 2.7 (A) An autoradiograph of the posterior region of the primary visual cortex from the left side of a macaque monkey brain. The pattern is a radioactive trace of the activity evoked by an image like that in figure 2.6B. The vertical lines correspond to circles at eccentricities of 1°, 2.3°, and 5.4°, and the horizontal lines (from top to bottom) represent radial lines in the visual image at a values of $-90°$, $-45°$, $0°$, $45°$, and $90°$. Only the part of cortex corresponding to the central region of the visual field on one side is shown. (B) The mathematical map from the visual coordinates ϵ and a to the cortical coordinates X and Y described by equations 2.15 and 2.17. (A adapted from Tootell et al., 1982.)

Figure 2.7A shows a dramatic illustration of the retinotopic map in the primary visual cortex of a monkey. The pattern on the cortex seen in figure 2.7A was produced by imaging a radioactive analogue of glucose that was taken up by active neurons while a monkey viewed a visual image consisting of concentric circles and radial lines, similar to the pattern in figure 2.6B. The vertical lines correspond to the circles in the image, and the roughly horizontal lines are due to the activity evoked by the radial lines. The fovea is represented at the leftmost pole of this piece of cortex, and eccentricity increases toward the right. Azimuthal angles are positive in the lower half of the piece of cortex shown, and negative in the upper half.

To describe the map in figure 2.7A mathematically, we write the horizontal and vertical coordinates, X and Y, describing points on the cortical surface as functions of the eccentricity ϵ and azimuth a of the corresponding points in the visual field, $X(\epsilon, a)$, $Y(\epsilon, a)$. This map is characterized

cortical magnification factor

by local factors, each called a cortical magnification factor, that indicate the relationship between small displacements ΔX, ΔY across the cortex and the corresponding small image displacements $\Delta \epsilon$, Δa. In general, these factors can be derived from the four elements of the Jacobian matrix, $\partial X/\partial \epsilon$, $\partial X/\partial a$, $\partial Y/\partial \epsilon$, and $\partial Y/\partial a$. However, few experiments characterize all four elements, and it is common to include additional constraints.

Figure 2.7B shows an example of such a map. Here, we assume that $X(\epsilon)$ is only a function of eccentricity, and that $Y(\epsilon, a)$ is proportional to the azimuth angle a. Further, we assume that purely radial ($\Delta a = 0$) and purely meridional ($\Delta\epsilon = 0$) displacements have the same cortical magnification factor $M(\epsilon)$, which itself is only a function of eccentricity. Consider two nearby image points with coordinates ϵ, a and $\epsilon + \Delta\epsilon, a$, separated by an angular distance $\Delta\epsilon$. The distance separating the activity evoked by these two image points on the cortex is ΔX. By the definition of the cortical magnification factor, these two quantities satisfy $\Delta X = M(\epsilon)\Delta\epsilon$ or, taking the limit as ΔX and $\Delta\epsilon$ go to 0,

$$\frac{dX}{d\epsilon} = M(\epsilon). \tag{2.13}$$

For the macaque monkey, results such as figure 2.7A suggest that

$$M(\epsilon) = \frac{\lambda}{\epsilon_0 + \epsilon}, \tag{2.14}$$

with $\lambda \approx 12$ mm and $\epsilon_0 \approx 1°$. Integrating equation 2.13 and defining $X = 0$ to be the point representing $\epsilon = 0$, we find

$$X = \lambda \ln(1 + \epsilon/\epsilon_0). \tag{2.15}$$

For purely meridional displacements, the angular distance between two points at eccentricity ϵ with an azimuthal angle difference Δa is $\Delta a\epsilon\pi/180°$. Here, the factor of ϵ corrects for the increase of arc length as a function of eccentricity, and the factor of $\pi/180°$ converts ϵ from degrees to radians. The separation on the cortex, ΔY, corresponding to these points has a magnitude given by the cortical amplification times this distance. Taking the limit $\Delta a \to 0$, we find that

$$\frac{dY}{da} = -\frac{\epsilon\pi}{180°}M(\epsilon). \tag{2.16}$$

The minus sign in this relationship appears because the visual field is inverted on the cortex. Solving equation 2.16 gives

$$Y = -\frac{\lambda\epsilon a\pi}{(\epsilon_0 + \epsilon)180°}. \tag{2.17}$$

The map defined by equations 2.15 and 2.17 is only approximate, particularly for small eccentricities. It is also not isotropic, which means that the magnification factor $M(\epsilon)$ only describes displacements for which either $\Delta\epsilon = 0$ or $\Delta a = 0$. Nevertheless, figure 2.7B shows that these coordinates agree fairly well with the map in figure 2.7A.

For eccentricities appreciably greater than $1°$, equations 2.15 and 2.17 reduce to $X \approx \lambda \ln(\epsilon/\epsilon_0)$ and $Y \approx -\lambda\pi a/180°$. These two formulas can be combined by defining the complex numbers $Z = X + iY$ and $z = (\epsilon/\epsilon_0)$ $\exp(-i\pi a/180°)$ (with i equal to the square root of -1) and writing $Z = \lambda \ln(z)$. For this reason, the cortical map is sometimes called a complex logarithmic map. For an image scaled radially by a factor γ, eccentricities change according to $\epsilon \to \gamma\epsilon$ while a is unaffected. Scaling of the eccentricity produces a shift $X \to X + \lambda \ln(\gamma)$ over the range of values where the simple logarithmic form of the map is valid. The logarithmic transformation thus causes images that are scaled radially outward on the retina to be represented at locations on the cortex translated in the X direction.

complex logarithmic map

Visual Stimuli

Earlier in this chapter, we used the function $s(t)$ to characterize a time-dependent stimulus. The description of visual stimuli is more complex. Gray-scale images appearing on a two-dimensional surface, such as a video monitor, can be described by giving the luminance, or light intensity, at each point on the screen. These pixel locations are parameterized by Cartesian coordinates x and y, as in the lower panel of figure 2.6A. However, pixel-by-pixel light intensities are not a useful way of parameterizing a visual image for the purposes of characterizing neuronal responses. This is because visually responsive neurons, like many sensory neurons, adapt to the overall level of screen illumination. To avoid dealing with adaptation effects, we describe the stimulus by a function $s(x, y, t)$ that is proportional to the difference between the luminance at the point (x, y) at time t and the average or background level of luminance. Often $s(x, y, t)$ is also divided by the background luminance level, making it dimensionless. The *contrast* resulting quantity is called the contrast.

During recordings, visual neurons are usually stimulated by images that vary over both space and time. A commonly used stimulus, the counter-*counterphase* phase sinusoidal grating, is described by
sinusoidal grating

$$s(x, y, t) = A \cos \left(Kx \cos \Theta + Ky \sin \Theta - \Phi \right) \cos(\omega t) . \qquad (2.18)$$

Figure 2.8 shows a similar grating (a spatial square wave is drawn rather than a sinusoid) and illustrates the significance of the parameters K, Θ, *spatial frequency K* Φ, and ω. K and ω are the spatial and temporal frequencies of the grat-*frequency ω* ing (these are angular frequencies), Θ is its orientation, Φ is its spatial *orientation Θ* phase, and A is its contrast amplitude. This stimulus oscillates in both *spatial phase Φ* space and time. At any fixed time, it oscillates in the direction perpendic-*amplitude A* ular to the orientation angle Θ as a function of position, with wavelength $2\pi/K$ (figure 2.8A). At any fixed position, it oscillates in time with period $2\pi/\omega$ (figure 2.8B). For convenience, Θ is measured relative to the y axis rather than the x axis, so that a stimulus with $\Theta = 0$ varies in the x, but not in the y, direction. Φ determines the spatial location of the light and dark stripes of the grating. Changing Φ by an amount $\Delta\Phi$ shifts the grating in the direction perpendicular to its orientation by a fraction $\Delta\Phi/2\pi$ of its wavelength. The contrast amplitude A controls the maximum degree of difference between light and dark areas. Because x and y are measured in degrees, K is expressed in the rather unusual units of radians per degree and $K/2\pi$ is typically reported in units of cycles per degree. Φ has units of radians, ω is in radians/s, and $\omega/2\pi$ is in Hz.

Experiments that consider reverse correlation and spike-triggered averages use various types of random and white-noise stimuli in addition to bars and gratings. A white-noise stimulus, in this case, is one that is *white-noise image* uncorrelated in both space and time so that

$$\frac{1}{T} \int_0^T dt \, s(x, y, t) s(x', y', t + \tau) = \sigma_s^2 \delta(\tau) \delta(x - x') \delta(y - y') . \qquad (2.19)$$

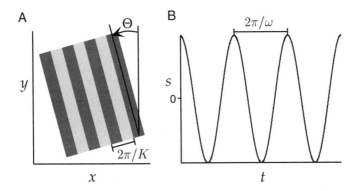

Figure 2.8 A counterphase grating. (A) A portion of a square-wave grating analogous to the sinusoidal grating of equation 2.18. The lighter stripes are regions where $s > 0$, and $s < 0$ within the darker stripes. K determines the wavelength of the grating and Θ, its orientation. Changing its spatial phase, Φ, shifts the entire light-dark pattern in the direction perpendicular to the stripes. (B) The light-dark intensity at any point of the spatial grating oscillates sinusoidally in time with period $2\pi/\omega$.

Of course, in practice a discrete approximation of such a stimulus must be used by dividing the image space into pixels and time into small bins. In addition, more structured random sets of images (randomly oriented bars, for example) are sometimes used to enhance the responses obtained during stimulation.

The Nyquist Frequency

Many factors limit the maximal spatial frequency that can be resolved by the visual system, but one interesting effect arises from the size and spacing of individual photoreceptors on the retina. The region of the retina with the highest resolution is the fovea at the center of the visual field. Within the macaque or human fovea, cone photoreceptors are densely packed in a regular array. Along any direction in the visual field, a regular array of tightly packed photoreceptors of size Δx samples points at locations $m\Delta x$ for $m = 1, 2, \ldots$. The (angular) frequency that defines the resolution of such an array is called the Nyquist frequency and is given by   *Nyquist frequency*

$$K_{\text{nyq}} = \frac{\pi}{\Delta x}. \tag{2.20}$$

To understand the significance of the Nyquist frequency, consider sampling two cosine gratings with spatial frequencies of K and $2K_{\text{nyq}} - K$, with $K < K_{\text{nyq}}$. These are described by $s = \cos(Kx)$ and $s = \cos((2K_{\text{nyq}} - K)x)$. At the sampled points, these functions are identical because $\cos((2K_{\text{nyq}} - K)m\Delta x) = \cos(2\pi m - Km\Delta x) = \cos(-Km\Delta x) = \cos(Km\Delta x)$ by the periodicity and evenness of the cosine function (see figure 2.9). As a result, these two gratings cannot be distinguished by examining them only at the sampled points. Any two spatial frequencies $K < K_{\text{nyq}}$ and $2K_{\text{nyq}} - K$

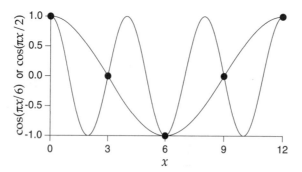

Figure 2.9 Aliasing and the Nyquist frequency. The two curves are the functions $\cos(\pi x/6)$ and $\cos(\pi x/2)$ plotted against x, and the dots show points sampled with a spacing of $\Delta x = 3$. The Nyquist frequency in this case is $\pi/3$, and the two cosine curves match at the sampled points because their spatial frequencies satisfy the relation $2\pi/3 - \pi/6 = \pi/2$.

can be confused with one another in this way, a phenomenon known as aliasing. Conversely, if an image is constructed solely of frequencies less than K_{nyq}, it can be reconstructed perfectly from the finite set of samples provided by the array. There are 120 cones per degree at the fovea of the macaque retina, which makes $K_{nyq}/(2\pi) = 1/(2\Delta x) = 60$ cycles per degree. In this result, we have divided the right side of equation 2.20, which gives K_{nyq} in units of radians per degree, by 2π to convert the answer to cycles per degree.

2.4   Reverse-Correlation Methods: Simple Cells

The spike-triggered average for visual stimuli is defined, as in chapter 1, as the average over trials of stimuli evaluated at times $t_i - \tau$, where t_i for $i = 1, 2, \ldots, n$ are the spike times. Because the light intensity of a visual image depends on location as well as time, the spike-triggered average stimulus is a function of three variables,

$$C(x, y, \tau) = \frac{1}{\langle n \rangle} \left\langle \sum_{i=1}^{n} s(x, y, t_i - \tau) \right\rangle .$$ (2.21)

Here, as in chapter 1, the brackets denote trial averaging, and we have used the approximation $1/n \approx 1/\langle n \rangle$. $C(x, y, \tau)$ is the average value of the visual stimulus at the point (x, y) a time τ before a spike was fired. Similarly, we can define the correlation function between the firing rate at time t and the stimulus at time $t + \tau$, for trials of duration T, as

$$Q_{rs}(x, y, \tau) = \frac{1}{T} \int_0^T dt\, r(t) s(x, y, t + \tau) .$$ (2.22)

The spike-triggered average is related to the reverse-correlation function, as discussed in chapter 1, by

$$C(x, y, \tau) = \frac{Q_{rs}(x, y, -\tau)}{\langle r \rangle},$$ (2.23)

where $\langle r \rangle$ is, as usual, the average firing rate over the entire trial, $\langle r \rangle = \langle n \rangle / T$.

To estimate the firing rate of a neuron in response to a particular image, we add a function of the output of a linear filter of the stimulus to the background firing rate r_0, as in equation 2.8, $r_{est}(t) = r_0 + F(L(t))$. As in equation 2.7, the linear estimate $L(t)$ is obtained by integrating over the past history of the stimulus with a kernel acting as the weighting function. Because visual stimuli depend on spatial location, we must decide how contributions from different image locations are to be combined to determine $L(t)$. The simplest assumption is that the contributions from *linear response* different spatial points sum linearly, so that $L(t)$ is obtained by integrating *estimate* over all x and y values:

$$L(t) = \int_0^\infty d\tau \int dx dy \, D(x, y, \tau) s(x, y, t - \tau).$$ (2.24)

The kernel $D(x, y, \tau)$ determines how strongly, and with what sign, the visual stimulus at the point (x, y) and at time $t - \tau$ affects the firing rate of the neuron at time t. As in equation 2.6, the optimal kernel is given in terms of the firing rate-stimulus correlation function, or the spike-triggered average, for a white-noise stimulus with variance parameter σ_s^2 by

$$D(x, y, \tau) = \frac{Q_{rs}(x, y, -\tau)}{\sigma_s^2} = \frac{\langle r \rangle C(x, y, \tau)}{\sigma_s^2}.$$ (2.25)

The kernel $D(x, y, \tau)$ defines the space-time receptive field of a neuron. Because $D(x, y, \tau)$ is a function of three variables, it can be difficult to *space-time* measure and visualize. For some neurons, the kernel can be written as *receptive field* a product of two functions, one that describes the spatial receptive field and the other, the temporal receptive field,

$$D(x, y, \tau) = D_s(x, y) D_t(\tau).$$ (2.26)

Such neurons are said to have separable space-time receptive fields. Separability requires that the spatial structure of the receptive field not *separable* change over time except by an overall multiplicative factor. When *receptive field* $D(x, y, \tau)$ cannot be written as the product of two terms, the neuron is said to have a nonseparable space-time receptive field. Given the freedom *nonseparable* in equation 2.8 to set the scale of D (by suitably adjusting the function F), *receptive field* we typically normalize D_s so that its integral is 1, and use a similar rule for the components from which D_t is constructed. We begin our analysis by studying first the spatial and then the temporal components of a separable space-time receptive field, and then proceed to the nonseparable case. For simplicity, we ignore the possibility that cells can have slightly different receptive fields for the two eyes, which underlies the disparity tuning considered in chapter 1.

Spatial Receptive Fields

Figures 2.10A and C show the spatial structure of spike-triggered average stimuli for two simple cells in the primary visual cortex of a cat (area 17) with approximately separable space-time receptive fields. These receptive fields are elongated in one direction. There are some regions within the receptive field where D_s is positive, called ON regions, and others where it is negative, called OFF regions. The integral of the linear kernel times the stimulus can be visualized by noting how the OFF (black) and ON (white) regions overlap the image (see figure 2.11). The response of a neuron is enhanced if ON regions are illuminated ($s > 0$) or if OFF regions are darkened ($s < 0$) relative to the background level of illumination. Conversely, they are suppressed by darkening ON regions or illuminating OFF regions. As a result, the neurons of figures 2.10A and C respond most vigorously to light-dark edges positioned along the border between the ON and OFF regions, and oriented parallel to this border and to the elongated direction of the receptive fields (figure 2.11). Figures 2.10 and 2.11 show receptive fields with two major subregions. Simple cells are found with from one to five subregions. Along with the ON-OFF patterns we have seen, another typical arrangement is a three-lobed receptive field with OFF-ON-OFF or ON-OFF-ON subregions.

Gabor function

A mathematical approximation of the spatial receptive field of a simple cell is provided by a Gabor function, which is a product of a Gaussian function and a sinusoidal function. Gabor functions are by no means the only functions used to fit spatial receptive fields of simple cells. For example, gradients of Gaussians are sometimes used. However, we will stick to Gabor functions, and to simplify the notation, we choose the coordinates x and y so that the borders between the ON and OFF regions are parallel to the y axis. We also place the origin of the coordinates at the center of the receptive field. With these choices, we can approximate the observed receptive field structures using the Gabor function

$$D_s(x, y) = \frac{1}{2\pi\sigma_x\sigma_y} \exp\left(-\frac{x^2}{2\sigma_x^2} - \frac{y^2}{2\sigma_y^2}\right) \cos(kx - \phi). \qquad (2.27)$$

rf size σ_x, σ_y

preferred spatial frequency k

preferred spatial phase ϕ

The parameters in this function determine the properties of the spatial receptive field: σ_x and σ_y determine its extent in the x and y directions, respectively; k, the preferred spatial frequency, determines the spacing of light and dark bars that produce the maximum response (the preferred spatial wavelength is $2\pi/k$); and ϕ is the preferred spatial phase, which determines where the ON-OFF boundaries fall within the receptive field. For this spatial receptive field, the sinusoidal grating described by equation 2.18 that produces the maximum response for a fixed value of A has $K = k$, $\Phi = \phi$, and $\Theta = 0$.

Figures 2.10B and 2.10D show Gabor functions chosen specifically to match the data in figures 2.10A and 2.10C. Figure 2.12 shows x and y plots of a variety of Gabor functions with different parameter values. As

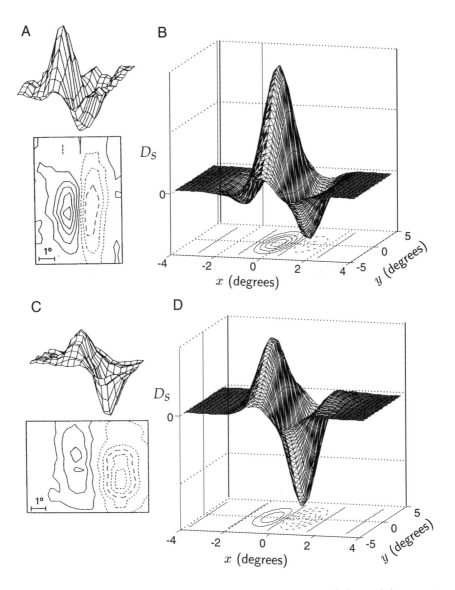

Figure 2.10 Spatial receptive field structure of simple cells. (A) and (C) Spatial structure of the receptive fields of two neurons in cat primary visual cortex determined by averaging stimuli between 50 ms and 100 ms prior to an action potential. The upper plots are three-dimensional representations, with the horizontal dimensions acting as the x-y plane and the vertical dimension indicating the magnitude and sign of $D_s(x, y)$. The lower contour plots represent the x-y plane. Regions with solid contour curves are ON areas where $D_s(x, y) > 0$, and regions with dashed contours are OFF areas where $D_s(x, y) < 0$. (B) and (D) Gabor functions (equation 2.27) with $\sigma_x = 1°$, $\sigma_y = 2°$, $1/k = 0.56°$, and $\phi = 1 - \pi/2$ (B) or $\phi = 1 - \pi$ (D), chosen to match the receptive fields in A and C. (A and C adapted from Jones and Palmer, 1987a.)

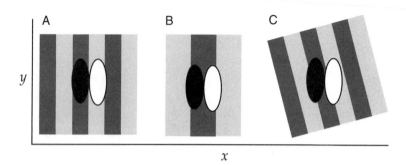

Figure 2.11 Grating stimuli superimposed on spatial receptive fields similar to those shown in figure 2.10. The receptive field is shown as two oval regions, one dark to represent an OFF area where $D_s < 0$ and one white to denote an ON region where $D_s > 0$. (A) A grating with the spatial wavelength, orientation, and spatial phase shown produces a high firing rate because a dark band completely overlaps the OFF area of the receptive field and a light band overlaps the ON area. (B) The grating shown is nonoptimal due to a mismatch in both the spatial phase and frequency, so that the ON and OFF regions each overlap both light and dark stripes. (C) The grating shown is at a nonoptimal orientation because each region of the receptive field overlaps both light and dark stripes.

seen in figure 2.12, Gabor functions can have various types of symmetry, and variable numbers of significant oscillations (or subregions) within the Gaussian envelope. The number of subregions within the receptive field is determined by the product $k\sigma_x$ and is typically expressed in terms *bandwidth* of a quantity known as the bandwidth b. The bandwidth is defined as $b = \log_2(K_+/K_-)$, where $K_+ > k$ and $K_- < k$ are the spatial frequencies of gratings that produce one-half the response amplitude of a grating with $K = k$. High bandwidths correspond to low values of $k\sigma_x$, meaning that the receptive field has few subregions and poor spatial frequency selectivity. Neurons with more subfields are more selective to spatial frequency, and they have smaller bandwidths and larger values of $k\sigma_x$.

The bandwidth is the width of the spatial frequency tuning curve measured in octaves. The spatial frequency tuning curve as a function of K for a Gabor receptive field with preferred spatial frequency k and receptive field width σ_x is proportional to $\exp(-\sigma_x^2(k - K)^2/2)$ (see equation 2.34 below). The values of K_+ and K_- needed to compute the bandwidth are thus determined by the condition $\exp(-\sigma_x^2(k - K_\pm)^2/2) = 1/2$. Solving this equation gives $K_\pm = k \pm (2\ln(2))^{1/2}/\sigma_x$, from which we obtain

$$b = \log_2\left(\frac{k\sigma_x + \sqrt{2\ln(2)}}{k\sigma_x - \sqrt{2\ln(2)}}\right) \quad \text{or} \quad k\sigma_x = \sqrt{2\ln(2)}\,\frac{2^b + 1}{2^b - 1}. \qquad (2.28)$$

Bandwidth is defined only if $k\sigma_x > (2\ln(2))^{1/2}$, but this is usually the case. Bandwidths typically range from about 0.5 to 2.5, corresponding to $k\sigma_x$ between 1.7 and 6.9.

The response characterized by equation 2.27 is maximal if light-dark edges are parallel to the y axis, so the preferred orientation angle is 0. An arbitrary preferred orientation θ can be generated by rotating the coordi-

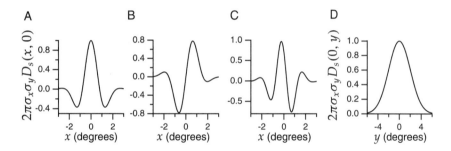

Figure 2.12 Gabor functions of the form given by equation 2.27. For convenience we plot the dimensionless function $2\pi\sigma_x\sigma_y D_s$. (A) A Gabor function with $\sigma_x = 1°$, $1/k = 0.5°$, and $\phi = 0$ plotted as a function of x for $y = 0$. This function is symmetric about $x = 0$. (B) A Gabor function with $\sigma_x = 1°$, $1/k = 0.5°$, and $\phi = \pi/2$ plotted as a function of x for $y = 0$. This function is antisymmetric about $x = 0$ and corresponds to using a sine instead of a cosine function in equation 2.27. (C) A Gabor function with $\sigma_x = 1°$, $1/k = 0.33°$, and $\phi = \pi/4$ plotted as a function of x for $y = 0$. This function has no particular symmetry properties with respect to $x = 0$. (D) The Gabor function of equation 2.27 with $\sigma_y = 2°$ plotted as a function of y for $x = 0$. This function is simply a Gaussian.

nates, making the substitutions $x \rightarrow x\cos(\theta) + y\sin(\theta)$ and $y \rightarrow y\cos(\theta) - x\sin(\theta)$ in equation 2.27. This produces a spatial receptive field that is maximally responsive to a grating with $\Theta = \theta$. Similarly, a receptive field centered at the point (x_0, y_0) rather than at the origin can be constructed by making the substitutions $x \rightarrow x - x_0$ and $y \rightarrow y - y_0$.

preferred
orientation θ

rf center x_0, y_0

Temporal Receptive Fields

Figure 2.13 reveals the temporal development of the space-time receptive field of a neuron in the cat primary visual cortex through a series of snapshots of its spatial receptive field. More than 300 ms prior to a spike, there is little correlation between the visual stimulus and the upcoming spike. Around 210 ms before the spike ($\tau = 210$ ms), a two-lobed OFF-ON receptive field, similar to the ones in figure 2.10, is evident. As τ decreases (recall that τ measures time in a reversed sense), this structure first fades away and then reverses, so that the receptive field 75 ms before a spike has the opposite sign from what appeared at $\tau = 210$ ms. Due to latency effects, the spatial structure of the receptive field is less significant for $\tau < 75$ ms. The stimulus preferred by this cell is thus an appropriately aligned dark-light boundary that reverses to a light-dark boundary over time.

Reversal effects like those seen in figure 2.13 are a common feature of space-time receptive fields. Although the magnitudes and signs of the different spatial regions in figure 2.13 vary over time, their locations and shapes remain fairly constant. This indicates that the neuron has, to a good approximation, a separable space-time receptive field. When a space-time receptive field is separable, the reversal can be described by a function $D_t(\tau)$ that rises from 0, becomes positive, then negative, and ultimately

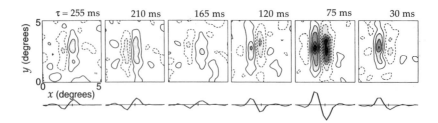

Figure 2.13 Temporal evolution of a spatial receptive field. Each panel is a plot of $D(x, y, \tau)$ for a different value of τ. As in figure 2.10, regions with solid contour curves are areas where $D(x, y, \tau) > 0$ and regions with dashed contours have $D(x, y, \tau) < 0$. The curves below the contour diagrams are one-dimensional plots of the receptive field as a function of x alone. The receptive field is maximally different from 0 for $\tau = 75$ ms with the spatial receptive field reversed from what it was at $\tau = 210$ ms. (Adapted from DeAngelis et al., 1995.)

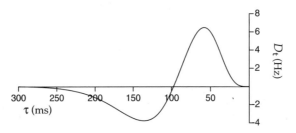

Figure 2.14 Temporal structure of a receptive field. The function $D_t(\tau)$ of equation 2.29 with $\alpha = 1/(15 \text{ ms})$.

returns to 0 as τ increases. Adelson and Bergen (1985) proposed the function shown in figure 2.14,

$$D_t(\tau) = \alpha \exp(-\alpha\tau) \left(\frac{(\alpha\tau)^5}{5!} - \frac{(\alpha\tau)^7}{7!} \right) , \qquad (2.29)$$

for $\tau \geq 0$, and $D_t(\tau) = 0$ for $\tau < 0$. Here, α is a constant that sets the scale for the temporal development of the function. Single-phase responses are also seen for V1 neurons, and these can be described by eliminating the second term in equation 2.29. Three-phase responses, which are sometimes seen, must be described by a more complicated function.

Response of a Simple Cell to a Counterphase Grating

The response of a simple cell to a counterphase grating stimulus (equation 2.18) can be estimated by computing the function $L(t)$. For the separable receptive field given by the product of the spatial factor in equation 2.27 and the temporal factor in 2.29, the linear estimate of the response can be written as the product of two terms,

$$L(t) = L_s L_t(t) , \qquad (2.30)$$

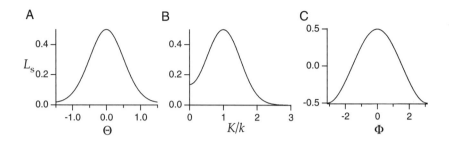

Figure 2.15 Selectivity of a Gabor filter with $\theta = \phi = 0$, $\sigma_x = \sigma_y = \sigma$, and $k\sigma = 2$ acting on a cosine grating with $A = 1$. (A) L_s as a function of stimulus orientation Θ for a grating with the preferred spatial frequency and phase, $K = k$ and $\Phi = 0$. (B) L_s as a function of the ratio of the stimulus spatial frequency to its preferred value, K/k, for a grating oriented in the preferred direction $\Theta = 0$ and with the preferred phase $\Phi = 0$. (C) L_s as a function of stimulus spatial phase Φ for a grating with the preferred spatial frequency and orientation, $K = k$ and $\Theta = 0$.

where

$$L_s = \int dxdy \, D_s(x, y) A \cos\left(Kx\cos(\Theta) + Ky\sin(\Theta) - \Phi\right) \qquad (2.31)$$

and

$$L_t(t) = \int_0^\infty d\tau \, D_t(\tau) \cos\left(\omega(t - \tau)\right). \qquad (2.32)$$

The reader is invited to compute these integrals for the case $\sigma_x = \sigma_y = \sigma$. To show the selectivity of the resulting spatial receptive fields, we plot (in figure 2.15) L_s as functions of the parameters Θ, K, and Φ that determine the orientation, spatial frequency, and spatial phase of the stimulus. It is also instructive to write out L_s for various special parameter values. First, if the spatial phase of the stimulus and the preferred spatial phase of the receptive field are 0 ($\Phi = \phi = 0$), we find that

$$L_s = A\exp\left(-\frac{\sigma^2(k^2 + K^2)}{2}\right)\cosh\left(\sigma^2 kK\cos(\Theta)\right), \qquad (2.33)$$

which determines the orientation and spatial frequency tuning for an optimal spatial phase. Second, for a grating with the preferred orientation $\Theta = 0$ and a spatial frequency that is not too small, the full expression for L_s can be simplified by noting that $\exp(-\sigma^2 kK) \approx 0$ for the values of $k\sigma$ normally encountered (for example, if $K = k$ and $k\sigma = 2$, $\exp(-\sigma^2 kK) = 0.02$). Using this approximation, we find

$$L_s = \frac{A}{2}\exp\left(-\frac{\sigma^2(k - K)^2}{2}\right)\cos(\phi - \Phi), \qquad (2.34)$$

which reveals a Gaussian dependence on spatial frequency and a cosine dependence on spatial phase.

The amplitude of the sinusoidally oscillating linear response estimate (equation 2.32) is plotted as a function of the temporal frequency of the

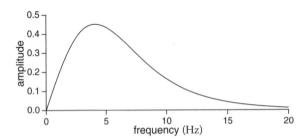

Figure 2.16 Frequency response of a model simple cell based on the temporal kernel of equation 2.29. The amplitude of the sinusoidal oscillations of $L_t(t)$ produced by a counterphase grating is plotted as a function of the temporal oscillation frequency, $\omega/2\pi$.

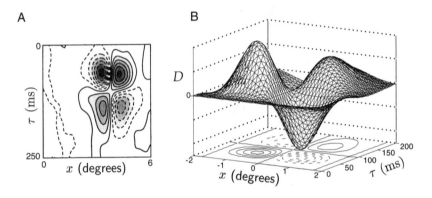

Figure 2.17 A separable space-time receptive field. (A) An x-τ plot of an approximately separable space-time receptive field from cat primary visual cortex. OFF regions are shown with dashed contour lines and ON regions with solid contour lines. The receptive field has side-by-side OFF and ON regions that reverse as a function of τ. (B) Mathematical description of the space-time receptive field in A constructed by multiplying a Gabor function (evaluated at $y = 0$) with $\sigma_x = 1°$, $1/k = 0.56°$, and $\phi = \pi/2$ by the temporal kernel of equation 2.29 with $1/\alpha = 15$ ms. (A adapted from DeAngelis et al., 1995.)

stimulus ($\omega/2\pi$ rather than the angular frequency ω) in figure 2.16. The peak value around 4 Hz and roll-off above 10 Hz are typical for V1 neurons and for cortical neurons in other primary sensory areas as well.

Space-Time Receptive Fields

To display the function $D(x, y, \tau)$ in a space-time plot rather than as a sequence of spatial plots (as in figure 2.13), we suppress the y dependence and plot an x-τ projection of the space-time kernel. Figure 2.17A shows a space-time plot of the receptive field of a simple cell in the cat primary visual cortex. This receptive field is approximately separable, and it has OFF and ON subregions that reverse to ON and OFF subregions as a function of τ, similar to the reversal seen in figure 2.13. Figure 2.17B shows an

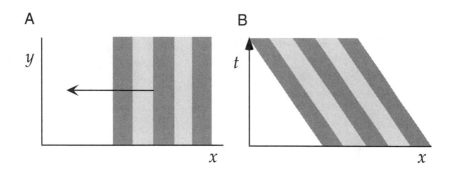

Figure 2.18 Space and space-time diagrams of a moving grating. (A) A vertically oriented grating moves to the left on a two-dimensional screen. (B) The space-time diagram of the image in A. The *x* location of the dark and light bands moves to the left as time progresses upward, representing the motion of the grating.

x-τ plot of a separable space-time kernel, similar to the one in figure 2.17A, generated by multiplying a Gabor function by the temporal kernel of equation 2.29.

We can also plot the visual stimulus in a space-time diagram, suppressing the y coordinate by assuming that the image does not vary as a function of y. For example, figure 2.18A shows a grating of vertically oriented stripes moving to the left on an x-y plot. In the x-t plot of figure 2.18B, this image appears as a series of sloped dark and light bands. These represent the projection of the image in figure 2.18A onto the x axis evolving as a function of time. The leftward slope of the bands corresponds to the leftward movement of the image.

Most neurons in primary visual cortex do not respond strongly to static images, but respond vigorously to flashed and moving bars and gratings. The receptive field structure of figure 2.17 reveals why this is the case, as is shown in figures 2.19 and 2.20. The image in figures 2.19A-C is a dark bar that is flashed on for a brief period of time. To describe the linear response estimate at different times, we consider a space-time receptive field similar to the one in figure 2.17A. The receptive field is positioned at three different times in figures 2.19A, B, and C. The height of the horizontal axis of the receptive field diagram indicates the time when the estimation is being made. Figure 2.19A corresponds to an estimate of $L(t)$ at the moment when the image first appears. At this time, $L(t) = 0$. As time progresses, the receptive field diagram moves upward. Figure 2.19B generates an estimate at the moment of maximum response, when the dark image overlaps the OFF area of the space-time receptive field, producing a positive contribution to $L(t)$. Figure 2.19C shows a later time when the dark image overlaps an ON region, generating a negative $L(t)$. The response for this flashed image is thus transient firing followed by suppression, as shown in Figure 2.19D.

Figures 2.19E and 2.19F show why a static dark bar is an ineffective stimulus. The static bar overlaps both the OFF region for small τ and the re-

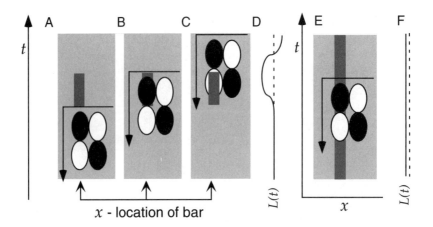

Figure 2.19 Responses to dark bars estimated from a separable space-time receptive field. Dark ovals in the receptive field diagrams are OFF regions and light circles are ON regions. The linear estimate of the response at any time is determined
by positioning the receptive field diagram so that its horizontal axis matches the
time of response estimation and noting how the OFF and ON regions overlap with
the image. (A-C) The image is a dark bar that is flashed on for a short interval of
time. There is no response (A) until the dark image overlaps the OFF region (B)
when $L(t) > 0$. The response is later suppressed when the dark bar overlaps the
ON region (C) and $L(t) < 0$. (D) A plot of $L(t)$ versus time corresponding to the
responses generated in A-C. Time runs vertically in this plot, and $L(t)$ is plotted
horizontally with the dashed line indicating the zero axis and positive values plotted to the left. (E) The image is a static dark bar. The bar overlaps both an OFF and
an ON region, generating opposing positive and negative contributions to $L(t)$. (F)
The weak response corresponding to E, plotted as in D.

versed ON region for large τ, generating opposing positive and negative
contributions to $L(t)$. The flashed dark bar of figures 2.19A-C is a more
effective stimulus because there is a time when it overlaps only the OFF
region.

Figure 2.20 shows why a moving grating is a particularly effective stimulus. The grating moves to the left in 2.20A-C. At the time corresponding to
the positioning of the receptive field diagram in 2.20A, a dark band stimulus overlaps both OFF regions and light bands overlap both ON regions.
Thus, all four regions contribute positive amounts to $L(t)$. As time progresses and the receptive field moves upward in the figure, the alignment
will sometimes be optimal, as in 2.20A, and sometimes nonoptimal, as in
2.20B. This produces an $L(t)$ that oscillates as a function of time between
positive and negative values (2.20C). Figures 2.20D-F show that a neuron
with this receptive field responds equally to a grating moving to the right.
Like the left-moving grating in figures 2.20A-C, the right-moving grating
can overlap the receptive field in an optimal manner (2.20D), producing
a strong response, or in a maximally negative manner (2.20E), producing
strong suppression of response, again resulting in an oscillating response
(2.20F). Separable space-time receptive fields can produce responses that
are maximal for certain speeds of grating motion, but they are not sensitive
to the direction of motion.

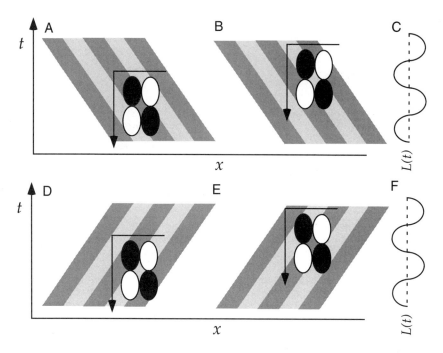

Figure 2.20 Responses to moving gratings estimated from a separable space-time receptive field. The receptive field is the same as in figure 2.19. (A-C) The stimulus is a grating moving to the left. At the time corresponding to A, OFF regions overlap with dark bands and ON regions with light bands, generating a strong response. At the time of the estimate in B, the alignment is reversed, and $L(t)$ is negative. (C) A plot of $L(t)$ versus time corresponding to the responses generated in A-B. Time runs vertically in this plot and $L(t)$ is plotted horizontally, with the dashed line indicating the zero axis and positive values plotted to the left. (D-F) The stimulus is a grating moving to the right. The responses are identical to those in A-C.

Nonseparable Receptive Fields

Many neurons in primary visual cortex are selective for the direction of motion of an image. Accounting for direction selectivity requires nonseparable space-time receptive fields. An example of a nonseparable receptive field is shown in figure 2.21A. This neuron has a three-lobed OFF-ON-OFF spatial receptive field, and these subregions shift to the left as time moves forward (and τ decreases). This means that the optimal stimulus for this neuron has light and dark areas that move toward the left. One way to describe a nonseparable receptive field structure is to use a separable function constructed from a product of a Gabor function for D_s and equation 2.29 for D_t, but to write these as functions of a mixture or rotation of the x and τ variables. The rotation of the space-time receptive field, as seen in figure 2.21B, is achieved by mixing the space and time coordinates, using the transformation

$$D(x, y, \tau) = D_s(x', y)D_t(\tau') \tag{2.35}$$

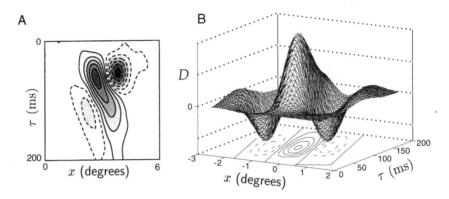

Figure 2.21 A nonseparable space-time receptive field. (A) An x-τ plot of the space-time receptive field of a neuron from cat primary visual cortex. OFF regions are shown with dashed contour lines and ON regions with solid contour lines. The receptive field has a central ON region and two flanking OFF regions that shift to the left over time. (B) Mathematical description of the space-time receptive field in A constructed from equations 2.35 - 2.37. The Gabor function used (evaluated at $y = 0$) had $\sigma_x = 1°$, $1/k = 0.5°$, and $\phi = 0$. D_t is given by the expression in equation 2.29 with $\alpha = 20$ ms, except that the second term, with the seventh power function, was omitted because the receptive field does not reverse sign in this example. The x-τ rotation angle used was $\psi = \pi/9$, and the conversion factor was $c = 0.02 °/$ms. (A adapted from DeAngelis et al., 1995.)

with

$$x' = x\cos(\psi) - c\tau\sin(\psi) \qquad (2.36)$$

and

$$\tau' = \tau\cos(\psi) + \frac{x}{c}\sin(\psi). \qquad (2.37)$$

The factor c converts between the units of time (ms) and space (degrees), and ψ is the space-time rotation angle. The rotation operation is not the only way to generate nonseparable space-time receptive fields. They are often constructed by adding together two or more separable space-time receptive fields with different spatial and temporal characteristics.

Figure 2.22 shows how a nonseparable space-time receptive field can produce a response that is sensitive to the direction of motion of a grating. Figures 2.22A-C show a left-moving grating and, in 2.22A, the receptive field is positioned at a time when a light area of the image overlaps the central ON region and dark areas overlap the flanking OFF regions. This produces a large positive $L(t)$. At other times, the alignment is nonoptimal (2.22B), and over time, $L(t)$ oscillates between large positive and negative values (2.22C). The nonseparable space-time receptive field does not overlap optimally with the right-moving grating of figures 2.22D-F at any time, and the response is correspondingly weaker (2.22F). Thus, a neuron with *direction selectivity* a nonseparable space-time receptive field can be selective for the direction *preferred velocity* of motion of a grating and for its velocity, responding most vigorously to an optimally spaced grating moving at a velocity given, in terms of the parameters in equation 2.36, by $c\tan(\psi)$.

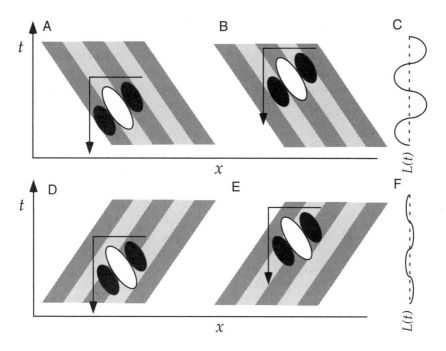

Figure 2.22 Responses to moving gratings estimated from a nonseparable space-time receptive field. Dark areas in the receptive field diagrams represent OFF regions and light areas, ON regions. (A-C) The stimulus is a grating moving to the left. At the time corresponding to A, OFF regions overlap with dark bands and the ON region overlaps a light band, generating a strong response. At the time of the estimate in B, the alignment is reversed, and $L(t)$ is negative. (C) A plot of $L(t)$ versus time corresponding to the responses generated in A and B. Time runs vertically in this plot, and $L(t)$ is plotted horizontally with the dashed line indicating the zero axis. (D-F) The stimulus is a grating moving to the right. Because of the tilt of the space-time receptive field, the alignment with the right-moving grating is never optimal and the response is weak (F).

Static Nonlinearities: Simple Cells

Once the linear response estimate $L(t)$ has been computed, the firing rate of a visually responsive neuron can be approximated by using equation 2.8, $r_{\text{est}}(t) = r_0 + F(L(t))$, where F is an appropriately chosen static nonlinearity. The simplest choice for F consistent with the positive nature of firing rates is rectification, $F = G[L]_+$, with G set to fit the magnitude of the measured firing rates. However, this choice makes the firing rate a linear function of the contrast amplitude, which does not match the data on the contrast dependence of visual responses. Neural responses saturate as the contrast of the image increases, and are more accurately described *contrast saturation* by $r \propto A^n/(A_{1/2}^n + A^n)$ where n is near 2, and $A_{1/2}$ is a parameter equal to the contrast amplitude that produces a half-maximal response. This led Heeger (1992) to propose that an appropriate static nonlinearity to use is

$$F(L) = \frac{G[L]_+^2}{A_{1/2}^2 + G[L]_+^2} \, , \tag{2.38}$$

because this reproduces the observed contrast dependence. A number of variants and extensions of this idea have also been considered, including, for example, that the denominator of this expression should include L factors for additional neurons with nearby receptive fields. This can account for the effects of visual stimuli outside the "classical" receptive field. Discussion of these effects is beyond the scope of this chapter.

2.5   Static Nonlinearities: Complex Cells

Recall that neurons in primary visual cortex are characterized as simple or complex. While linear methods, such as spike-triggered averages, are useful for revealing the properties of simple cells, at least to a first approximation, complex cells display features that are fundamentally incompatible with a linear description. The spatial receptive fields of complex cells cannot be divided into separate ON and OFF regions that sum linearly to generate the response. Areas where light and dark images excite the neuron overlap, making it difficult to measure and interpret spike-triggered average stimuli. Nevertheless, like simple cells, complex cells are selective to the spatial frequency and orientation of a grating. However, unlike simple cells, complex cells respond to bars of light or dark no matter where they are placed within the overall receptive field. Likewise, the responses of complex cells to grating stimuli show little dependence on spatial phase. *spatial-phase* Thus, a complex cell is selective for a particular type of image independent *invariance* of its exact spatial position within the receptive field. This may represent an early stage in the visual processing that ultimately leads to position-invariant object recognition.

Complex cells also have temporal response characteristics that distinguish them from simple cells. Complex cell responses to moving gratings are approximately constant, not oscillatory as in figures 2.20 and 2.22. The firing rate of a complex cell responding to a counterphase grating oscillating with frequency ω has both a constant component and an oscillatory component with a frequency of 2ω, a phenomenon known as frequency *frequency doubling* doubling.

Even though spike-triggered average stimuli and reverse-correlation functions fail to capture the response properties of complex cells, complex-cell responses can be described, to a first approximation, by a relatively straightforward extension of the reverse-correlation approach. The key observation comes from equation 2.34, which shows how the linear response estimate of a simple cell depends on spatial phase for an optimally oriented grating with K not too small. Consider two such responses, labeled L_1 and L_2, with preferred spatial phases ϕ and $\phi - \pi/2$. Including both the spatial and the temporal response factors, we find, for preferred spatial phase ϕ,

$$L_1 = AB(\omega, K) \cos(\phi - \Phi) \cos(\omega t - \delta), \qquad (2.39)$$

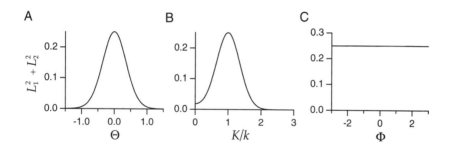

Figure 2.23 Selectivity of a complex cell model in response to a sinusoidal grating. The width and preferred spatial frequency of the Gabor functions underlying the estimated firing rate satisfy $k\sigma = 2$. (A) The complex cell response estimate, $L_1^2 + L_2^2$, as a function of stimulus orientation Θ for a grating with the preferred spatial frequency $K = k$. (B) $L_1^2 + L_2^2$ as a function of the ratio of the stimulus spatial frequency to its preferred value, K/k, for a grating oriented in the preferred direction $\Theta = 0$. (C) $L_1^2 + L_2^2$ as a function of stimulus spatial phase Φ for a grating with the preferred spatial frequency and orientation, $K = k$ and $\Theta = 0$.

where $B(\omega, K)$ is a temporal and spatial frequency-dependent amplitude factor. We do not need the explicit form of $B(\omega, K)$ here, but the reader is urged to derive it. For preferred spatial phase $\phi - \pi/2$,

$$L_2 = AB(\omega, K)\sin(\phi - \Phi)\cos(\omega t - \delta) \qquad (2.40)$$

because $\cos(\phi - \pi/2 - \Phi) = \sin(\phi - \Phi)$. If we square and add these two terms, we obtain a result that does not depend on Φ,

$$L_1^2 + L_2^2 = A^2 B^2(\omega, K)\cos^2(\omega t - \delta), \qquad (2.41)$$

because $\cos^2(\phi - \Phi) + \sin^2(\phi - \Phi) = 1$. Thus, we can describe the spatial-phase-invariant response of a complex cell by writing

$$r(t) = r_0 + G\left(L_1^2 + L_2^2\right), \qquad (2.42)$$

for some constant G. The selectivities of such a response estimate to grating orientation, spatial frequency, and spatial phase are shown in figure 2.23. The response of the model complex cell is tuned to orientation and spatial frequency, but the spatial phase dependence, illustrated for a simple cell in figure 2.15C, is absent. In computing the curve for figure 2.23C, we used the exact expressions for L_1 and L_2 from the integrals in equations 2.31 and 2.32, not the approximation used in equation 2.34 to simplify the previous discussion. Although it is not visible in the figure, there is a weak dependence on Φ when the exact expressions are used.

The complex cell response given by equations 2.41 and 2.42 reproduces the frequency-doubling effect seen in complex cell responses because the factor $\cos^2(\omega t - \delta)$ oscillates with frequency 2ω. This follows from the identity

$$\cos^2(\omega t - \delta) = \frac{1}{2}\cos\left(2(\omega t - \delta)\right) + \frac{1}{2}. \qquad (2.43)$$

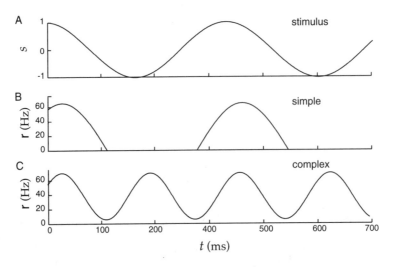

Figure 2.24 Temporal responses of model simple and complex cells to a counter-phase grating. (A) The stimulus $s(x, y, t)$ at a given point (x, y) plotted as a function of time. (B) The rectified linear response estimate of a model simple cell to this grating with a temporal kernel given by equation 2.29 with $\alpha = 1/(15 \text{ ms})$. (C) The frequency-doubled response of a model complex cell with the same temporal kernel but with the estimated rate given by a squaring operation rather than recti-fication. The background firing rate is $r_0 = 5$ Hz. Note the temporal phase shift of both B and C relative to A.

In addition, the last term on the right side of this equation generates the constant component of the complex cell response to a counterphase grat-ing. Figure 2.24 shows a comparison of model simple and complex cell responses to a counterphase grating, and illustrates this phenomenon.

energy model

The description of a complex cell response that we have presented is called an "energy" model because of its resemblance to the equation for the en-ergy of a simple harmonic oscillator. The pair of linear filters used, with preferred spatial phases separated by $\pi/2$, is called a quadrature pair. Be-cause of rectification, the terms L_1^2 and L_2^2 cannot be constructed by squar-ing the outputs of single simple cells. However, they can each be con-structed by summing the squares of rectified outputs from two simple cells with preferred spatial phases separated by π. Thus, we can write the com-plex cell response as the sum of the squares of four rectified simple cell responses,

$$r(t) = r_0 + G \left([L_1]_+^2 + [L_2]_+^2 + [L_3]_+^2 + [L_4]_+^2 \right), \qquad (2.44)$$

where the different $[L]_+$ terms represent the responses of simple cells with preferred spatial phases ϕ, $\phi + \pi/2$, $\phi + \pi$, and $\phi + 3\pi/2$. While such a construction is possible, it should not be interpreted too literally because complex cells receive input from many sources, including the LGN and other complex cells. Rather, this model should be viewed as purely de-scriptive. Mechanistic models of complex cells are described at the end of this chapter and in chapter 7.

2.6   Receptive Fields in the Retina and LGN

We end this discussion of the visual system by returning to the initial stages of the visual pathway and briefly describing the receptive field properties of neurons in the retina and LGN. Retinal ganglion cells display a wide variety of response characteristics, including nonlinear and direction-selective responses. However, a class of retinal ganglion cells (X cells in the cat or P cells in the monkey retina and LGN) can be described by a linear model built using reverse-correlation methods. The receptive fields of this class of retinal ganglion cells and an analogous type of LGN relay neurons are similar, so we do not treat them separately. The spatial structure of the receptive fields of these neurons has a center-surround structure consisting either of a circular central ON region surrounded by an annular OFF region, or the opposite arrangement of a central OFF region surrounded by an ON region. Such receptive fields are called ON-center and OFF-center, respectively. Figure 2.25A shows the spatial receptive fields of an ON-center cat LGN neuron.

The spatial structure of retinal ganglion and LGN receptive fields is well captured by a difference-of-Gaussians model in which the spatial receptive field is expressed as

difference of Gaussians

$$D_s(x, y) = \pm \left(\frac{1}{2\pi\sigma_{cen}^2} \exp\left(-\frac{x^2 + y^2}{2\sigma_{cen}^2} \right) - \frac{B}{2\pi\sigma_{sur}^2} \exp\left(-\frac{x^2 + y^2}{2\sigma_{sur}^2} \right) \right).$$

(2.45)

Here the center of the receptive field has been placed at $x = y = 0$. The first Gaussian function in equation 2.45 describes the center, and the second, the surround. The size of the central region is determined by the parameter σ_{cen}, while σ_{sur}, which is greater than σ_{cen}, determines the size of the surround. B controls the balance between center and surround contributions. The $\pm$ sign allows both ON-center (+) and OFF-center ($-$) cases to be represented. Figure 2.25B shows a spatial receptive field formed from the difference of two Gaussians that approximates the receptive field structure in figure 2.25A.

Figure 2.25C shows that the spatial structure of the receptive field reverses over time with, in this case, a central ON region reversing to an OFF region as τ increases. Similarly, the OFF surround region changes to an ON region with increasing τ, although the reversal and the onset are slower for the surround than for the central region. Because of the difference between the time course of the center and of the surround regions, the space-time receptive field is not separable, although the center and surround components are individually separable. The basic features of LGN neuron space-time receptive fields are captured by

$$D(x, y, \tau) = \pm \left(\frac{D_t^{cen}(\tau)}{2\pi\sigma_{cen}^2} \exp\left(-\frac{x^2 + y^2}{2\sigma_{cen}^2} \right) - \frac{BD_t^{sur}(\tau)}{2\pi\sigma_{sur}^2} \exp\left(-\frac{x^2 + y^2}{2\sigma_{sur}^2} \right) \right).$$

(2.46)

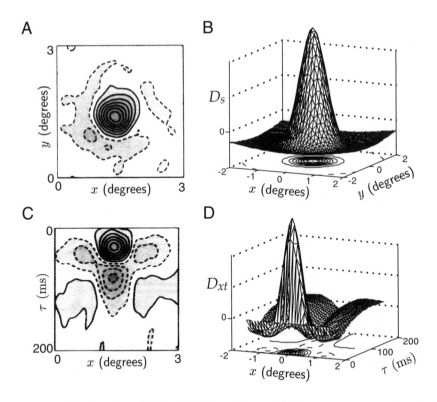

Figure 2.25 Receptive fields of LGN neurons. (A) The center-surround spatial structure of the receptive field of a cat LGN X cell. This has a central ON region (solid contours) and a surrounding OFF region (dashed contours). (B) A fit of the receptive field shown in A using a difference-of-Gaussians function (equation 2.45) with $\sigma_{cen} = 0.3°$, $\sigma_{sur} = 1.5°$, and $B = 5$. (C) The space-time receptive field of a cat LGN X cell. Note that the center and surround regions both reverse sign as a function of τ and that the temporal evolution is slower for the surround than for the center. (D) A fit of the space-time receptive field in C using equation 2.46 with the same parameters for the Gaussian functions as in B, and temporal factors given by equation 2.47 with $1/\alpha_{cen} = 16$ ms for the center, $1/\alpha_{sur} = 32$ ms for the surround, and $1/\beta_{cen} = 1/\beta_{sur} = 64$ ms. (A and C adapted from DeAngelis et al., 1995.)

Separate functions of time multiply the center and surround, but they can both be described by the same functions, using two sets of parameters,

$$D_t^{cen,sur}(\tau) = \alpha_{cen,sur}^2 \tau \exp(-\alpha_{cen,sur}\tau) - \beta_{cen,sur}^2 \tau \exp(-\beta_{cen,sur}\tau). \quad (2.47)$$

The parameters α_{cen} and α_{sur} control the latency of the response in the center and surround regions, respectively, and β_{cen} and β_{sur} affect the time of the reversal. This function has characteristics similar to the function in equation 2.29, but the latency effect is less pronounced. Figure 2.25D shows the space-time receptive field of equation 2.46 with parameters chosen to match figure 2.25C.

Figure 2.26 shows the results of a direct test of a reverse-correlation model of an LGN neuron. The kernel needed to describe a particular LGN cell was first extracted by using a white-noise stimulus. This, together with

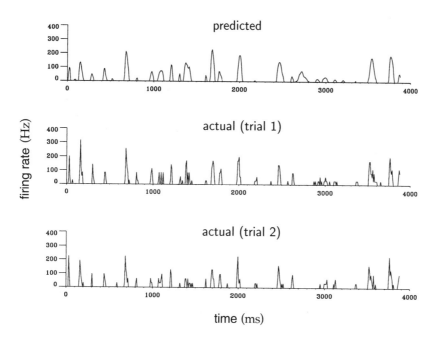

Figure 2.26 Comparison of predicted and measured firing rates for a cat LGN neuron responding to a video movie. The top panel is the rate predicted by integrating the product of the video image intensity and a linear filter obtained for this neuron from a spike-triggered average of a white-noise stimulus. The resulting linear prediction was rectified. The middle and lower panels are measured firing rates extracted from two different sets of trials. (Adapted from Dan et al., 1996.)

a rectifying static nonlinearity, was used to predict the firing rate of the neuron in response to a video movie. The top panel in figure 2.26 shows the resulting prediction, and the middle and lower panels show the actual firing rates extracted from two different groups of trials. The correlation coefficient between the predicted and actual firing rates was 0.5, which was very close to the correlation coefficient between firing rates extracted from different groups of trials. This means that the error of the prediction was no worse than the variability of the neural response itself.

2.7 Constructing V1 Receptive Fields

The models of visual receptive fields we have been discussing are purely descriptive, but they provide an important framework for studying how the circuits of the retina, LGN, and primary visual cortex generate neural responses. In an example of a more mechanistic model, Hubel and Wiesel (1962) proposed that the oriented receptive fields of cortical neurons could be generated by summing the input from appropriately selected LGN neurons. Their construction, shown in figure 2.27A, consists of alternating rows of ON-center and OFF-center LGN cells providing convergent input to a cortical simple cell. The left side of figure 2.27A shows the spatial ar-

Hubel-Wiesel simple cell model

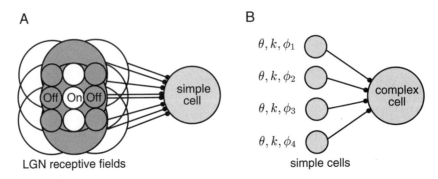

Figure 2.27 (A) The Hubel-Wiesel model of orientation selectivity. The spatial arrangement of the receptive fields of nine LGN neurons are shown, with a row of three ON-center fields flanked on either side by rows of three OFF-center fields. White areas denote ON fields and gray areas, OFF fields. In the model, the converging LGN inputs are summed by the simple cell. This arrangement produces a receptive field oriented in the vertical direction. (B) The Hubel-Wiesel model of a complex cell. Inputs from a number of simple cells with similar orientation and spatial frequency preferences (θ and k), but different spatial phase preferences (ϕ_1, ϕ_2, ϕ_3, and ϕ_4), converge on a complex cell and are summed. This produces a complex cell output that is selective for orientation and spatial frequency, but not for spatial phase. The figure shows four simple cells converging on a complex cell, but additional simple cells can be included to give a more complete coverage of spatial phase.

rangement of LGN receptive fields that, when summed, form bands of ON and OFF regions resembling the receptive field of an oriented simple cell. This model accounts for the selectivity of a simple cell purely on the basis of feedforward input from the LGN. We leave the study of this model as an exercise for the reader. Other models, which we discuss in chapter 7, include the effects of recurrent intracortical connections as well.

In a previous section, we showed how the properties of complex cell responses could be accounted for by using a squaring static nonlinearity. While this provides a good description of complex cells, there is little indication that complex cells actually square their inputs. Models of complex cells can be constructed without introducing a squaring nonlinearity. One such example is another model proposed by Hubel and Wiesel (1962), *Hubel-Wiesel* which is depicted in figure 2.27B. Here the phase-invariant response of a *complex cell model* complex cell is produced by summing together the responses of several simple cells with similar orientation and spatial frequency tuning, but different preferred spatial phases. In this model, the complex cell inherits its orientation and spatial frequency preference from the simple cells that drive it, but spatial phase selectivity is reduced because the outputs of simple cells with a variety of spatial phase selectivities are summed. Analysis of this model is left as an exercise. While the model generates complex cell responses, there are indications that complex cells in primary visual cortex are not driven exclusively by simple cell input. An alternative model is considered in chapter 7.

2.8   Chapter Summary

We continued from chapter 1 our study of the ways that neurons encode information, focusing on reverse-correlation analysis, particularly as applied to neurons in the retina, visual thalamus (LGN), and primary visual cortex. We used the tools of systems identification, especially the linear filter, Wiener kernel, and static nonlinearity, to build descriptive linear and nonlinear models of the transformation from dynamic stimuli to time-dependent firing rates. We discussed the complex logarithmic map governing the way that neighborhood relationships in the retina are transformed into cortex, Nyquist sampling in the retina, and Gabor functions as descriptive models of separable and nonseparable receptive fields. Models based on Gabor filters and static nonlinearities were shown to account for the basic response properties of simple and complex cells in primary visual cortex, including selectivity for orientation, spatial frequency and phase, velocity, and direction. Retinal ganglion cell and LGN responses were modeled using a difference-of-Gaussians kernel. We briefly described simple circuit models of simple and complex cells.

2.9   Appendices

A: The Optimal Kernel

Using equation 2.1 for the estimated firing rate, the expression in equation 2.3 to be minimized is

$$E = \frac{1}{T} \int_0^T dt \left(r_0 + \int_0^\infty d\tau \, D(\tau) s(t-\tau) - r(t) \right)^2 . \qquad (2.48)$$

The minimum is obtained by setting the derivative of E with respect to the function D to 0. A quantity, such as E, that depends on a function, D in this case, is called a functional, and the derivative we need is a functional derivative. Finding the extrema of functionals is the subject of a branch of mathematics called the calculus of variations. A simple way to define a functional derivative is to introduce a small time interval Δt and evaluate all functions at integer multiples of Δt. We define $r_i = r(i\Delta t)$, $D_k = D(k\Delta t)$, and $s_{i-k} = s((i-k)\Delta t)$. If Δt is small enough, the integrals in equation 2.48 can be approximated by sums, and we can write   *functional derivative*

$$E = \frac{\Delta t}{T} \sum_{i=0}^{T/\Delta t} \left(r_0 + \Delta t \sum_{k=0}^\infty D_k s_{i-k} - r_i \right)^2 . \qquad (2.49)$$

E is minimized by setting its derivative with respect to D_j for all values of j to 0,

$$\frac{\partial E}{\partial D_j} = 0 = \frac{2\Delta t}{T} \sum_{i=0}^{T/\Delta t} \left(r_0 + \Delta t \sum_{k=0}^\infty D_k s_{i-k} - r_i \right) s_{i-j} \Delta t . \qquad (2.50)$$

Rearranging and simplifying this expression gives the condition

$$\Delta t \sum_{k=0}^{\infty} D_k \left(\frac{\Delta t}{T} \sum_{i=0}^{T/\Delta t} s_{i-k} s_{i-j} \right) = \frac{\Delta t}{T} \sum_{i=0}^{T/\Delta t} (r_i - r_0) s_{i-j}. \qquad (2.51)$$

If we take the limit $\Delta t \to 0$ and make the replacements $i\Delta t \to t$, $j\Delta t \to \tau$, and $k\Delta t \to \tau'$, the sums in equation 2.51 turn back into integrals, the indexed variables become functions, and we find

$$\int_0^{\infty} d\tau' \, D(\tau') \left(\frac{1}{T} \int_0^T dt \, s(t - \tau') s(t - \tau) \right) = \frac{1}{T} \int_0^T dt \, (r(t) - r_0) s(t - \tau).$$
$$(2.52)$$

The term proportional to r_0 on the right side of this equation can be dropped because the time integral of s is 0. The remaining term is the firing rate-stimulus correlation function evaluated at $-\tau$, $Q_{rs}(-\tau)$. The term in large parentheses on the left side of 2.52 is the stimulus autocorrelation function. By shifting the integration variable $t \to t + \tau$, we find that it is $Q_{ss}(\tau - \tau')$, so 2.52 can be re-expressed in the form of equation 2.4.

Equation 2.6 provides the solution to equation 2.4 only for a white-noise stimulus. For an arbitrary stimulus, equation 2.4 can easily be solved by the method of Fourier transforms if we ignore causality and allow the estimated rate at time t to depend on the stimulus at times later than t, so that

$$r_{est}(t) = r_0 + \int_{-\infty}^{\infty} d\tau \, D(\tau) s(t - \tau). \qquad (2.53)$$

The estimate written in this acausal form satisfies a slightly modified version of equation 2.4,

$$\int_{-\infty}^{\infty} d\tau' \, Q_{ss}(\tau - \tau') D(\tau') = Q_{rs}(-\tau). \qquad (2.54)$$

We define the Fourier transforms (see the Mathematical Appendix)

$$\tilde{D}(\omega) = \int_{-\infty}^{\infty} dt \, D(t) \exp(i\omega t) \quad \text{and} \quad \tilde{Q}_{ss}(\omega) = \int_{-\infty}^{\infty} d\tau \, Q_{ss}(\tau) \exp(i\omega \tau),$$
$$(2.55)$$

as well as $\tilde{Q}_{rs}(\omega)$ defined analogously to $\tilde{Q}_{ss}(\omega)$.

Equation 2.54 is solved by taking the Fourier transform of both sides. The integral of the product of two functions that appears on the left side of equation 2.54 is called a convolution. To evaluate its Fourier transform, we make use of an important theorem stating that the Fourier transform of a convolution is the product of the Fourier transforms of the two functions involved (see the Mathematical Appendix),

$$\int_{-\infty}^{\infty} d\tau \, \exp(i\omega \tau) \int_{-\infty}^{\infty} d\tau' \, Q_{ss}(\tau - \tau') D(\tau') = \tilde{D}(\omega) \tilde{Q}_{ss}(\omega). \qquad (2.56)$$

In terms of the Fourier transforms, equation 2.54 then becomes

$$\tilde{D}(\omega)\tilde{Q}_{ss}(\omega) = \tilde{Q}_{rs}(-\omega), \qquad (2.57)$$

which can be solved directly to obtain $\tilde{D}(\omega) = \tilde{Q}_{rs}(-\omega)/\tilde{Q}_{ss}(\omega)$. The inverse Fourier transform from which $D(\tau)$ is recovered is (Mathematical Appendix)

$$D(\tau) = \frac{1}{2\pi}\int_{-\infty}^{\infty} d\omega\, \tilde{D}(\omega)\exp(-i\omega\tau), \qquad (2.58)$$

so the optimal acausal kernel when the stimulus is temporally correlated is given by

$$D(\tau) = \frac{1}{2\pi}\int_{-\infty}^{\infty} d\omega\, \frac{\tilde{Q}_{rs}(-\omega)}{\tilde{Q}_{ss}(\omega)}\exp(-i\omega\tau). \qquad (2.59)$$

B: The Most Effective Stimulus

We seek the stimulus that produces the maximum predicted responses at time t subject to the fixed energy constraint

$$\int_0^T dt'\, \left(s(t')\right)^2 = \text{constant}. \qquad (2.60)$$

We impose this constraint by the method of Lagrange multipliers (see the Mathematical Appendix), which means that we must find the unconstrained maximum value with respect to s of

$$r_{est}(t) + \lambda \int_0^T dt'\, s^2(t') = r_0 + \int_0^\infty d\tau\, D(\tau)s(t-\tau) + \lambda \int_0^T dt'\, \left(s(t')\right)^2, \qquad (2.61)$$

where λ is the Lagrange multiplier. Setting the derivative of this expression with respect to the function s to 0 (using the same methods used in appendix A) gives

$$D(\tau) = -2\lambda s(t-\tau). \qquad (2.62)$$

The value of λ (which is less than 0) is determined by requiring that condition 2.60 is satisfied, but the precise value is not important for our purposes. The essential result is the proportionality between the optimal stimulus and $D(\tau)$.

C: Bussgang's Theorem

Bussgang (1952, 1975) proved that an estimate based on the optimal kernel for linear estimation can still be self-consistent (although not necessarily optimal) when nonlinearities are present. The self-consistency condition is that when the nonlinear estimate $r_{est} = r_0 + F(L(t))$ is substituted into equation 2.6, the relationship between the linear kernel and the firing rate-stimulus correlation function should still hold. In other words, we require that

$$D(\tau) = \frac{1}{\sigma_s^2 T} \int_0^T dt\, r_{est}(t) s(t - \tau) = \frac{1}{\sigma_s^2 T} \int_0^T dt\, F(L(t)) s(t - \tau). \quad (2.63)$$

We have dropped the r_0 term because the time integral of s is 0. In general, equation 2.63 does not hold, but if the stimulus used to extract D is Gaussian white noise, equation 2.63 reduces to a simple normalization condition on the function F. This result is based on the identity, valid for a Gaussian white-noise stimulus,

$$\frac{1}{\sigma_s^2 T} \int_0^T dt\, F(L(t)) s(t - \tau) = \frac{D(\tau)}{T} \int_0^T dt\, \frac{dF(L(t))}{dL}. \quad (2.64)$$

For the right side of this equation to be $D(\tau)$, the remaining expression, involving the integral of the derivative of F, must be equal to 1. This can be achieved by appropriate scaling of F. The critical identity 2.64 is based on integration by parts for a Gaussian weighted integral. A simplified proof is presented as a problem on the exercise web site.

2.10   Annotated Bibliography

Marmarelis & Marmarelis (1978), **Rieke et al. (1997)**, and **Gabbiani & Koch (1998)** provide general discussions of reverse-correlation methods. A useful reference relevant to our presentation of their application to the visual system is **Carandini et al. (1996)**. Volterra and Wiener functional expansions are discussed in **Wiener (1958)** and **Marmarelis & Marmarelis (1978)**.

General introductions to the visual system include **Hubel & Wiesel (1962, 1977)**, **Orban (1984)**, **Hubel (1988)**, **Wandell (1995)**, and **De Valois & De Valois (1990)**. Our treatment follows **Dowling (1987)** on processing in the retina, and Schwartz (1977), Van Essen et al. (1984), and Rovamo & Virsu (1984) on aspects of the retinotopic map from the eye to the brain. Properties of this map are used to account for aspects of visual hallucinations in Ermentrout & Cowan (1979). We also follow **Movshon et al. (1978a, 1978b)** for definitions of simple and complex cells, Daugman (1985) and Jones & Palmer (1987b) on the use of Gabor functions (Gabor, 1946) to describe visual receptive fields, and **DeAngelis et al. (1995)** on space-time receptive fields. Our description of the energy model of complex cells is based on Adelson & Bergen (1985), which is related to work by Pollen & Ronner (1982), Van Santen & Sperling (1984), and Watson & Ahumada (1985), and to earlier ideas of Reichardt (1961) and Barlow & Levick (1965). Heeger's (1992, 1993) model of contrast saturation is reviewed in **Carandini et al.**

(1996), and has been applied in an approach more closely related to the representational learning models of chapter 10 by Simoncelli & Schwartz (1999). The difference-of-Gaussians model for retinal and LGN receptive fields is due to Rodieck (1965) and Enroth-Cugell and Robson (1966). A useful reference on modeling of the early visual system is Wörgötter & Koch (1991). The issue of linearity and nonlinearity in early visual processing is reviewed by **Ferster (1994)**.

3 Neural Decoding

3.1 Encoding and Decoding

In chapters 1 and 2, we considered the problem of predicting neural responses to known stimuli. The nervous system faces the reverse problem, determining what is going on in the real world from neuronal spiking patterns. It is interesting to attempt such computations ourselves, using the responses of one or more neurons to identify a particular stimulus or to extract the value of a stimulus parameter. We will assess the accuracy with which this can be done primarily by using optimal decoding techniques, regardless of whether the computations involved seem biologically plausible. Some biophysically realistic implementations are discussed in chapter 7. Optimal decoding allows us to determine limits on the accuracy and reliability of neuronal encoding. In addition, it is useful for estimating the information content of neuronal spike trains, an issue addressed in chapter 4.

As we discuss in chapter 1, neural responses, even to a single repeated stimulus, are typically described by stochastic models due to their inherent variability. In addition, stimuli themselves are often described stochastically. For example, the stimuli used in an experiment might be drawn randomly from a specified probability distribution. Natural stimuli can also be modeled stochastically as a way of capturing the statistical properties of complex environments.

Given this twofold stochastic model, encoding and decoding are related through a basic identity of probability theory called Bayes theorem. Let $\mathbf{r}$ represent the response of a neuron or a population of neurons to a stimulus characterized by a parameter s. Throughout this chapter, $\mathbf{r} = (r_1, r_2, \dots, r_N)$ for N neurons is a list of spike-count firing rates, although, for the present discussion, it could be any other set of parameters describing the neuronal response. Several different probabilities and conditional probabilities enter into our discussion. A conditional probability is just an ordinary probability of an event occurring, except that its occurrence is subject to an additional condition. The conditional probability of event A occurring subject to the condition B is denoted by $P[A|B]$. The probabilities we need are:

conditional probability

prior probability
- $P[s]$, the probability of stimulus s being presented, often called the prior probability

- $P[\mathbf{r}]$, the probability of response $\mathbf{r}$ being recorded

joint probability
- $P[\mathbf{r}, s]$, the probability of stimulus s being presented and response $\mathbf{r}$ being recorded. This is called the joint probability

- $P[\mathbf{r}|s]$, the conditional probability of evoking response $\mathbf{r}$, given that stimulus s was presented

- $P[s|\mathbf{r}]$, the conditional probability that stimulus s was presented, given that response $\mathbf{r}$ was recorded.

Note that $P[\mathbf{r}|s]$ is the probability of observing the rates $\mathbf{r}$, given that the stimulus took the value s, while $P[\mathbf{r}]$ is the probability of the rates taking the values $\mathbf{r}$ independent of what stimulus was used. $P[\mathbf{r}]$ can be computed from $P[\mathbf{r}|s]$ by summing over all stimulus values weighted by their probabilities,

$$P[\mathbf{r}] = \sum_s P[\mathbf{r}|s]P[s] \text{ and similarly } P[s] = \sum_{\mathbf{r}} P[s|\mathbf{r}]P[\mathbf{r}]. \quad (3.1)$$

An additional relationship between the probabilities listed above can be derived by noticing that the joint probability $P[\mathbf{r}, s]$ can be expressed as either the conditional probability $P[\mathbf{r}|s]$ times the probability of the stimulus, or as $P[s|\mathbf{r}]$ times the probability of the response,

$$P[\mathbf{r}, s] = P[\mathbf{r}|s]P[s] = P[s|\mathbf{r}]P[\mathbf{r}]. \quad (3.2)$$

Bayes theorem
This is the basis of Bayes theorem relating $P[s|\mathbf{r}]$ to $P[\mathbf{r}|s]$:

$$P[s|\mathbf{r}] = \frac{P[\mathbf{r}|s]P[s]}{P[\mathbf{r}]}, \quad (3.3)$$

assuming that $P[\mathbf{r}] \neq 0$. Encoding is characterized by the set of probabilities $P[\mathbf{r}|s]$ for all stimuli and responses. Decoding a response, on the other hand, amounts to determining the probabilities $P[s|\mathbf{r}]$. According to Bayes theorem, $P[s|\mathbf{r}]$ can be obtained from $P[\mathbf{r}|s]$, but the stimulus probability $P[s]$ is also needed. As a result, decoding requires knowledge of the statistical properties of experimentally or naturally occurring stimuli.

In the above discussion, we have assumed that both the stimulus and the response are characterized by discrete values so that ordinary probabilities, not probability densities, are used to describe their distributions. For example, firing rates obtained by counting spikes over the duration of a trial take discrete values and can be described by a probability. However, we sometimes treat the response firing rates or the stimulus values as continuous variables. In this case, the probabilities listed must be replaced by the corresponding probability densities, $p[\mathbf{r}]$, $p[\mathbf{r}|s]$, etc. Nevertheless, the relationships discussed above are equally valid.

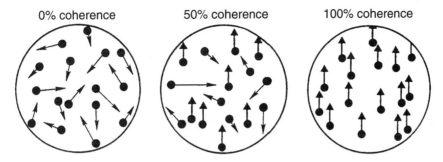

Figure 3.1 The moving random-dot stimulus for different levels of coherence. The visual image consists of randomly placed dots that jump every 45 ms according to the scheme described in the text. At 0% coherence the dots move randomly. At 50% coherence, half the dots move randomly and half move together (upward in this example). At 100% coherence all the dots move together. (Adapted from Britten et al., 1992.)

In the following sections, we present examples of decoding that involve both single neurons and neuronal populations. We first study a restricted case of single-cell decoding, discrimination between two different stimulus values. We then consider extracting the value of a parameter that characterizes a static stimulus from the responses of a population of neurons. As a final example, we return to single neurons and discuss spike-train decoding, in which an estimate of a time-varying stimulus is constructed from the spike train it evokes.

3.2   Discrimination

To introduce the notion of discriminability and the receiver operating characteristic that lie at the heart of discrimination analysis, we will discuss a fascinating study performed by Britten et al. (1992). In their experiments, a monkey was trained to discriminate between two directions of motion of a visual stimulus. The stimulus was a pattern of dots on a video monitor that jump from random initial locations to new locations every 45 ms. To introduce a sense of directed movement at a particular velocity, a percentage of the dots move together by a fixed amount in a fixed direction (figure 3.1). The coherently moving dots are selected randomly at each time step, and the remaining dots move to random new locations. The percentage of dots that move together in the fixed direction is called the coherence level. At 0% coherence, the image appears chaotic with no sense of any particular direction of motion. As the coherence increases, a sense of movement in a particular direction appears in the image until, at 100% coherence, the entire array of dots moves together on the monitor. By varying the degree of coherence, the task of detecting the movement direction can be made more or less difficult.

The experiments combined neural recording with behavioral measurements. In the behavioral part, the monkey had to report the direction

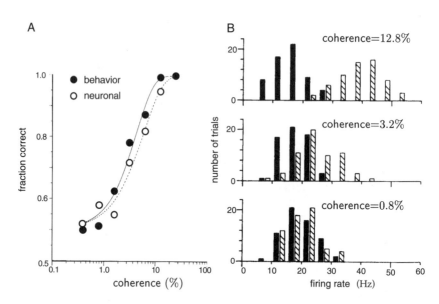

Figure 3.2 Behavioral and electrophysiological results from a random-dot motion-discrimination task. (A) The filled circles show the fraction of correct discriminations made by a monkey as a function of the degree of coherence of the motion. The open circles show the discrimination accuracy that an ideal observer could achieve on the analogous two-alternative forced-choice discrimination task, given the neural responses. (B) Firing-rate histograms for three different levels of coherence. Hatched rectangles show the results for motion in the plus direction, and solid rectangles, for motion in the minus direction. The histograms have been thinned for clarity so that not all the bins are shown. (Adapted from Britten et al., 1992.)

of motion in the random dot images. During the same task, recordings were made from neurons in area MT. Only two possible directions of coherent movement of the dots were used while a particular neuron was being recorded; either the direction that produced the maximum response in that neuron, or the opposite direction. The monkey's task was to discriminate between these two directions. The filled circles and solid curve in figure 3.2A show the proportion of correct responses in a typical experiment. Below 1% coherence, the responses were near chance (fraction correct = 0.5), but the monkey approached perfect performance (fraction correct = 1) above 10% coherence.

Figure 3.2B shows histograms of average firing rates in response to different levels of movement coherence. The firing rates plotted are the number of spikes recorded during the 2 s period that the stimulus was presented, divided by 2 s. The neuron shown tended to fire more spikes when the motion was in its preferred direction, which we will call the plus (or +) direction (hatched histogram), than in the other, minus (or −) direction (solid histogram). At high coherence levels, the firing-rate distributions corresponding to the two directions are fairly well separated, while at low coherence levels, they merge. Although spike count rates take only discrete values, it is more convenient to treat r as a continuous variable for our discussion. Treated as probability densities, these two distributions are approximately Gaussian with the same variance, σ_r^2, but different means, $\langle r \rangle_+$ for the plus direction and $\langle r \rangle_-$ for the minus direction. A convenient measure of the separation between the distributions is the discriminability

discriminability d'

$$d' = \frac{\langle r \rangle_+ - \langle r \rangle_-}{\sigma_r} \,, \tag{3.4}$$

which is the distance between the means in units of their common standard deviation. The larger d', the more separated the distributions.

In the example we are considering, decoding involves using the neural response to determine in which of the two possible directions the stimulus moved. A simple decoding procedure is to determine the firing rate r during a trial and compare it to a threshold number z. If $r \geq z$, we report "plus"; otherwise we report "minus". Figure 3.2B suggests that if we choose z to lie somewhere between the two distributions, this procedure will give the correct answer at high coherence, but will have difficulty distinguishing the two directions at low coherence. This difficulty is clearly related to the degree to which the two distributions in figure 3.2B overlap, and thus to the discriminability.

The probability that the procedure outlined in the previous paragraph will generate the correct answer (called a hit) when the stimulus is moving in the plus direction is the conditional probability that $r \geq z$ given a plus stimulus, $P[r \geq z|+]$. The probability that it will give the answer "plus" when the stimulus is actually moving in the minus direction (called a false alarm) is similarly $P[r \geq z|-]$. These two probabilities completely determine the performance of the decoding procedure because the probabilities for the other two cases (reporting "minus" when the correct answer is "plus", and reporting "minus" when the correct answer is "minus") are $1 - P[r \geq z|+]$ and $1 - P[r \geq z|-]$, respectively. In signal detection theory, the quantity used to perform the discrimination, r in our case, is called the test, and the two probabilities corresponding to reporting a "plus" answer have specific names:

test size and power or false alarm and hit rate

$$\begin{aligned}\alpha(z) &= P[r \geq z|-] \quad \text{is the size or false alarm rate of the test} \\ \beta(z) &= P[r \geq z|+] \quad \text{is the power or hit rate of the test.}\end{aligned} \tag{3.5}$$

The following table shows how the probabilities of the test giving correct and incorrect answers in the different cases depend on α and β.

stimulus	probability	
	correct	incorrect
$+$	β	$1 - \beta$
$-$	$1 - \alpha$	α

The performance of the decoding procedure we have been discussing depends critically on the value of the threshold z to which the rate r is compared. Obviously, we would like to use a threshold for which the size is near 0 and the power near 1. In general, it is impossible to choose the threshold so that both the size and the power of the test are optimized; a compromise must be made. A logical optimization criterion is to maximize the probability of getting a correct answer, which is equal to $(\beta(z) + 1 - \alpha(z))/2$ if the plus and minus stimuli occur with equal probability. Although this is a possible approach for the experiment we are studying, the analysis we present introduces a powerful technique that makes better use of the full range of recorded data and can be generalized to tasks where the optimal strategy is unknown. This approach makes use of ROC curves, which indicate how the size and power of a test trade off as the threshold is varied.

ROC Curves

receiver operating
characteristic, ROC

The receiver operating characteristic (ROC) curve provides a way of evaluating how test performance depends on the choice of the threshold z. Each point on an ROC curve corresponds to a different value of z. The x coordinate of the point is α, the size of the test for this value of z, and the y coordinate is β, its power. As the threshold is varied continuously, these points trace out the ROC plot. If $z = 0$, the firing rate will always be greater than or equal to z, so the decoding procedure will always give the answer "plus". Thus, for $z = 0$, $\alpha = \beta = 1$, producing a point at the upper-right corner of the ROC plot. At the other extreme, if z is very large, r will always be less than z, the test will always report "minus", and $\alpha = \beta = 0$. This produces a point at the bottom-left corner of the plot. Between these extremes, a curve is traced out as a function of z.

Figure 3.3 shows ROC curves computed by Britten et al. for several different values of the stimulus coherence. At high coherence levels, when the task is easy, the ROC curve rises rapidly from $\alpha(z) = 0$, $\beta(z) = 0$ as the threshold is lowered from a high value, and the probability $\beta(z)$ of a correct "plus" answer quickly approaches 1 without a concomitant increase in $\alpha(z)$. As the threshold is lowered further, the probability of giving the answer "plus" when the correct answer is "minus" also rises, and $\alpha(z)$ increases. When the task is difficult, the curve rises more slowly as z is lowered; and if the task is impossible, in that the test merely gives random answers, the curve will lie along the diagonal $\alpha = \beta$, because the probabilities of answers being correct and incorrect are equal. This is exactly the trend of the ROC curves at different coherence levels shown in figure 3.3.

Examination of figure 3.3 suggests a relationship between the area under the ROC curve and the level of performance on the task. When the ROC curve in figure 3.3 lies along the diagonal, the area underneath it is 1/2, which is the probability of a correct answer in this case (given any threshold). When the task is easy and the ROC curve hugs the left axis and upper limit in figure 3.3, the area under it approaches 1, which is again the probability of a correct answer (given an appropriate threshold). However, the precise relationship between task performance and the area under the

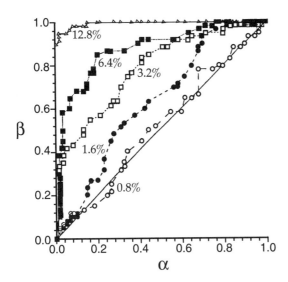

Figure 3.3 ROC curves for a variety of motion coherence levels. Each curve is the locus of points $(\alpha(z), \beta(z))$ for all z values. The values of α and β were computed from histograms such as those in figure 3.2B. The diagonal line is the ROC curve for random guessing. (Adapted from Britten et al., 1992.)

ROC curve is complicated by the fact that different threshold values can be used. This ambiguity can be removed by considering a slightly different task, called two-alternative forced choice. Here, the stimulus is presented twice, once with motion in the plus direction and once in the minus direction. The task is to decide which presentation corresponded to the plus direction, given the firing rates on both trials, r_1 and r_2. A natural extension of the test procedure we have been discussing is to answer trial 1 if $r_1 \geq r_2$ and otherwise answer trial 2. This removes the threshold variable from consideration.

two-alternative forced choice

In the two-alternative force-choice task, the value of r on one trial serves as the threshold for the other trial. For example, if the order of stimulus presentation is plus, then minus, the comparison procedure we have outlined will report the correct answer if $r_1 \geq z$ where $z = r_2$, and this has probability $P[r_1 \geq z|+] = \beta(z)$ with $z = r_2$. To determine the probability of getting the correct answer in a two-alternative forced-choice task, we need to integrate this probability over all possible values of r_2 weighted by their probability of occurrence. For small Δz, the probability that r_2 takes a value in the range between z and $z + \Delta z$ when the second trial has a minus stimulus is $p[z|-]\Delta z$, where $p[z|-]$ is the conditional firing-rate probability density for a firing rate $r = z$. Integrating over all values of z gives the probability of getting the correct answer,

$$P[\text{correct}] = \int_0^\infty dz\, p[z|-]\beta(z)\,. \tag{3.6}$$

Because the two-alternative forced-choice test is symmetric, this is also the probability of being correct if the order of the stimuli is reversed.

The probability that $r \geq z$ for a minus stimulus, which is just $\alpha(z)$, can be written as an integral of the conditional firing-rate probability density $p[r|-]$,

$$\alpha(z) = \int_z^\infty dr\, p[r|-]\,. \tag{3.7}$$

Taking the derivative of this equation with respect to z, we find that

$$\frac{d\alpha}{dz} = -p[z|-]\,. \tag{3.8}$$

This allows us to make the replacement $dz\, p[z|-] \to -d\alpha$ in the integral of equation 3.6 and to change the integration variable from z to α. Noting that $\alpha = 1$ when $z = 0$ and $\alpha = 0$ when $z = \infty$, we find

$$P[\text{correct}] = \int_0^1 d\alpha\, \beta\,. \tag{3.9}$$

The ROC curve is just β plotted as a function of α, so this integral is the area under the ROC curve. Thus, the area under the ROC curve is the probability of responding correctly in the two-alternative forced-choice test.

Suppose that $p[r|+]$ and $p[r|-]$ are both Gaussian functions with means $\langle r \rangle_+$ and $\langle r \rangle_-$, and a common variance σ_r^2. The reader is invited to show that, in this case,

$$P[\text{correct}] = \frac{1}{2}\text{erfc}\left(\frac{\langle r \rangle_- - \langle r \rangle_+}{2\sigma_r}\right) = \frac{1}{2}\text{erfc}\left(-\frac{d'}{2}\right), \tag{3.10}$$

complementary error function

where d' is the discriminability defined in equation 3.4 and $\text{erfc}(x)$ is the complementary error function (which is an integral of a Gaussian distribution) defined as

$$\text{erfc}(x) = \frac{2}{\sqrt{\pi}}\int_x^\infty dy\, \exp(-y^2)\,. \tag{3.11}$$

In the case where the distributions are equal-variance Gaussians, the relationship between the discriminability and the area under the ROC curve is invertible because the complementary error function is monotonic. It is common to quote d' values even for non-Gaussian distributions by inverting the relationship between $P[\text{correct}]$ and d' in equation 3.10.

ROC Analysis of Motion Discrimination

To interpret their experiment as a two-alternative forced-choice task, Britten et al. imagined that, in addition to being given the firing rate of the recorded neuron during stimulus presentation, the observer is given the firing rate of a hypothetical "anti-neuron" having response characteristics exactly opposite from the recorded neuron. In reality, the responses of this anti-neuron to a plus stimulus were just those of the recorded neuron to a minus stimulus, and vice versa. The idea of using the responses of a single neuron to opposite stimuli as if they were the simultaneous responses of two different neurons also reappears in our discussion of spike-train decoding. An observer predicting motion directions on the basis of just these two neurons at a level equal to the area under the ROC curve is termed an ideal observer.

Figure 3.2A shows a typical result for the performance of an ideal observer using one recorded neuron and its anti-neuron partner. The open circles in figure 3.2A were obtained by calculating the areas under the ROC curves for this neuron. Amazingly, the ability of the ideal observer to perform the discrimination task using a single neuron/anti-neuron pair is equal to the ability of the monkey to do the task. Although the choices of the ideal observer and the monkey do not necessarily match on a trial-to-trial basis, their performances are comparable when averaged over trials. This seems remarkable because the monkey presumably has access to a large population of neurons, while the ideal observer uses only two. One speculation is that correlations in the response variability between neurons limit the performance of the monkey.

The Likelihood Ratio Test

The discrimination test we have considered compares the firing rate to a threshold value. Could an observer do better than this already remarkable performance by comparing some other function of the firing rate to a threshold? What is the best test function to use for this purpose? The Neyman-Pearson lemma (proved in appendix A) shows that it is impossible to do better than to choose as the test function the ratio of probability densities (or, where appropriate, probabilities),

Neyman-Pearson lemma

$$l(r) = \frac{p[r|+]}{p[r|-]}, \tag{3.12}$$

which is known as the likelihood ratio. The test function r used above is not equal to the likelihood ratio. However, if the likelihood ratio is a monotonically increasing function of r, as it is for the data of Britten et al., the firing-rate threshold test is equivalent to using the likelihood ratio and is indeed optimal. Similarly, any monotonic function of the likelihood ratio will provide as good a test as the likelihood itself, and the logarithm is frequently used.

likelihood ratio

There is a direct relationship between the likelihood ratio and the ROC curve. As in equations 3.7 and 3.8, we can write

$$\beta(z) = \int_z^\infty dr\, p[r|+] \quad \text{so} \quad \frac{d\beta}{dz} = -p[z|+]. \tag{3.13}$$

Combining this result with 3.8, we find that

$$\frac{d\beta}{d\alpha} = \frac{d\beta}{dz}\frac{dz}{d\alpha} = \frac{p[z|+]}{p[z|-]} = l(z),\qquad(3.14)$$

so the slope of the ROC curve is equal to the likelihood ratio.

Another way of seeing that comparing the likelihood ratio to a threshold value is an optimal decoding procedure for discrimination uses a Bayesian approach based on associating a cost or penalty with getting the wrong answer. Suppose that the penalty associated with answering "minus" when the correct answer is "plus" is quantified by the loss parameter L_-. Similarly, quantify the loss for answering "plus" when the correct answer is "minus" as L_+. For convenience, we assume that there is neither loss nor gain for answering correctly. The probabilities that the correct answer is "plus" or "minus", given the firing rate r, are $P[+|r]$ and $P[-|r]$ respectively. These probabilities are related to the conditional firing-rate probability densities by Bayes theorem,

$$P[+|r] = \frac{p[r|+]P[+]}{p[r]}\quad\text{and}\quad P[-|r] = \frac{p[r|-]P[-]}{p[r]}.\qquad(3.15)$$

The average loss expected for a "plus" answer when the firing rate is r is the loss associated with being wrong times the probability of being wrong, $\text{Loss}_+ = L_+ P[-|r]$. Similarly, the expected loss when answering "minus" is $\text{Loss}_- = L_- P[+|r]$. A reasonable strategy is to cut the losses, answering "plus" if $\text{Loss}_+ \le \text{Loss}_-$ and "minus" otherwise. Using equation 3.15, we find that this strategy gives the response "plus" if

$$l(r) = \frac{p[r|+]}{p[r|-]} \ge \frac{L_+}{L_-}\frac{P[-]}{P[+]}.\qquad(3.16)$$

This shows that the strategy of comparing the likelihood ratio to a threshold is a way of minimizing the expected loss. The right side of this inequality gives an explicit formula for the value of the threshold that should be used, and reflects two factors. One is the relative losses for the two sorts of possible errors. The other is the prior probabilities that the stimulus is plus or minus. Interestingly, it is possible to change the thresholds that human subjects use in discrimination tasks by manipulating these two factors.

If the conditional probability densities $p[r|+]$ and $p[r|-]$ are Gaussians with means r_+ and r_- and identical variances σ_r^2, and $P[+] = P[-] = 1/2$, the probability $P[+|r]$ is a sigmoidal function of r,

$$P[+|r] = \frac{1}{1 + \exp(-d'(r - r_{\text{ave}})/\sigma_r)},\qquad(3.17)$$

where $r_{\text{ave}} = (r_+ + r_-)/2$. This provides an alternate interpretation of the parameter d' that is often used in the psychophysics literature; it determines the slope of a sigmoidal function fitted to $P[+|r]$.

We have thus far considered discriminating between two quite distinct stimulus values, plus and minus. Often we are interested in discriminating between two stimulus values $s + \Delta s$ and s that are very close to one another. In this case, the likelihood ratio is

$$\frac{p[r|s+\Delta s]}{p[r|s]} \approx \frac{p[r|s] + \Delta s\, \partial p[r|s]/\partial s}{p[r|s]}$$

$$= 1 + \Delta s \frac{\partial \ln p[r|s]}{\partial s}. \qquad (3.18)$$

For small Δs, a test that compares

$$Z(r) = \frac{\partial \ln p[r|s]}{\partial s} \qquad (3.19)$$

to a threshold $(z - 1)/\Delta s$ is equivalent to the likelihood ratio test. The function $Z(r)$ is sometimes called the score.   *score $Z(r)$*

3.3 Population Decoding

The use of large numbers of neurons to represent information is a basic operating principle of many nervous systems. Population coding has a number of advantages, including reduction of uncertainty due to neuronal variability and the ability to represent a number of different stimulus attributes simultaneously. Individual neurons in such a population typically have different but overlapping selectivities, so that many neurons, but not necessarily all, respond to a given stimulus. In the previous section, we discussed discrimination between stimuli on the basis of the response of a single neuron. The responses of a population of neurons can also be used for discrimination, with the only essential difference being that terms such as $p[r|s]$ are replaced by $p[\mathbf{r}|s]$, the conditional probability density of the population response $\mathbf{r}$. ROC analysis, likelihood ratio tests, and the Neyman-Pearson lemma continue to apply in exactly the same way. Discrimination is a special case of decoding in which only a few different stimulus values are considered. A more general problem is the extraction of a continuous stimulus parameter from one or more neuronal responses. In this section, we study how the value of a continuous parameter associated with a static stimulus can be decoded from the spike-count firing rates of a population of neurons.

Encoding and Decoding Direction

The cercal system of the cricket, which senses the direction of incoming air currents as a warning of approaching predators, is an interesting example of population coding involving a relatively small number of neurons. Crickets and related insects have two appendages called cerci extending

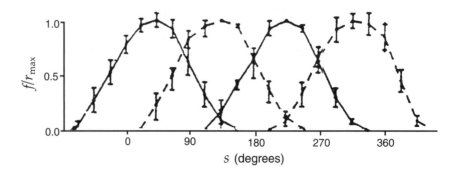

Figure 3.4 Tuning curves for the four low-velocity interneurons of the cricket cercal system plotted as a function of the wind direction s. Each neuron responds with a firing rate that is closely approximated by a half-wave rectified cosine function. The preferred directions of the neurons are located 90° from each other, and r_{max} values are typically around 40 Hz. Error bars show standard deviations. (Adapted from Theunissen and Miller, 1991.)

from their hind ends. These are covered with hairs that are deflected by air currents. Each hair is attached to a neuron that fires when the hair is deflected. Thousands of these primary sensory neurons send axons to a set of interneurons that relay the sensory information to the rest of the cricket's nervous system. No single interneuron of the cercal system responds to all wind directions, and multiple interneurons respond to any given wind direction. This implies that the interneurons encode the wind direction collectively as a population.

Theunissen and Miller (1991) measured both the mean and the variance of responses of cercal interneurons while blowing air currents at the cerci. At low wind velocities, information about wind direction is encoded by just four interneurons. Figure 3.4 shows average firing-rate tuning curves for the four relevant interneurons as a function of wind direction. These neurons are sensitive primarily to the angle of the wind around the vertical axis and not to its elevation above the horizontal plane. Wind speed was held constant in these experiments, so we do not discuss how it is encoded. The interneuron tuning curves are well approximated by half-wave rectified cosine functions. Neuron a (where $a = 1, 2, 3, 4$) responds with a maximum average firing rate when the angle of the wind direction is s_a, the preferred-direction angle for that neuron. The tuning curve for interneuron a in response to wind direction s, $\langle r_a \rangle = f_a(s)$, normalized to

cosine tuning   its maximum, can be written as

$$\left(\frac{f(s)}{r_{max}}\right)_a = [(\cos(s - s_a)]_+ , \qquad (3.20)$$

where the half-wave rectification eliminates negative firing rates. Here r_{max}, which may be different for each neuron, is a constant equal to the maximum average firing rate. The fit can be improved somewhat by introducing a small offset rate, but the simple cosine is adequate for our purposes.

To determine the wind direction from the firing rates of the cercal interneurons, it is useful to change the notation somewhat. In place of the angle
s, we can represent wind direction by a spatial vector $\vec{v}$ pointing parallel
to the wind velocity and having unit length $|\vec{v}| = 1$ (we use over-arrows
to denote spatial vectors). Similarly, we can represent the preferred wind
direction for each interneuron by a vector $\vec{c}_a$ of unit length pointing in the
direction specified by the angle s_a. In this case, we can use the vector dot   *dot product*
product to write $\cos(s - s_a) = \vec{v} \cdot \vec{c}_a$. In terms of these vectors, the average
firing rate is proportional to a half-wave rectified projection of the wind
direction vector onto the preferred-direction axis of the neuron,

$$\left(\frac{f(s)}{r_{\max}} \right)_a = [\vec{v} \cdot \vec{c}_a]_+ \ . \tag{3.21}$$

Decoding the cercal system is particularly easy because of the close relationship between the representation of wind direction it provides and a
two-dimensional Cartesian coordinate system. In a Cartesian system, vectors are parameterized by their projections onto x and y axes, v_x and v_y.
These projections can be written as dot products of the vector being repre-
sented, $\vec{v}$, with vectors of unit length $\vec{x}$ and $\vec{y}$ lying along the x and y axes,
$v_x = \vec{v} \cdot \vec{x}$ and $v_y = \vec{v} \cdot \vec{y}$. Except for the half-wave rectification, these equations are identical to equation 3.21. Furthermore, the preferred directions
of the four interneurons, like the x and y axes of a Cartesian coordinate
system, lie along two perpendicular directions (figure 3.5A). Four neurons
are required, rather than two, because firing rates cannot represent negative projections. The cricket discovered the Cartesian coordinate system
long before Descartes did, but failed to invent negative numbers! Perhaps credit should also be given to the leech, for Lewis and Kristan (1998)
have shown that the direction of touch sensation in its body segments is
encoded by four neurons in a virtually identical arrangement.

A vector $\vec{v}$ can be reconstructed from its Cartesian components through the
component-weighted vector sum $\vec{v} = v_x \vec{x} + v_y \vec{y}$. Because the firing rates of
the cercal interneurons we have been discussing are proportional to the
Cartesian components of the wind direction vector, a similar sum should
allow us to reconstruct the wind direction from a knowledge of the interneuron firing rates, except that four, not two, terms must be included.
If r_a is the spike-count firing rate of neuron a, an estimate of the wind direction on any given trial can be obtained from the direction of the vector

$$\vec{v}_{\text{pop}} = \sum_{a=1}^{4} \left(\frac{r}{r_{\max}} \right)_a \vec{c}_a \ . \tag{3.22}$$

This vector is known as the population vector, and the associated decoding   *population vector*
method is called the vector method. This decoding scheme works quite   *vector method*
well. Figure 3.5B shows the root-mean-square difference between the direction determined by equation 3.22 and the actual wind direction that
evoked the firing rates. The difference between the decoded and actual
wind directions is around $6°$ except for dips at the angles corresponding
to the preferred directions of the neurons. These dips are not due to the

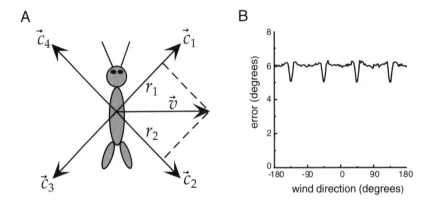

Figure 3.5 (A) Preferred directions of four cercal interneurons in relation to the cricket's body. The firing rate of each neuron for a fixed wind speed is proportional to the projection of the wind velocity vector $\vec{v}$ onto the preferred-direction axis of the neuron. The projection directions $\vec{c}_1$, $\vec{c}_2$, $\vec{c}_3$, and $\vec{c}_4$ for the four neurons are separated by 90°, and they collectively form a Cartesian coordinate system. (B) The root-mean-square error in the wind direction determined by vector decoding of the firing rates of four cercal interneurons. These results were obtained through simulation by randomly generating interneuron responses to a variety of wind directions, with the average values and trial-to-trial variability of the firing rates matched to the experimental data. The generated rates were then decoded using equation 3.22 and compared to the wind direction used to generate them. (B adapted from Salinas and Abbott, 1994.)

fact that one of the neurons responds maximally; rather, they arise because the two neurons with tuning curves adjacent to the maximally responding neuron are most sensitive to wind direction at these points.

As discussed in chapter 1, tuning curves of certain neurons in the primary motor cortex (M1) of the monkey can be described by cosine functions of arm movement direction. Thus, a vector decomposition similar to that of the cercal system appears to take place in M1. Many M1 neurons have nonzero offset rates, r_0, so they can represent the cosine function over most or all of its range. When an arm movement is made in the direction represented by a vector of unit length, $\vec{v}$, the average firing rates for such an M1 neuron, labeled by an index a (assuming that it fires over the entire range of angles), can be written as

$$\left(\frac{\langle r \rangle - r_0}{r_{max}}\right)_a = \left(\frac{f(s) - r_0}{r_{max}}\right)_a = \vec{v} \cdot \vec{c}_a, \qquad (3.23)$$

where $\vec{c}_a$ is the preferred-direction vector that defines the selectivity of the neuron. Because these firing rates represent the full cosine function, it would, in principle, be possible to encode all movement directions in three dimensions using just three neurons. Instead, many thousands of M1 neurons have arm-movement-related tuning curves, resulting in a highly redundant representation. Of course, these neurons encode additional movement-related quantities; for example, their firing rates depend on the initial position of the arm relative to the body as well as on movement velocity and acceleration. This complicates the interpretation of their activity as reporting movement direction in a particular coordinate system.

Unlike the cercal interneurons, M1 neurons do not have orthogonal preferred directions that form a Cartesian coordinate system. Instead, the preferred directions of the neurons appear to point in all directions with roughly equal probability. If the projection axes are not orthogonal, the Cartesian sum of equation 3.22 is not the correct way to reconstruct $\vec{v}$. Nevertheless, if the preferred directions point uniformly in all directions and the number of neurons N is sufficiently large, the population vector

$$\vec{v}_{\text{pop}} = \sum_{a=1}^{N} \left(\frac{r - r_0}{r_{\max}} \right)_a \vec{c}_a \tag{3.24}$$

will, on average, point in a direction parallel to the arm movement direction vector $\vec{v}$. If we average equation 3.24 over trials and use equation 3.23, we find

$$\langle \vec{v}_{\text{pop}} \rangle = \sum_{a=1}^{N} (\vec{v} \cdot \vec{c}_a) \vec{c}_a . \tag{3.25}$$

We leave as an exercise the proof that $\langle \vec{v}_{\text{pop}} \rangle$ is approximately parallel to $\vec{v}$ if a large enough number of neurons is included in the sum, and if their preferred-direction vectors point randomly in all directions with equal probability. Later in this chapter, we discuss how corrections can be made if the distribution of preferred directions is not uniform or the number of neurons is not large. The population vectors constructed from equation 3.24 on the basis of responses of neurons in primary motor cortex, recorded while a monkey performed a reaching task, are compared with the actual directions of arm movements in figure 3.6.

Optimal Decoding Methods

The vector method is a simple decoding method that can perform quite well in certain cases, but it is neither a general nor an optimal way to reconstruct a stimulus from the firing rates of a population of neurons. In this section, we discuss two methods that can, by some measure, be considered optimal. These are called Bayesian inference and maximum a posteriori (MAP) inference. We also discuss a special case of MAP called maximum likelihood (ML) inference. The Bayesian approach involves finding the minimum of a loss function that expresses the cost of estimation errors. MAP inference and ML inference generally produce estimates that are as accurate, in terms of the variance of the estimate, as any that can be achieved by a wide class of estimation methods (so-called unbiased estimates), at least when large numbers of neurons are used in the decoding. Bayesian and MAP estimates use the conditional probability that a stimulus parameter takes a value between s and $s + \Delta s$, given that the set of N encoding neurons fired at rates given by $\mathbf{r}$. The probability density

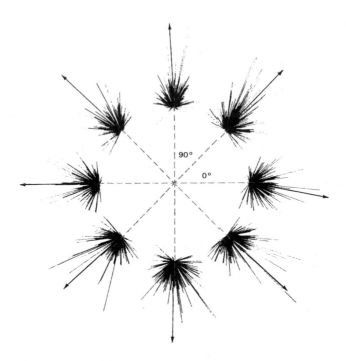

Figure 3.6 Comparison of population vectors with actual arm movement directions. Results are shown for eight different movement directions. Actual arm movement directions are radially outward at angles that are multiples of 45°. The groups of lines without arrows show the preferred-direction vectors of the recorded neurons multiplied by their firing rates. Vector sums of these terms for each movement direction are indicated by the arrows. The fact that the arrows point approximately radially outward shows that the population vector reconstructs the actual movement direction fairly accurately. (Figure adapted from Kandel et al., 1991, based on data from Kalaska et al., 1983.)

needed for a continuous stimulus parameter, $p[s|\mathbf{r}]$, can be obtained from the encoding probability density $p[\mathbf{r}|s]$ by the continuous version of Bayes theorem (equation 3.3),

$$p[s|\mathbf{r}] = \frac{p[\mathbf{r}|s]p[s]}{p[\mathbf{r}]} \,. \qquad (3.26)$$

A disadvantage of these methods is that extracting $p[s|\mathbf{r}]$ from experimental data can be difficult. In contrast, the vector method only requires us to know the preferred stimulus values of the encoding neurons.

Bayesian inference   As mentioned in the previous paragraph, Bayesian inference is based on the minimization of a particular loss function $L(s, s_{\text{bayes}})$ that quantifies the "cost" of reporting the estimate s_{bayes} when the correct answer is s. The loss function provides a way of defining the optimality criterion for decoding analogous to the loss computation discussed previously for optimal discrimination. The value of s_{bayes} is chosen to minimize the expected loss averaged over all stimuli for a given set of rates, that is, to minimize the function $\int ds\, L(s, s_{\text{bayes}}) p[s|\mathbf{r}]$. If the loss function is the squared differ-

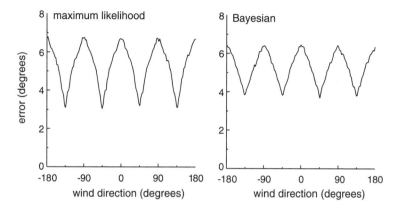

Figure 3.7 Maximum likelihood and Bayesian estimation errors for the cricket cercal system. ML and Bayesian estimates of the wind direction were compared with the actual stimulus value for a large number of simulated firing rates. Firing rates were generated as for figure 3.5B. The error shown is the root-mean-squared difference between the estimated and actual stimulus angles. (Adapted from Salinas and Abbott, 1994.)

ence between the estimate and the true value, $L(s, s_{\text{bayes}}) = (s - s_{\text{bayes}})^2$, the estimate that minimizes the expected loss is the mean

$$s_{\text{bayes}} = \int ds\, p[s|\mathbf{r}]s \ . \qquad (3.27)$$

If the loss function is the absolute value of the difference, $L(s, s_{\text{bayes}}) = |s - s_{\text{bayes}}|$, then s_{bayes} is the median rather than the mean of the distribution $p[s|\mathbf{r}]$.

Maximum a posteriori (MAP) inference does not involve a loss function but instead simply chooses the stimulus value, s_{MAP}, that maximizes the conditional probability density of the stimulus, $p[s_{\text{MAP}}|\mathbf{r}]$. The MAP approach is thus to choose as the estimate s_{MAP} the most likely stimulus value for a given set of rates. If the prior or stimulus probability density $p[s]$ is independent of s, then $p[s|\mathbf{r}]$ and $p[\mathbf{r}|s]$ have the same dependence on s, because the factor $p[s]/p[\mathbf{r}]$ in equation 3.26 is independent of s. In this case, the MAP algorithm is equivalent to maximizing the likelihood function, that is, choosing s_{ML} to maximize $p[\mathbf{r}|s_{\text{ML}}]$, which is called maximum likelihood (ML) inference.

MAP inference

ML inference

Previously we applied the vector decoding method to the cercal system of the cricket. Figure 3.7 shows the root-mean-squared difference between the true and estimated wind directions for the cercal system, using ML and Bayesian methods. For the cercal interneurons, the response probability density $p[\mathbf{r}|s]$ is a product of four Gaussians with means and variances given by the data points and error bars in figure 3.4. The Bayesian estimate in figure 3.7 is based on the squared-difference loss function. Both estimates use a constant stimulus probability density $p[s]$, so the ML and MAP estimates are identical. The maximum likelihood estimate is either more or less accurate than the Bayesian estimate, depending on the angle.

The Bayesian result has a slightly smaller average error across all angles. The dips in the error curves in figure 3.7, as in the curve of figure 3.5B, appear at angles where one tuning curve peaks and two others rise from threshold (see figure 3.4). As in figure 3.5B, these dips are due to the two neurons responding near threshold, not to the maximally responding neuron. They occur because neurons are most sensitive at points where their tuning curves have maximum slopes, which in this case is near threshold (see figure 3.11).

Comparing these results with figure 3.5B shows the improved performance of these methods relative to the vector method. The vector method performs extremely well for this system, so the degree of improvement is not large. This is because the cercal responses are well described by cosine functions and their preferred directions are 90° apart. Much more dramatic differences occur when the tuning curves are not cosines or the preferred stimulus directions are not perpendicular.

Up to now, we have considered the decoding of a direction angle. We now turn to the more general case of decoding an arbitrary continuous stimulus parameter. An instructive example is provided by an array of N neurons with preferred stimulus values distributed uniformly across the full range of possible stimulus values. An example of such an array for Gaussian tuning curves,

$$f_a(s) = r_{\max} \exp\left(-\frac{1}{2}\left(\frac{s - s_a}{\sigma_a}\right)^2\right), \qquad (3.28)$$

is shown in figure 3.8. In this example, each neuron has a tuning curve with a different preferred value s_a and potentially a different width σ_a (although all the curves in figure 3.8 have the same width). If the tuning curves are evenly and densely distributed across the range of s values, the sum of all tuning curves $\sum f_a(s)$ is approximately independent of s. The roughly flat line in figure 3.8 is proportional to this sum. The constancy of the sum over tuning curves will be useful in the following analysis.

Tuning curves give the mean firing rates of the neurons across multiple trials. In any single trial, measured firing rates will vary from their mean values. To implement the Bayesian, MAP, or ML approach, we need to know the conditional firing-rate probability density $p[\mathbf{r}|s]$ that describes this variability. We assume that the firing rate r_a of neuron a is determined by counting n_a spikes over a trial of duration T (so that $r_a = n_a/T$), and that the variability can be described by the homogeneous Poisson model discussed in chapter 1. In this case, the probability of stimulus s evoking $n_a = r_a T$ spikes, when the average firing rate is $\langle r_a \rangle = f_a(s)$, is given by (see chapter 1)

$$P[r_a|s] = \frac{(f_a(s)T)^{r_a T}}{(r_a T)!} \exp(-f_a(s)T). \qquad (3.29)$$

If we assume that each neuron fires independently, the firing-rate proba-

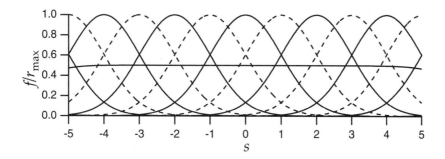

Figure 3.8 An array of Gaussian tuning curves spanning stimulus values from -5 to 5. The peak values of the tuning curves fall on the integer values of s and the tuning curves all have $\sigma_a = 1$. For clarity, the curves are drawn alternately with dashed and solid lines. The approximately flat curve with value near 0.5 is $1/5$ the sum of the tuning curves shown, indicating that this sum is approximately independent of s.

bility for the population is the product of the individual probabilities,

$$P[\mathbf{r}|s] = \prod_{a=1}^{N} \frac{(f_a(s)T)^{r_a T}}{(r_a T)!} \exp(-f_a(s)T). \tag{3.30}$$

The assumption of independence simplifies the calculations considerably.

The filled circles in figure 3.9 show a set of randomly generated firing rates for the array of Gaussian tuning curves in figure 3.8 for $s = 0$. This figure also illustrates a useful way of visualizing population responses: plotting the responses as a function of the preferred stimulus values. The dashed curve in figure 3.9 is the tuning curve for the neuron with $s_a = 0$. Because the tuning curves are functions of $|s - s_a|$, the values of the dashed curve at $s_a = -5, -4, \ldots, 5$ are the mean activities of the cells with preferred values at those locations for a stimulus at $s = 0$.

To apply the ML estimation algorithm, we only need to consider the terms in $P[\mathbf{r}|s]$ that depend on s. Because equation 3.30 involves a product, it is convenient to take its logarithm and write

$$\ln P[\mathbf{r}|s] = T \sum_{a=1}^{N} r_a \ln\left(f_a(s)\right) + \ldots, \tag{3.31}$$

where the ellipsis represents terms that are independent or approximately independent of s, including, as discussed above, $\sum f_a(s)$. Because maximizing a function and maximizing its logarithm are equivalent, we can use the logarithm of the conditional probability in place of the actual probability in ML decoding.

The ML estimated stimulus, s_{ML}, is the stimulus that maximizes the right side of equation 3.31. Setting the derivative to 0, we find that s_{ML} is deter-

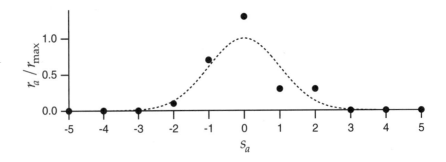

Figure 3.9 Simulated responses of 11 neurons with the Gaussian tuning curves shown in figure 3.8 to a stimulus value of 0. Firing rates for a single trial, generated using the Poisson model, are plotted as a function of the preferred-stimulus values of the different neurons in the population (filled circles). The dashed curve shows the tuning curve for the neuron with $s_a = 0$. Its heights at integer values of s_a are the average responses of the corresponding cells. It is possible to have $r_a > r_{max}$ (point at $s_a = 0$) because r_{max} is the maximum average firing rate, not the maximum firing rate.

mined by

$$\sum_{a=1}^{N} r_a \frac{f'_a(s_{ML})}{f_a(s_{ML})} = 0, \tag{3.32}$$

where the prime denotes a derivative. If the tuning curves are the Gaussians of equation 3.28, this equation can be solved explicitly using the result $f'_a(s)/f_a(s) = (s_a - s)/\sigma_a^2$,

$$s_{ML} = \frac{\sum r_a s_a/\sigma_a^2}{\sum r_a/\sigma_a^2} . \tag{3.33}$$

If all the tuning curves have the same width, this reduces to

$$s_{ML} = \frac{\sum r_a s_a}{\sum r_a}, \tag{3.34}$$

which is a simple estimation formula with an intuitive interpretation as the firing-rate weighted average of the preferred values of the encoding neurons. The numerator of this expression is reminiscent of the population vector.

Although equation 3.33 gives the ML estimate for a population of neurons with Poisson variability, it has some undesirable properties as a decoding algorithm. Consider a neuron with a preferred stimulus value s_a that is much greater than the actual stimulus value s. Because $s_a \gg s$, the average firing rate of this neuron is essentially 0. For a Poisson distribution, zero rate implies zero variability. If, however, this neuron fires one or more spikes on a trial due to a non-Poisson source of variability, this will cause a large error in the estimate because of the large weighting factor s_a.

The MAP estimation procedure is similar in spirit to the ML approach, but the MAP estimate, s_{MAP}, may differ from s_{ML} if the probability density $p[s]$ depends on s. The MAP algorithm allows us to include prior knowledge about the distribution of stimulus values in the decoding estimate. As noted above, if the $p[s]$ is constant, the MAP and ML estimates are identical. In addition, if many neurons are observed, or if a small number of neurons is observed over a long trial period, even a nonconstant stimulus distribution has little effect and $s_{MAP} \approx s_{ML}$.

The MAP estimate is computed from the distribution $p[s|\mathbf{r}]$ determined by Bayes theorem. In terms of the logarithms of the probabilities, $\ln p[s|\mathbf{r}] = \ln P[\mathbf{r}|s] + \ln p[s] - \ln P[\mathbf{r}]$. The last term in this expression is independent of s and can be absorbed into the ignored s-independent terms, so we can write, as in equation 3.31,

$$\ln p[s|\mathbf{r}] = T \sum_{a=1}^{N} r_a \ln \left(f_a(s) \right) + \ln p[s] + \dots. \tag{3.35}$$

Maximizing this determines the MAP estimate,

$$T \sum_{a=1}^{N} \frac{r_a f_a'(s_{MAP})}{f_a(s_{MAP})} + \frac{p'[s_{MAP}]}{p[s_{MAP}]} = 0. \tag{3.36}$$

If the stimulus or prior distribution is itself Gaussian with mean s_{prior} and variance σ_{prior}, and we use the Gaussian array of tuning curves, equation 3.36 yields

$$s_{MAP} = \frac{T \sum r_a s_a / \sigma_a^2 + s_{prior}/\sigma_{prior}^2}{T \sum r_a / \sigma_a^2 + 1/\sigma_{prior}^2}. \tag{3.37}$$

Figure 3.10 compares the conditional stimulus probability densities $p[s|\mathbf{r}]$ for a constant stimulus distribution (solid curve) and for a Gaussian stimulus distribution with $s_{prior} = -2$ and $\sigma_{prior} = 1$, using the firing rates given by the filled circles in figure 3.9. If the stimulus distribution is constant, $p[s|\mathbf{r}]$ peaks near the true stimulus value of 0. The effect of a nonconstant stimulus distribution is to shift the curve toward the value -2, where the stimulus probability density has its maximum, and to decrease its width by a small amount. The estimate is shifted to the left because the prior distribution suggests that the stimulus is more likely to take negative values than positive ones, independent of the evoked response. The decreased width is due to the added information that the prior distribution provides. The curves in figure 3.10 can be computed from equations 3.28 and 3.35 as Gaussians with variances $1/(T \sum r_a/\sigma_a^2)$ (constant prior) and $1/(T \sum r_a/\sigma_a^2 + 1/\sigma_{prior}^2)$ (Gaussian prior).

The accuracy with which an estimate s_{est} describes a stimulus s can be characterized by two important quantities, its bias $b_{est}(s)$ and its variance $\sigma_{est}^2(s)$. The bias is the difference between the average of s_{est} across trials that use the stimulus s and the true value of the stimulus (i.e., s), *bias*

$$b_{est}(s) = \langle s_{est} \rangle - s. \tag{3.38}$$

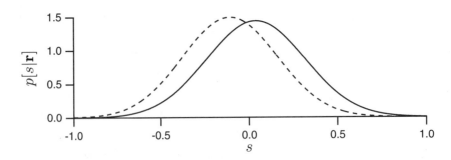

Figure 3.10 Probability densities for the stimulus, given the firing rates shown in figure 3.9 and assuming the tuning curves of figure 3.8. The solid curve is $p[s|\mathbf{r}]$ when the prior distribution of stimulus values is constant and the true value of the stimulus is $s = 0$. The dashed curve is for a Gaussian prior distribution with a mean of -2 and variance of 1, again with the true stimulus being $s = 0$. The peaks of the solid and dashed curves are at $s = 0.0385$ and $s = -0.107$, respectively.

Note that the bias depends on the true value of the stimulus. An estimate is termed unbiased if $b_{\text{est}}(s) = 0$ for all stimulus values.

variance The variance of the estimator, which quantifies how much the estimate varies about its mean value, is defined as

$$\sigma^2_{\text{est}}(s) = \langle (s_{\text{est}} - \langle s_{\text{est}} \rangle)^2 \rangle . \tag{3.39}$$

The bias and variance can be used to compute the trial-average squared estimation error, $\langle (s_{\text{est}} - s)^2 \rangle$. This is a measure of the spread of the estimated values about the true value of the stimulus. Because $s = \langle s_{\text{est}} \rangle - b_{\text{est}}(s)$, we *estimation error* can write the squared estimation error as

$$\langle (s_{\text{est}} - s)^2 \rangle = \langle (s_{\text{est}} - \langle s_{\text{est}} \rangle + b_{\text{est}}(s))^2 \rangle = \sigma^2_{\text{est}}(s) + b^2_{\text{est}}(s) . \tag{3.40}$$

In other words, the average squared estimation error is the sum of the variance and the square of the bias. For an unbiased estimate, the average squared estimation error is equal to the variance of the estimator.

Fisher Information

Decoding can be used to limit the accuracy with which a neural system encodes the value of a stimulus parameter because the encoding accuracy cannot exceed the accuracy of an optimal decoding method. Of course, we must be sure that the decoding technique used to establish such a bound is truly optimal, or else the result will reflect the limitations of the decoding procedure, not bounds on the neural system being studied. The Fisher information is a quantity that provides one such measure of encoding accuracy. Through a bound known as the Cramér-Rao bound, the Fisher information limits the accuracy with which any decoding scheme can extract an estimate of an encoded quantity.

Cramér-Rao bound The Cramér-Rao bound limits the variance of any estimate s_{est} according to (appendix B)

$$\sigma^2_{\text{est}}(s) \geq \frac{\left(1 + b'_{\text{est}}(s)\right)^2}{I_{\text{F}}(s)}, \tag{3.41}$$

where $b'_{\text{est}}(s)$ is the derivative of $b_{\text{est}}(s)$. If we assume here that the firing rates take continuous values and that their distribution in response to a stimulus s is described by the conditional probability density $p[\mathbf{r}|s]$, the quantity $I_{\text{F}}(s)$ in equation 3.41 is the Fisher information of the firing-rate distribution, which is related to $p[\mathbf{r}|s]$ (assuming the latter is sufficiently smooth) by

Fisher information

$$I_{\text{F}}(s) = \left\langle -\frac{\partial^2 \ln p[\mathbf{r}|s]}{\partial s^2} \right\rangle = \int d\mathbf{r}\, p[\mathbf{r}|s] \left(-\frac{\partial^2 \ln p[\mathbf{r}|s]}{\partial s^2} \right). \tag{3.42}$$

The reader can verify that the Fisher information can also be written as

$$I_{\text{F}}(s) = \left\langle \left(\frac{\partial \ln p[\mathbf{r}|s]}{\partial s} \right)^2 \right\rangle = \int d\mathbf{r}\, p[\mathbf{r}|s] \left(\frac{\partial \ln p[\mathbf{r}|s]}{\partial s} \right)^2. \tag{3.43}$$

The Cramér-Rao bound sets a limit on the accuracy of any unbiased estimate of the stimulus. When $b_{\text{est}}(s) = 0$, equation 3.40 indicates that the average squared estimation error is equal to σ^2_{est} and, by equation 3.41, this satisfies the bound $\sigma^2_{\text{est}} \geq 1/I_{\text{F}}(s)$. Provided that we restrict ourselves to unbiased decoding schemes, the Fisher information sets an absolute limit on decoding accuracy, and it thus provides a useful limit on encoding accuracy. Although imposing zero bias on the decoding estimate seems reasonable, the restriction is not trivial. In general, minimizing the decoding error in equation 3.40 involves a trade-off between minimizing the bias and minimizing the variance of the estimator. In some cases, biased schemes may produce more accurate results than unbiased ones. For a biased estimator, the average squared estimation error and the variance of the estimate are not equal, and the estimation error can be either larger or smaller than $1/I_{\text{F}}(s)$.

The limit on decoding accuracy set by the Fisher information can be attained by a decoding scheme we have studied, the maximum likelihood method. In the limit of large numbers of encoding neurons, and for most firing-rate distributions, the ML estimate is unbiased and saturates the Cramér-Rao bound. In other words, the variance of the ML estimate is given asymptotically (for large N) by $\sigma^2_{\text{ML}}(s) = 1/I_{\text{F}}(s)$. Any unbiased estimator that saturates the Cramér-Rao lower bound is called efficient. Furthermore, $I_{\text{F}}(s)$ grows linearly with N, and the ML estimate obeys a central limit theorem, so that $N^{1/2}(s_{\text{ML}} - s)$ is Gaussian distributed with a variance that is independent of N in the large N limit. Finally, in the limit $N \to \infty$, the ML estimate is asymptotically consistent, in the sense that $P[|s_{\text{ML}} - s| > \epsilon] \to 0$ for any $\epsilon > 0$.

efficiency

asymptotic consistency

As equation 3.42 shows, the Fisher information is a measure of the expected curvature of the log likelihood at the stimulus value s. Curvature is important because the likelihood is expected to be at a maximum near the true stimulus value s that caused the responses. If the likelihood is very curved, and thus the Fisher information is large, responses typical for the stimulus s are much less likely to occur for slightly different stimuli. Therefore, the typical response provides a strong indication of the value of the stimulus. If the likelihood is fairly flat, and thus the Fisher information is small, responses common for s are likely to occur for slightly different stimuli as well. Thus, the response does not as clearly determine the stimulus value. The Fisher information is purely local in the sense that it does not reflect the existence of stimulus values completely different from s that are likely to evoke the same responses as those evoked by s itself. However, this does not happen for the sort of simple population codes we consider. Shannon's mutual information measure, discussed in chapter 4, takes such possibilities into account.

The Fisher information for a population of neurons with uniformly arrayed tuning curves (the Gaussian array in figure 3.8, for example) and Poisson statistics can be computed from the conditional firing-rate probability in equation 3.30. Because the spike-count rate is described here by a probability rather than a probability density, we use the discrete analog of equation 3.42,

$$I_F(s) = \left\langle -\frac{\partial^2 \ln P[\mathbf{r}|s]}{\partial s^2} \right\rangle = T \sum_{a=1}^{N} \left(\langle r_a \rangle \left(\left(\frac{f_a'(s)}{f_a(s)} \right)^2 - \frac{f_a''(s)}{f_a(s)} \right) + f_a''(s) \right).$$

$$(3.44)$$

Note that we have used the full expression, equation 3.30, in deriving this result, not the truncated form of $\ln P[\mathbf{r}|s]$ in equation 3.31. We next make the replacement $\langle r_a \rangle = f_a(s)$, producing the final result

$$I_F(s) = T \sum_{a=1}^{N} \frac{\left(f_a'(s) \right)^2}{f_a(s)}.$$

$$(3.45)$$

In this expression, each neuron contributes an amount to the Fisher information proportional to the square of its tuning curve slope and inversely proportional to the average firing rate for the particular stimulus value being estimated. Highly sloped tuning curves give firing rates that are sensitive to the precise value of the stimulus. Figure 3.11 shows the contribution to the sum in equation 3.45 from a single neuron with a Gaussian tuning curve, the neuron with $s_a = 0$ in figure 3.8. For comparison purposes, a dashed curve proportional to the tuning curve is also plotted. Note that the Fisher information vanishes for the stimulus value that produces the maximum average firing rate, because $f_a'(s) = 0$ at this point. The firing rate of a neuron at the peak of its tuning curve is relatively unaffected by small changes in the stimulus. Individual neurons carry the most Fisher information in regions of their tuning curves where average firing rates are rapidly varying functions of the stimulus value, not where the firing rate is highest.

The Fisher information can be used to derive an interesting result on the optimal widths of response tuning curves. Consider a population of neu-

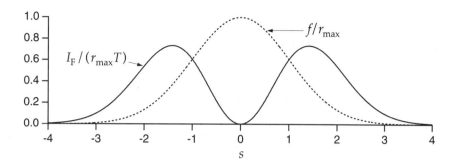

Figure 3.11 The Fisher information for a single neuron with a Gaussian tuning curve with $s=0$ and $\sigma_a=1$, and Poisson variability. The Fisher information (solid curve) has been divided by $r_{max}T$, the peak firing rate of the tuning curve times the duration of the trial. The dashed curve shows the tuning curve scaled by r_{max}. Note that the Fisher information is greatest where the slope of the tuning curve is highest, and vanishes at $s=0$, where the tuning curve peaks.

rons with tuning curves of identical shapes, distributed evenly over a range of stimulus values as in figure 3.8. Equation 3.45 indicates that the Fisher information will be largest if the tuning curves of individual neurons are rapidly varying (making the square of their derivatives large), and if many neurons respond (making the sum over neurons large). For typical neuronal response tuning curves, these two requirements are in conflict with one another. If the population of neurons has narrow tuning curves, individual neural responses are rapidly varying functions of the stimulus, but few neurons respond. Broad tuning curves allow many neurons to respond, but the individual responses are not as sensitive to the stimulus value. To determine whether narrow or broad tuning curves produce the more accurate encodings, we consider a dense distribution of Gaussian tuning curves, all with $\sigma_a = \sigma_r$. Using such curves in equation 3.45, we find

$$I_F(s) = T \sum_{a=1}^{N} \frac{r_{max}(s-s_a)^2}{\sigma_r^4} \exp\left(-\frac{1}{2}\left(\frac{s-s_a}{\sigma_r}\right)^2\right). \qquad (3.46)$$

This expression can be approximated by replacing the sum over neurons with an integral over their preferred stimulus values and multiplying by a density factor ρ_s. The factor ρ_s is the density with which the neurons cover the range of stimulus values, and it is equal to the number of neurons with preferred stimulus values lying within a unit range of s values. Replacing the sum over a with an integral over a continuous preferred stimulus parameter ξ (which replaces s_a), we find

sums→integrals

$$I_F(s) \approx \rho_s T \int_{-\infty}^{\infty} d\xi \, \frac{r_{max}(s-\xi)^2}{\sigma_r^4} \exp\left(-\frac{1}{2}\left(\frac{s-\xi}{\sigma_r}\right)^2\right)$$

$$= \frac{\sqrt{2\pi}\rho_s\sigma_r r_{max}T}{\sigma_r^2}. \qquad (3.47)$$

We have expressed the final result in this form because the number of neurons that respond to a given stimulus value is roughly $\rho_s\sigma_r$, and the Fisher information is proportional to this number divided by the square of the tuning curve width. Combining these factors, the Fisher information is inversely proportional to σ_r, and the encoding accuracy increases with narrower tuning curves.

The advantage of using narrow tuning curves goes away if the stimulus is characterized by more than one parameter. Consider a stimulus with D parameters and suppose that the response tuning curves are products of identical Gaussians for each of these parameters. If the tuning curves cover the D-dimensional space of stimulus values with a uniform density ρ_s, the number of responding neurons for any stimulus value is proportional to $\rho_s \sigma_r^D$ and, using the same integral approximation as in equation 3.47, the Fisher information is

$$I_F = \frac{(2\pi)^{D/2} \rho_s \sigma_r^D r_{max} T}{D \sigma_r^2} = \frac{(2\pi)^{D/2} \rho_s \sigma_r^{D-2} r_{max} T}{D}. \qquad (3.48)$$

This equation, which reduces to the result given above if $D = 1$, allows us to examine the effect of tuning curve width on encoding accuracy. The trade-off between the encoding accuracy of individual neurons and the number of responding neurons depends on the dimension of the stimulus space. Narrowing the tuning curves (making σ_r smaller) increases the Fisher information for $D = 1$, decreases it for $D > 2$, and has no impact if $D = 2$.

Optimal Discrimination

In the first part of this chapter, we considered discrimination between two values of a stimulus. An alternative to the procedures discussed there is simply to decode the responses and discriminate on the basis of the estimated stimulus values. Consider the case of discriminating between s and $s + \Delta s$ for small Δs. For large N, the average value of the difference between the ML estimates for the two stimulus values is equal to Δs (because the estimate is unbiased) and the variance of each estimate (for small *ML* Δs) is $1/I_F(s)$. Thus, the discriminability, defined in equation 3.4, for the *discriminability* ML-based test is

$$d' = \Delta s \sqrt{I_F(s)}. \qquad (3.49)$$

The larger the Fisher information, the higher the discriminability. We leave as an exercise the proof that for small Δs, this discriminability is the same as that of the likelihood ratio test $Z(\mathbf{r})$ defined in equation 3.19.

Discrimination by ML estimation requires maximizing the likelihood, and this may be computationally challenging. The likelihood ratio test described previously may be simpler, especially for Poisson variability, because, for small Δs, the likelihood ratio test Z defined in equation 3.19 is a linear function of the firing rates,

$$Z = T \sum_{a=1}^{N} r_a \frac{f_a'(s)}{f_a(s)}. \qquad (3.50)$$

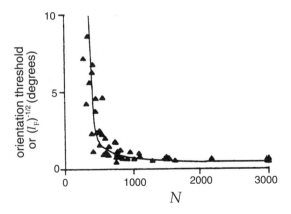

Figure 3.12 Comparison of Fisher information and discrimination thresholds for orientation tuning. The solid curve is the minimum standard deviation of an estimate of orientation angle from the Cramér-Rao bound, plotted as a function of the number of neurons (N) involved in the estimation. The triangles are data points from an experiment that determined the threshold for discrimination of the orientation of line images by human subjects as a function of line length and eccentricity. An effective number of neurons involved in the task was estimated for the different line lengths and eccentricities, using the cortical magnification factor discussed in chapter 2. (Adapted from Paradiso, 1988.)

Figure 3.12 shows an interesting comparison of the Fisher information for orientation tuning in the primary visual cortex with human orientation discrimination thresholds. Agreement like this can occur for difficult tasks, like discrimination at threshold, where the performance of a subject may be limited by basic constraints on neuronal encoding accuracy.

3.4   Spike-Train Decoding

The decoding methods we have considered estimate or discriminate static stimulus values on the basis of spike-count firing rates. Spike-count firing rates do not provide sufficient information for reconstructing a stimulus that varies during the course of a trial. Instead, we can estimate such a stimulus from the sequence of firing times t_i for $i = 1, 2, \ldots, n$ of the spikes that it evokes. One method for doing this is similar to the Wiener kernel approach used to estimate the firing rate from the stimulus in chapter 2, and to approximate a firing rate using a sliding window function in chapter 1. For simplicity, we restrict our discussion to the decoding of a single neuron. We assume, as we did in chapter 2, that the time average of the stimulus being estimated is 0.

In spike-train decoding, we attempt to construct an estimate of the stimulus at time t from the sequence of spikes evoked up to that time. There are paradoxical aspects of this procedure. The firing of an action potential at time t_i is only affected by the stimulus $s(t)$ prior to that time, $t < t_i$, and yet, in spike decoding, we attempt to extract information from this

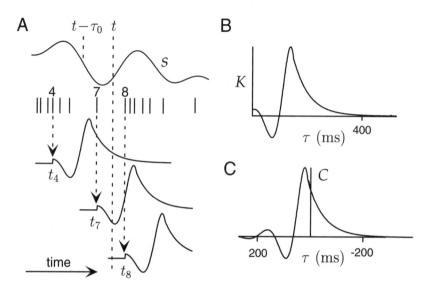

Figure 3.13 Illustration of spike-train decoding. (A) The top trace denotes a stimulus that evokes the spike train appearing below it (second trace from top). At time t an estimate is being made of the stimulus at time $t - \tau_0$. The estimate is obtained by summing the values of the kernels where they cross the dashed line labeled t, for spikes up to and including spike 7. Two such kernels are shown in the third and fourth traces from the top. The real estimate is obtained by summing similar contributions from all of the spikes. The kernel is 0 for negative values of its argument, so spikes for $i \geq 8$ do not contribute to the estimate at this time (e.g., fifth trace from top). (B) The kernel used in A. This has been truncated to zero value for negative values of τ. (C) The spike-triggered average corresponding to the kernel in B, assuming no spike-train correlations. Note that C has been plotted with the τ axis reversed, following the convention established in chapter 1. With this convention, K in panel B is simply a shifted and truncated version of the curve appearing here. In this case $\tau_0 = 160$ ms.

action potential about the value of the stimulus at a later time $t > t_i$. That is, the evoked spikes tell us about the past behavior of the stimulus and, in spike decoding, we attempt to use this information to predict the current stimulus value. Clearly, this requires that the stimulus have some form of temporal correlation so that past behavior provides information about the current stimulus value. To make the decoding task easier, we

prediction delay τ_0   can introduce a prediction delay, τ_0, and attempt to construct, from spikes occurring prior to time t, an estimate of the stimulus at time $t - \tau_0$ (see figure 3.13A). Such a delayed estimate uses a combination of spikes that could have been fired in response to the stimulus $s(t - \tau_0)$ being estimated (those for which $t - \tau_0 < t_i < t$; spike 7 in figure 3.13A), and spikes that occurred too early to be affected by the value of $s(t - \tau_0)$ (those for which $t_i < t - \tau_0$; spikes 1-6 in figure 3.13A), but that can contribute to its estimation on the basis of stimulus correlations. The estimation task gets easier as τ_0 is increased, but this delays the decoding and makes the result less behaviorally relevant. We will consider decoding with an arbitrary delay and later discuss how to set a specific value for τ_0.

stimulus estimate   The stimulus estimate is constructed as a linear sum over all spikes. A spike occurring at time t_i contributes a kernel $K(t - t_i)$, and the total estimate is obtained by summing over all spikes,

$$s_{est}(t - \tau_0) = \sum_{i=1}^{n} K(t - t_i) - \langle r \rangle \int_{-\infty}^{\infty} d\tau \, K(\tau) \,. \qquad (3.51)$$

The last term, with $\langle r \rangle = \langle n \rangle / T$ the average firing rate over the trial, is included to impose the condition that the time average of s_{est} is 0, in agreement with the time-average condition on s. The sum in equation 3.51 includes all spikes, so the constraint that only those spikes occurring prior to the time t (spikes 1-7 in figure 3.13A) should be included must be imposed by requiring $K(t - t_i) = 0$ for $t - t_i \leq 0$. A kernel satisfying this constraint is termed causal. We ignore the causality constraint for now and construct an acausal kernel, but we will return to issues of causality later in the discussion. Figure 3.13A shows how spikes contribute to a stimulus estimate, using the kernel shown in figure 3.13B.

Equation 3.51 can be written in a compact way by using the neural response function $\rho(t) = \sum \delta(t - t_i)$ introduced in chapter 1,

$$s_{est}(t - \tau_0) = \int_{-\infty}^{\infty} d\tau \, (\rho(t - \tau) - \langle r \rangle) \, K(\tau) \,. \qquad (3.52)$$

Using this form of the estimate, the construction of the optimal kernel K proceeds very much like the construction of the optimal kernel for predicting firing rates in chapter 2. We choose K so that the squared difference between the estimated stimulus and the actual stimulus, averaged over both time and trials,

$$\frac{1}{T} \int_0^T dt \left\langle \left(\int_{-\infty}^{\infty} d\tau \, (\rho(t - \tau) - \langle r \rangle) \, K(\tau) - s(t - \tau_0) \right)^2 \right\rangle, \qquad (3.53)$$

is minimized. The calculation proceeds as in appendix A of chapter 2, and the result is that K obeys the equation

$$\int_{-\infty}^{\infty} d\tau' \, Q_{\rho\rho}(\tau - \tau') K(\tau') = Q_{rs}(\tau - \tau_0) \,, \qquad (3.54)$$

where $Q_{\rho\rho}$ is the spike-train autocorrelation function,

$$Q_{\rho\rho}(\tau - \tau') = \frac{1}{T} \int_0^T dt \, \langle (\rho(t - \tau) - \langle r \rangle)(\rho(t - \tau') - \langle r \rangle) \rangle \,, \qquad (3.55)$$

as defined in chapter 1. Q_{rs} is the correlation of the firing rate and the stimulus, which is related to the spike-triggered average C, both introduced in chapter 1,

$$Q_{rs}(\tau - \tau_0) = \langle r \rangle C(\tau_0 - \tau) = \frac{1}{T} \left\langle \sum_{i=1}^{n} s(t_i + \tau - \tau_0) \right\rangle. \qquad (3.56)$$

At this point in the derivation of the optimal linear kernel for firing-rate prediction in chapter 2, we chose the stimulus to be uncorrelated so that an integral equation similar to 3.54 would be simplified. This could always be done because we have complete control over the stimulus in this type of experiment. However, we do not have similar control of the neuron, and must deal with whatever spike-train autocorrelation function it gives us. If the spike train is uncorrelated, which tends to happen at low rates,

$$Q_{\rho\rho}(\tau) = \langle r \rangle \delta(\tau), \tag{3.57}$$

and we find from equation 3.54 that

$$K(\tau) = \frac{1}{\langle r \rangle} Q_{rs}(\tau - \tau_0) = C(\tau_0 - \tau) = \frac{1}{\langle n \rangle} \left\langle \sum_{i=1}^{n} s(t_i + \tau - \tau_0) \right\rangle. \tag{3.58}$$

This is the average value of the stimulus at time $\tau - \tau_0$ relative to the appearance of a spike. Because $\tau - \tau_0$ can be either positive or negative, stimulus estimation, unlike firing-rate estimation, involves both forward and backward correlation and the average values of the stimulus both before and after a spike. Decoding in this way follows a simple rule: every time a spike appears, we replace it with the average stimulus surrounding a spike, shifted by an amount τ_0 (figure 3.13).

The need for either stimulus correlations or a nonzero prediction delay is clear from equation 3.58. Correlations between a spike and subsequent stimuli can arise, in a causal system, only from correlations between the stimulus and itself. If these are absent, as for white noise, $K(\tau)$ will be 0 for $\tau > \tau_0$. For causal decoding, we must also have $K(\tau) = 0$ for $\tau < 0$. Thus, if $\tau_0 = 0$ and the stimulus is uncorrelated, $K(\tau) = 0$ for all values of τ.

optimal kernel   When the spike-train autocorrelation function is not a δ function, an acausal solution for K can be expressed as an inverse Fourier transform,

$$K(\tau) = \frac{1}{2\pi} \int d\omega \, \tilde{K}(\omega) \exp(-i\omega\tau), \tag{3.59}$$

where, as shown in appendix C,

$$\tilde{K}(\omega) = \frac{\tilde{Q}_{rs}(\omega) \exp(i\omega\tau_0)}{\tilde{Q}_{\rho\rho}(\omega)}. \tag{3.60}$$

Here $\tilde{Q}_{rs}$ and $\tilde{Q}_{\rho\rho}$ are the Fourier transforms of Q_{rs} and $Q_{\rho\rho}$. The numerator in this expression reproduces the expression $Q_{rs}(\tau - \tau_0)$ in equation 3.58. The role of the denominator is to correct for any autocorrelations in the response spike train. Such correlations introduce a bias in the decoding, and the denominator in equation 3.60 corrects for this bias.

If we ignore the constraint of causality, then, because the occurrence of a spike cannot depend on the behavior of a stimulus in the very distant past, we can expect $K(\tau)$ from equation 3.58 to vanish for sufficiently negative values of $\tau - \tau_0$. For most neurons, this will occur for $\tau - \tau_0$ more negative than minus a few hundred ms. The decoding kernel can therefore be made small for negative values of τ by choosing τ_0 large enough, but this may require a fairly large prediction delay. We can force exact adherence to the causality constraint for $\tau < 0$ by replacing $K(\tau)$ by $\Theta(\tau)K(\tau)$, where   *causality constraint* $\Theta(\tau)$ is defined such that $\Theta(\tau) = 1$ for $\tau > 0$ and $\Theta(\tau) = 0$ for $\tau < 0$. The causality constraint was imposed in this way in figure 3.13B. When it is multiplied by $\Theta(\tau)$, the restricted K is no longer the optimal decoding kernel, but it may be close to optimal.

Another way of imposing causality on the decoding kernel is to expand $K(\tau)$ as a weighted sum of causal basis functions (functions that vanish for negative arguments and span the space of functions satisfying the causal constraint). The optimal weights are then determined by minimizing the estimation error. This approach has the advantage of producing a truly optimal kernel for any desired value of τ_0. A simpler but nonoptimal approach is to consider a fixed functional form for $K(\tau)$ that vanishes for $\tau \leq 0$ and is characterized by a number of free parameters that can be determined by minimizing the decoding error. Finally, the optimal causal kernel, also called the Wiener-Hopf filter, can be obtained by a technique that involves so-called spectral factorization of $\tilde{Q}_{\rho\rho}(\omega)$.

Figure 3.14 shows an example of spike-train decoding for the H1 neuron of the fly discussed in chapter 2. The top panel gives two reconstruction kernels, one acausal and one causal, and the bottom panel compares the reconstructed stimulus velocity with the actual stimulus velocity. The middle panel in figure 3.14 points out one further wrinkle in the procedure. Flies have two H1 neurons, one on each side of the body, that respond to motion in opposite directions. As is often the case, half-wave rectification prevents a single neuron from encoding both directions of motion. In the experiment described in the figure, rather than recording from both H1 neurons, Bialek et al. (1991) recorded from a single H1 neuron, but presented both the stimulus $s(t)$ and its negative, $-s(t)$. The two rows of spikes in the middle panel show sample traces for each of these presentations. This procedure provides a reasonable approximation of recording both H1 neurons, and produces a neuron/anti-neuron pair of recordings similar to the one that we discussed in connection with motion discrimination from area MT neurons. The stimulus is then decoded by summing the kernel $K(t - t_i)$ for all spike times t_i of the recorded H1 neuron and summing $-K(t - t_j)$ for all spike times t_j of its anti-neuron partner.

The fly has only two H1 neurons from which it must extract information about visual motion, so it seems reasonable that stimulus reconstruction using the spike-train decoding technique can produce quite accurate results (figure 3.14). It is perhaps more surprising that accurate decoding, at least in the sense of percent correct discriminations, can be obtained from single neurons out of the large population of MT neurons responding to visual motion in the monkey. Of course, the reconstruction of a time-dependent stimulus from H1 responses is more challenging than the binary discrimination done with MT neurons. Furthermore, it is worth remembering that in all the examples we have considered, including decoding wind direction from the cercal system and arm movement direction

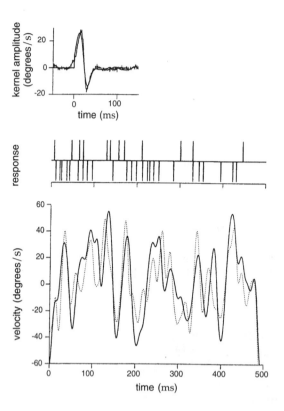

Figure 3.14 Decoding the stimulus from an H1 neuron of the fly. The upper panel is the decoding kernel. The jagged curve is the optimal acausal filter, and the smooth curve is a kernel obtained by expanding in a causal set of basis functions. In both cases, the kernels are shifted by $\tau_0 = 40$ ms. The middle panel shows typical responses of the H1 neuron to the stimuli $s(t)$ (upper trace) and $-s(t)$ (bottom trace). The dashed line in the lower panel shows the actual stimulus, and the solid line is the estimated stimulus from the optimal linear reconstruction using the acausal filter. (Adapted from Rieke et al., 1997.)

from a population of M1 neurons, the stimuli used are extremely simple compared with the naturally occurring stimuli that must be interpreted during normal behavior.

3.5   Chapter Summary

We have considered the decoding of stimulus characteristics from the responses they evoke, including discrimination between stimulus values, the decoding of static stimuli on the basis of population responses, and the decoding of dynamic stimulus parameters from spike trains. Discrimination was studied using the receiver operating characteristic, likelihood ratio tests, and the Neyman-Pearson lemma. For static parameter decoding we introduced the vector method; Bayesian, maximum a posteriori, and maximum likelihood inference; the Fisher information; and the Cramér-Rao lower bound. We also showed how to use ideas from Wiener filtering to reconstruct an approximation of a time-varying stimulus from the spike trains it evokes.

3.6 Appendices

A: The Neyman-Pearson Lemma

Consider the difference $\Delta \beta$ in the power of two tests that have identical sizes α. One uses the likelihood ratio $l(r)$, and the other uses a different test function $h(r)$. For the test $h(r)$ using the threshold z_h,

$$\alpha_h(z_h) = \int dr \, p[r|-]\Theta(h(r) - z_h) \text{ and } \beta_h(z_h) = \int dr \, p[r|+]\Theta(h(r) - z_h).$$

$$(3.61)$$

Similar equations hold for the $\alpha_l(z_l)$ and $\beta_l(z_l)$ values for the test $l(r)$ using the threshold z_l. We use the Θ function, which is 1 for positive and 0 for negative values of its argument, to impose the condition that the test is greater than the threshold. Comparing the β values for the two tests, we find

$$\Delta \beta = \beta_l(z_l) - \beta_h(z_h) = \int dr \, p[r|+]\Theta(l(r) - z_l) - \int dr \, p[r|+]\Theta(h(r) - z_h).$$

$$(3.62)$$

The range of integration where $l(r) \geq z_l$ and also $h(r) \geq z_h$ cancels between these two integrals, so, in a more compact notation, we can write

$$\Delta \beta = \int dr \, p[r|+] \left(\Theta(l(r) - z_l)\Theta(z_h - h(r)) - \Theta(z_l - l(r))\Theta(h(r) - z_h) \right).$$

$$(3.63)$$

Using the definition $l(r) = p[r|+]/p[r|-]$, we can replace $p[r|+]$ with $l(r)p[r|-]$ in this equation, giving

$$\Delta \beta = \int dr \, l(r)p[r|-]\left(\Theta(l(r) - z_l)\Theta(z_h - h(r)) - \Theta(z_l - l(r))\Theta(h(r) - z_h) \right).$$

$$(3.64)$$

Then, due to the conditions imposed on $l(r)$ by the Θ functions within the integrals, replacing $l(r)$ by z can neither decrease the value of the integral resulting from the first term in the large parentheses, nor increase the value arising from the second. This leads to the inequality

$$\Delta \beta \geq z \int dr \, p[r|-] \left(\Theta(l(r) - z_l)\Theta(z_h - h(r)) - \Theta(z_l - l(r))\Theta(h(r) - z_h) \right).$$

$$(3.65)$$

Putting back the region of integration that cancels between these two terms (for which $l(r) \geq z_l$ and $h(r) \geq z_h$), we find

$$\Delta\beta \geq z\left[\int dr\, p[r|-]\Theta(l(r)-z_l) - \int dr\, p[r|-]\Theta(h(r)-z_h)\right]. \quad (3.66)$$

By definition, these integrals are the sizes of the two tests, which are equal by hypothesis. Thus $\Delta\beta \geq 0$, showing that no test can be better than the likelihood ratio $l(r)$, at least in the sense of maximizing the power for a given size.

B: The Cramér-Rao Bound

Cauchy-Schwarz inequality

The Cramér-Rao lower bound for an estimator s_{est} is based on the Cauchy-Schwarz inequality, which states that for any two quantities A and B,

$$\langle A^2\rangle\langle B^2\rangle \geq \langle AB\rangle^2. \quad (3.67)$$

To prove this inequality, note that

$$\left\langle\left((\langle B^2\rangle A - \langle AB\rangle B\right)^2\right\rangle \geq 0 \quad (3.68)$$

because it is the average value of a square. Computing the square gives

$$\langle B^2\rangle^2\langle A^2\rangle - \langle AB\rangle^2\langle B^2\rangle \geq 0, \quad (3.69)$$

from which the inequality follows directly.

Consider the inequality of equation 3.67 with $A = \partial\ln p/\partial s$ and $B = s_{est} - \langle s_{est}\rangle$. From equations 3.43 and 3.39, we have $\langle A^2\rangle = I_F$ and $\langle B^2\rangle = \sigma_{est}^2$. The Cauchy-Schwarz inequality then gives

$$\sigma_{est}^2(s)I_F \geq \left\langle\frac{\partial\ln p[r|s]}{\partial s}(s_{est} - \langle s_{est}\rangle)\right\rangle^2. \quad (3.70)$$

To evaluate the expression on the right side of the inequality 3.70, we differentiate the defining equation for the bias (equation 3.38),

$$s + b_{est}(s) = \langle s_{est}\rangle = \int d\mathbf{r}\, p[\mathbf{r}|s]s_{est}, \quad (3.71)$$

with respect to s to obtain

$$\begin{aligned}
1 + b'_{est}(s) &= \int d\mathbf{r}\,\frac{\partial p[\mathbf{r}|s]}{\partial s}s_{est}\\
&= \int d\mathbf{r}\, p[\mathbf{r}|s]\frac{\partial\ln p[\mathbf{r}|s]}{\partial s}s_{est}\\
&= \int d\mathbf{r}\, p[\mathbf{r}|s]\frac{\partial\ln p[\mathbf{r}|s]}{\partial s}(s_{est} - \langle s_{est}\rangle). \quad (3.72)
\end{aligned}$$

The last equality follows from the identity

$$\int d\mathbf{r}\, p[\mathbf{r}|s] \frac{\partial \ln p[\mathbf{r}|s]}{\partial s} \langle s_{\mathrm{est}} \rangle = \langle s_{\mathrm{est}} \rangle \int d\mathbf{r}\, \frac{\partial p[\mathbf{r}|s]}{\partial s} = 0 \qquad (3.73)$$

because $\int d\mathbf{r}\, p[\mathbf{r}|s] = 1$. The last line of equation 3.72 is just another way of writing the expression being squared on the right side of the inequality 3.70, so combining this result with the inequality gives

$$\sigma_{\mathrm{est}}^2(s) I_{\mathrm{F}} \geq (1 + b_{\mathrm{est}}'(s))^2 \,, \qquad (3.74)$$

which, when rearranged, is the Cramér-Rao bound of equation 3.41.

C: The Optimal Spike-Decoding Filter

The optimal linear kernel for spike-train decoding is determined by solving equation 3.54. This is done by taking the Fourier transform of both sides of the equation, that is, multiplying both sides by $\exp(i\omega\tau)$ and integrating over τ,

$$\int_{-\infty}^{\infty} d\tau \exp(i\omega\tau) \int_{-\infty}^{\infty} d\tau'\, Q_{\rho\rho}(\tau - \tau')K(\tau') = \int_{-\infty}^{\infty} d\tau \exp(i\omega\tau) Q_{rs}(\tau - \tau_0) \,. \qquad (3.75)$$

By making the replacement of integration variable $\tau \to \tau + \tau_0$, we find that the right side of this equation is

$$\exp(i\omega\tau_0) \int_{-\infty}^{\infty} d\tau \exp(i\omega\tau) Q_{rs}(\tau) = \exp(i\omega\tau_0)\tilde{Q}_{rs}(\omega) \,, \qquad (3.76)$$

where $\tilde{Q}_{rs}(\omega)$ is the Fourier transform of $Q_{rs}(\tau)$. The integral of the product of two functions that appears on the left side of equations 3.54 and 3.75 is a convolution. As a result of the theorem on the Fourier transforms of convolutions (see the Mathematical Appendix),

$$\int_{-\infty}^{\infty} d\tau \exp(i\omega\tau) \int_{-\infty}^{\infty} d\tau'\, Q_{\rho\rho}(\tau - \tau')K(\tau') = \tilde{Q}_{\rho\rho}(\omega)\tilde{K}(\omega) \,, \qquad (3.77)$$

where $\tilde{Q}_{\rho\rho}(\omega)$ and $\tilde{K}(\omega)$ are the Fourier transforms of $Q_{\rho\rho}(\tau)$ and $K(\tau)$ respectively:

$$\tilde{Q}_{\rho\rho}(\omega) = \int_{-\infty}^{\infty} d\tau \exp(i\omega\tau) Q_{\rho\rho}(\tau) \quad \text{and} \quad \tilde{K}(\omega) = \int_{-\infty}^{\infty} d\tau \exp(i\omega\tau) K(\tau) \,. \qquad (3.78)$$

Putting the left and right sides of equation 3.75 together as we have evaluated them, we find that

$$\tilde{Q}_{\rho\rho}(\omega)\tilde{K}(\omega) = \exp(i\omega\tau_0)\tilde{Q}_{rs}(\omega) \,. \qquad (3.79)$$

Equation 3.60 follows directly from this result, and equation 3.59 then determines $K(\tau)$ as the inverse Fourier transform of $\tilde{K}(\omega)$.

3.7   Annotated Bibliography

Statistical analysis of discrimination, various forms of decoding, the Neyman-Pearson lemma, the Fisher information, and the Cramér-Rao lower bound can be found in **Cox & Hinckley (1974)**. Receiver operator characteristics and signal detection theory are described comprehensively in **Green & Swets (1966)** and Graham (1989). Our account of spiketrain decoding follows that of **Rieke et al. (1997)**. Spectral factorization is discussed in Poor (1994). **Newsome et al. (1989)** and **Salzman et al. (1992)** present important results concerning visual motion discrimination and recordings from area MT, and **Shadlen et al. (1996)** provides a theoretically oriented review.

The vector method of population decoding has been considered in the context of a number of systems, and references include Humphrey et al. (1970), Georgopoulos et al. (1986 & 1988), van Gisbergen et al. (1987), and Lee et al. (1988). Various theoretical aspects of population decoding, such as vector and ML decoding and the Fisher information, that comprise our account were developed by Paradiso (1988), Baldi and Heiligenberg (1988), Vogels (1990), Snippe & Koenderink (1992), Zohary (1992), Seung & Sompolinsky (1993), Touretzky et al. (1993), Salinas & Abbott (1994), Sanger (1994, 1996), Snippe (1996), and Oram et al. (1998). Population codes are also known as coarse codes in the connectionist literature (Hinton, 1981). In our discussion of the effect of tuning curve widths on the Fisher information, we followed Zhang and Sejnowski (1999), but see also Snippe & Koenderink (1992) and Hinton (1984).

4 Information Theory

4.1 Entropy and Mutual Information

Neural encoding and decoding focus on the question "What does the response of a neuron tell us about a stimulus?" In this chapter we consider a related but different question "How much does the neural response tell us about a stimulus?" The techniques of information theory allow us to answer this question in a quantitative manner. Furthermore, we can use them to ask what forms of neural response are optimal for conveying information about natural stimuli. Information theoretic principles play an important role in many of the unsupervised learning methods that are discussed in chapters 8 and 10.

Shannon invented information theory as a general framework for quantifying the ability of a coding scheme or a communication channel (such as the optic nerve) to convey information. It is assumed that the code involves a number of symbols (such as different neuronal responses), and that the coding and transmission processes are stochastic and noisy. The quantities we consider in this chapter, the entropy and the mutual information, depend on the probabilities with which these symbols, or combinations of them, are used. Entropy is a measure of the theoretical capacity of a code to convey information. Mutual information measures how much of that capacity is actually used when the code is employed to describe a particular set of data. Communication channels, if they are noisy, have only limited capacities to convey information. The techniques of information theory are used to evaluate these limits and find coding schemes that saturate them.

In neuroscience applications, the symbols we consider are neuronal responses, and the data sets they describe are stimulus characteristics. In the most complete analyses, which are considered at the end of the chapter, the neuronal response is characterized by a list of action potential firing times. The symbols being analyzed in this case are sequences of action potentials. Computing the entropy and mutual information for spike sequences can be difficult because the frequency of occurrence of many different spike sequences must be determined. This typically requires a large amount of data. For this reason, many information theory analyses use simplified descriptions of the response of a neuron that reduce the number of possible "symbols" (i.e., responses) that need to be considered. We discuss cases in which the symbols consist of responses described by spike-count firing rates. We also consider the extension to continuous-valued firing rates. Because a reduced description of a spike train can carry no more information than the full spike train itself, this approach provides a lower bound on the actual information carried by the spike train.

Entropy

Entropy is a quantity that, roughly speaking, measures how "interesting" or "surprising" a set of responses is. Suppose that we are given a set of neural responses. If each response is identical, or if only a few different responses appear, we might conclude that this data set is relatively uninteresting. A more interesting set might show a larger range of different responses, perhaps in a highly irregular and unpredictable sequence. How can we quantify this intuitive notion of an interesting set of responses?

We begin by characterizing the responses in terms of their spike-count firing rates (i.e., the number of spikes divided by the trial duration), which can take a discrete set of different values. The methods we discuss are based on the probabilities $P[r]$ of observing a response with a spike-count rate r. The most widely used measure of entropy, due to Shannon, expresses the "surprise" associated with seeing a response rate r as a function of the probability of getting that response, $h(P[r])$, and quantifies the entropy as the average of $h(P[r])$ over all possible responses. The function $h(P[r])$, which acts as a measure of surprise, is chosen to satisfy a number of conditions. First, $h(P[r])$ should be a decreasing function of $P[r]$ because low probability responses are more surprising than high probability responses. Further, the surprise measure for a response that consists of two independent spike counts should be the sum of the measures for each spike count separately. This assures that the entropy and information measures we ultimately obtain will be additive for independent sources. Suppose we record rates r_1 and r_2 from two neurons that respond independently of each other. Because the responses are independent, the probability of getting this pair of responses is the product of their individual probabilities, $P[r_1]P[r_2]$, so the additivity condition requires that

$$h(P[r_1]P[r_2]) = h(P[r_1]) + h(P[r_2]) . \tag{4.1}$$

The logarithm is the only function that satisfies such an identity for all P. Thus, it only remains to decide what base to use for the logarithm. By convention, base 2 logarithms are used so that information can be compared easily with results for binary systems. To indicate that the base 2 logarithm *bits* is being used, information is reported in units of "bits", with

$$h(P[r]) = -\log_2 P[r] . \tag{4.2}$$

The minus sign makes h a decreasing function of its argument, as required. Note that information is really a dimensionless number. The bit, like the radian for angles, is not a dimensional unit but a reminder that a particular system is being used.

Expression (4.2) quantifies the surprise or unpredictability associated with a particular response. Shannon's entropy is just this measure averaged over all responses,

$$H = -\sum_r P[r] \log_2 P[r].$$ (4.3)

entropy

In the sum that determines the entropy, the factor $h = -\log_2 P[r]$ is multiplied by the probability that the response with rate r occurs. Responses with extremely low probabilities may contribute little to the total entropy, despite having large h values, because they occur so rarely. In the limit when $P[r] \to 0$, $h \to \infty$, but an event that does not occur does not contribute to the entropy because the problematic expression $-0\log_2 0$ is evaluated as $-\epsilon \log_2 \epsilon$ in the limit $\epsilon \to 0$, which is 0. Very high probability responses also contribute little because they have $h \approx 0$. The responses that contribute most to the entropy have high enough probabilities so that they appear with a fair frequency, but not high enough to make h too small.

Computing the entropy in some simple cases helps provide a feel for what it measures. First, imagine the least interesting situation: when a neuron responds every time by firing at the same rate. In this case, all of the probabilities $P[r]$ are 0, except for one of them, which is 1. This means that every term in the sum of equation (4.3) is 0 because either $P[r] = 0$ or $\log_2 1 = 0$. Thus, a set of identical responses has zero entropy. Next, imagine that the neuron responds in only two possible ways, either with rate r_+ or r_-. In this case, there are only two nonzero terms in equation (4.3), and, using the fact that $P[r_-] = 1 - P[r_+]$, the entropy is

$$H = -(1 - P[r_+]) \log_2 (1 - P[r_+]) - P[r_+] \log_2 P[r_+].$$ (4.4)

This entropy, plotted in figure 4.1A, takes its maximum value of 1 bit when $P[r_-] = P[r_+] = 1/2$. Thus, a code consisting of two equally likely responses has one bit of entropy.

Mutual Information

To convey information about a set of stimuli, neural responses must be different for different stimuli. Entropy is a measure of response variability, but it does not tell us anything about the source of that variability. A neuron can provide information about a stimulus only if its response variability is correlated with changes in that stimulus, rather than being purely random or correlated with other unrelated factors. One way to determine whether response variability is correlated with stimulus variability is to compare the responses obtained using a different stimulus on every trial with those measured in trials involving repeated presentations of the same

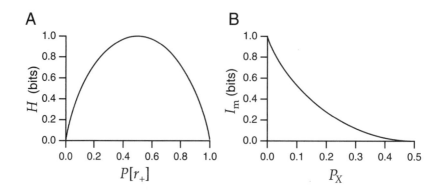

Figure 4.1 (A) The entropy of a binary code. $P[r_+]$ is the probability of a response at rate r_+, and $P[r_-] = 1 - P[r_+]$ is the probability of the other response, r_-. The entropy is maximum when $P[r_-] = P[r_+] = 1/2$. (B) The mutual information for a binary encoding of a binary stimulus. P_X is the probability of an incorrect response being evoked. The plot shows only $P_X \leq 1/2$ because values of $P_X > 1/2$ correspond to an encoding in which the relationship between the two responses and the two stimuli is reversed and the error probability is $1 - P_X$.

stimulus. Responses that are informative about the identity of the stimulus should exhibit larger variability for trials involving different stimuli than for trials that use the same stimulus repetitively. Mutual information is an entropy-based measure related to this idea.

The mutual information is the difference between the total response entropy and the average response entropy on trials that involve repetitive presentation of the same stimulus. Subtracting the entropy when the stimulus does not change removes from the total entropy the contribution from response variability that is not associated with the identity of the stimulus. When the responses are characterized by a spike-count rate, the total response entropy is given by equation 4.3. The entropy of the responses evoked by repeated presentations of a given stimulus s is computed using the conditional probability $P[r|s]$, the probability of a response at rate r given that stimulus s was presented, instead of the response probability $P[r]$ in equation 4.3. The entropy of the responses to a given stimulus is thus

$$H_s = - \sum_r P[r|s] \log_2 P[r|s]. \tag{4.5}$$

noise entropy

If we average this quantity over all the stimuli, we obtain a quantity called the noise entropy

$$H_{\text{noise}} = \sum_s P[s] H_s = - \sum_{s,r} P[s] P[r|s] \log_2 P[r|s]. \tag{4.6}$$

This is the entropy associated with that part of the response variability that is not due to changes in the stimulus, but arises from other sources. The mutual information is obtained by subtracting the noise entropy from the full response entropy, which from equations 4.3 and 4.6 gives

$$I_m = H - H_{\text{noise}} = -\sum_r P[r] \log_2 P[r] + \sum_{s,r} P[s]P[r|s] \log_2 P[r|s]. \quad (4.7)$$

The probability of a response r is related to the conditional probability $P[r|s]$ and the probability $P[s]$ that stimulus s is presented by the identity (chapter 3),

$$P[r] = \sum_s P[s]P[r|s]. \quad (4.8)$$

Using this, and writing the difference of the two logarithms in equation 4.7 as the logarithm of the ratio of their arguments, we can rewrite the mutual   *mutual information* information as

$$I_m = \sum_{s,r} P[s]P[r|s] \log_2 \left(\frac{P[r|s]}{P[r]} \right). \quad (4.9)$$

Recall from chapter 3 that

$$P[r,s] = P[s]P[r|s] = P[r]P[s|r], \quad (4.10)$$

where $P[r,s]$ is the joint probability of stimulus s appearing and response r being evoked. Equation 4.10 can be used to derive yet another form for the mutual information,

$$I_m = \sum_{s,r} P[r,s] \log_2 \left(\frac{P[r,s]}{P[r]P[s]} \right). \quad (4.11)$$

This equation reveals that the mutual information is symmetric with respect to interchange of s and r, which means that the mutual information that a set of responses conveys about a set of stimuli is identical to the mutual information that the set of stimuli conveys about the responses. To see this explicitly, we apply equation 4.10 again to write

$$I_m = -\sum_s P[s] \log_2 P[s] + \sum_{s,r} P[r]P[s|r] \log_2 P[s|r]. \quad (4.12)$$

This result is the same as equation 4.7, except that the roles of the stimulus and the response have been interchanged. Equation 4.12 shows how response variability limits the ability of a spike train to carry information. The second term on the right side, which is negative, is the average uncertainty about the identity of the stimulus given the response, and reduces the total stimulus entropy represented by the first term.

To provide some concrete examples, we compute the mutual information for a few simple cases. First, suppose that the responses of the neuron are completely unaffected by the identity of the stimulus. In this case, $P[r|s] = P[r]$, and from equation 4.9 it follows immediately that $I_m = 0$. At the other extreme, suppose that each stimulus s produces a unique and distinct response r_s. Then, $P[r_s] = P[s]$ and $P[r|s]$ is 1 if $r = r_s$ and 0 otherwise. This causes the sum over r in equation 4.9 to collapse to just one term, and the mutual information becomes

$$I_m = \sum_s P[s] \log_2 \left(\frac{1}{P[r_s]} \right) = -\sum_s P[s] \log_2 P[s] . \qquad (4.13)$$

The last expression, which follows from the fact that $P[r_s] = P[s]$, is the entropy of the stimulus. Thus, with no variability and a one-to-one map from stimulus to response, the mutual information is equal to the full stimulus entropy.

Finally, imagine that there are only two possible stimulus values, which we label $+$ and $-$, and that the neuron responds with just two rates, r_+ and r_-. We associate the response r_+ with the $+$ stimulus, and the response r_- with the $-$ stimulus, but the encoding is not perfect. The probability of an incorrect response is P_X, meaning that for the correct responses $P[r_+|+] = P[r_-|-] = 1 - P_X$, and for the incorrect responses $P[r_+|-] = P[r_-|+] = P_X$. We assume that the two stimuli are presented with equal probability so that $P[r_+] = P[r_-] = 1/2$, which, from equation 4.4, makes the full response entropy 1 bit. The noise entropy is $-(1 - P_X) \log_2 (1 - P_X) - P_X \log_2 P_X$. Thus, the mutual information is

$$I_m = 1 + (1 - P_X) \log_2 (1 - P_X) + P_X \log_2 P_X . \qquad (4.14)$$

This is plotted in figure 4.1B. When the encoding is error-free ($P_X = 0$), the mutual information is 1 bit, which is equal to both the full response entropy and the stimulus entropy. When the encoding is random ($P_X = 1/2$), the mutual information goes to 0.

It is instructive to consider this example from the perspective of decoding. We can think of the neuron as being a communication channel that reports noisily on the stimulus. From this perspective, we want to know the probability that a $+$ was presented, given that the response r_+ was recorded. By Bayes theorem, this is $P[+|r_+] = P[r_+|+]P[+]/P[r_+] = 1 - P_X$. Before the response is recorded, the expectation was that $+$ and $-$ were equally likely. If the response r_+ is recorded, this expectation changes to $1 - P_X$. The mutual information measures the corresponding reduction in uncertainty or, equivalently, the tightening of the posterior distribution due to the response.

KL divergence   The mutual information is related to a measure used in statistics called the Kullback-Leibler (KL) divergence. The KL divergence between one probability distribution $P[r]$ and another distribution $Q[r]$ is

$$D_{KL}(P, Q) = \sum_r P[r] \log_2 \left(\frac{P[r]}{Q[r]} \right) . \qquad (4.15)$$

The KL divergence has a property normally associated with a distance measure, $D_{KL}(P, Q) \geq 0$ with equality if and only if $P = Q$ (proven in appendix A). However, unlike a distance, it is not symmetric with respect to interchange of P and Q. Comparing the definition 4.15 with equation 4.11, we see that the mutual information is the KL divergence between the distributions $P[r, s]$ and $P[r]P[s]$. If the stimulus and the response were independent of one another, $P[r, s]$ would be equal to $P[r]P[s]$. Thus, the mutual information is the KL divergence between the actual probability distribution $P[r, s]$ and the value it would take if the stimulus and response were independent. The fact that $D_{KL} \geq 0$ proves that the mutual information cannot be negative. In addition, it can never be larger than either the full response entropy or the entropy of the stimulus set.

Entropy and Mutual Information for Continuous Variables

Up to now we have characterized neural responses using discrete spike-count rates. As in chapter 3, it is often convenient to treat these rates instead as continuous variables. There is a complication associated with entropies that are defined in terms of continuous response variables. If we could measure the value of a continuously defined firing rate with unlimited accuracy, it would be possible to convey an infinite amount of information using the endless sequence of decimal digits of this single variable. Of course, practical considerations always limit the accuracy with which a firing rate can be measured or conveyed.

To define the entropy associated with a continuous measure of a neural response, we must include some limit on the measurement accuracy. The effects of this limit typically cancel in computations of mutual information because the mutual information is the difference between two entropies. In this section, we show how entropy and mutual information are computed for responses characterized by continuous firing rates. For completeness, we also treat the stimulus parameter s as a continuous variable. This means that the probability $P[s]$ is replaced by the probability density $p[s]$, and sums over s are replaced by integrals.

For a continuously defined firing rate, the probability of the firing rate lying in the range between r and $r + \Delta r$, for small Δr, is expressed in terms of a probability density as $p[r]\Delta r$. Summing over discrete bins of size Δr, we find, by analogy with equation (4.3),

$$H = -\sum p[r]\Delta r \log_2(p[r]\Delta r)$$
$$= -\sum p[r]\Delta r \log_2 p[r] - \log_2 \Delta r. \qquad (4.16)$$

To extract the last term we have expressed the logarithm of a product as the sum of two logarithms and used the fact that the sum of the response probabilities is 1. We would now like to take the limit $\Delta r \to 0$ but we cannot, because the $\log_2 \Delta r$ term diverges in this limit. This divergence reflects the fact that a continuous variable measured with perfect accuracy has infinite entropy. However, for reasonable (i.e., Riemann integrable) $p[r]$, everything works out fine for the first term because the sum becomes an integral in the limit $\Delta r \to 0$. In this limit, we can write

$$\lim_{\Delta r \to 0} \{ H + \log_2 \Delta r \} = - \int dr \, p[r] \log_2 p[r] . \qquad (4.17)$$

continuous entropy Δr is best thought of as a limit on the resolution with which the firing rate can be measured. Unless this limit is known, the entropy of a probability density for a continuous variable can be determined only up to an additive constant. However, if two entropies computed with the same resolution are subtracted, the troublesome term involving Δr cancels, and we can proceed without knowing its precise value. All of the cases where we use equation 4.17 are of this form. The integral on the right side of *differential entropy* equation 4.17 is sometimes called the differential entropy.

The noise entropy, for a continuous variable like the firing rate, can be written in a manner similar to the response entropy 4.17, except that the *continuous noise* conditional probability density $p[r|s]$ is used:
entropy

$$\lim_{\Delta r \to 0} \{ H_{\text{noise}} + \log_2 \Delta r \} = - \int ds \int dr \, p[s] p[r|s] \log_2 p[r|s] . \qquad (4.18)$$

The mutual information is the difference between the expressions in equa-
continuous mutual tions 4.17 and 4.18,
information

$$I_{\text{m}} = \int ds \int dr \, p[s] p[r|s] \log_2 \left(\frac{p[r|s]}{p[r]} \right) . \qquad (4.19)$$

Note that the factor of $\log_2 \Delta r$ cancels in the expression for the mutual information because both entropies are evaluated at the same resolution.

In chapter 3, we described the Fisher information as a local measure of how tightly the responses determine the stimulus. The Fisher information is local because it depends on the expected curvature of the likelihood $P[\mathbf{r}|s]$ (typically for the responses of many cells) evaluated at the true stimulus value. The mutual information is a global measure in the sense that it depends on the average overall uncertainty in the decoding distribution $p[s|\mathbf{r}]$, including values of s both close to and far from the true stimulus. If the decoding distribution $p[s|\mathbf{r}]$ has a single peak about the true stimulus, the Fisher information and the mutual information are closely related. In particular, for large numbers of neurons, the maximum likelihood estimator tends to have a sharply peaked Gaussian distribution, as discussed in chapter 3. In this case, the mutual information is, up to an additive constant, the logarithm of the Fisher information averaged over the distribution of stimuli.

4.2   Information and Entropy Maximization

Entropy and mutual information are useful quantities for characterizing the nature and efficiency of neural encoding and selectivity. Often, in addition to such characterizations, we seek to understand the computational implications of an observed response selectivity. For example, we might ask whether neural responses to natural stimuli are optimized to convey as much information as possible. This hypothesis can be tested by computing the response characteristics that maximize the mutual information conveyed about naturally occurring stimuli and comparing the results with responses observed experimentally.

Because the mutual information is the full response entropy minus the noise entropy, maximizing the information involves a compromise. We must make the response entropy as large as possible without allowing the noise entropy to get too big. If the noise entropy is small, maximizing the response entropy, subject to an appropriate constraint, maximizes the mutual information to a good approximation. We therefore begin our discussion by studying how response entropy can be maximized. Later in the discussion, we will consider the effects of noise entropy.

Constraints play a crucial role in this analysis. We have already seen that the theoretical information-carrying capacity associated with a continuous firing rate is limited only by the resolution with which the firing rate can be defined. Even with a finite resolution, a firing rate could convey an infinite amount of information if it could take arbitrarily high values. Thus, we must impose some constraint that limits the firing rate to a realistic range. Possible constraints include limiting the maximum allowed firing rate or holding the average firing rate or its variance fixed.

Entropy Maximization for a Single Neuron

To maximize the response entropy, we must find a probability density $p[r]$ that makes the integral in equation 4.17 as large as possible while satisfying whatever constraints we impose. During the maximization process, the resolution Δr is held fixed, so the $\log_2 \Delta r$ term remains constant, and it can be ignored. As a result, it will not generally appear in the following equations. One constraint that always applies in entropy maximization is that the integral of the probability density must be 1. Suppose that the neuron in question has a maximum firing rate of r_{max}. Then, the integrals in question extend from 0 to r_{max}. To find the $p[r]$ producing the maximum entropy, we must maximize

$$-\int_0^{r_{max}} dr\, p[r] \log_2 p[r]\,, \qquad (4.20)$$

subject to the constraint

$$\int_0^{r_{max}} dr\, p[r] = 1\,. \qquad (4.21)$$

The result, computed using Lagrange multipliers (see the Mathematical Appendix), is that the probability density that maximizes the entropy subject to this constraint is a constant,

$$p[r] = \frac{1}{r_{max}}, \tag{4.22}$$

independent of r. The entropy for this probability density, for finite firing-rate resolution Δr, is

$$H = \log_2 r_{max} - \log_2 \Delta r = \log_2 \left(\frac{r_{max}}{\Delta r} \right). \tag{4.23}$$

histogram
equalization

Equation 4.22 is the basis of a signal-processing technique called histogram equalization. Applied to neural responses, this is a procedure for tailoring the neuronal selectivity so that $p[r] = 1/r_{max}$ in response to a set of stimuli over which the entropy is to be maximized. Suppose a neuron responds to a stimulus characterized by the parameter s by firing at a rate $r = f(s)$. For small Δs, the probability that the continuous stimulus variable falls in the range between s and $s + \Delta s$ is given in terms of the stimulus probability density by $p[s]\Delta s$. This produces a response that falls in the range between $f(s + \Delta s)$ and $f(s)$. If the response probability density takes its optimal value, $p[r] = 1/r_{max}$, the probability that the response falls within this range is $|f(s + \Delta s) - f(s)|/r_{max}$. Setting these two probabilities equal to each other, we find that $|f(s + \Delta s) - f(s)|/r_{max} = p[s]\Delta s$.

Consider the case of a monotonically increasing response so that $f(s + \Delta s) > f(s)$ for positive Δs. Then, in the limit $\Delta s \to 0$, the equalization condition becomes

$$\frac{df}{ds} = r_{max}p[s], \tag{4.24}$$

which has the solution

$$f(s) = r_{max} \int_{s_{min}}^{s} ds' \, p[s'], \tag{4.25}$$

where s_{min} is the minimum value of s, which is assumed to generate no response. Thus, entropy maximization requires that the average firing rate of the responding neuron be proportional to the integral of the probability density of the stimulus.

Laughlin (1981) has provided evidence that responses of the large monopolar cell (LMC) in the visual system of the fly satisfy the entropy-maximizing condition. The LMC responds to contrast, and Laughlin measured the probability distribution of contrasts of natural scenes in habitats where the flies he studied live. The solid curve in figure 4.2 is the integral of this measured distribution. The data points in figure 4.2 are LMC responses as a function of contrast. These responses are measured as membrane potential fluctuation amplitudes, not as firing rates, but the analysis presented can be applied without modification. As figure 4.2 indicates, the response as a function of contrast is very close to the integrated probability density, suggesting that the LMC is using a maximum entropy encoding.

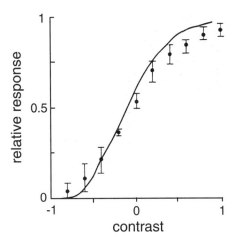

Figure 4.2 Contrast response of the fly LMC (data points) compared to the integral of the natural contrast probability distribution (solid curve). The relative response is the amplitude of the membrane potential fluctuation produced by the onset of a light or dark image with a given level of contrast divided by the maximum response. Contrast is defined relative to the background level of illumination. (Adapted from Laughlin, 1981.)

Even though neurons have maximum firing rates, the constraint $r \leq r_{max}$ may not always be the factor that limits the entropy. For example, the average firing rate of the neuron may be constrained to values much less than r_{max}, or the variance of the firing rate might be constrained. The reader is invited to show that the entropy-maximizing probability density, if the average firing rate is constrained to a fixed value, is an exponential. A related calculation shows that the probability density that maximizes the entropy subject to constraints on the firing rate and its variance is a Gaussian.

Populations of Neurons

When a population of neurons encodes a stimulus, optimizing their individual response properties will not necessarily lead to an optimized population response. Optimizing individual responses could result in a highly redundant population representation in which different neurons encode the same information. Entropy maximization for a population requires that the neurons convey independent pieces of information (i.e., they must have different response selectivities). Let the vector $\mathbf{r}$ with components r_a for $a = 1, 2, \ldots, N$ denote the firing rates for a population of N neurons, measured with resolution Δr. If $p[\mathbf{r}]$ is the probability of evoking a population response characterized by the vector $\mathbf{r}$, the entropy for the entire population response is

$$H = -\int d\mathbf{r}\, p[\mathbf{r}] \log_2 p[\mathbf{r}] - N \log_2 \Delta r. \qquad (4.26)$$

Along with the full population entropy of Equation 4.26, we can also consider the entropy associated with individual neurons within the population. If $p[r_a] = \int \prod_{b \neq a} dr_b\, p[\mathbf{r}]$ is the probability density for response r_a from neuron a, its entropy is

$$H_a = -\int dr_a\, p[r_a] \log_2 p[r_a] - \log_2 \Delta r = -\int d\mathbf{r}\, p[\mathbf{r}] \log_2 p[r_a] - \log_2 \Delta r.$$

(4.27)

The true population entropy can never be greater than the sum of these individual neuron entropies over the entire population,

$$H \leq \sum_a H_a.$$

(4.28)

To prove this, we note that the difference between the full entropy and the sum of individual neuron entropies is

$$\sum_a H_a - H = \int d\mathbf{r}\, p[\mathbf{r}] \log_2 \left(\frac{p[\mathbf{r}]}{\prod_a p_a[r_a]} \right) \geq 0.$$

(4.29)

The inequality follows from the fact that the middle expression is the KL divergence between the probability distributions $p[\mathbf{r}]$ and $\prod_a p[r_a]$, and a KL divergence is always nonnegative. Equality holds only if

$$p[\mathbf{r}] = \prod_a p[r_a],$$

(4.30)

that is, if the responses of the neurons are statistically independent. Thus, the full response entropy is never greater than the sum of the entropies of the individual neurons in the population, and it reaches the limiting value when equation 4.30 is satisfied. A code that satisfies this condition is called a *factorial code* because the probability factorizes into a product of single neuron probabilities. When the population-response probability density factorizes, this implies that the individual neurons respond independently. The entropy difference in equation 4.29 has been suggested as a measure of *redundancy*.

Combining this result with the results of the previous section, we conclude that the maximum population-response entropy can be achieved by satisfying two conditions. First, the individual neurons must respond independently, which means that $p[\mathbf{r}] = \prod_a p[r_a]$ must factorize. Second, they must all have response probabilities that are optimal for whatever constraints are imposed (e.g., flat, exponential, or Gaussian). If the same constraint is imposed on every neuron, the second condition implies that every neuron must have the same response probability density. In other words, $p[r_a]$ must be the same for all a values, a property called *probability equalization*. This does not imply that all the neurons respond identically to every stimulus. Indeed, the conditional probabilities $p[r_a|s]$ must be different for different neurons if they are to act independently. We proceed by considering factorization and probability equalization as general principles of entropy maximization, without imposing explicit constraints.

Exact factorization and probability equalization are difficult to achieve, especially if the form of the neural response is restricted. These goals are likely to be impossible to achieve, for example, if the neural responses are modeled as having a linear relation to the stimulus. A more modest goal is to require that the lowest-order moments of the population-response probability distribution match those of a fully factorized and equalized distribution. If the individual response probability distributions are equal, the average firing rates and firing rate variances will be the same for all neurons, $\langle r_a \rangle = \langle r \rangle$ and $\langle (r_a - \langle r \rangle)^2 \rangle = \sigma_r^2$ for all a. Furthermore, the covariance matrix for a factorized and probability-equalized population distribution is proportional to the identity matrix,

$$Q_{ab} = \int d\mathbf{r}\, p[\mathbf{r}](r_a - \langle r \rangle)(r_b - \langle r \rangle) = \sigma_r^2 \delta_{ab}\,. \tag{4.31}$$

Finding response distributions that satisfy only the decorrelation and variance equalization condition of equation 4.31 is usually tractable. In the following examples, we restrict ourselves to this easier task. This maximizes the entropy only if the statistics of the responses are Gaussian, but it is a reasonable procedure even in a non-Gaussian case, because it typically reduces the redundancy in the population code and spreads the load of information transmission equally among the neurons.

decorrelation and variance equalization

Application to Retinal Ganglion Cell Receptive Fields

Entropy and information maximization have been used to explain properties of visual receptive fields in the retina, LGN, and primary visual cortex. The basic assumption is that these receptive fields serve to maximize the amount of information that the associated neural responses convey about natural visual scenes in the presence of noise. Information theoretical analyses are sensitive to the statistical properties of the stimuli being represented, so the statistics of natural scenes play an important role in these studies. Natural scenes exhibit substantial spatial and temporal redundancy. Maximizing the information conveyed requires removing this redundancy from the neural responses.

It should be kept in mind that the information maximization approach sets limited goals and requires strong assumptions about the nature of the constraints relevant to the nervous system. In addition, the approach analyzes only the representational properties of neural responses and ignores the computational goals of the visual system, such as object recognition or target tracking. Finally, maximizing other measures of performance, different from the mutual information, may give similar results. Nevertheless, the principle of information maximization is quite successful at accounting for properties of receptive fields early in the visual pathway.

In chapter 2, a visual image was defined by a contrast function $s(x, y, t)$ with a trial-averaged value of 0. For the calculations we present here, it is more convenient to express the x and y coordinates for locations on the viewing screen in terms of a single vector $\vec{x} = (x, y)$, or sometimes $\vec{y} = (x, y)$. Using this notation, the linear estimate of the response of a visual neuron discussed in chapter 2 can be written as

$$L(t) = \int_0^\infty d\tau \int d\vec{x}\, D(\vec{x}, \tau) s(\vec{x}, t - \tau). \qquad (4.32)$$

If the space-time receptive field $D(\vec{x}, \tau)$ is separable, $D(\vec{x}, \tau) = D_s(\vec{x}) D_t(\tau)$, and we can rewrite $L(t)$ as the product of integrals involving temporal and spatial filters. To keep the notation simple, we assume that the stimulus can also be separated, so that $s(\vec{x}, t) = s_s(\vec{x}) s_t(t)$. Then, $L(t) = L_s L_t(t)$ where

$$L_s = \int d\vec{x}\, D_s(\vec{x}) s_s(\vec{x}) \qquad (4.33)$$

and

$$L_t(t) = \int_0^\infty d\tau\, D_t(\tau) s_t(t - \tau). \qquad (4.34)$$

In the following, we analyze the spatial and temporal components, D_s and D_t, separately by considering the information-carrying capacity of L_s and L_t. We study the spatial receptive fields of retinal ganglion cells in this section, and the temporal response properties of LGN cells in the next. Later, we discuss the application of information maximization ideas to primary visual cortex.

To derive appropriately optimal spatial filters, we consider an array of retinal ganglion cells with receptive fields covering a small patch of the retina. We assume that the statistics of the input are spatially (and temporally) stationary or translation-invariant. This means that all locations and directions in space (and all times), at least within the patch we consider, are equivalent. This equivalence allows us to give all of the receptive fields the same spatial structure, with the receptive fields of different cells merely being shifted to different points within the visual field. As a result, we write the spatial kernel describing a retinal ganglion cell with receptive field centered at the point $\vec{a}$ as $D_s(\vec{x} - \vec{a})$. The linear response of this cell is then

$$L_s(\vec{a}) = \int d\vec{x}\, D_s(\vec{x} - \vec{a}) s_s(\vec{x}). \qquad (4.35)$$

Note that we are labeling the neurons by the locations $\vec{a}$ of the centers of their receptive fields rather than by an integer index such as i. This is a convenient labeling scheme that allows sums over neurons to be replaced by sums over parameters describing their receptive fields. The vectors $\vec{a}$ for the different neurons take on discrete values corresponding to the different neurons in the population. If many neurons are being considered, these discrete vectors may fill the range of receptive field locations quite densely. In this case, it is reasonable to approximate the large but discrete set of $\vec{a}$ values with a vector $\vec{a}$ that is allowed to vary continuously. In other words, as an approximation, we proceed as if there were a neuron corresponding to every continuous value of $\vec{a}$. This allows us to treat $L(\vec{a})$ as a function of $\vec{a}$ and to replace sums over neurons with integrals over $\vec{a}$. In the case we are considering, the receptive fields of retinal ganglion cells cover the retina densely, with many receptive fields overlapping each point on the retina, so the replacement of discrete sums over neurons with continuous integrals over $\vec{a}$ is quite accurate.

The Whitening Filter

We will not attempt a complete entropy maximization for the case of retinal ganglion cells. Instead, we follow the approximate procedure of setting the correlation matrix between different neurons within the population proportional to the identity matrix (equation 4.31). The relevant correlation is the average, over all stimuli, of the product of the linear responses of two cells, with receptive fields centered at $\vec{a}$ and $\vec{b}$,

$$Q_{LL}(\vec{a}, \vec{b}) = \langle L_s(\vec{a}) L_s(\vec{b}) \rangle = \int d\vec{x}\, d\vec{y}\, D_s(\vec{x} - \vec{a}) D_s(\vec{y} - \vec{b}) \langle s_s(\vec{x}) s_s(\vec{y}) \rangle \,.$$

(4.36)

The average here, denoted by angle brackets, is not over trials but over the set of natural scenes for which we believe the receptive field is optimized. By analogy with equation 4.31, decorrelation and variance equalization of the different retinal ganglion cells, when $\vec{a}$ and $\vec{b}$ are taken to be continuous variables, require that we set this correlation function proportional to a δ function,

$$Q_{LL}(\vec{a}, \vec{b}) = \sigma_L^2 \delta(\vec{a} - \vec{b}) \,.$$

(4.37)

This is the continuous variable analog of making a discrete correlation matrix proportional to the identity matrix (equation 4.31). The δ function with vector arguments is nonzero only when all of the components of $\vec{a}$ and $\vec{b}$ are identical.

The quantity $\langle s_s(\vec{x}) s_s(\vec{y}) \rangle$ in equation 4.36 is the correlation function of the stimulus averaged over natural scenes. Our assumption of homogeneity implies that this quantity is only a function of the vector difference $\vec{x} - \vec{y}$ (actually, if all directions are equivalent, it is only a function of the magnitude $|\vec{x} - \vec{y}|$), and we write it as

$$Q_{ss}(\vec{x} - \vec{y}) = \langle s_s(\vec{x}) s_s(\vec{y}) \rangle \,.$$

(4.38)

To determine the form of the receptive field filter that is optimal, we must solve equation 4.37 for D_s. This is done by expressing D_s and Q_{ss} in terms of their Fourier transforms $\tilde{D}_s$ and $\tilde{Q}_{ss}$,

$$D_s(\vec{x} - \vec{a}) = \frac{1}{4\pi^2} \int d\vec{\kappa}\, \exp\left(-i\vec{\kappa} \cdot (\vec{x} - \vec{a})\right) \tilde{D}_s(\vec{\kappa})$$

(4.39)

$$Q_{ss}(\vec{x} - \vec{y}) = \frac{1}{4\pi^2} \int d\vec{\kappa}\, \exp\left(-i\vec{\kappa} \cdot (\vec{x} - \vec{y})\right) \tilde{Q}_{ss}(\vec{\kappa}) \,.$$

(4.40)

$\tilde{Q}_{ss}$, which is real and nonnegative, is also called the stimulus power spectrum (see chapter 1). In terms of these Fourier transforms, equation 4.37 becomes

$$|\tilde{D}_s(\vec{\kappa})|^2 \tilde{Q}_{ss}(\vec{\kappa}) = \sigma_L^2 , \qquad (4.41)$$

from which we find

$$|\tilde{D}_s(\vec{\kappa})| = \frac{\sigma_L}{\sqrt{\tilde{Q}_{ss}(\vec{\kappa})}} . \qquad (4.42)$$

whitening filter

The linear kernel described by equation 4.42 exactly compensates for whatever dependence the Fourier transform of the stimulus correlation function has on the spatial frequency $\vec{\kappa}$, making the product $\tilde{Q}_{ss}(\vec{\kappa})|\tilde{D}_s(\vec{\kappa})|^2$ independent of $\vec{\kappa}$. This product is the power spectrum of L. The output of the optimal filter has a power spectrum that is independent of spatial frequency, and therefore has the same characteristics as white noise. Therefore, the kernel in equation 4.42 is called a whitening filter. Different spatial frequencies act independently in a linear system, so decorrelation and variance equalization require them to be utilized at equal signal strength.

The calculation we have performed determines only the amplitude $|\tilde{D}_s(\vec{\kappa})|$, and not $\tilde{D}_s(\vec{\kappa})$ itself. Thus, decorrelation and variance equalization do not uniquely specify the form of the linear kernel. We study some consequences of the freedom to choose different linear kernels satisfying equation 4.42 later in the chapter.

The spatial correlation function for natural scenes has been measured, with the result that $\tilde{Q}_{ss}(\vec{\kappa})$ is proportional to $1/|\vec{\kappa}|^2$ over the range it has been evaluated. The behavior near $\vec{\kappa} = 0$ is not well established, but the divergence of $1/|\vec{\kappa}|^2$ near $\vec{\kappa} = 0$ can be removed by setting $\tilde{Q}_{ss}(\vec{\kappa})$ proportional to $1/(|\vec{\kappa}|^2 + \kappa_0^2)$ where κ_0 is a constant. The stimuli of interest in the calculation of retinal ganglion receptive fields are natural images as they appear on the retina, not in the photographs from which the natural scenes statistics are measured. An additional factor must be included in $\tilde{Q}_{ss}(\vec{\kappa})$ to account for filtering introduced by the optics of the eye (the optical modulation transfer function). A simple model of the optical modulation transfer function results in an exponential correction to the stimulus correlation function,

optical modulation transfer function

$$\tilde{Q}_{ss}(\vec{\kappa}) \propto \frac{\exp(-\alpha|\vec{\kappa}|)}{|\vec{\kappa}|^2 + \kappa_0^2} , \qquad (4.43)$$

with α a parameter. Substituting this into equation 4.42 gives the rather peculiar result that the amplitude $|\tilde{D}_s(\vec{\kappa})|$, being proportional to the inverse of the square root of $\tilde{Q}_{ss}$, is predicted to grow exponentially for large $|\vec{\kappa}|$. Whitening filters maximize entropy by equalizing the distribution of response power over the entire spatial frequency range. High spatial frequency components of images are relatively rare in natural scenes and, even if they occur, are greatly attenuated by the eye. The whitening filter compensates for this by boosting the responses to high spatial frequencies. Although this is the result of the entropy maximization calculation, it is not a good strategy to use in an unrestricted way for visual processing. Real inputs to retinal ganglion cells involve a mixture of true signal and noise coming from biophysical sources in the retina. At high spatial frequencies, for which the true signal is weak, inputs to retinal ganglion cells are likely to be dominated by noise, especially in low-light conditions. Boosting the amplitude of this noise-dominated input and transmitting it to the brain is not an efficient visual encoding strategy.

The problem of excessive boosting of responses at high spatial frequency arises in the entropy maximization calculation because no distinction has been made between the entropy coming from true signals and that coming from noise. To correct this problem, we should maximize the information transmitted by the retinal ganglion cells about natural scenes, rather than maximize the entropy. A full information-maximization calculation of the receptive field properties of retinal ganglion cells can be performed, but this requires introducing a number of assumptions about the constraints that are relevant, and it is not entirely obvious what these constraints should be. Instead, we will follow an approximate procedure that pre-filters the input to eliminate as much noise as possible, and then uses the results of this section to maximize the entropy of a linear filter acting on the prefiltered input signal.

Filtering Input Noise

Suppose that the visual stimulus on the retina is the sum of the true stimulus $s_s(\vec{x})$ that should be conveyed to the brain and a noise term $\eta(\vec{x})$ that reflects image distortion, photoreceptor noise, and other signals that are not worth conveying beyond the retina. To deal with such a mixed input signal, we express the Fourier transform of the linear kernel $\tilde{D}_s(\vec{\kappa})$ as a product of two terms: a noise filter, $\tilde{D}_\eta(\vec{\kappa})$, that eliminates as much of the noise as possible; and a whitening filter, $\tilde{D}_w(\vec{\kappa})$, that satisfies equation 4.42. The Fourier transform of the complete filter is then $\tilde{D}_s(\vec{\kappa}) = \tilde{D}_w(\vec{\kappa})\tilde{D}_\eta(\vec{\kappa})$.

To determine the form of the noise filter, we demand that when it is applied to the total input $s_s(\vec{x}) + \eta(\vec{x})$, the result is as close to the signal part of the input, $s_s(\vec{x})$, as possible. The problem of minimizing the average squared difference between the filtered noisy signal and the true signal is formally the same as the problems we solved in chapter 2 (appendix A) and chapter 3 (appendix C) to determine the optimal kernels for rate prediction and for spike decoding (also see the Mathematical Appendix). The general solution is that the Fourier transform of the optimal filter is the Fourier transform of the cross-correlation between the quantity being filtered and the quantity being approximated divided by the Fourier transform of the autocorrelation of the quantity being filtered. In the present example, there is a slight complication that the integral in equation 4.35 is not in the form of a convolution, because D_s is written as a function of $\vec{x} - \vec{a}$ rather than $\vec{a} - \vec{x}$. However, in the case we consider, this ultimately makes no difference to the final answer.

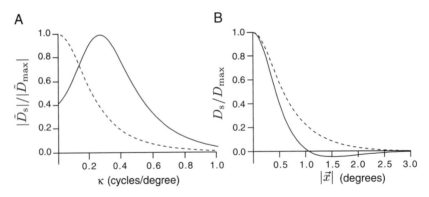

Figure 4.3 Receptive field properties predicted by entropy maximization and noise suppression of responses to natural images. (A) The amplitude of the predicted Fourier-transformed linear filters for low (solid curve) and high (dashed curve) input noise. $|\tilde{D}_s(\vec{\kappa})|$ is plotted relative to its maximum value. (B) The linear kernel as a function of the distance from the center of the receptive field for low (solid curve) and high (dashed curve) input noise. Observe the center-surround structure at low noise. $\tilde{D}_s(\vec{\kappa})$ is taken to be real, and $D_s(|\vec{x}|)$ is plotted relative to its maximum value. Parameter values used were $1/\alpha = 0.16$ cycles/degree, $k_0 = 0.16$ cycles/degree, and $\tilde{Q}_{\eta\eta}/\tilde{Q}_{ss}(0) = 0.05$ for the low-noise case and 1 for the high-noise case.

The calculation simplifies because we assume that the signal and noise terms are uncorrelated, so that $\langle s_s(\vec{x})\eta(\vec{y})\rangle = 0$. Then, the relevant cross-correlation for this problem is

$$\langle (s_s(\vec{x}) + \eta(\vec{x}))s_s(\vec{y})\rangle = Q_{ss}(\vec{x} - \vec{y}), \qquad (4.44)$$

and the autocorrelation is

$$\langle (s_s(\vec{x}) + \eta(\vec{x}))(s_s(\vec{y}) + \eta(\vec{y}))\rangle = Q_{ss}(\vec{x} - \vec{y}) + Q_{\eta\eta}(\vec{x} - \vec{y}), \qquad (4.45)$$

where Q_{ss} and $Q_{\eta\eta}$ are, respectively, the stimulus and noise autocorrelations functions. These results imply that the optimal noise filter is real and

noise filter   given, in terms of the Fourier transforms of Q_{ss} and $Q_{\eta\eta}$, by

$$\tilde{D}_\eta(\vec{\kappa}) = \frac{\tilde{Q}_{ss}(\vec{\kappa})}{\tilde{Q}_{ss}(\vec{\kappa}) + \tilde{Q}_{\eta\eta}(\vec{\kappa})}. \qquad (4.46)$$

Because the noise filter is designed so that its output matches the signal as closely as possible, we make the approximation of using the same whitening filter as before (equation 4.42). Combining the two, we find that

$$|\tilde{D}_s(\vec{\kappa})| \propto \frac{\sigma_L\sqrt{\tilde{Q}_{ss}(\vec{\kappa})}}{\tilde{Q}_{ss}(\vec{\kappa}) + \tilde{Q}_{\eta\eta}(\vec{\kappa})}. \qquad (4.47)$$

Linear kernels resulting from equation 4.47, using equation 4.43 for the stimulus correlation function, are plotted in figure 4.3. For this figure, we have assumed that the input noise is white so that $\tilde{Q}_{\eta\eta}$ is independent of $\vec{\kappa}$. Both the amplitude of the Fourier transform of the kernel (figure 4.3A) and the actual spatial kernel $D_s(\vec{x})$ (figure 4.3B) are plotted under conditions of low and high noise. The linear kernels in figure 4.3B have been constructed by assuming that $\tilde{D}_s(\vec{\kappa})$ satisfies equation 4.47 and is real, which minimizes the spatial extent of the resulting receptive field. The resulting function $D_s(\vec{x})$ is radially symmetric, so it depends only on the distance $|\vec{x}|$ from the center of the receptive field to the point $\vec{x}$, and this radial dependence is plotted in figure 4.3B. Under low noise conditions (solid lines in figure 4.3), the linear kernel has a bandpass character and the predicted receptive field has a center-surround structure, which matches the retinal ganglion receptive fields shown in chapter 2. This structure eliminates one major source of redundancy in natural scenes: the strong similarity of neighboring inputs owing to the predominance of low spatial frequencies in images.

When the noise level is high (dashed lines in figure 4.3), the structure of the optimal receptive field is different. In spatial frequency terms, the filter is now low-pass, and the receptive field loses its surround. This structure averages over neighboring pixels to extract the true signal obscured by the uncorrelated noise. In the retina, we expect the signal-to-noise ratio to be controlled by the level of ambient light, with low levels of illumination corresponding to the high-noise case. The predicted change in the receptive fields at low illumination (high noise) matches what actually happens in the retina. At low light levels, circuitry changes within the retina remove the opposing surrounds from retinal ganglion cell receptive fields.

Temporal Processing in the LGN

Natural images tend to change relatively slowly over time. This means that there is substantial redundancy in the succession of natural images, suggesting an opportunity for efficient temporal filtering to complement efficient spatial filtering. An analysis similar to that of the previous section can be performed to account for the temporal receptive fields of visually responsive neurons early in the visual pathway. Recall that the predicted linear temporal response is given by $L_t(t)$, as expressed in equation 4.34. The analog of equation 4.37 for temporal decorrelation and variance equalization is

$$\langle L_t(t)L_t(t')\rangle = \sigma_L^2 \delta(t-t')\,. \tag{4.48}$$

This is mathematically identical to equation 4.37 except that the role of the spatial variables $\vec{a}$ and $\vec{b}$ has been replaced by the temporal variables t and t'. The analysis proceeds exactly as above, and the optimal filter is the product of a noise filter and a temporal whitening filter, as before. The temporal linear kernel $D_t(\tau)$ is written in terms of its Fourier transform,

$$D_t(\tau) = \frac{1}{2\pi}\int d\omega\, \exp(-i\omega\tau)\tilde{D}_t(\omega)\,, \tag{4.49}$$

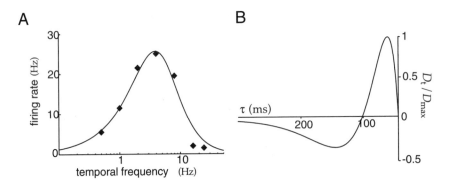

Figure 4.4 (A) Predicted (curve) and actual (diamonds) selectivity of an LGN cell as a function of temporal frequency. The predicted curve is based on the optimal linear filter $\tilde{D}_t(\omega)$ with $\omega_0 = 5.5$ Hz. (B) Causal, minimum phase, temporal form of the optimal filter. (Adapted from Dong and Atick, 1995; data in A from Saul and Humphrey, 1990.)

and $\tilde{D}_t(\omega)$ is given by an equation similar to 4.47,

$$|\tilde{D}_t(\omega)| \propto \frac{\sigma_L \sqrt{\tilde{Q}_{ss}(\omega)}}{\tilde{Q}_{ss}(\omega) + \tilde{Q}_{\eta\eta}(\omega)} \, . \tag{4.50}$$

In this case, $\tilde{Q}_{ss}(\omega)$ and $\tilde{Q}_{\eta\eta}(\omega)$ are the power spectra of the signal and the noise in the temporal domain.

Dong and Atick (1995) analyzed temporal receptive fields in the LGN in this way, under the assumption that a substantial fraction of the temporal redundancy of visual stimuli is removed in the LGN rather than in the retina. They determined that the temporal power spectrum of natural scenes has the form

$$\tilde{Q}_{ss}(\omega) \propto \frac{1}{\omega^2 + \omega_0^2} \, , \tag{4.51}$$

where ω_0 is a constant. The resulting filter, in both the temporal frequency and the time domains, is plotted in figure 4.4. Figure 4.4A shows the predicted and actual frequency responses of an LGN cell. This is similar to the plot in figure 4.3A, except that the result has been normalized to a realistic response level so that it can be compared with data. Because the optimization procedure determines only the amplitude of the Fourier transform of the linear kernel, $D_t(\tau)$ is not uniquely specified. To determine the temporal kernel, we require it to be causal ($D_t(\tau) = 0$ for $\tau < 0$) and impose a technical condition known as minimum phase, which assures that the output changes as rapidly as possible when the stimulus varies. Figure 4.4B shows the resulting form of the temporal filter. The space-time receptive fields shown in chapter 2 tend to change sign as a function of τ. The temporal filter in figure 4.4B has exactly this property.

An interesting test of the notion of optimal coding was carried out by Dan, Atick, and Reid (1996). They used both natural scene and white-noise stimuli while recording cat LGN cells. Figure 4.5A shows the power

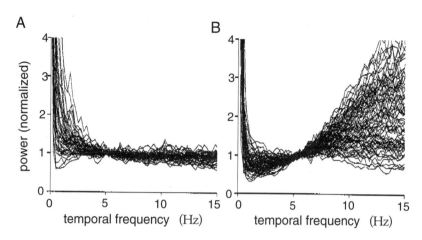

Figure 4.5 (A) Power spectra of the spike trains of 51 cat LGN cells in response to presentation of the movie *Casablanca*, normalized to their own values between 5 and 6 Hz. B) Equivalently normalized power spectra of the spike trains of 75 LGN cells in response to white-noise stimuli. (Adapted from Dan et al., 1996.)

spectra of spike trains of cat LGN cells in response to natural scenes (the movie *Casablanca*), and figure 4.5B shows power spectra in response to white-noise stimuli. The power spectra of the responses to natural scenes are quite flat above about $\omega = 3$ Hz. In response to white noise, on the other hand, they rise with ω. This is exactly what we would expect if LGN cells are acting as temporal whitening filters. In the case of natural stimuli, the whitening filter evenly distributes the output power over a broad frequency range. Responses to white-noise stimuli increase at high frequencies due to the boosting of inputs at these frequencies by the whitening filter.

Cortical Coding

Computational concerns beyond mere linear information transfer are likely to be relevant at the level of cortical processing of visual images. For one thing, the primary visual cortex has many more neurons than the LGN, yet they can collectively convey no more information about the visual world than they receive. As we saw in chapter 2, neurons in primary visual cortex are selective for quantities, such as spatial frequency and orientation, that are of particular importance in relation to object recognition but not for information transfer. Nevertheless, the methods described in the previous section can be used to understand restricted aspects of receptive fields of neurons in primary visual cortex, namely, the way that their multiple selectivities are collectively assigned. For example, cells that respond best at high spatial frequencies tend to respond more to low temporal frequency components of images, and vice versa.

The stimulus power spectrum written as a function of both spatial and temporal frequency has been estimated as $\tilde{Q}_{ss}(\vec{\kappa}, \omega) \propto 1/\left(|\vec{\kappa}|^2 + \alpha^2\omega^2\right)$,

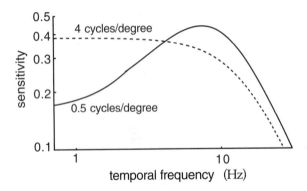

Figure 4.6 Dependence of temporal frequency tuning on preferred spatial frequency for space-time receptive fields derived from information maximization in the presence of noise. The curves show a transition from partial whitening in temporal frequency for low preferred spatial frequency (solid curve, 0.5 cycles/degree) to temporal summation for high preferred spatial frequency (dashed curve, 4 cycles/degree). (Adapted from Li, 1996.)

where $\alpha = 0.4$ cycle seconds/degree. This correlation function decreases both for high spatial and high temporal frequencies. Figure 4.6 shows how temporal selectivity for a combined noise and whitening filter, constructed using this stimulus power spectrum, changes for different preferred spatial frequencies. The basic idea is that components with fairly low stimulus power are boosted by the whitening filter, while those with very low stimulus power get suppressed by the noise filter. As shown by Li (1996), if a cell is selective for high spatial frequencies, the input signal rapidly falls below the noise (treated as white) as the temporal frequency of the input is increased. As a result, the noise filter of equation 4.46 causes the temporal response to be largest at 0 temporal frequency (dashed curve of figure 4.6). If instead the cell is selective for low spatial frequencies, the signal dominates the noise up to higher temporal frequencies, and the whitening filter causes the response to increase as a function of temporal frequency up to a maximum value where the noise filter begins to suppress the response (solid curve in figure 4.6). Receptive fields with preference for high spatial frequency thus act as low-pass temporal filters, and receptive fields with selectivity for low spatial frequency act as bandpass temporal filters.

Similar conclusions can be drawn concerning other joint selectivities. For example, color-selective (chrominance) cells tend to be selective for low temporal frequencies, because their input signal-to-noise ratio is lower than that for broadband (luminance) cells. There is also an interesting predicted relationship between ocular dominance and spatial frequency tuning due to the nature of the correlations between the two eyes. Optimal receptive fields with low spatial frequency tuning (for which the input signal-to-noise ratio is high) have enhanced sensitivity to differences between inputs coming from the two eyes. Receptive fields tuned to intermediate and high spatial frequencies suppress ocular differences.

4.3   Entropy and Information for Spike Trains

Computing the entropy or information content of a neuronal response characterized by spike times is much more difficult than computing these quantities for responses described by firing rates. Nevertheless, these computations are important, because firing rates are incomplete descriptions that can lead to serious underestimates of the entropy and information. In this section, we discuss how the entropy and mutual information can be computed for spike trains. Extensive further discussion can be found in the book by Rieke et al. (1997).

Spike-train entropy calculations are typically based on the study of long-duration recordings consisting of many action potentials. The entropy or mutual information typically grows linearly with the length of the spike train being considered. For this reason, the entropy and mutual information of spike trains are reported as entropy or information rates. These are the total entropy or information divided by the duration of the spike train. We write the entropy rate as $\dot{H}$ rather than H. Alternatively, entropy and mutual information can be divided by the total number of action potentials and reported as bits per spike rather than bits per second.

*entropy and
information rates*

To compute entropy and information rates for a spike train, we need to determine the probabilities that various temporal patterns of action potentials appear. These probabilities replace the factors $P[r]$ or $p[r]$ that occur when discrete or continuous firing rates are used to characterize a neural response. The temporal pattern of a group of action potentials can be specified by listing either the individual spike times or the sequence of intervals between successive spikes. The entropy and mutual information calculations we present are based on a spike-time description, but as an initial example we consider an approximate computation of entropy using interspike intervals.

The probability of an interspike interval falling in the range between τ and $\tau + \Delta\tau$ is given in terms of the interspike interval probability density by $p[\tau]\Delta\tau$. Because the interspike interval is a continuous variable, we must specify a resolution $\Delta\tau$ with which it is measured to define the entropy. If the different interspike intervals are statistically independent, the entropy associated with the interspike intervals in a spike train of average rate $\langle r \rangle$ and of duration T is the number of intervals, $\langle r \rangle T$, times the integral over τ of $-p[\tau]\log_2(p[\tau]\Delta\tau)$. The entropy rate is obtained by dividing this result by T, and the entropy per spike requires dividing by the number of spikes, $\langle r \rangle T$. The assumption of independent interspike intervals is critical for obtaining the spike-train entropy solely in terms of $p[\tau]$. Correlations between different interspike intervals reduce the total entropy, so the result obtained by assuming independent intervals provides an upper bound on the true entropy of a spike train. Thus, in general, the entropy rate $\dot{H}$ for a spike train with interspike interval distribution $p[\tau]$ and average rate $\langle r \rangle$

satisfies

$$\dot{H} \leq -\langle r \rangle \int_0^\infty d\tau\, p[\tau] \log_2(p[\tau]\Delta\tau) . \tag{4.52}$$

If a spike train is described by a homogeneous Poisson process with rate $\langle r \rangle$, we have $p[\tau] = \langle r \rangle \exp(-\langle r \rangle \tau)$, and the interspikes are statistically in-

Poisson entropy rate

dependent (chapter 1). Equation 4.52 is then an equality and, performing the integrals,

$$\dot{H} = \frac{\langle r \rangle}{\ln(2)} \left(1 - \ln(\langle r \rangle \Delta\tau)\right) . \tag{4.53}$$

We now turn to a more general calculation of the spike-train entropy. To make entropy calculations practical, a long spike train is broken into statistically independent subunits, and the total entropy is written as the sum of the entropies for the individual subunits. In the case of equation 4.52, the subunit was the interspike interval. If interspike intervals are not independent, and we wish to compute a result and not merely a bound, we must work with larger subunit descriptions. Strong et al. (1998) proposed a scheme that uses spike sequences of duration T_s as these basic subunits. Note that the variable T_s is used here to denote the duration of the spike sequence being considered, while T, which is much larger than T_s, is the duration of the entire spike train.

The time that a spike occurs is a continuous variable, so, as in the case of interspike intervals, a resolution must be specified when spike train entropies are computed. This can be done by dividing time into discrete bins of size Δt. We assume that the bins are small enough so that not more than one spike appears in a bin. Depending on whether or not a spike occurred within it, each bin is labeled by a 0 (no spike) or a 1 (spike). A spike sequence defined over a block of duration T_s is thus represented by a string of $T_s/\Delta t$ zeros and ones. We denote such a sequence by $B(t)$, where B is a $T_s/\Delta t$ bit binary number, and t specifies the time of the first bin in the sequence being considered. Both T_s and t are integer multiples of the bin size Δt.

The probability of a sequence B occurring at any time during the entire response is denoted by $P[B]$. This can be obtained by counting the number of times the sequence B occurs anywhere within the spike trains being analyzed (including overlapping cases). The spike-train entropy rate implied by this distribution is

$$\dot{H} = -\frac{1}{T_s} \sum_B P[B] \log_2 P[B] , \tag{4.54}$$

where the sum is over all the sequences B found in the data set, and we have divided by the duration T_s of a single sequence to obtain an entropy rate.

If the spike sequences in nonoverlapping intervals of duration T_s are independent, the full spike-train entropy rate is also given by equation 4.54.

However, any correlations between successive intervals (if $B(t+T_s)$ is correlated with $B(t)$, for example) reduce the total spike-train entropy, causing equation 4.54 to overestimate the true entropy rate. Thus, for finite T_s, this equation provides an upper bound on the true entropy rate. If T_s is too small, $B(t+T_s)$ and $B(t)$ are likely to be correlated, and the overestimate may be severe. As T_s increases, we expect the correlations to get smaller, and equation 4.54 should provide a more accurate value. For any finite data set, T_s cannot be increased past a certain point, because there will not be enough spike sequences of duration T_s in the data set to determine their probabilities. Thus, in practice, T_s must be increased until the point where the extraction of probabilities becomes problematic, and some form of extrapolation to $T_s \rightarrow \infty$ must be made.

Statistical mechanics arguments suggest that the difference between the entropy for finite T_s and the true entropy for $T_s \rightarrow \infty$ should be proportional to $1/T_s$ for large T_s. Therefore, the true entropy can be estimated, as in figure 4.7, by linearly extrapolating a plot of the entropy rate versus $1/T_s$ to the point $1/T_s = 0$. In figure 4.7 (upper line), this has been done for data from the motion-sensitive H1 neuron of the fly visual system. The plotted points show entropy rates computed for different values of $1/T_s$, and they vary linearly over most of the range of the plot. However, when $1/T_s$ goes below about $20/s$ (or $T_s > 50$ ms), the dependence suddenly changes. This is the point at which the amount of data is insufficient to extract even an overestimate of the entropy. By linearly extrapolating the linear part of the series of computed points in figure 4.7, Strong et al. estimated that the H1 spike trains had an entropy rate of 157 bits/s when the resolution was $\Delta t = 3$ ms.

To compute the mutual information rate for a spike train, we must subtract the noise entropy rate from the full spike-train entropy rate. The noise entropy rate is determined from the probabilities of finding various sequences B, given that they were evoked by the same stimulus. This is done by considering sequences $B(t)$ that start at a fixed time t. If the same stimulus is used in repeated trials, sequences that begin at time t in every trial are generated by the same stimulus. Therefore, the conditional probability of the response, given the stimulus, is in this case the distribution $P[B(t)]$ for response sequences beginning at time t. This is obtained by determining the fraction of trials on which $B(t)$ was evoked. Note that $P[B(t)]$ is the probability of finding a given sequence at time t within a set of spike trains obtained on trials using the same stimulus. In contrast, $P[B]$, used in the spike-train entropy calculation, is the probability of finding the sequence B at any time within these trains. Determining $P[B(t)]$ for a sufficient number of spike sequences may take a large number of trials using the same stimulus.

The full noise entropy is computed by averaging the noise entropy at time t over all t values. The average over t plays the role of the average over

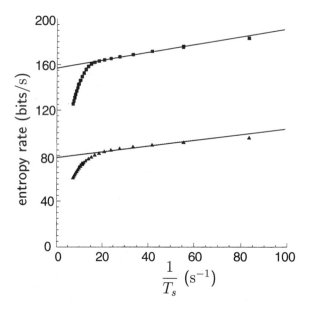

Figure 4.7 Entropy and noise entropy rates for the H1 visual neuron in the fly responding to a randomly moving visual image. The filled circles in the upper trace show the full spike-train entropy rate computed for different values of $1/T_s$. The straight line is a linear extrapolation to $1/T_s = 0$, which corresponds to $T_s \to \infty$. The lower trace shows the spike train noise entropy rate for different values of $1/T_s$. The straight line is again an extrapolation to $1/T_s = 0$. Both entropy rates increase as functions of $1/T_s$, and the true spike-train and noise entropy rates are overestimated at large values of $1/T_s$. At $1/T_s \approx 20/s$, there is a sudden shift in the dependence. This occurs when there is insufficient data to compute the spike sequence probabilities. The difference between the y intercepts of the two straight lines plotted is the mutual information rate. The resolution is $\Delta t = 3$ ms. (Adapted from Strong et al., 1998.)

stimuli in equation 4.6. The result is

$$\dot{H}_{\text{noise}} = -\frac{\Delta t}{T} \sum_t \left(\frac{1}{T_s} \sum_B P[B(t)] \log_2 P[B(t)] \right) , \qquad (4.55)$$

where $T/\Delta t$ is the number of different t values being summed.

If equation 4.55 is based on finite-length spike sequences, it provides an upper bound on the noise entropy rate. The true noise entropy rate is estimated by performing a linear extrapolation in $1/T_s$ to $1/T_s = 0$, as was done for the spike-train entropy rate. The result, shown in figure 4.7, is a noise entropy of 79 bits/s for $\Delta t = 3$ ms. The information rate is obtained by taking the difference between the extrapolated values for the spike-train and noise entropy rates. The result for the fly H1 neuron used in figure 4.7 is an information rate of 157 - 79 = 78 bits/s or 1.8 bits/spike. Values in the range 1 to 3 bits/spike are typical results of such calculations for a variety of preparations.

Both the spike-train and noise entropy rates depend on Δt. The leading dependence, coming from the $\log_2 \Delta t$ term discussed previously, cancels in the computation of the information rate, but the information can still depend on Δt through nondivergent terms. This reflects the fact that more information can be extracted from accurately measured spike times than from poorly measured spike times. Thus, we expect the information rate to increase with decreasing Δt, at least over some range of Δt values. At some critical value of Δt that matches the natural degree of noise jitter in the spike timings, we expect the information rate to stop increasing. This value of Δt is interesting because it tells us about the degree of spike timing accuracy in neural encoding.

The information conveyed by spike trains can be used to compare responses to different stimuli and thereby reveal stimulus-specific aspects of neural encoding. For example, Rieke et al. (1995) compared the information conveyed by single neurons in a peripheral auditory organ (the amphibian papilla) of the bullfrog in response to broadband noise or to noise filtered to have an amplitude spectrum close to that of natural bullfrog calls (although the phases for each frequency component were chosen randomly). They determined that the cells conveyed on average of 46 bits per second (1.4 bits per spike) for broadband noise and 133 bits per second (7.8 bits per spike) for stimuli with call-like spectra, despite the fact that the broadband noise had a higher entropy. The spike trains in response to the call-like stimuli conveyed information with near maximal efficiency.

4.4 Chapter Summary

Shannon's information theory can be used to determine how much a neural response tells both us and, presumably, the animal in which the neuron lives, about a stimulus. Entropy is a measure of the uncertainty or surprise associated with a stochastic variable, such as a stimulus. Mutual information quantifies the reduction in uncertainty associated with the observation of another variable, such as a response. The mutual information is related to the Kullback-Leibler divergence between two probability distributions. We defined the response and noise entropies for probability distributions of discrete and continuous firing rates, and considered how the information transmitted about a set of stimuli might be optimized. The principles of entropy and information maximization were used to account for features of the receptive fields of cells in the retina, LGN, and primary visual cortex. This analysis introduced probability factorization and equalization, and whitening and noise filters. Finally, we discussed how the information conveyed about dynamic stimuli by spike sequences can be estimated.

4.5   Appendix

Positivity of the Kullback-Leibler Divergence

Jensen's inequality   The logarithm is a concave function, which means that $\log_2 \langle z \rangle \geq \langle \log_2 z \rangle$, where the angle brackets denote averaging with respect to some probability distribution and z is any positive quantity. The equality holds only if z is a constant. If we consider this relation, known as Jensen's inequality, with $z = Q[r]/P[r]$ and the average defined over the probability distribution $P[r]$, we find

$$-D_{\mathrm{KL}}(P, Q) = \sum_r P[r] \log_2 \left(\frac{Q[r]}{P[r]} \right) \leq \log_2 \left(\sum_r P[r] \frac{Q[r]}{P[r]} \right) = 0 . \quad (4.56)$$

The last equality holds because $Q[r]$ is a probability distribution and thus satisfies $\sum_r Q[r] = 1$. Equation 4.56 implies that $D_{\mathrm{KL}}(P, Q) \geq 0$, with equality holding if and only if $P[r] = Q[r]$. A similar result holds for the Kullback-Leibler divergence between two probability densities,

$$D_{\mathrm{KL}}(p, q) = \int dr \, p[r] \log_2 \left(\frac{p[r]}{q[r]} \right) \geq 0 . \quad (4.57)$$

4.6   Annotated Bibliography

Information theory was created by Shannon (see **Shannon & Weaver, 1949**) largely as a way of understanding communication in the face of noise. **Cover & Thomas (1991)** provides a review, and **Rieke et al. (1997)** gives a treatment specialized to neural coding. Information theory and theories inspired by it, such as histogram equalization, were adopted in neuroscience and psychology as a way of understanding sensory transduction and coding, as discussed by **Barlow (1961)** and **Uttley (1979)**. We followed a more recent set of studies, inspired by Linkser (1988) and Barlow (1989), which have particularly focused on optimal coding in early vision; Atick & Redlich (1990), Plumbley (1991), Atick et al. (1992), **Atick (1992)**, van Hateren (1992; 1993), Li & Atick (1994a), Dong & Atick (1995), and Dan et al. (1996). Li & Atick (1994b) discuss the extension to joint selectivities of cells in V1; and Li & Atick (1994a) and Li (1996) treat stereo and motion sensitivities as examples.

The statistics of natural sensory inputs is reviewed by **Field (1987)**. Campbell & Gubisch (1966) estimated the optimal modulation transfer function.

We followed the technique of Strong et al. (1998) for computing the mutual information about a dynamical stimulus in spike trains. Bialek et al. (1993) presents an earlier approach based on stimulus reconstruction.

II Neurons and Neural Circuits

5 Model Neurons I: Neuroelectronics

5.1 Introduction

A great deal is known about the biophysical mechanisms responsible for generating neuronal activity, and this knowledge provides a basis for constructing neuron models. Such models range from highly detailed descriptions involving thousands of coupled differential equations to greatly simplified caricatures useful for studying large interconnected networks. In this chapter, we discuss the basic electrical properties of neurons and the mathematical models by which they are described. We present a simple but nevertheless useful model neuron, the integrate-and-fire model, in a basic version and with added membrane and synaptic conductances. We also discuss the Hodgkin-Huxley model, which describes the conductances responsible for generating action potentials. In chapter 6, we continue by presenting more complex models, in terms of their conductances and morphology. Circuits and networks of model neurons are discussed in chapter 7. This chapter makes use of basic concepts of electrical circuit theory, which are reviewed in the Mathematical Appendix.

5.2 Electrical Properties of Neurons

Like other cells, neurons are packed with a huge number and variety of ions and molecules. A cubic micron of cytoplasm might contain, for example, 10^{10} water molecules, 10^8 ions, 10^7 small molecules such as amino acids and nucleotides, and 10^5 proteins. Many of these molecules carry charges, either positive or negative. Most of the time, there is an excess concentration of negative charge inside a neuron. Excess charges that are mobile, like ions, repel each other and build up on the inside surface of the cell membrane. Electrostatic forces attract an equal density of positive ions from the extracellular medium to the outside surface of the membrane.

The cell membrane is a lipid bilayer 3 to 4 nm thick that is essentially impermeable to most charged molecules. This insulating feature causes the cell membrane to act as a capacitor by separating the charges lying

cell membrane

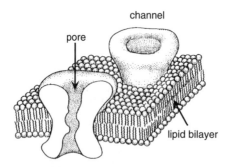

Figure 5.1 A schematic diagram of a section of the lipid bilayer that forms the cell membrane with two ion channels embedded in it. The membrane is 3 to 4 nm thick and the ion channels are about 10 nm long. (Adapted from Hille, 1992.)

ion channels

along its interior and exterior surfaces. Numerous ion-conducting channels embedded in the cell membrane (figure 5.1) lower the effective membrane resistance for ion flow to a value about 10,000 times smaller than that of a pure lipid bilayer. The resulting membrane conductance depends on the density and types of ion channels. A typical neuron may have a dozen or more different types of channels, anywhere from a few to hundreds of channels in a square micron of membrane, and hundreds of thousands to millions of channels in all. Many, but not all, channels are highly selective,

channel selectivity

allowing only a single type of ion to pass through them (to an accuracy of about 1 ion in 10^4). The capacity of channels for conducting ions across the cell membrane can be modified by many factors, including the membrane potential (voltage-dependent channels), the internal concentration of various intracellular messengers (Ca^{2+}-dependent channels, for example), and the extracellular concentration of neurotransmitters or neuromodulators (synaptic receptor channels, for example). The membrane also contains

ion pumps

selective pumps that expend energy to maintain differences in the concentrations of ions inside and outside the cell.

membrane potential

By convention, the potential of the extracellular fluid outside a neuron is defined to be 0. When a neuron is inactive, the excess internal negative charge causes the potential inside the cell membrane to be negative. This potential is an equilibrium point at which the flow of ions into the cell matches that out of the cell. The potential can change if the balance of ion flow is modified by the opening or closing of ion channels. Under normal conditions, neuronal membrane potentials vary over a range from about -90 to +50 mV. The order of magnitude of these potentials can be estimated from basic physical principles.

Membrane potentials are small enough to allow neurons to take advantage of thermal energy to help transport ions across the membrane, but are large enough so that thermal fluctuations do not swamp the signaling capabilities of the neuron. These conditions imply that potential differences across the cell membrane must lie in a range such that the energy gained or lost by an ion traversing the membrane is the same order of magnitude as its thermal energy. The thermal energy of an ion is about $k_B T$ where k_B

is the Boltzmann constant and T is the temperature on an absolute Kelvin scale. For chemists and biologists (though not for physicists), it is more customary to discuss moles of ions rather than single ions. A mole of ions has Avogadro's number times as much thermal energy as a single ion, or RT, where R is the universal gas constant, equal to 8.31 joules/mol K° = 1.99 cal/mol K°. RT is about 2500 joules/mol or 0.6 kCal/mol at normal temperatures.

To estimate the size of typical membrane potentials, we equate the thermal energy of a mole of ions to the energy gained or lost when a mole of ions crosses a membrane with a potential difference V_T across it. This energy is FV_T, where F is the Faraday constant, $F = 96{,}480$ coulombs/mol, equal to Avogadro's number times the charge of a single proton, q. Setting $FV_T = RT$ gives

V_T

$$V_T = \frac{RT}{F} = \frac{k_B T}{q} \ . \tag{5.1}$$

This is an important parameter that enters into a number of calculations. V_T is between 24 and 27 mV for the typical temperatures of cold- and warm-blooded animals. This sets the overall scale for membrane potentials across neuronal membranes, which range from about -3 to +2 times V_T.

Intracellular Resistance

Membrane potentials measured at different places within a neuron can take different values. For example, the potentials in the soma, dendrite, and axon can all be different. Potential differences between different parts of a neuron cause ions to flow within the cell, which tends to equalize these differences. The intracellular medium provides a resistance to such flow. This resistance is highest for long, narrow stretches of dendritic or axonal cable, such as the segment shown in figure 5.2. The longitudinal current I_L flowing along such a cable segment can be computed from Ohm's law. For the cylindrical segment of dendrite shown in figure 5.2, the longitudinal current flowing from right to left satisfies $V_2 - V_1 = I_L R_L$. Here, R_L is the longitudinal resistance, which grows in proportion to the length of the segment (long segments have higher resistances than short ones) and is inversely proportional to the cross-sectional area of the segment (thin segments have higher resistances than fat ones). The constant of proportionality, called the intracellular resistivity, r_L, typically falls in a range from 1 to 3 kΩ mm. The longitudinal resistance of the segment in figure 5.2 is r_L times the length L divided by the cross-sectional area πa^2, $R_L = r_L L / \pi a^2$. A segment 100 μm long with a radius of 2 μm has a longitudinal resistance of about 8 MΩ. A voltage difference of 8 mV would be required to force 1 nA of current down such a segment.

longitudinal current I_L

longitudinal resistance R_L

intracellular resistivity r_L

We can also use the intracellular resistivity to estimate crudely the conductance of a single channel. The conductance, being the inverse of a resistance, is equal to the cross-sectional area of the channel pore divided by

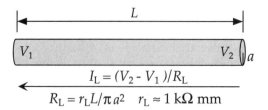

Figure 5.2 The longitudinal resistance of a cylindrical segment of neuronal cable with length L and radius a. The difference between the membrane potentials at the ends of this segment is related to the longitudinal current within the segment by Ohm's law, with R_L the longitudinal resistance of the segment. The arrow indicates the direction of positive current flow. The constant r_L is the intracellular resistivity, and a typical value is given.

single-channel conductance

its length and by r_L We approximate the channel pore as a tube of length 6 nm and opening area 0.15 nm^2. This gives an estimate of 0.15 nm$^2/(1$ kΩ mm $\times$ 6 nm) $\approx$ 25 pS, which is the right order of magnitude for a channel conductance.

Membrane Capacitance and Resistance

electrotonic compactness

The intracellular resistance to current flow can cause substantial differences in the membrane potential measured in different parts of a neuron, especially during rapid transient excursions of the membrane potential from its resting value, such as action potentials. Neurons that have few of the long, narrow cable segments that produce high longitudinal resistances may have relatively uniform membrane potentials across their surfaces. Such neurons are termed electrotonically compact. For electrotonically compact neurons, or for less compact neurons in situations where spatial variations in the membrane potential are not thought to play an important functional role, the entire neuron may be adequately described by a single membrane potential. Here, we discuss the membrane capacitance and resistance using such a description. An analysis for the case of spatially varying membrane potentials is presented in chapter 6.

membrane capacitance C_m

specific membrane capacitance c_m

We have mentioned that there is typically an excess negative charge on the inside surface of the cell membrane of a neuron, and a balancing positive charge on its outside surface (figure 5.3). In this arrangement, the cell membrane creates a capacitance C_m, and the voltage across the membrane V and the amount of this excess charge Q are related by the standard equation for a capacitor, $Q = C_m V$. The membrane capacitance is proportional to the total amount of membrane or, equivalently, to the surface area of the cell. The constant of proportionality, called the specific membrane capacitance, is the capacitance per unit area of membrane, and it is approximately the same for all neurons, $c_m \approx 10$ nF/mm^2. The total capacitance C_m is the membrane surface area A times the specific capacitance, $C_m = c_m A$. Neuronal surface areas tend to be in the range 0.01 to 0.1 mm^2, so the membrane capacitance for a whole neuron is typically 0.1

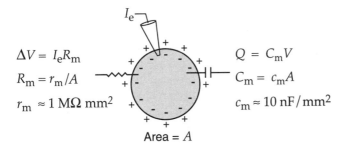

$$\Delta V = I_e R_m$$

$$R_m = r_m/A$$

$$r_m \approx 1\ \text{M}\Omega\ \text{mm}^2$$

$$Q = C_m V$$

$$C_m = c_m A$$

$$c_m \approx 10\ \text{nF/mm}^2$$

Area = A

Figure 5.3 The capacitance and membrane resistance of a neuron considered as a single compartment. The membrane capacitance determines how the membrane potential V and excess internal charge Q are related. The membrane resistance R_m determines the size of the membrane potential deviation ΔV caused by a small current I_e entering through an electrode, for example. Equations relating the total membrane capacitance and resistance, C_m and R_m, to the specific membrane capacitance and resistance, c_m and r_m, are given along with typical values of c_m and r_m. The value of r_m may vary considerably under different conditions and for different neurons.

to 1 nF. For a neuron with a total membrane capacitance of 1 nF, 7×10^{-11} coulomb or about 10^9 singly charged ions are required to produce a resting potential of -70 mV. This is about $1/100{,}000$ of the total number of ions in a neuron and is the amount of charge delivered by a 0.7 nA current in 100 ms.

We can use the membrane capacitance to determine how much current is required to change the membrane potential at a given rate. The time derivative of the basic equation relating the membrane potential and charge,

$$C_m \frac{dV}{dt} = \frac{dQ}{dt}\,, \tag{5.2}$$

plays an important role in the mathematical modeling of neurons. The time derivative of the charge dQ/dt is equal to the current passing into the cell, so the amount of current needed to change the membrane potential of a neuron with a total capacitance C_m at a rate dV/dt is $C_m dV/dt$. For example, 1 nA will change the membrane potential of a neuron with a capacitance of 1 nF at a rate of 1 mV/ms.

The capacitance of a neuron determines how much current is required to make the membrane potential change at a given rate. Holding the membrane potential steady at a level different from its resting value also requires current, but this current is determined by the membrane resistance rather than by the capacitance of the cell. For example, if a small constant current I_e is injected into a neuron through an electrode, as in figure 5.3, the membrane potential will shift away from its resting value by an amount ΔV given by Ohm's law, $\Delta V = I_e R_m$. R_m is known as the membrane or input resistance. The restriction to small currents and small ΔV is required because membrane resistances can vary as a function of voltage, whereas Ohm's law assumes R_m is constant over the range ΔV.

membrane resistance R_m

*membrane
conductance*

*specific membrane
resistance* r_m

The membrane resistance is the inverse of the membrane conductance, and, like the capacitance, the conductance of a piece of cell membrane is proportional to its surface area. The constant of proportionality is the membrane conductance per unit area, but we write it as $1/r_m$, where r_m is called the specific membrane resistance. Conversely, the membrane resistance R_m is equal to r_m divided by the surface area. When a neuron is in a resting state, the specific membrane resistance is around 1 MΩ mm^2. This number is much more variable than the specific membrane capacitance. Membrane resistances vary considerably among cells, and under different conditions and at different times for a given neuron, depending on the number, type, and state of its ion channels. For total surface areas between 0.01 and 0.1 mm^2 the membrane resistance is typically in the range 10 to 100 MΩ. With a 100 MΩ membrane resistance, a constant current of 0.1 nA is required to hold the membrane potential 10 mV away from its resting value.

*membrane time
constant* τ_m

The product of the membrane capacitance and the membrane resistance is a quantity with the units of time called the membrane time constant, $\tau_m = R_m C_m$. Because C_m and R_m have inverse dependences on the membrane surface area, the membrane time constant is independent of area and equal to the product of the specific membrane capacitance and resistance, $\tau_m = r_m c_m$. The membrane time constant sets the basic time scale for changes in the membrane potential and typically falls in the range between 10 and 100 ms.

Equilibrium and Reversal Potentials

Electric forces and diffusion are responsible for driving ions through channel pores. Voltage differences between the exterior and interior of the cell produce forces on ions. Negative membrane potentials attract positive ions into the neuron and repel negative ions. In addition, ions diffuse through channels because the ion concentrations differ inside and outside the neuron. These differences are maintained by the ion pumps within the cell membrane. The concentrations of Na$^+$ and Ca^{2+} are higher outside the cell than inside, so these ions are driven into the neuron by diffusion. K$^+$ is more concentrated inside the neuron than outside, so it tends to diffuse out of the cell.

*equilibrium
potential*

It is convenient to characterize the current flow due to diffusion in terms of an equilibrium potential.This is defined as the membrane potential at which current flow due to electric forces cancels the diffusive flow. For channels that conduct a single type of ion, the equilibrium potential can be computed easily. The potential difference across the cell membrane biases the flow of ions into or out of a neuron. Consider, for example, a positively charged ion and a negative membrane potential. In this case, the membrane potential opposes the flow of ions out of the cell. Ions can cross the membrane and leave the interior of the cell only if their thermal energy suffices to overcome the energy barrier produced by the membrane potential. If the ion has an electric charge zq, where q is the charge of one proton, it must have a thermal energy of at least $-zqV$ to cross the membrane (this is a positive energy for $z > 0$ and $V < 0$). The probability that an ion has a thermal energy greater than or equal to $-zqV$, when the temperature (on an absolute scale) is T, is $\exp(zqV/k_BT)$. This is determined by integrating the Boltzmann distribution for energies greater than or equal to $-zqV$. In molar units, this result can be written as $\exp(zFV/RT)$, which is equal to $\exp(zV/V_T)$ by equation 5.1.

The biasing effect of the electrical potential can be overcome by an opposing concentration gradient. A concentration of ions inside the cell, [inside], that is sufficiently greater than the concentration outside the cell, [outside], can compensate for the Boltzmann probability factor. The rate at which ions flow into the cell is proportional to [outside]. The flow of ions out of the cell is proportional to [inside] times the Boltzmann factor, because in this direction only those ions that have sufficient thermal energy can leave the cell. The net flow of ions will be 0 when the inward and outward flows are equal. We use the letter E to denote the particular potential that satisfies this balancing condition, which is then

$$[\text{outside}] = [\text{inside}] \exp(zE/V_T) . \tag{5.3}$$

Solving this equation for E, we find   *Nernst equation*

$$E = \frac{V_T}{z} \ln \left(\frac{[\text{outside}]}{[\text{inside}]} \right) . \tag{5.4}$$

Equation 5.4 is the Nernst equation. The reader can check that if the result is derived for either sign of ionic charge or membrane potential, the result is identical to 5.4, which thus applies in all cases. The equilibrium potential for a K^+ conducting channel, labeled E_K, typically falls in the range between -70 and -90 mV. The Na^+ equilibrium potential, E_{Na}, is 50 mV or higher, and E_{Ca}, for Ca^{2+} channels, is higher still, around 150 mV. Finally, Cl^- equilibrium potentials are typically around -60 to -65 mV, near the resting potential of many neurons.

The Nernst equation (5.4) applies when the channels that generate a particular conductance allow only one type of ion to pass through them. Some channels are not so selective, and in this case the potential E is not determined by equation 5.4. Instead, it takes a value intermediate between the equilibrium potentials of the individual ion types that it conducts. An approximate formula, known as the Goldman equation (see Tuckwell, 1988; or Johnston and Wu, 1995), can be used to estimate E for such conductances. In this case, E is often called a reversal potential, rather than an equilibrium potential, because the direction of current flow through the channel switches as the membrane potential passes through E.   *Goldman equation* / *reversal potential*

A conductance with an equilibrium or reversal potential E tends to move the membrane potential of the neuron toward the value E. When $V > E$, this means that positive current will flow outward, and when $V < E$, positive current will flow inward. Because Na^+ and Ca^{2+} conductances have

depolarization   positive reversal potentials, they tend to depolarize a neuron (make its
                      membrane potential less negative). K^+ conductances, with their nega-
hyperpolarization   tive E values, normally hyperpolarize a neuron (make its membrane po-
                      tential more negative). Cl^- conductances, with reversal potentials near
                      the resting potential, may pass little net current. Instead, their primary
                      impact is to change the membrane resistance of the cell. Such conduc-
shunting   tances are sometimes called shunting, although all conductances "shunt",
conductances   that is, increase the total conductance of a neuron. Synaptic conductances
                      are also characterized by reversal potentials and are termed excitatory or
                      inhibitory on this basis. Synapses with reversal potentials less than the
inhibitory and   threshold for action potential generation are typically called inhibitory,
excitatory synapses and those with reversal potentials above the action potential threshold are
                      called excitatory.

The Membrane Current

The total current flowing across the membrane through all of its ion channels is called the membrane current of the neuron. By convention, the
membrane current is defined as positive when positive ions leave the neuron and negative when positive ions enter the neuron. The total membrane
current is determined by summing currents due to all of the different types
of channels within the cell membrane, including voltage-dependent and
synaptic channels. To facilitate comparisons between neurons of different sizes, it is convenient to use the membrane current per unit area of cell
membrane current   membrane, which we call i_m. The total membrane current is obtained from
per unit area i_m i_m by multiplying it by A, the total surface area of the cell.

We label the different types of channels in a cell membrane with an index
i. As discussed in the last section, the current carried by a set of channels
of type i with reversal potential E_i, vanishes when the membrane potential satisfies $V = E_i$. For many types of channels, the current increases
or decreases approximately linearly when the membrane potential devi-
driving force   ates from this value. The difference $V - E_i$ is called the driving force, and
specific   the membrane current per unit area due to the type i channels is written
conductance g_i as $g_i(V - E_i)$. The factor g_i is the conductance per unit area, or specific
                     conductance, due to these channels. Summing over the different types of
membrane current   channels, we obtain the total membrane current,

$$i_m = \sum_i g_i(V - E_i)\,. \tag{5.5}$$

Sometimes a more complicated expression called the Goldman-Hodgkin-
Katz formula is used to relate the membrane current to g_i and membrane
potential (see Tuckwell, 1988; or Johnston and Wu, 1995), but we will restrict our discussion to the simpler relationship used in equation 5.5.

Much of the complexity and richness of neuronal dynamics arises because
membrane conductances change over time. However, some of the factors that contribute to the total membrane current can be treated as relatively constant, and these are typically grouped together into a single term

called the leakage current. The currents carried by ion pumps that main-   *leakage current*
tain the concentration gradients that make equilibrium potentials nonzero
typically fall into this category. For example, one type of pump uses the
energy of ATP hydrolysis to move three Na^+ ions out of the cell for every
two K^+ ions it moves in.

It is normally assumed that ion pumps work at relatively steady rates
so that the currents they generate can be included in a time-independent
leakage conductance. Sometimes, this assumption is dropped and explicit
pump currents are modeled. In either case, all of the time-independent
contributions to the membrane current can be lumped together into a single leakage term $\overline{g}_L (V - E_L)$. Because this term hides many sins, its reversal potential E_L is not usually equal to the equilibrium potential of any
specific ion. Instead, it is often kept as a free parameter and adjusted to
make the resting potential of the model neuron match that of the cell being   *resting potential*
modeled. Similarly, $\overline{g}_L$ is adjusted to match the membrane conductance at
rest. The line over the parameter $\overline{g}_L$ is used to indicate that it has constant value. A similar notation is used later in this chapter to distinguish
variable conductances from the fixed parameters that describe them. The
leakage conductance is called a passive conductance to distinguish it from
variable conductances that are termed active.

5.3   Single-Compartment Models

Models that describe the membrane potential of a neuron by a single variable V are called single-compartment models. This chapter deals exclusively with such models. Multi-compartment models, which can describe
spatial variations in the membrane potential, are considered in chapter 6.
The equations for single-compartment models, like those of all neuron
models, describe how charges flow into and out of a neuron and affect
its membrane potential.

Equation 5.2 provides the basic relationship that determines the membrane potential for a single-compartment model. This equation states that
the rate of change of the membrane potential is proportional to the rate
at which charge builds up inside the cell. The rate of charge buildup is,
in turn, equal to the total amount of current entering the neuron. The
relevant currents are those arising from all the membrane and synaptic
conductances plus, in an experimental setting, any current injected into
the cell through an electrode. From equation 5.2, the sum of these currents
is equal to $C_m dV/dt$, the total capacitance of the neuron times the rate of
change of the membrane potential. Because the membrane current is usually characterized as a current per unit area, i_m, it is more convenient to
divide this relationship by the surface area of the neuron. Then, the total
current per unit area is equal to $c_m dV/dt$, where $c_m = C_m/A$ is the specific membrane capacitance. One complication in this procedure is that the
electrode current, I_e, is not typically expressed as a current per unit area,
so we must divide by the total surface area of the neuron, A. Putting all

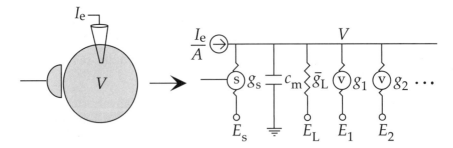

Figure 5.4 The equivalent circuit for a one-compartment neuron model. The neuron is represented, at the left, by a single compartment of surface area A with a synapse and a current-injecting electrode. At right is the equivalent circuit. The circled s indicates a synaptic conductance that depends on the activity of a presynaptic neuron. A single synaptic conductance g_s is indicated, but in general there may be several different types. The circled v indicates a voltage-dependent conductance, and I_e is the current passing through the electrode. The dots stand for possible additional membrane conductances.

this together, the basic equation for all single-compartment models is

$$c_m \frac{dV}{dt} = -i_m + \frac{I_e}{A}.$$ (5.6)

By convention, current that enters the neuron through an electrode is defined as positive-inward, whereas membrane current is defined as positive-outward. This explains the different signs for the currents in equation 5.6. The membrane current in equation 5.6 is determined by equation 5.5 and additional equations that specify the conductance variables g_i. The structure of such a model is the same as that of an electrical circuit, called the *equivalent circuit*, consisting of a capacitor and a set of variable and nonvariable resistors corresponding to the different membrane conductances. Figure 5.4 shows the equivalent circuit for a generic one-compartment model.

5.4   Integrate-and-Fire Models

A neuron will typically fire an action potential when its membrane potential reaches a threshold value of about -55 to -50 mV. During the action potential, the membrane potential follows a rapid, stereotyped trajectory and then returns to a value that is hyperpolarized relative to the threshold potential. As we will see, the mechanisms by which voltage-dependent K$^+$ and Na$^+$ conductances produce action potentials are well understood and can be modeled quite accurately. On the other hand, neuron models can be simplified and simulations can be accelerated dramatically if the biophysical mechanisms responsible for action potentials are not explicitly included in the model. Integrate-and-fire models do this by stipulating that an action potential occurs whenever the membrane potential of the model neuron reaches a threshold value V_{th}. After the action potential, the potential is reset to a value V_{reset} below the threshold potential, $V_{\text{reset}} < V_{\text{th}}$.

The basic integrate-and-fire model was proposed by Lapicque in 1907, long before the mechanisms that generate action potentials were understood. Despite its age and simplicity, the integrate-and-fire model is still an extremely useful description of neuronal activity. By avoiding a biophysical description of the action potential, integrate-and-fire models are left with the simpler task of modeling only subthreshold membrane potential dynamics. This can be done with various levels of rigor. In the simplest version of these models, all active membrane conductances are ignored, including, for the moment, synaptic inputs, and the entire membrane conductance is modeled as a single passive leakage term, $i_{\text{m}} = \overline{g}_{\text{L}}(V - E_{\text{L}})$. This version is called the passive or leaky integrate-and-fire model. For *passive* small fluctuations about the resting membrane potential, neuronal con- *integrate-and-fire* ductances are approximately constant, and the passive integrate-and-fire *model* model assumes that this constancy holds over the entire subthreshold range. For some neurons this is a reasonable approximation, and for others it is not. With these approximations, the model neuron behaves like an electric circuit consisting of a resistor and a capacitor in parallel (figure 5.4), and the membrane potential is determined by equation 5.6 with $i_{\text{m}} = \overline{g}_{\text{L}}(V - E_{\text{L}})$,

$$c_{\text{m}}\frac{dV}{dt} = -\overline{g}_{\text{L}}(V - E_{\text{L}}) + \frac{I_{\text{e}}}{A}. \tag{5.7}$$

It is convenient to multiply equation 5.7 by the specific membrane resistance r_{m}, which in this case is given by $r_{\text{m}} = 1/\overline{g}_{\text{L}}$. This cancels the factor of $\overline{g}_{\text{L}}$ on the right side of the equation and leaves a factor $c_{\text{m}}r_{\text{m}} = \tau_{\text{m}}$ on the left side, where τ_{m} is the membrane time constant of the neuron. The electrode current ends up being multiplied by r_{m}/A, which is the total membrane resistance R_{m}. Thus, the basic equation of the passive integrate-and-fire models is

$$\tau_{\text{m}}\frac{dV}{dt} = E_{\text{L}} - V + R_{\text{m}}I_{\text{e}}. \tag{5.8}$$

To generate action potentials in the model, equation 5.8 is augmented by the rule that whenever V reaches the threshold value V_{th}, an action potential is fired and the potential is reset to V_{reset}. Equation 5.8 indicates that when $I_{\text{e}} = 0$, the membrane potential relaxes exponentially with time constant τ_{m} to $V = E_{\text{L}}$. Thus, E_{L} is the resting potential of the model cell.

The membrane potential for the passive integrate-and-fire model is determined by integrating equation 5.8 (a numerical method for doing this is described in appendix A) and applying the threshold and reset rule for action potential generation. The response of a passive integrate-and-fire model neuron to a time-varying electrode current is shown in figure 5.5.

The firing rate of an integrate-and-fire model in response to a constant injected current can be computed analytically. When I_{e} is independent of

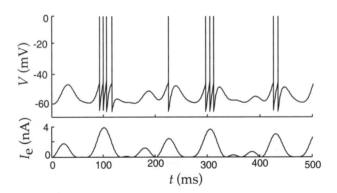

Figure 5.5 A passive integrate-and-fire model driven by a time-varying electrode current. The upper trace is the membrane potential, and the bottom trace the driving current. The action potentials in this figure are simply pasted onto the membrane potential trajectory whenever it reaches the threshold value. The parameters of the model are $E_L = V_{reset} = -65$ mV, $V_{th} = -50$ mV, $\tau_m = 10$ ms, and $R_m = 10$ MΩ.

time, the subthreshold potential $V(t)$ can easily be computed by solving equation 5.8, and is

$$V(t) = E_L + R_m I_e + (V(0) - E_L - R_m I_e)\exp(-t/\tau_m),\qquad (5.9)$$

where $V(0)$ is the value of V at time $t = 0$. This solution can be checked by substituting it into equation 5.8. It is valid for the integrate-and-fire model only as long as V stays below the threshold. Suppose that at $t = 0$, the neuron has just fired an action potential and is thus at the reset potential, so that $V(0) = V_{reset}$. The next action potential will occur when the membrane potential reaches the threshold, that is, at a time $t = t_{isi}$ when

$$V(t_{isi}) = V_{th} = E_L + R_m I_e + (V_{reset} - E_L - R_m I_e)\exp(-t_{isi}/\tau_m).\quad (5.10)$$

By solving this for t_{isi}, the time of the next action potential, we can determine the interspike interval for constant I_e, or equivalently its inverse, which we call the interspike-interval firing rate of the neuron,

$$r_{isi} = \frac{1}{t_{isi}} = \left(\tau_m \ln\left(\frac{R_m I_e + E_L - V_{reset}}{R_m I_e + E_L - V_{th}}\right)\right)^{-1}.\qquad (5.11)$$

This expression is valid if $R_m I_e > V_{th} - E_L$; otherwise $r_{isi} = 0$. For sufficiently large values of I_e, we can use the linear approximation of the logarithm ($\ln(1 + z) \approx z$ for small z) to show that

$$r_{isi} \approx \left[\frac{E_L - V_{th} + R_m I_e}{\tau_m(V_{th} - V_{reset})}\right]_+,\qquad (5.12)$$

which shows that the firing rate grows linearly with I_e for large I_e.

Figure 5.6A compares r_{isi} as a function of I_e, using appropriate parameter values, with data from current injection into a cortical neuron in vivo. The firing rate of the cortical neuron in figure 5.6A has been defined as the

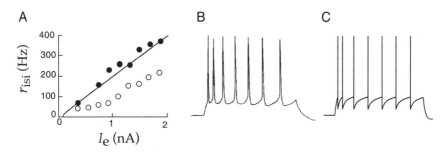

Figure 5.6 (A) Comparison of interspike-interval firing rates as a function of injected current for an integrate-and-fire model and a cortical neuron measure in vivo. The line gives r_{isi} for a model neuron with $\tau_m = 30$ ms, $E_L = V_{reset} = -65$ mV, $V_{th} = -50$ mV, and $R_m = 90$ MΩ. The data points are from a pyramidal cell in the primary visual cortex of a cat. The filled circles show the inverse of the interspike interval for the first two spikes fired, and the open circles show the steady-state interspike-interval firing rate after spike-rate adaptation. (B) A recording of the firing of a cortical neuron under constant current injection, showing spike-rate adaptation. (C) Membrane voltage trajectory and spikes for an integrate-and-fire model with an added current, with $r_m \Delta g_{sra} = 0.06$, $\tau_{sra} = 100$ ms, and $E_K = -70$ mV (see equations 5.13 and 5.14). (Data in A from Ahmed et al., 1998; B from McCormick, 1990.)

inverse of the interval between pairs of spikes. The rates determined in this way, using the first two spikes fired by the neuron in response to the injected current (filled circles in figure 5.6A), agree fairly well with the results of the integrate-and-fire model described in the figure caption. However, the real neuron exhibits spike-rate adaptation, in that the interspike intervals lengthen over time when a constant current is injected into the cell (figure 5.6B), before settling to a steady-state value. The steady-state firing rate in figure 5.6A (open circles) could also be fitted by an integrate-and-fire model, but not using the same parameters that were used to fit the initial spikes. Spike-rate adaptation is a common feature of cortical pyramidal cells, and consideration of this phenomenon allows us to show how an integrate-and-fire model can be modified to incorporate more complex dynamics.

spike-rate adaptation

Spike-Rate Adaptation and Refractoriness

The passive integrate-and-fire model that we have described thus far is based on two separate approximations, a highly simplified description of the action potential and a linear approximation for the total membrane current. If details of the action-potential generation process are not important for a particular modeling goal, the first approximation can be retained while the membrane current is modeled in as much detail as is necessary. We will illustrate this process by developing a heuristic description of spike-rate adaptation using a model conductance that has characteristics similar to measured neuronal conductances known to play important roles in producing this effect.

We model spike-rate adaptation by including an additional current in the model,

$$\tau_m \frac{dV}{dt} = E_L - V - r_m g_{sra}(V - E_K) + R_m I_e.$$ (5.13)

The spike-rate adaptation conductance g_{sra} has been modeled as a K^+ conductance so, when activated, it will hyperpolarize the neuron, slowing any spiking that may be occurring. We assume that this conductance relaxes to 0 exponentially with time constant τ_{sra} through the equation

$$\tau_{sra} \frac{dg_{sra}}{dt} = -g_{sra}.$$ (5.14)

Whenever the neuron fires a spike, g_{sra} is increased by an amount Δg_{sra}, that is, $g_{sra} \rightarrow g_{sra} + \Delta g_{sra}$. During repetitive firing, the current builds up in a sequence of steps causing the firing rate to adapt. Figures 5.6B and 5.6C compare the adapting firing pattern of a cortical neuron with the output of the model.

As discussed in chapter 1, the probability that a neuron fires is significantly reduced for a short period of time after the appearance of an action potential. Such a refractory effect is not included in the basic integrate-and-fire model. The simplest way of including an absolute refractory period in the model is to add a condition to the basic threshold crossing rule that forbids firing for a period of time immediately after a spike. Refractoriness can be incorporated in a more realistic way by adding a conductance similar to the spike-rate adaptation conductance discussed above, but with a faster recovery time and a larger conductance increment following an action potential. With a large increment, the current can essentially clamp the neuron to E_K following a spike, temporarily preventing further firing and producing an absolute refractory period. As this conductance relaxes back to 0, firing will be possible but initially less likely, producing a relative refractory period. When recovery is completed, normal firing can resume.

Another scheme that is sometimes used to model refractory effects is to raise the threshold for action-potential generation following a spike and then allow it to relax back to its normal value. Spike-rate adaptation can also be described by using an integrated version of the integrate-and-fire model known as the spike-response model, in which membrane potential waveforms are determined by summing precomputed postsynaptic potentials and after-spike hyperpolarizations. Finally, spike-rate adaptation and other effects can be incorporated into the integrate-and-fire framework by allowing the parameters $\bar{g}_L$ and E_L in equation 5.7 to vary with time.

5.5 Voltage-Dependent Conductances

Most of the interesting electrical properties of neurons, including their ability to fire and propagate action potentials, arise from nonlinearities

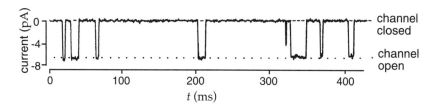

Figure 5.7 Recording of the current passing through a single ion channel. This is a synaptic receptor channel sensitive to the neurotransmitter acetylcholine. A small amount of acetylcholine was applied to the preparation to produce occasional channel openings. In the open state, the channel passes 6.6 pA at a holding potential of -140 mV. This is equivalent to more than 10^7 charges per second passing through the channel, and corresponds to an open channel conductance of 47 pS. (From Hille, 1992.)

associated with active membrane conductances. Recordings of the current flowing through single channels indicate that channels fluctuate rapidly between open and closed states in a stochastic manner (figure 5.7). Models of membrane and synaptic conductances must describe how the probability that a channel is in an open, ion-conducting state at any given time depends on the membrane potential (for a voltage-dependent conductance), the presence or absence of a neurotransmitter (for a synaptic conductance), or a number of other factors, such as the concentration of Ca^{2+} or other messenger molecules inside the cell. In this chapter, we consider two classes of active conductances, voltage-dependent membrane conductances and transmitter-dependent synaptic conductances. An additional type, the Ca^{2+}-dependent conductance, is considered in chapter 6.

stochastic channel

voltage-dependent, synaptic, and Ca^{2+}-dependent conductances

In a later section of this chapter, we discuss stochastic models of individual channels based on state diagrams and transition rates. However, most neuron models use deterministic descriptions of the conductances arising from many channels of a given type. This is justified because of the large number of channels of each type in the cell membrane of a typical neuron. If large numbers of channels are present, and if they fluctuate independently of each other (which they do, to a good approximation), then, from the law of large numbers, the fraction of channels open at any given time is approximately equal to the probability that any one channel is in an open state. This allows us to move between single-channel probabilistic formulations and macroscopic deterministic descriptions of membrane conductances.

We have denoted the conductance per unit area of membrane due to a set of ion channels of type i by g_i. The value of g_i at any given time is determined by multiplying the conductance of an open channel by the density of channels in the membrane and by the fraction of channels that are open at that time. The product of the first two factors is a constant called the maximal conductance that is denoted by $\bar{g}_i$. It is the conductance per unit area of membrane if all the channels of type i are open. Maximal conductance parameters tend to range from $\mu S/mm^2$ to mS/mm^2. The fraction of channels in the open state is equivalent to the probability of finding any *open probability P_i*   given channel in the open state, and it is denoted by P_i. Thus, $g_i = \bar{g}_i P_i$. The dependence of a conductance on voltage, transmitter concentration, or other factors arises through effects on the open probability.

The open probability of a voltage-dependent conductance depends, as its name suggests, on the membrane potential of the neuron. In this chapter, we discuss models of two such conductances, the so-called delayed-rectifier K$^+$ and fast Na$^+$ conductances. The formalism we present, which is almost universally used to describe voltage-dependent conductances, was developed by Hodgkin and Huxley (1952) as part of their pioneering work showing how these conductances generate action potentials in the squid giant axon. Other conductances are modeled in chapter 6.

Persistent Conductances

Figure 5.8 shows cartoons of the mechanisms by which voltage-dependent channels open and close as a function of membrane potential. Channels are depicted for two different types of conductances, termed persistent (figure 5.8A) and transient (figure 5.8B). We begin by discussing persistent conductances. Figure 5.8A shows a swinging gate attached to a voltage

channel gate   sensor that can open or close the pore of the channel. In reality, channel gating mechanisms involve complex changes in the conformational structure of the channel, but the simple swinging gate picture is sufficient if we are interested only in the current-carrying capacity of the channel. A channel that acts as if it had a single type of gate (although, as we will see, this is actually modeled as a number of identical subgates), like the channel in figure 5.8A, produces what is called a persistent or noninactivating

activation   conductance. Opening of the gate is called activation of the conductance,
deactivation   and gate closing is called deactivation. For this type of channel, the probability that the gate is open, P_K, increases when the neuron is depolarized and decreases when it is hyperpolarized. The delayed-rectifier K$^+$ conductance that is responsible for repolarizing a neuron after an action potential is such a persistent conductance.

The opening of the gate that describes a persistent conductance may involve a number of conformational changes. For example, the delayed-rectifier K$^+$ conductance is constructed from four identical subunits, and it appears that all four must undergo a structural change for the channel to open. In general, if k independent, identical events are required for a channel to open, P_K can be written as

$$P_K = n^k,\tag{5.15}$$

where n is the probability that any one of the k independent gating events has occurred. Here, n, which varies between 0 and 1, is called a gating
activation   or an activation variable, and a description of its voltage and time depen-
variable n   dence amounts to a description of the conductance. We can think of n as the probability of an individual subunit gate being open, and $1 - n$ as the probability that it is closed.

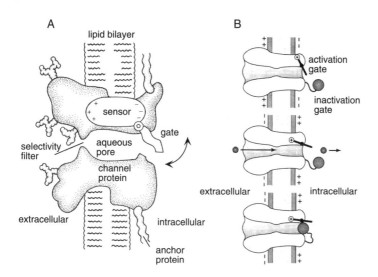

Figure 5.8 Gating of membrane channels. In both figures, the interior of the neuron is to the right of the membrane, and the extracellular medium is to the left. (A) A cartoon of gating of a persistent conductance. A gate is opened and closed by a sensor that responds to the membrane potential. The channel also has a region that selectively allows ions of a particular type to pass through the channel, for example, K$^+$ ions for a potassium channel. (B) A cartoon of the gating of a transient conductance. The activation gate is coupled to a voltage sensor (denoted by a circled +) and acts like the gate in A. A second gate, denoted by the ball, can block that channel once it is open. The top figure shows the channel in a deactivated (and deinactivated) state. The middle panel shows an activated channel, and the bottom panel shows an inactivated channel. Only the middle panel corresponds to an open, ion-conducting state. (A from Hille, 1992; B from Kandel et al., 1991.)

Although using the value of $k = 4$ is consistent with the four-subunit structure of the delayed-rectifier conductance, in practice k is an integer chosen to fit the data, and should be interpreted as a functional definition of a subunit rather than a reflection of a realistic structural model of the channel. Indeed, the structure of the channel was not known at the time that Hodgkin and Huxley chose the form of equation 5.15 and suggested that $k = 4$.

We describe the transition of each subunit gate by a simple kinetic scheme in which the gating transition closed → open occurs at a voltage-dependent rate $\alpha_n(V)$, and the reverse transition, open → closed, occurs at a voltage-dependent rate $\beta_n(V)$. The probability that a subunit gate opens over a short interval of time is proportional to the probability of finding the gate closed, $1 - n$, multiplied by the opening rate $\alpha_n(V)$. Likewise, the probability that a subunit gate closes during a short time interval is proportional to the probability of finding the gate open, n, multiplied by the closing rate $\beta_n(V)$. The rate at which the open probability for a subunit gate changes is given by the difference of these two terms,

channel kinetics

opening rate $\alpha_n(V)$

closing rate $\beta_n(V)$

$$\frac{dn}{dt} = \alpha_n(V)(1 - n) - \beta_n(V)n. \tag{5.16}$$

The first term in equation 5.16 describes the opening process, and the second term the closing process (hence the minus sign) that lowers the probability of being in the configuration with an open subunit gate. Equation 5.16 can be written in another useful form by dividing through by

gating equation $\alpha_n(V) + \beta_n(V)$,

$$\tau_n(V)\frac{dn}{dt} = n_\infty(V) - n, \tag{5.17}$$

$\tau_n(V)$ where

$$\tau_n(V) = \frac{1}{\alpha_n(V) + \beta_n(V)} \tag{5.18}$$

$n_\infty(V)$ and

$$n_\infty(V) = \frac{\alpha_n(V)}{\alpha_n(V) + \beta_n(V)}. \tag{5.19}$$

Equation 5.17 indicates that for a fixed voltage V, n approaches the limiting value $n_\infty(V)$ exponentially with time constant $\tau_n(V)$.

The key elements in the equation that determines n are the opening and closing rate functions $\alpha_n(V)$ and $\beta_n(V)$. These are obtained by fitting experimental data. It is useful to discuss the form that we expect these rate functions to take on the basis of thermodynamic arguments. The state transitions described by α_n, for example, are likely to be rate-limited by barriers requiring thermal energy. These transitions involve the movement of charged components of the gate across part of the membrane, so the height of these energy barriers should be affected by the membrane potential. The transition requires the movement of an effective charge, which we denote by qB_α, through the potential V. This requires an energy $qB_\alpha V$. The constant B_α reflects both the amount of charge being moved and the distance over which it travels. The probability that thermal fluctuations will provide enough energy to surmount this energy barrier is proportional to the Boltzmann factor, $\exp(-qB_\alpha V/k_B T)$. Based on this argument, we expect α_n to be of the form

$$\alpha_n(V) = A_\alpha \exp(-qB_\alpha V/k_B T) = A_\alpha \exp(-B_\alpha V/V_T) \tag{5.20}$$

for some constant A_α. The closing rate β_n should be expressed similarly, except with different constants A_β and B_β. From equation 5.19, we then find that $n_\infty(V)$ is expected to be a sigmoidal function

$$n_\infty(V) = \frac{1}{1 + (A_\beta/A_\alpha)\exp((B_\alpha - B_\beta)V/V_T)}. \tag{5.21}$$

For a voltage-activated conductance, depolarization causes n to grow toward 1, and hyperpolarization causes n to shrink toward 0. Thus, we expect that the opening rate, α_n, should be an increasing function of V (and thus $B_\alpha < 0$), and β_n should be a decreasing function of V (and thus $B_\beta > 0$). Examples of the functions we have discussed are plotted in figure 5.9.

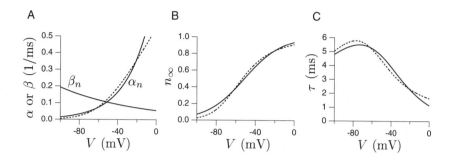

Figure 5.9 Generic voltage-dependent gating functions compared with Hodgkin-Huxley results for the delayed-rectifier K$^+$ conductance. (A) The exponential α_n and β_n functions expected from thermodynamic arguments are indicated by the solid curves. Parameter values used were $A_\alpha = 1.22$ ms^{-1}, $A_\beta = 0.056$ ms^{-1}, $B_\alpha/V_T = -0.04$/mV, and $B_\beta/V_T = 0.0125$/mV. The fit of Hodgkin and Huxley for β_n is identical to the solid curve shown. The Hodgkin-Huxley fit for α_n is the dashed curve. (B) The corresponding function $n_\infty(V)$ of equation 5.21 (solid curve). The dashed curve is obtained using the α_n and β_n functions of the Hodgkin-Huxley fit (equation 5.22). (C) The corresponding function $\tau_n(V)$, obtained from equation 5.18 (solid curve). Again the dashed curve is the result of using the Hodgkin-Huxley rate functions.

While thermodynamic arguments support the forms we have presented, they rely on simplistic assumptions. Not surprisingly, the resulting functional forms do not always fit the data, and various alternatives are often employed. The data upon which these fits are based are typically obtained using a technique called voltage clamping. In this technique, an amplifier *voltage clamping* is configured to inject the appropriate amount of electrode current to hold the membrane potential at a constant value. By current conservation, this current is equal to the membrane current of the cell. Hodgkin and Huxley fitted the rate functions for the delayed-rectifier K$^+$ conductance they studied, using the equations

$$\alpha_n = \frac{.01(V+55)}{1 - \exp(-.1(V+55))} \quad \text{and} \quad \beta_n = 0.125 \exp(-0.0125(V+65)),$$

$$(5.22)$$

where V is expressed in mV, and α_n and β_n are both expressed in units of 1/ms. The fit for β_n is exactly the exponential form we have discussed, with $A_\beta = 0.125 \exp(-0.0125 \cdot 65)$ ms^{-1} and $B_\beta/V_T = 0.0125$ mV^{-1}, but the fit for α_n uses a different functional form. The dashed curves in figure 5.9 plot the formulas of equation 5.22.

Transient Conductances

Some channels only open transiently when the membrane potential is depolarized because they are gated by two processes with opposite voltage dependences. Figure 5.8B is a schematic of a channel that is controlled by two gates and generates a transient conductance. The swinging gate in figure 5.8B behaves exactly like the gate in figure 5.8A. The probability that it

*activation
variable m*

is open is written as m^k, where m is an activation variable similar to n and k is an integer. Hodgkin and Huxley used $k = 3$ for their model of the fast Na^+ conductance. The ball in figure 5.8B acts as the second gate. The probability that the ball does not block the channel pore is written as h and is called the inactivation variable. The activation and inactivation variables m and h are distinguished by having opposite voltage dependences. Depolarization causes m to increase and h to decrease, and hyperpolarization decreases m while increasing h.

*inactivation
variable h*

For the channel in figure 5.8B to conduct, both gates must be open, and assuming the two gates act independently, this has probability

$$P_{Na} = m^k h. \tag{5.23}$$

This is the general form used to describe the open probability for a transient conductance. We could raise the h factor in this expression to an arbitrary power, as we did for m, but we omit this complication to streamline the discussion. The activation m and inactivation h, like all gating variables, vary between 0 and 1. They are described by equations identical to 5.16, except that the rate functions α_n and β_n are replaced by either α_m and β_m, or α_h and β_h. These rate functions were fitted by Hodgkin and Huxley using the equations (in units of $1/ms$ with V in mV)

$$\alpha_m = \frac{.1(V+40)}{1-\exp(-.1(V+40))} \qquad \beta_m = 4\exp(-.0556(V+65))$$

$$\alpha_h = .07\exp(-.05(V+65)) \qquad \beta_h = 1/(1+\exp(-.1(V+35))). \tag{5.24}$$

Functions $m_\infty(V)$ and $h_\infty(V)$ describing the steady-state activation and inactivation levels, and voltage-dependent time constants for m and h can be defined as in equations 5.19 and 5.18. These are plotted in figure 5.10. For comparison, $n_\infty(V)$ and $\tau_n(V)$ for the K^+ conductance are also plotted. Note that $h_\infty(V)$, because it corresponds to an inactivation variable, is flipped relative to $m_\infty(V)$ and $n_\infty(V)$, so that it approaches 1 at hyperpolarized voltages and 0 at depolarized voltages.

deinactivation

The presence of two factors in equation (5.23) gives a transient conductance some interesting properties. To turn on a transient conductance maximally, it may first be necessary to hyperpolarize the neuron below its resting potential and then to depolarize it. Hyperpolarization raises the value of the inactivation h, a process called deinactivation. The second step, depolarization, increases the value of m, which is activation. Only when m and h are both nonzero is the conductance turned on. Note that the conductance can be reduced in magnitude by either decreasing m or h. Decreasing h is called inactivation to distinguish it from decreasing m, which is deactivation.

inactivation

Hyperpolarization-Activated Conductances

Persistent currents act as if they are controlled by an activation gate, while transient currents act as if they have both an activation and an inactivation

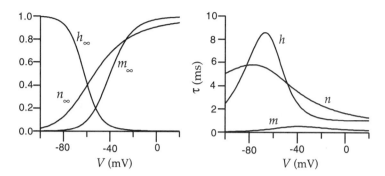

Figure 5.10 The voltage-dependent functions of the Hodgkin-Huxley model. The left panel shows $m_\infty(V)$, $h_\infty(V)$, and $n_\infty(V)$, the steady-state levels of activation and inactivation of the Na^+ conductance, and activation of the K^+ conductance. The right panel shows the voltage-dependent time constants that control the rates at which these steady-state levels are approached for the three gating variables.

gate. Another class of conductances, the hyperpolarization-activated conductances, behave as if they are controlled solely by an inactivation gate. They are thus persistent conductances, but they open when the neuron is hyperpolarized rather than depolarized. The opening probability for such channels is written solely in terms of an inactivation variable similar to h. Strictly speaking, these conductances deinactivate when they turn on and inactivate when they turn off. However, most people cannot bring themselves to say "deinactivate" all the time, so they say instead that these conductances are activated by hyperpolarization.

5.6 The Hodgkin-Huxley Model

The Hodgkin-Huxley model for the generation of the action potential, in its single-compartment form, is constructed by writing the membrane current in equation 5.6 as the sum of a leakage current, a delayed-rectified K^+ current, and a transient Na^+ current,

$$i_m = \overline{g}_L (V - E_L) + \overline{g}_K n^4 (V - E_K) + \overline{g}_{Na} m^3 h (V - E_{Na}). \tag{5.25}$$

The maximal conductances and reversal potentials used in the model are $\overline{g}_L = 0.003$ mS/mm², $\overline{g}_K = 0.36$ mS/mm², $\overline{g}_{Na} = 1.2$ mS/mm², $E_L = -54.387$ mV, $E_K = -77$ mV and $E_{Na} = 50$ mV. The full model consists of equation 5.6 with equation 5.25 for the membrane current, and equations of the form 5.17 for the gating variables n, m, and h. These equations can be integrated numerically, using the methods described in appendices A and B.

The temporal evolution of the dynamic variables of the Hodgkin-Huxley model during a single action potential is shown in figure 5.11. The initial rise of the membrane potential, prior to the action potential, seen in the upper panel of figure 5.11, is due to the injection of a positive electrode current into the model starting at $t = 5$ ms. When this current drives the

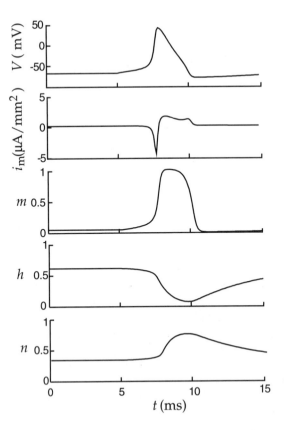

Figure 5.11 The dynamics of V, m, h, and n in the Hodgkin-Huxley model during the firing of an action potential. The upper-most trace is the membrane potential, the second trace is the membrane current produced by the sum of the Hodgkin-Huxley K$^+$ and Na$^+$ conductances, and subsequent traces show the temporal evolution of m, h, and n. Current injection was initiated at $t = 5$ ms.

membrane potential up to about -50 mV, the m variable that describes activation of the Na$^+$ conductance suddenly jumps from nearly 0 to a value near 1. Initially, the h variable, expressing the degree of inactivation of the Na$^+$ conductance, is around 0.6. Thus, for a brief period both m and h are significantly different from 0. This causes a large influx of Na$^+$ ions, producing the sharp downward spike of inward current shown in the second trace from the top. The inward current pulse causes the membrane potential to rise rapidly to around 50 mV (near the Na$^+$ equilibrium potential). The rapid increase in both V and m is due to a positive feedback effect. Depolarization of the membrane potential causes m to increase, and the resulting activation of the Na$^+$ conductance makes V increase. The rise in the membrane potential causes the Na$^+$ conductance to inactivate by driving h toward 0. This shuts off the Na$^+$ current. In addition, the rise in V activates the K$^+$ conductance by driving n toward 1. This increases the K$^+$ current, which drives the membrane potential back down to negative values. The final recovery involves the readjustment of m, h, and n to their initial values.

The Hodgkin-Huxley model can also be used to study propagation of an action potential down an axon, but for this purpose a multi-compartment model must be constructed. Methods for building such a model, and results from it, are described in chapter 6.

5.7   Modeling Channels

In previous sections, we described the Hodgkin-Huxley formalism for describing voltage-dependent conductances arising from a large number of channels. With the advent of single-channel studies, microscopic descriptions of the transitions between the conformational states of channel molecules have been developed. Because these models describe complex molecules, they typically involve many states and transitions. Here, we discuss simple versions of these models that capture the spirit of single-channel modeling without getting mired in the details.

Models of single channels are based on state diagrams that indicate the possible conformational states that the channel can assume. Typically, one of the states in the diagram is designated as open and ion-conducting, while the other states are nonconducting. The current conducted by the channel is written as $\overline{g}P(V - E)$, where E is the reversal potential, $\overline{g}$ is the single-channel open conductance, and P is 1 whenever the open state is occupied, and 0 otherwise. Channel models can be instantiated directly from state diagrams simply by keeping track of the state of the channel and allowing stochastic changes of state to occur at appropriate transition rates. If the model is updated in short time steps of duration Δt, the probability that the channel makes a given transition during an update interval is the transition rate times Δt.

Figure 5.12 shows the state diagram and simulation results for a model of a single delayed-rectifier K^+ channel that is closely related to the Hodgkin-Huxley description of the macroscopic delayed-rectifier conductance. The factors α_n and β_n in the transition rates shown in the state diagram of figure 5.12 are the voltage-dependent rate functions of the Hodgkin-Huxley model. The model uses the same four subunit structure assumed in the Hodgkin-Huxley model. We can think of state 1 in this diagram as a state in which all the subunit gates are closed. States 2, 3, 4, and 5 have 1, 2, 3, and 4 open subunit gates, respectively. State 5 is the sole open state. The factors of 1, 2, 3, and 4 in the transition rates in figure 5.12 correspond to the number of subunit gates that can make a given transition. For example, the transition rate from state 1 to state 2 is four times faster than the rate from state 4 to state 5. This is because any one of the four subunit gates can open to get from state 1 to state 2, but the transition from state 4 to state 5 requires the single remaining closed subunit gate to open.

The lower panels in figure 5.12 show simulations of this model involving 1, 10, and 100 channels. The sum of currents from all of these channels is compared with the current predicted by the Hodgkin-Huxley model

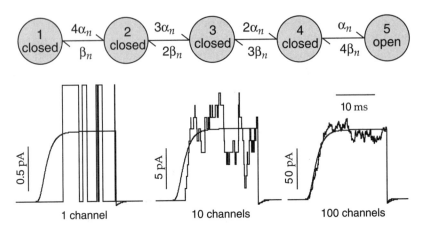

Figure 5.12 A model of the delayed-rectifier K^+ channel. The upper diagram shows the states and transition rates of the model. In the simulations shown in the lower panels, the membrane potential was initially held at -100 mV, then held at 10 mV for 20 ms, and finally returned to a holding potential of -100 mV. The smooth curves in these panels show the membrane current predicted by the Hodgkin-Huxley model in this situation. The left panel shows a simulation of a single channel that opened several times during the depolarization. The middle panel shows the total current from 10 simulated channels, and the right panel corresponds to 100 channels. As the number of channels increases, the Hodgkin-Huxley model provides a more accurate description of the current.

(scaled by the appropriate maximal conductance). For each channel, the pattern of opening and closing is random, but when enough channels are summed, the total current matches that of the Hodgkin-Huxley model quite well.

To see how the channel model in figure 5.12 reproduces the results of the Hodgkin-Huxley model when the currents from many channels are summed, we must consider a probabilistic description of the channel model. We denote the probability that a channel is in state a of figure 5.12 by p_a, with $a = 1, 2, \ldots, 5$. Dynamic equations for these probabilities are easily derived by setting the rate of change for a given p_a equal to the probability per unit time of entry into state a from other states minus the rate for leaving state a. The entry probability per unit time is the product of the appropriate transition rate times the probability that the state making the transition is occupied. The probability per unit time for leaving is p_a times the sum of all the rates for possible transitions out of the state. Following this reasoning, the equations for the state probabilities are (using the notation $\dot{p} = dp/dt$)

$$\dot{p}_1 = \beta_n p_2 - 4\alpha_n p_1$$
$$\dot{p}_2 = 4\alpha_n p_1 + 2\beta_n p_3 - (\beta_n + 3\alpha_n)p_2$$
$$\dot{p}_3 = 3\alpha_n p_2 + 3\beta_n p_4 - (2\beta_n + 2\alpha_n)p_3$$
$$\dot{p}_4 = 2\alpha_n p_3 + 4\beta_n p_5 - (3\beta_n + \alpha_n)p_4$$
$$\dot{p}_5 = \alpha_n p_4 - 4\beta_n p_5 .$$

$$(5.26)$$

A solution for these equations can be constructed if we recall that, in the

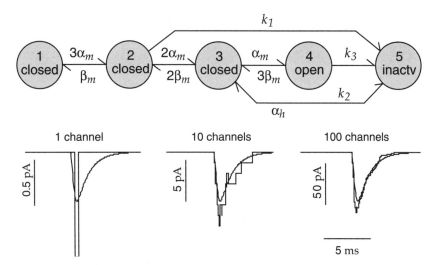

Figure 5.13 A model of the fast Na^+ channel. The upper diagram shows the states and transitions rates of the model. The values $k_1 = 0.24$/ms, $k_2 = 0.4$/ms, and $k_3 = 1.5$/ms were used in the simulations shown in the lower panels. For these simulations, the membrane potential was initially held at -100 mV, then held at 10 mV for 20 ms, and finally returned to a holding potential of -100 mV. The smooth curves in these panels show the current predicted by the Hodgkin-Huxley model in this situation. The left panel shows a simulation of a single channel that opened once during the depolarization. The middle panel shows the total current from 10 simulated channels, and the right panel corresponds to 100 channels. As the number of channels increases, the Hodgkin-Huxley model provides a fairly accurate description of the current, but it is not identical to the channel model in this case.

Hodgkin-Huxley model, n is the probability of a subunit gate being in the open state and $1 - n$ is the probability of it being closed. If we use that same notation here, state 1 has four closed subunit gates, and thus $p_1 = (1 - n)^4$. State 5, the open state, has four open subunit gates, so $p_5 = n^4 = P$. State 2 has one open subunit gate, which can be any one of the four subunit gates, and three closed states, making $p_2 = 4n(1 - n)^3$. Similar arguments yield $p_3 = 6n^2(1 - n)^2$ and $p_4 = 4n^3(1 - n)$. These expressions generate a solution to the above equations, provided that n satisfies equation 5.16, as the reader can verify.

In the Hodgkin-Huxley model of the Na^+ conductance, the activation and inactivation processes are assumed to act independently. The schematic in figure 5.8B, which cartoons the mechanism believed to be responsible for inactivation, suggests that this assumption is incorrect. The ball that inactivates the channel is located inside the cell membrane, where it cannot be affected directly by the potential across the membrane. Furthermore, in this scheme the ball cannot occupy the channel pore until the activation gate has opened, making the two processes interdependent.

The state diagram in figure 5.13 reflects this by having a state-dependent, voltage-independent inactivation mechanism. This diagram is a simplified version of an Na^+ channel model due to Patlak (1991). The sequence of transitions that lead to channel opening through states 1, 2, 3, and 4 is

state-dependent inactivation

identical to that of the Hodgkin-Huxley model, with transition rates determined by the Hodgkin-Huxley functions $\alpha_m(V)$ and $\beta_m(V)$ and appropriate combinatoric factors. State 4 is the open state. The transition to the inactivated state 5, however, is quite different from the inactivation process in the Hodgkin-Huxley model. Inactivation transitions to state 5 can occur only from states 2, 3, and 4, and the corresponding transition rates k_1, k_2, and k_3 are constants, independent of voltage. The deinactivation process occurs at the Hodgkin-Huxley rate $\alpha_h(V)$ from state 5 to state 3.

Figure 5.13 shows simulations of this Na^+ channel model. In contrast to the K^+ channel model shown in figure 5.12, this model does not reproduce exactly the results of the Hodgkin-Huxley model when large numbers of channels are summed. Nevertheless, the two models agree quite well, as seen in the lower right panel of figure 5.13. The agreement, despite the different mechanisms of inactivation, is due to the speed of the activation process for the Na^+ conductance. The inactivation rate function $\beta_h(V)$ in the Hodgkin-Huxley model has a sigmoidal form similar to the asymptotic activation function $m_\infty(V)$ (see equation 5.24). This is indicative of the actual dependence of inactivation on m and not on V. However, the activation variable m of the Hodgkin-Huxley model reaches its voltage-dependent asymptotic value $m_\infty(V)$ so rapidly that it is difficult to distinguish inactivation processes that depend on m from those that depend on V. Differences between the two models are apparent only during a submillisecond time period while the conductance is activating. Experiments that can resolve this time scale support the channel model over the original Hodgkin-Huxley description.

5.8   Synaptic Conductances

Synaptic transmission at a spike-mediated chemical synapse begins when an action potential invades the presynaptic terminal and activates voltage-dependent Ca^{2+} channels, leading to a rise in the concentration of Ca^{2+} within the terminal. This causes vesicles containing transmitter molecules to fuse with the cell membrane and release their contents into the synaptic cleft between the pre- and postsynaptic sides of the synapse. The transmitter molecules then diffuse across the cleft and bind to receptors on the postsynaptic neuron. Binding of transmitter molecules leads to the opening of ion channels that modify the conductance of the postsynaptic neuron, completing the transmission of the signal from one neuron to the other. Postsynaptic ion channels can be activated directly by binding to the transmitter, or indirectly when the transmitter binds to a distinct receptor that affects ion channels through an intracellular second-messenger signaling pathway.

As with a voltage-dependent conductance, a synaptic conductance can be written as the product of a maximal conductance and an open channel probability, $g_s = \bar{g}_s P$. The open probability for a synaptic conductance can be expressed as a product of two terms that reflect processes occurring on the pre- and postsynaptic sides of the synapse, $P = P_s P_{rel}$. The factor P_s is the probability that a postsynaptic channel opens, given that the transmitter was released by the presynaptic terminal. Because there are typically many postsynaptic channels, this can also be taken as the fraction of channels opened by the transmitter.

synaptic open probability P_s

P_{rel} is related to the probability that transmitter is released by the presynaptic terminal following the arrival of an action potential. This reflects the fact that transmitter release is a stochastic process. Release of transmitter at a presynaptic terminal does not necessarily occur every time an action potential arrives and, conversely, spontaneous release can occur even in the absence of the depolarization due to an action potential. The interpretation of P_{rel} is a bit subtle because a synaptic connection between neurons may involve multiple anatomical synapses, and each of these may have multiple independent transmitter release sites. The factor P_{rel}, in our discussion, is the average of the release probabilities at each release site. If there are many release sites, the total amount of transmitter released by all the sites is proportional to P_{rel}. If there is a single release site, P_{rel} is the probability that it releases transmitter. We will restrict our discussion to these two interpretations of P_{rel}. For a modest number of release sites with widely varying release probabilities, the current we discuss describes only an average over multiple trials.

transmitter release probability P_{rel}

Synapses can exert their effects on the soma, dendrites, axon spike-initiation zone, or presynaptic terminals of their postsynaptic targets. There are two broad classes of synaptic conductances that are distinguished by whether the transmitter binds to the synaptic channel and activates it directly, or the transmitter binds to a distinct receptor that activates the conductance indirectly through an intracellular signaling pathway. The first class is called ionotropic and the second, metabotropic. Ionotropic conductances activate and deactivate more rapidly than metabotropic conductances. Metabotropic receptors can, in addition to opening channels, cause long-lasting changes inside a neuron. They typically operate through pathways that involve G-protein-mediated receptors and various intracellular signaling molecules known as second messengers. Many neuromodulators, including serotonin, dopamine, norepinephrine, and acetylcholine, act through metabotropic receptors. These have a wide variety of important effects on the functioning of the nervous system.

ionotropic receptor

metabotropic receptor

Glutamate and GABA (γ-aminobutyric acid) are the major excitatory and inhibitory transmitters in the brain. Both act ionotropically and metabotropically. The principal ionotropic receptor types for glutamate are called AMPA and NMDA. Both AMPA and NMDA receptors produce mixed-cation conductances with reversal potentials around 0 mV. The AMPA current activates and deactivates rapidly. The NMDA receptor is somewhat slower to activate and deactivates considerably more slowly. In addition, NMDA receptors have an unusual voltage dependence that we discuss in a later section, and are more permeable to Ca^{2+} than AMPA receptors.

glutamate, GABA

AMPA, NMDA

GABA$_A$, GABA$_B$ GABA activates two important inhibitory synaptic conductances in the brain. GABA$_A$ receptors produce a relatively fast ionotropic Cl$^-$ conductance. GABA$_B$ receptors are metabotropic, and act to produce a slower and longer-lasting K$^+$ conductance.

gap junctions   In addition to chemical synapses, neurons can be coupled through electrical synapses (gap junctions) that produce a synaptic current proportional to the difference between the pre- and postsynaptic membrane potentials. Some gap junctions rectify so that positive and negative current flows are not equal for potential differences of the same magnitude.

The Postsynaptic Conductance

In a simple model of a directly activated receptor channel, the transmitter interacts with the channel through a binding reaction in which k transmitter molecules bind to a closed receptor and open it. In the reverse reaction, the transmitter molecules unbind from the receptor and it closes. These processes are analogous to the opening and closing involved in the gating of a voltage-dependent channel, and the same type of equation is used to describe how the open probability P_s changes with time,

$$\frac{dP_s}{dt} = \alpha_s(1 - P_s) - \beta_s P_s . \tag{5.27}$$

Here, β_s determines the closing rate of the channel and is usually assumed to be a constant. The opening rate, α_s, on the other hand, depends on the concentration of transmitter available for binding to the receptor. If the concentration of transmitter at the site of the synaptic channel is [transmitter], the probability of finding k transmitter molecules within binding range of the channel is proportional to [transmitter]^k, and α_s is some constant of proportionality times this factor.

When an action potential invades the presynaptic terminal, the transmitter concentration rises and α_s grows rapidly, causing P_s to increase. Following the release of transmitter, diffusion out of the cleft, enzyme-mediated degradation, and presynaptic uptake mechanisms can all contribute to a rapid reduction of the transmitter concentration. This sets α_s to 0, and P_s follows suit by decaying exponentially with a time constant $1/\beta_s$. Typically, the time constant for channel closing is considerably larger than the opening time.

As a simple model of transmitter release, we assume that the transmitter concentration in the synaptic cleft rises extremely rapidly after vesicle release, remains at a high value for a period of duration T, and then falls rapidly to 0. Thus, the transmitter concentration is modeled as a square pulse. While the transmitter concentration is nonzero, α_s takes a constant value much greater than β_s, otherwise $\alpha_s = 0$. Suppose that vesicle release occurs at time $t = 0$ and that the synaptic channel open probability takes the value $P_s(0)$ at this time. While the transmitter concentration in the cleft

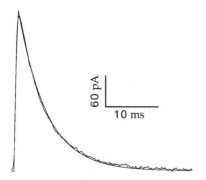

Figure 5.14 A fit of the model discussed in the text to the average EPSC (excitatory postsynaptic current) recorded from mossy fiber input to a CA3 pyramidal cell in a hippocampal slice preparation. The smooth line is the theoretical curve and the wiggly line is the result of averaging recordings from a number of trials. (Adapted from Destexhe et al., 1994.)

is nonzero, α_s is so much larger than β_s that we can ignore the term involving β_s in equation 5.27. Integrating equation 5.27 under this assumption, we find that

$$P_s(t) = 1 + (P_s(0) - 1)\exp(-\alpha_s t) \quad \text{for} \quad 0 \le t \le T. \qquad (5.28)$$

The open probability takes its maximum value at time $t = T$ and then, for $t \ge T$, decays exponentially at a rate determined by the constant β_s,

$$P_s(t) = P_s(T)\exp(-\beta_s(t - T)) \quad \text{for} \quad t \ge T. \qquad (5.29)$$

If $P_s(0) = 0$, as it will if there is no synaptic release immediately before the release at $t = 0$, equation 5.28 simplifies to $P_s(t) = 1 - \exp(-\alpha_s t)$ for $0 \le t \le T$, and this reaches a maximum value $P_{max} = P_s(T) = 1 - \exp(-\alpha_s T)$. In terms of this parameter, a simple manipulation of equation 5.28 shows that we can write, in the general case,

$$P_s(T) = P_s(0) + P_{max}(1 - P_s(0)). \qquad (5.30)$$

Figure 5.14 shows a fit to a recorded postsynaptic current using this formalism. In this case, β_s was set to 0.19 ms^{-1}. The transmitter concentration was modeled as a square pulse of duration $T = 1$ ms during which $\alpha_s = 0.93$ ms^{-1}. Inverting these values, we find that the time constant determining the rapid rise seen in figure 5.14A is 0.9 ms, while the fall of the current is an exponential with a time constant of 5.26 ms.

For a fast synapse like the one shown in figure 5.14, the rise of the conductance following a presynaptic action potential is so rapid that it can be approximated as instantaneous. In this case, the synaptic conductance due to a single presynaptic action potential occurring at $t = 0$ is often written as an exponential, $P_s = P_{max}\exp(-t/\tau_s)$ (see the AMPA trace in figure 5.15A), where from equation 5.29, $\tau_s = 1/\beta_s$. The synaptic conductance due to a

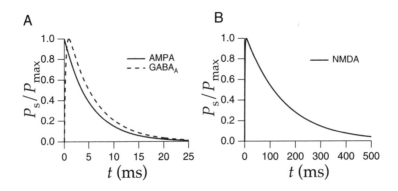

Figure 5.15 Time-dependent open probabilities fitted to match AMPA, GABA$_A$, and NMDA synaptic conductances. (A) The AMPA curve is a single exponential described by equation 5.31 with $\tau_s = 5.26$ ms. The GABA$_A$ curve is a difference of exponentials with $\tau_1 = 5.6$ ms and $\tau_{\text{rise}} = 0.3$ ms. (B) The NMDA curve is the differences of two exponentials with $\tau_1 = 152$ ms and $\tau_{\text{rise}} = 1.5$ ms. (Parameters are from Destexhe et al., 1994.)

sequence of action potentials at arbitrary times can be modeled by allowing P_s to decay exponentially to 0 according to the equation

$$\tau_s \frac{dP_s}{dt} = -P_s \,, \tag{5.31}$$

and, on the basis of the equation 5.30, making the replacement

$$P_s \rightarrow P_s + P_{\max}(1 - P_s) \tag{5.32}$$

immediately after each presynaptic action potential.

Equations 5.28 and 5.29 can also be used to model synapses with slower rise times, but other functional forms are often used. One way of describing both the rise and the fall of a synaptic conductance is to express P_s as the difference of two exponentials (see the GABA$_A$ and NMDA traces in figure 5.15). For an isolated presynaptic action potential occurring at $t = 0$, the synaptic conductance is written as

$$P_s = P_{\max} B \left(\exp(-t/\tau_1) - \exp(-t/\tau_2)\right) \,, \tag{5.33}$$

where $\tau_1 > \tau_2$, and B is a normalization factor that assures that the peak value of P_s is equal to $P_{\max}$,

$$B = \left(\left(\frac{\tau_2}{\tau_1}\right)^{\tau_{\text{rise}}/\tau_1} - \left(\frac{\tau_2}{\tau_1}\right)^{\tau_{\text{rise}}/\tau_2}\right)^{-1} . \tag{5.34}$$

The rise time of the synapse is determined by $\tau_{\text{rise}} = \tau_1 \tau_2 / (\tau_1 - \tau_2)$, while the fall time is set by τ_1. This conductance reaches its peak value $\tau_{\text{rise}} \ln(\tau_1/\tau_2)$ after the presynaptic action potential.

Another way of describing a synaptic conductance is to use the expression

$$P_s = \frac{P_{\max} t}{\tau_s} \exp(1 - t/\tau_s) \tag{5.35}$$

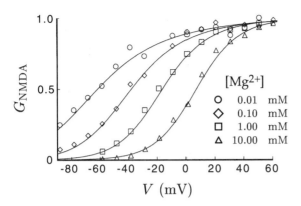

Figure 5.16 Dependence of the NMDA conductance on the membrane potential and extracellular Mg^{2+} concentration. Normal extracellular Mg^{2+} concentrations are in the range of 1 to 2 mM. The solid lines are the factors G_{NMDA} of equation 5.36 for different values of $[Mg^{2+}]$, and the symbols indicate the data points. (Adapted from Jahr and Stevens, 1990.)

for an isolated presynaptic release that occurs at time $t = 0$. This expression, called an alpha function, starts at 0, reaches its peak value at $t = \tau_s$, and then decays with a time constant τ_s.

alpha function

We mentioned earlier in this chapter that NMDA receptor conductance has an additional dependence on the postsynaptic potential not normally seen in other conductances. To incorporate this dependence, the current due to the NMDA receptor can be described using an additional factor that depends on the postsynaptic potential, V. The NMDA current is written as $\bar{g}_{NMDA} G_{NMDA}(V) P(V - E_{NMDA})$. P is the usual open probability factor. The factor $G_{NMDA}(V)$ describes an extra voltage dependence due to the fact that when the postsynaptic neuron is near its resting potential, NMDA receptors are blocked by Mg^{2+} ions. To activate the conductance, the postsynaptic neuron must be depolarized to knock out the blocking ions. Jahr and Stevens (1990) have fitted this dependence by (figure 5.16)

NMDA receptor

$$G_{NMDA} = \left(1 + \frac{[Mg^{2+}]}{3.57 \text{ mM}} \exp(-V/16.13 \text{ mV})\right)^{-1}. \qquad (5.36)$$

NMDA receptors conduct Ca^{2+} ions as well as monovalent cations. Entry of Ca^{2+} ions through NMDA receptors is a critical event for long-term modification of synaptic strength. The fact that the opening of NMDA channels requires both pre- and postsynaptic depolarization means NMDA receptors can act as coincidence detectors of simultaneous pre- and postsynaptic activity. This plays an important role in connection with the Hebb rule for synaptic modification discussed in chapter 8.

coincidence detection

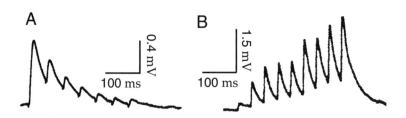

Figure 5.17 Depression and facilitation of excitatory intracortical synapses. (A) Depression of an excitatory synapse between two layer 5 pyramidal cells recorded in a slice of rat somatosensory cortex. Spikes were evoked by current injection into the presynaptic neuron, and postsynaptic potentials were recorded with a second electrode. (B) Facilitation of an excitatory synapse from a pyramidal neuron to an inhibitory interneuron in layer 2/3 of rat somatosensory cortex. (A from Markram and Tsodyks, 1996; B from Markram et al., 1998.)

Release Probability and Short-Term Plasticity

short-term plasticity

long-term plasticity

The probability of transmitter release and the magnitude of the resulting conductance change in the postsynaptic neuron can depend on the history of activity at a synapse. The effects of activity on synaptic conductances are termed short- and long-term. Short-term plasticity refers to a number of phenomena that affect the probability that a presynaptic action potential opens postsynaptic channels and that last anywhere from milliseconds to tens of seconds. The effects of long-term plasticity are extremely persistent, lasting, for example, as long as the preparation being studied can be kept alive. The modeling and implications of long-term plasticity are considered in chapter 8. Here we present a simple way of describing short-term synaptic plasticity as a modification in the release probability for synaptic transmission. Short-term modifications of synaptic transmission can involve other mechanisms than merely changes in the probability of transmission, but for simplicity we absorb all these effects into a modification of the factor P_{rel} introduced previously. Thus, P_{rel} can be interpreted more generally as a presynaptic factor affecting synaptic transmission.

depression
facilitation

Figure 5.17 illustrates two principal types of short-term plasticity, depression and facilitation. Figure 5.17A shows trial-averaged postsynaptic current pulses produced in one cortical pyramidal neuron by evoking a regular series of action potentials in a second pyramidal neuron presynaptic to the first. The pulses dramatically decrease in amplitude upon repeated activation of the synaptic conductance, revealing short-term synaptic depression. Figure 5.17B shows a similar series of averaged postsynaptic current pulses recorded in a cortical inhibitory interneuron when a sequence of action potentials was evoked in a presynaptic pyramidal cell. In this case, the amplitude of the pulses increases, and thus the synapse facilitates. In general, synapses can exhibit facilitation and depression over a variety of time scales, and multiple components of short-term plasticity can be found at the same synapse. To keep the discussion simple, we consider synapses that exhibit either facilitation or depression described by a single time constant.

Facilitation and depression can both be modeled as presynaptic processes that modify the probability of transmitter release. We describe them using a simple nonmechanistic model that has similarities to the model of P_s presented in the previous subsection. For both facilitation and depression, the release probability after a long period of presynaptic silence is $P_{rel} = P_0$. Activity at the synapse causes P_{rel} to increase in the case of facilitation and to decrease for depression. Between presynaptic action potentials, the release probability decays exponentially back to its "resting" value, P_0,

$$\tau_P \frac{dP_{rel}}{dt} = P_0 - P_{rel} \,. \tag{5.37}$$

The parameter τ_P controls the rate at which the release probability decays to P_0.

The models of facilitation and depression differ in how the release probability is changed by presynaptic activity. In the case of facilitation, P_{rel} is augmented by making the replacement $P_{rel} \to P_{rel} + f_F(1 - P_{rel})$ immediately after a presynaptic action potential (as in equation 5.32). The parameter f_F (with $0 \le f_F \le 1$) controls the degree of facilitation, and the factor $(1 - P_{rel})$ prevents the release probability from growing larger than 1. To model depression, the release probability is reduced after a presynaptic action potential by making the replacement $P_{rel} \to f_D P_{rel}$. In this case, the parameter f_D (with $0 \le f_D \le 1$) controls the amount of depression, and the factor P_{rel} prevents the release probability from becoming negative.

We begin by analyzing the effects of facilitation on synaptic transmission for a presynaptic spike train with Poisson statistics. In particular, we compute the average steady-state release probability, denoted by $\langle P_{rel} \rangle$. $\langle P_{rel} \rangle$ is determined by requiring that the facilitation that occurs after each presynaptic action potential is exactly canceled by the average exponential decrement that occurs between presynaptic spikes. Consider two presynaptic action potentials separated by an interval τ, and suppose that the release probability takes its average value $\langle P_{rel} \rangle$ at the time of the first spike. Immediately after this spike, it is augmented to $\langle P_{rel} \rangle + f_F(1 - \langle P_{rel} \rangle)$. By the time of the second spike, this will have decayed to $P_0 + (\langle P_{rel} \rangle + f_F(1 - \langle P_{rel} \rangle) - P_0) \exp(-\tau/\tau_P)$, which is obtained by integrating equation 5.37. The average value of the exponential decay factor in this expression is the integral over all positive τ values of $\exp(-\tau/\tau_P)$ times the probability density for a Poisson spike train with a firing rate r to produce an interspike interval of duration τ, which is $r \exp(-r\tau)$ (see chapter 1). Thus, the average exponential decrement is

$$r \int_0^\infty d\tau \, \exp(-r\tau - \tau/\tau_P) = \frac{r\tau_P}{1 + r\tau_P} \,. \tag{5.38}$$

In order for the release probability to return, on average, to its steady-state value between presynaptic spikes, we must therefore require that

$$\langle P_{rel} \rangle = P_0 + \left(\langle P_{rel} \rangle + f_F(1 - \langle P_{rel} \rangle) - P_0 \right) \frac{r\tau_P}{1 + r\tau_P} \,. \tag{5.39}$$

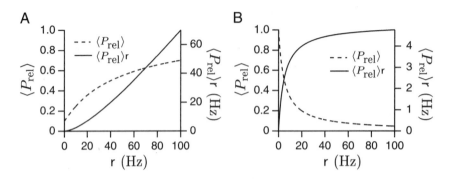

Figure 5.18 The effects of facilitation and depression on synaptic transmission. (A) Release probability and transmission rate for a facilitating synapse as a function of the firing rate of a Poisson presynaptic spike train. The dashed curve shows the rise of the average release probability as the presynaptic rate increases. The solid curve is the average rate of transmission, which is the average release probability times the presynaptic firing rate. The parameters of the model are $P_0 = 0.1$, $f_F = 0.4$, and $\tau_P = 50$ ms. (B) Same as A, but for the case of depression. The parameters of the model are $P_0 = 1$, $f_D = 0.4$, and $\tau_P = 500$ ms.

Solving for $\langle P_{rel} \rangle$ gives

$$\langle P_{rel} \rangle = \frac{P_0 + f_F r \tau_P}{1 + r f_F \tau_P} \,. \qquad (5.40)$$

This equals P_0 at low rates and rises toward the value 1 at high rates (figure 5.18A). As a result, isolated spikes in low-frequency trains are transmitted with lower probability than spikes occurring within high-frequency bursts. The synaptic transmission rate when the presynaptic neuron is firing at rate r is the firing rate times the release probability. This is approximately $P_0 r$ for small rates and approaches r at high rates (figure 5.18A).

The value of $\langle P_{rel} \rangle$ for a Poisson presynaptic spike train can also be computed in the case of depression. The only difference from the above derivation is that following a presynaptic spike, $\langle P_{rel} \rangle$ is decreased to $f_D \langle P_{rel} \rangle$. Thus, the consistency condition 5.39 is replaced by

$$\langle P_{rel} \rangle = P_0 + (f_D \langle P_{rel} \rangle - P_0) \frac{r \tau_P}{1 + r \tau_P} \,, \qquad (5.41)$$

giving

$$\langle P_{rel} \rangle = \frac{P_0}{1 + (1 - f_D) r \tau_P} \,. \qquad (5.42)$$

This equals P_0 at low rates and decreases as $1/r$ at high rates (figure 5.18B), which has some interesting consequences. As noted above, the average rate of successful synaptic transmissions is equal to $\langle P_{rel} \rangle$ times the presynaptic rate r. Because $\langle P_{rel} \rangle$ is proportional to $1/r$ at high rates, the average transmission rate is independent of r in this range. This can be seen by the flattening of the solid curve in figure 5.18B. As a result, synapses

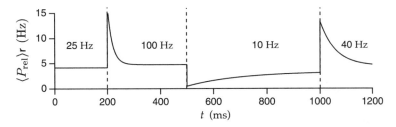

Figure 5.19 The average rate of transmission for a synapse with depression when the presynaptic firing rate changes in a sequence of steps. The firing rates were held constant at the values 25, 100, 10, and 40 Hz, except for abrupt changes at the times indicated by the dashed lines. The parameters of the model are $P_0 = 1$, $f_D = 0.6$, and $\tau_P = 500$ ms.

that depress do not convey information about the values of constant, high presynaptic firing rates to their postsynaptic targets. The presynaptic firing rate at which transmission starts to become independent of r is around $1/((1 - f_D)\tau_P)$.

Figure 5.19 shows the average transmission rate, $\langle P_{rel}\rangle r$, in response to a series of steps in the presynaptic firing rate. Note first that the steady-state transmission rates during the 25, 100, 10, and 40 Hz periods are quite similar. This is a consequence of the $1/r$ dependence of the average release probability, as discussed above. The largest transmission rates in the figure occur during the sharp upward transitions between different presynaptic rates. This illustrates the important point that depressing synapses amplify transient signals relative to steady-state inputs. The transients corresponding the 25 to 100 Hz transition and the 10 to 40 Hz transition are of roughly equal amplitudes, but the transient for the 10 to 40 Hz transition is broader than that for the 25 to 100 Hz transition.

The equality of amplitudes of the two upward transients in figure 5.19 is a consequence of the $1/r$ behavior of $\langle P_{rel}\rangle$. Suppose that the presynaptic firing rate makes a sudden transition from a steady value r to a new value $r + \Delta r$. Before the transition, the average release probability is given by equation 5.42. Immediately after the transition, before the release probability has had time to adjust to the new input rate, the average transmission rate will be this previous value of $\langle P_{rel}\rangle$ times the new rate $r + \Delta r$, which is $P_0(r + \Delta r)/(1 + (1 - f_D)r\tau_P)$. For sufficiently high rates, this is approximately proportional to $(r + \Delta r)/r$. The size of the change in the transmission rate is thus proportional to $\Delta r/r$, which means that depressing synapses not only amplify transient inputs, they transmit them in a scaled manner. The amplitude of the transient transmission rate is proportional to the fractional change, not the absolute change, in the presynaptic firing rate. The two transients seen in figure 5.19 have similar amplitudes because in both cases $\Delta r/r = 3$. The difference in the recovery time for the two upward transients in figure 5.19 is due to the fact that the effective time constant governing the recovery to a new steady-state level r is $\tau_P/(1 + (1 - f_D)\tau_P r)$.

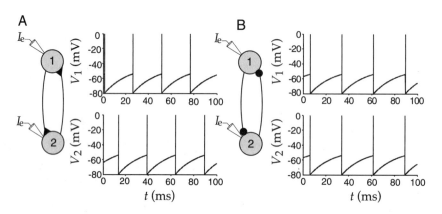

Figure 5.20 Two synaptically coupled integrate-and-fire neurons. (A) Excitatory synapses ($E_s = 0$ mV) produce an alternating, out-of-phase pattern of firing. (B) Inhibitory synapses ($E_s = -80$ mV) produce synchronous firing. Both model neurons have $E_L = -70$ mV, $V_{th} = -54$ mV, $V_{reset} = -80$ mV, $\tau_m = 20$ ms, $r_m \bar{g}_s = 0.05$, $P_{max} = 1$, $R_m I_e = 25$ mV, and $\tau_s = 10$ ms.

5.9   Synapses on Integrate-and-Fire Neurons

Synaptic inputs can be incorporated into an integrate-and-fire model by including synaptic conductances in the membrane current appearing in equation 5.8,

$$\tau_m \frac{dV}{dt} = E_L - V - r_m \bar{g}_s P_s (V - E_s) + R_m I_e. \tag{5.43}$$

For simplicity, we assume that $P_{rel} = 1$ in this example. The synaptic current is multiplied by r_m in equation 5.43 because equation 5.8 was multiplied by this factor. To model synaptic transmission, P_s changes whenever the presynaptic neuron fires an action potential using one of the schemes described previously.

Figures 5.20A and 5.20B show examples of two integrate-and-fire neurons driven by electrode currents and connected by identical excitatory or inhibitory synapses. The synaptic conductances in this example are described by the α function model. This means that the synaptic conductance a time t after the occurrence of a presynaptic action potential is given by equation 5.35. The figure shows a nonintuitive effect. When the synaptic time constant is sufficiently long ($\tau_s = 10$ ms in this exam-

synchronous and ple), excitatory connections produce a state in which the two neurons fire
asynchronous alternately, out of phase with one another, while inhibitory synapses pro-
firing duce synchronous firing. It is normally assumed that excitation produces synchrony. Actually, in some cases inhibitory connections can be more effective than excitatory connections at synchronizing neuronal firing.

Synapses have multiple effects on their postsynaptic targets. In equation 5.43, the term $r_m \bar{g}_s P_s E_s$ acts as a source of current to the neuron, while the term $r_m \overline{g}_s P_s V$ changes the membrane conductance. The effects of the latter term are referred to as shunting, and they can be identified most easily if we divide equation 5.43 by $1 + r_m \overline{g}_s P_s$ to obtain

$$\frac{\tau_m}{1 + r_m \overline{g}_s P_s} \frac{dV}{dt} = -V + \frac{E_L + r_m \overline{g}_s P_s E_s + R_m I_e}{1 + r_m \overline{g}_s P_s}. \tag{5.44}$$

The shunting effects of the synapse are seen in this equation as a decrease in the effective membrane time constant, and a divisive reduction in the impact of the leakage and synaptic reversal potentials and of the electrode current.

The shunting effects seen in equation 5.44 have been proposed as a possible basis for neural computations involving division. However, shunting has a divisive effect only on the membrane potential of an integrate-and-fire neuron; its effect on the firing rate is subtractive. To see this, assume that synaptic input is arriving at a sufficient rate to maintain a relatively constant value of P_s. In this case, shunting amounts to changing the value of the membrane resistance from R_m to $R_m/(1 + r_m \overline{g}_s P_s)$. Recalling equation 5.12 for the firing rate of the integrate-and-fire model, and the fact that $\tau_m = C_m R_m$, we can write the firing rate in a form that reveals its dependence on R_m,

$$r_{isi} \approx \left[\frac{E_L - V_{th}}{C_m R_m (V_{th} - V_{reset})} + \frac{I_e}{C_m (V_{th} - V_{reset})} \right]_+. \tag{5.45}$$

Changing R_m modifies only the constant term in this equation; it has no effect on the dependence of the firing rate on I_e.

Regular and Irregular Firing Modes

Integrate-and-fire models are useful for studying how neurons sum large numbers of synaptic inputs and how networks of neurons interact. One issue that has received considerable attention is the degree of variability in the firing output of integrate-and-fire neurons receiving synaptic input. This work has led to the realization that neurons can respond to multiple synaptic inputs in two different modes of operation depending on the balance that exists between excitatory and inhibitory contributions.

The two modes of operation are illustrated in figure 5.21, which shows membrane potentials of an integrate-and-fire model neuron responding to 1000 excitatory and 200 inhibitory inputs. Each input consists of an independent Poisson spike train driving a synaptic conductance. The upper panels of figure 5.21 show the membrane potential with the action potential generation mechanism of the model turned off, and figures 5.21A and 5.21B illustrate the two different modes of operation. In figure 5.21A, the effect of the excitatory inputs is strong enough, relative to that of the inhibitory inputs, to make the average membrane potential, when action

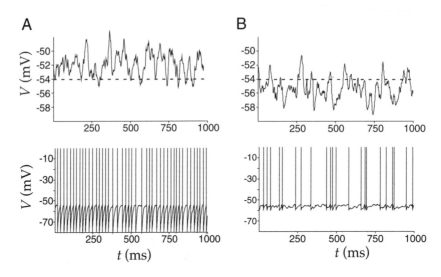

Figure 5.21 The regular and irregular firing modes of an integrate-and-fire model neuron. (A) The regular firing mode. Upper panel: The membrane potential of the model neuron when the spike generation mechanism is turned off. The average membrane potential is above the spiking threshold (dashed line). Lower panel: When the spike generation mechanism is turned on, it produces a regular spiking pattern. (B) The irregular firing mode. Upper panel: The membrane potential of the model neuron when the spike generation mechanism is turned off. The average membrane potential is below the spiking threshold (dashed line). Lower panel: When the spike generation mechanism is turned on, it produces an irregular spiking pattern. In order to keep the firing rates from differing too greatly between these two examples, the value of the reset voltage is higher in B than in A.

potential generation is blocked, more depolarized than the spiking threshold of the model (the dashed line in the figure). When the action potential mechanism is turned on (lower panel of figure 5.21A), this produces a fairly regular pattern of action potentials.

The irregularity of a spike train can be quantified using the coefficient of variation (C_V), the ratio of the standard deviation to the mean of the interspike intervals (see chapter 1). For the Poisson inputs being used in this example, $C_V = 1$, while for the spike train in the lower panel of figure 5.21A, $C_V = 0.3$. Thus, the output spike train is much more regular than the input trains. This is not surprising, because the model neuron effectively averages its many synaptic inputs. In the regular firing mode, the total synaptic input attempts to charge the neuron above the threshold, but every time the potential reaches the threshold, it gets reset and starts charging again. In this mode of operation, the timing of the action potentials is determined primarily by the charging rate of the cell, which is controlled by its membrane time constant.

Figure 5.21B shows the other mode of operation that produces an irregular firing pattern. In the irregular firing mode, the average membrane potential is more hyperpolarized than the threshold for action potential generation (upper panel of figure 5.21B). Action potentials are generated only when there is a fluctuation in the total synaptic input strong enough to make the membrane potential reach the threshold. This produces an irregular spike train, such as that seen in the lower panel of figure 5.21B, which has a C_V value of 0.84.

The high degree of variability seen in the spiking patterns of in vivo recordings of cortical neurons (see chapter 1) suggests that they are better approximated by an integrate-and-fire model operating in an irregular-firing mode. There are advantages to operating in the irregular-firing mode that may compensate for its increased variability. One is that neurons firing in the irregular mode reflect in their outputs the temporal properties of fluctuations in their total synaptic input. In the regular firing mode, the timing of output spikes is only weakly related to the temporal character of the input spike trains. In addition, neurons operating in the irregular firing mode can respond more quickly to changes in presynaptic spiking patterns and firing rates than those operating in the regular firing mode.

5.10   Chapter Summary

In this chapter, we considered the basic electrical properties of neurons, including their intracellular and membrane resistances, capacitances, and active voltage-dependent and synaptic conductances. We introduced the Nernst equation for equilibrium potentials and the formalism of Hodgkin and Huxley for describing persistent, transient, and hyperpolarization-activated conductances. Methods were introduced for modeling stochastic channel opening and stochastic synaptic transmission, including the effects of synaptic facilitation and depression. We discussed a number of ways of describing synaptic conductances following the release of a neurotransmitter. Two models of action potential generation were discussed, the simple integrate-and-fire scheme and the more realistic Hodgkin-Huxley model.

5.11   Appendices

A: Integrating the Membrane Potential

We begin by considering the numerical integration of equation 5.8. It is convenient to rewrite this equation in the form

$$\tau_V \frac{dV}{dt} = V_\infty - V \,, \tag{5.46}$$

where $\tau_V = \tau_m$ and $V_\infty = E_L + R_m I_e$. When the electrode current I_e is independent of time, the solution of this equation is

$$V(t) = V_\infty + (V(t_0) - V_\infty)\exp(-(t - t_0)/\tau_V) \,, \tag{5.47}$$

where t_0 is any time prior to t and $V(t_0)$ is the value of V at time t_0. Equation 5.9 is a special case of this result with $t_0 = 0$.

If I_e depends on time, the solution 5.47 is not valid. An analytic solution can still be written down in this case, but it is not particularly useful except in special cases. Over a small enough time period Δt, we can approximate $I_e(t)$ as constant and use the solution 5.47 to step from a time t to $t + \Delta t$. This requires replacing the variable t_0 in equation 5.47 with t, and t with $t + \Delta t$, so that

$$V(t + \Delta t) = V_\infty + (V(t) - V_\infty)\exp(-\Delta t/\tau_V).\qquad(5.48)$$

This equation provides an updating rule for the numerical integration of equation 5.46. Provided that Δt is sufficiently small, repeated application of the update rule 5.48 provides an accurate way of determining the membrane potential. Furthermore, this method is stable because if Δt is too large, it will only move V toward V_∞ and not, for example, make it grow without bound.

The equation for a general single-compartment conductance-based model, equation 5.6 with 5.5, can be written in the same form as equation 5.46 with

$$V_\infty = \frac{\sum_i g_i E_i + I_e/A}{\sum_i g_i}\qquad(5.49)$$

and

$$\tau_V = \frac{c_m}{\sum_i g_i}.\qquad(5.50)$$

Note that if c_m is in units of nF/mm^2 and the conductances are in the units μS/mm^2, τ_V comes out in ms units. Similarly, if the reversal potentials are given in units of mV, I_e is in nA, and A is in mm^2, V_∞ will be in mV units.

If we take the time interval Δt to be small enough so that the gating variables can be approximated as constant during this period, the membrane potential can again be integrated over one step, using equation 5.48. Of course, the gating variables are not fixed, so once V has been updated by this rule, the gating variables must be updated as well.

B: Integrating the Gating Variables

All the gating variables in a conductance-based model satisfy equations of the same form,

$$\tau_z \frac{dz}{dt} = z_\infty - z,\qquad(5.51)$$

where we use z to denote a generic variable. Note that this equation has the same form as equation 5.46, and it can be integrated in exactly the same way. We assume that Δt is sufficiently small so that V does not change appreciably over this time interval (and similarly [Ca^{2+}] is approximated as constant over this interval if any of the conductances are Ca^{2+}-dependent). Then, τ_z and z_∞, which are functions of V (and possibly $[Ca^{2+}]$) can be treated as constants over this period and z can be updated by a rule identical to 5.48,

$$z(t + \Delta t) = z_\infty + (z(t) - z_\infty) \exp(-\Delta t / \tau_z). \qquad (5.52)$$

An efficient integration scheme for conductance-based models is to alternate using rule (5.48) to update the membrane potential and rule (5.52) to update all the gating variables. It is important to alternate the updating of V with that of the gating variables, rather than doing them all simultaneously, as this keeps the method accurate to second order in Δt. If Ca^{2+}-dependent conductances are included, the intracellular Ca^{2+} concentration should be computed simultaneously with the membrane potential. By alternating the updating, we mean that the membrane potential is computed at times $0, \Delta t, 2\Delta t, \ldots$, while the gating variables are computed at times $\Delta t/2, 3\Delta t/2, 5\Delta t/2, \ldots$. A discussion of the second-order accuracy of this scheme is given in Mascagni and Sherman (1998).

5.12 Annotated Bibliography

Jack et al. (1975), **Tuckwell (1988)**, **Johnston & Wu (1995)**, **Koch & Segev (1998)**, and **Koch (1998)** cover much of the material in this chapter and in chapter 6. **Hille (1992)** provides a comprehensive treatment of ion channels. **Hodgkin & Huxley (1952)** presents the classic biophysical model of the action potential, and **Sakmann & Neher (1983)** describes patch clamp recording techniques allowing single channels to be studied electrophysiologically.

The integrate-and-fire model was introduced by Lapicque (1907). **Destexhe et al. (1994)** describes kinetic models of both ion channels and short-term postsynaptic effects at synapses. Marom & Abbott (1994) shows how the Na^+ channel model of Patlak (1991) can be reconciled with typical macroscopic conductance models. For a review of the spike-response model, the integrated version of the integrate-and-fire model, see **Gerstner (1998)**. Wang (1994) analyzes a spike-rate adaptation model similar to the one we presented, and Stevens & Zador (1998) introduces an integrate-and-fire model with time-dependent parameters.

The dynamic aspects of synaptic transmission are reviewed in **Magleby (1987)** and **Zucker (1989)**. Our presentation followed Abbott et al. (1997), Varela et al. (1997), and Tsodyks & Markram (1997). For additional implications of short-term synaptic plasticity for cortical processing, see Lisman (1997) and Chance et al. (1998). Wang & Rinzel (1992) notes that inhibitory synapses can synchronize coupled cells, and in our discussion we followed the treatment in van Vreeswijk et al. (1994). Our analysis of the regular and irregular firing mode regimes of integrate-and-fire cells was based on Troyer & Miller (1997). Numerical methods for integrating the equations of neuron models are discussed in **Mascagni & Sherman (1998)**.

6 Model Neurons II: Conductances and Morphology

6.1 Levels of Neuron Modeling

In modeling neurons, we must deal with two types of complexity: the intricate interplay of active conductances that makes neuronal dynamics so rich and interesting, and the elaborate morphology that allows neurons to receive and integrate inputs from so many other neurons. The first part of this chapter extends the material presented in chapter 5 by examining single-compartment models with a wider variety of voltage-dependent conductances, and hence a wider range of dynamic behaviors, than the Hodgkin-Huxley model. In the second part of the chapter, we introduce methods used to study the effects of morphology on the electrical characteristics of neurons. An analytic approach known as cable theory is presented first, followed by a discussion of multi-compartment models that permit numerical simulation of complex neuronal structures.

Model neurons range from greatly simplified caricatures to highly detailed descriptions involving thousands of differential equations. Choosing the most appropriate level of modeling for a given research problem requires a careful assessment of the experimental information available and a clear understanding of the research goals. Oversimplified models can, of course, give misleading results, but excessively detailed models can obscure interesting results beneath inessential and unconstrained complexity.

6.2 Conductance-Based Models

The electrical properties of neurons arise from membrane conductances with a wide variety of properties. The basic formalism developed by Hodgkin and Huxley to describe the Na^+ and K^+ conductances responsible for generating action potentials (discussed in chapter 5) is also used to represent most of the additional conductances encountered in neuron modeling. Models that treat these aspects of ionic conductances, known as conductance-based models, can reproduce the rich and complex dynamics of real neurons quite accurately. In this chapter, we discuss both single- and multi-compartment conductance-based models, beginning with the single-compartment case.

membrane potential equation

To review from chapter 5, the membrane potential of a single-compartment neuron model, V, is determined by integrating the equation

$$c_m \frac{dV}{dt} = -i_m + \frac{I_e}{A} \, , \tag{6.1}$$

with I_e the electrode current, A the membrane surface area of the cell, and i_m the membrane current. In the following subsections, we present expressions for the membrane current in terms of the reversal potentials, maximal conductance parameters, and gating variables of the different conductances of the models being considered. The gating variables and V comprise the dynamic variables of the model. All the gating variables are determined by equations of the form

gating equations

$$\tau_z(V) \frac{dz}{dt} = z_\infty(V) - z \, , \tag{6.2}$$

where z denotes a generic gating variable. The functions $\tau_z(V)$ and $z_\infty(V)$ are determined from experimental data. For some conductances, these are written in terms of the opening and closing rates $\alpha_z(V)$ and $\beta_z(V)$ (see chapter 5), as

$$\tau_z(V) = \frac{1}{\alpha_z(V) + \beta_z(V)} \quad \text{and} \quad z_\infty(V) = \frac{\alpha_z(V)}{\alpha_z(V) + \beta_z(V)} \, . \tag{6.3}$$

We have written $\tau_z(V)$ and $z_\infty(V)$ as functions of the membrane potential, but for Ca^{2+}-dependent currents they also depend on the internal Ca^{2+} concentration. We call $\alpha_z(V)$, $\beta_z(V)$, $\tau_z(V)$, and $z_\infty(V)$ gating functions. A method for numerically integrating equations 6.1 and 6.2 is described in the appendices of chapter 5.

In the following subsections, some basic features of conductance-based models are presented in a sequence of examples of increasing complexity. We do this to illustrate the effects of various conductances and combinations of conductances on neuronal activity. Different cells (and even the same cell held at different resting potentials) can have quite different response properties due to their particular combinations of conductances. Research on conductance-based models focuses on understanding how neuronal response dynamics arises from the properties of membrane and synaptic conductances, and how the characteristics of different neurons interact when they are coupled in networks.

The Connor-Stevens Model

The Hodgkin-Huxley model of action-potential generation, discussed in chapter 5, was developed on the basis of data from the giant axon of the squid, and we present a multi-compartment simulation of action-potential propagation using this model in a later section. The Connor-Stevens model (Connor and Stevens, 1971; Connor et al. 1977, which is the model we discuss) provides an alternative description of action-potential generation. Like the Hodgkin-Huxley model, it contains fast Na^+, delayed-rectifier K^+, and leakage conductances. The fast Na^+ and delayed-rectifier K^+ conductances have properties somewhat different from those of the Hodgkin-Huxley model, in particular faster kinetics, so the action potentials are briefer. In addition, the Connor-Stevens model contains an extra K^+ conductance, called the A-current, that is transient. K^+ conductances come in wide variety of different forms, and the Connor-Stevens model involves two of them.

A-type potassium current

The membrane current in the Connor-Stevens model is

$$i_m = \overline{g}_L(V - E_L) + \overline{g}_{Na}m^3h(V - E_{Na}) + \overline{g}_K n^4(V - E_K) + \overline{g}_A a^3 b(V - E_A),$$

(6.4)

where $\overline{g}_L = 0.003$ mS/mm^2 and E_L = -17 mV are the maximal conductance and reversal potential for the leak conductance; and $\overline{g}_{Na} = 1.2$ mS/mm^2, $\overline{g}_K = 0.2$ mS/mm^2, $\overline{g}_A = 0.477$ mS/mm^2, $E_{Na} = 55$ mV, E_K = -72 mV, and E_A = -75 mV (although the A-current is carried by K^+, the model does not require $E_A = E_K$). The gating variables, m, h, n, a, and b, are determined by equations of the form 6.2 with the gating functions given in appendix A.

The fast Na^+ and delayed-rectifier K^+ conductances generate action potentials in the Connor-Stevens model just as they do in the Hodgkin-Huxley model (see chapter 5). What is the role of the additional A-current? Figure 6.1 illustrates action-potential generation in the Connor-Stevens model. In the absence of an injected electrode current or synaptic input, the membrane potential of the model remains constant at a resting value of -68 mV. For a constant electrode current greater than a threshold value, the model neuron generates action potentials. Figure 6.1A shows how the firing rate of the model depends on the magnitude of the electrode current relative to the threshold value. The firing rate rises continuously from zero and then increases roughly linearly for currents over the range shown. Figure 6.1B shows an example of action-potential generation for one particular value of the electrode current.

Figure 6.1C shows the firing rate as a function of electrode current for the Connor-Stevens model with the maximal conductance of the A-current set to 0. The leakage conductance and reversal potential have been adjusted to keep the resting potential and membrane resistance the same as in the original model. The firing rate is clearly much higher with the A-current turned off. This is because the deinactivation rate of the A-current limits the rise time of the membrane potential between action potentials. In addition, the transition from no firing for currents less than the threshold value to firing with suprathreshold currents is different when the A-current is eliminated. Without the A-current, the firing rate jumps discontinuously to a nonzero value rather than rising continuously. Neurons with firing

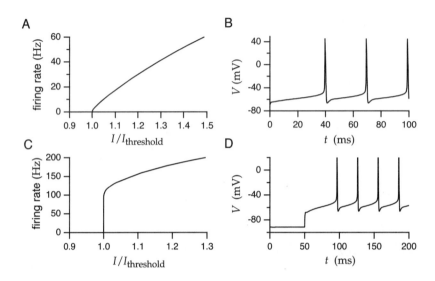

Figure 6.1 Firing of action potentials in the Connor-Stevens model. (A) Firing rate as a function of electrode current. The firing rate rises continuously from 0 as the current increases beyond the threshold value. (B) An example of action potentials generated by constant current injection. (C) Firing rate as a function of electrode current when the A-current is turned off. The firing rate now rises discontinuously from 0 as the current increases beyond the threshold value. (D) Delayed firing due to hyperpolarization. The neuron was held hyperpolarized for a prolonged period by injection of negative current. At $t = 50$ ms, the negative electrode current was switched to a positive value. The A-current delays the occurrence of the first action potential.

type I, type II

rates that rise continuously from 0 as a function of electrode current are called type I, and those with discontinuous jumps in their firing rates at threshold are called type II. An A-current is not the only mechanism that can produce a type I response but, as figures 6.1A and 6.1C show, it plays this role in the Connor-Stevens model. The Hodgkin-Huxley model produces a type II response.

Another effect of the A-current is illustrated in figure 6.1D. Here the model neuron was held hyperpolarized by negative current injection for an extended period of time, and then the current was switched to a positive value. While the neuron was hyperpolarized, the A-current deinactivated, that is, the variable b increased toward 1. When the electrode current switched sign and the neuron depolarized, the A-current first activated and then inactivated. This delayed the first spike following the change in the electrode current.

Postinhibitory Rebound and Bursting

transient Ca²⁺ conductance

The range of responses exhibited by the Connor-Stevens model neuron can be extended by including a transient Ca^{2+} conductance. The conductance we use was modeled by Huguenard and McCormick (1992) on the basis of

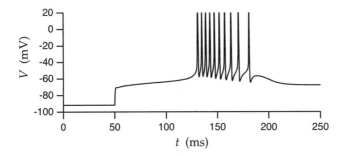

Figure 6.2 A burst of action potentials due to rebound from hyperpolarization. The model neuron was held hyperpolarized for an extended period (until the conductances came to equilibrium) by injection of constant negative electrode current. At $t = 50$ ms, the electrode current was set to 0, and a burst of Na$^+$ spikes was generated due to an underlying Ca^{2+} spike. The delay in the firing is caused by the presence of the A-current in the model.

data from thalamic relay cells. The membrane current due to the transient Ca^{2+} conductance is expressed as

$$i_{\text{CaT}} = \overline{g}_{\text{CaT}} M^2 H (V - E_{\text{Ca}}) \tag{6.5}$$

with, for the example given here, $\overline{g}_{\text{CaT}} = 0.013$ mS/mm^2 and $E_{\text{Ca}} = 120$ mV. The gating variables for the transient Ca^{2+} conductance are determined from the gating functions in appendix A.

Several different Ca^{2+} conductances are commonly expressed in neuronal membranes. These are categorized as L, T, N, and P types. L-type Ca^{2+} currents are persistent as far as their voltage dependence is concerned, and they activate at a relatively high threshold. They inactivate due to a Ca^{2+}-dependent rather than voltage-dependent process. T-type Ca^{2+} currents have lower activation thresholds and are transient. N- and P-type Ca^{2+} conductances have intermediate thresholds and are transient and persistent, respectively. They may be responsible for the Ca^{2+} entry that causes the release of transmitter at presynaptic terminals. Entry of Ca^{2+} into a neuron has many secondary consequences ranging from gating Ca^{2+}-dependent channels to inducing long-term modifications of synaptic conductances.

L, T, N and P type Ca^{2+} channels

A transient Ca^{2+} conductance acts, in many ways, like a slower version of the transient Na$^+$ conductance that generates action potentials. Instead of producing an action potential, a transient Ca^{2+} conductance generates a slower transient depolarization sometimes called a Ca^{2+} spike. This transient depolarization causes the neuron to fire a burst of action potentials, which are Na$^+$ spikes riding on the slower Ca^{2+} spike. Figure 6.2 shows such a burst and illustrates one way to produce it. In this example, the model neuron was hyperpolarized for an extended period and then released from hyperpolarization by setting the electrode current to 0. During the prolonged hyperpolarization, the transient Ca^{2+} conductance deinactivated. When the electrode current was set to 0, the resulting depolarization activated the transient Ca^{2+} conductance and generated a burst of

Ca^{2+} spike

burst

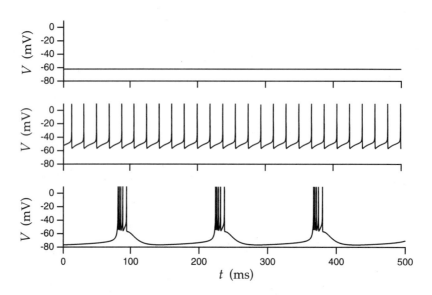

Figure 6.3 Three activity modes of a model thalamic neuron. Upper panel: with no electrode current, the model is silent. Middle panel: when a positive current is injected into the model neuron, it fires action potentials in a regular, periodic pattern. Lower panel: when negative current is injected into the model neuron, it fires action potentials in periodic bursts. (Adapted from Wang, 1994.)

action potentials. The burst in figure 6.2 is delayed due to the presence of the A-current in the Connor-Stevens model to which the Ca^{2+} conductance has been added, and it terminates when the Ca^{2+} conductance inactivates. Generation of action potentials in response to release from hyperpolar-

postinhibitory
rebound

ization is called postinhibitory rebound because, in a natural setting, the hyperpolarization would be caused by inhibitory synaptic input, not by current injection.

thalamic relay
neuron

The transient Ca^{2+} current is an important component of models of thalamic relay neurons. These neurons exhibit different firing patterns in sleep and wakeful states. Action potentials tend to appear in bursts during sleep. Figure 6.3 shows an example of three states of activity of a model thalamic relay cell due to Wang (1994) that has, in addition to fast Na^+, delayed-rectifier K^+, and transient Ca^{2+} conductances, a hyperpolarization-activated mixed-cation conductance and a persistent Na^+ conductance. The cell is silent or fires action potentials in a regular pattern or in bursts, depending on the level of current injection. In particular, injection of small amounts of negative current leads to bursting. This occurs because the hyperpolarization due to the current injection deinactivates the transient Ca^{2+} current and activates the hyperpolarization activated current. The regular firing mode of the middle plot of figure 6.3 is believed to be relevant during wakeful states, when the thalamus is faithfully reporting input from the sensory periphery to the cortex.

Neurons can fire action potentials either at a steady rate or in bursts even in the absence of current injection or synaptic input. Periodic bursting is a common feature of neurons in central pattern generators, which are neural circuits that produce periodic patterns of activity to drive rhythmic motor behaviors such as walking, running, or chewing. To illustrate periodic bursting, we consider a model constructed to match the activity of neurons in the crustacean stomatogastric ganglion (STG), a neuronal circuit that controls chewing and digestive rhythms in the foregut of lobsters and crabs. The STG is a model system for investigating the effects of neuromodulators, such as amines and neuropeptides, on the activity patterns of a neural network. Neuromodulators modify neuronal and network behavior by activating, deactivating, or otherwise altering the properties of membrane and synaptic channels. Neuromodulation has a major impact on virtually all neural networks, ranging from peripheral motor pattern generators like the STG to the sensory, motor, and cognitive circuits of the brain.

stomatogastric ganglion

neuromodulator

The model STG neuron contains fast Na^+, delayed-rectifier K^+, A-type K^+, and transient Ca^{2+} conductances similar to those discussed above, although the formulas and parameters used are somewhat different. In addition, the model has a Ca^{2+}-dependent K^+ conductance. Due to the complexity of the model, we do not provide complete descriptions of its conductances except for the Ca^{2+}-dependent K^+ conductance which plays a particularly significant role in the model.

The repolarization of the membrane potential after an action potential is often carried out both by the delayed-rectifier K^+ conductance and by a fast Ca^{2+}-dependent K^+ conductance. Ca^{2+}-dependent K^+ conductances may be voltage dependent, but they are activated primarily by a rise in the level of intracellular Ca^{2+}. A slow Ca^{2+}-dependent K^+ conductance called the after-hyperpolarization (AHP) conductance builds up during sequences of action potentials and typically contributes to the spike-rate adaptation discussed and modeled in chapter 5.

Ca^{2+}-dependent K^+ conductance

after-hyperpolarization conductance

The Ca^{2+}-dependent K^+ current in the model STG neuron is given by

$$i_{KCa} = \bar{g}_{KCa} c^4 (V - E_K), \qquad (6.6)$$

where c obeys an equation of the form 6.2, with c_∞ depending on both the membrane potential and the intracellular Ca^{2+} concentration, $[Ca^{2+}]$ (see appendix A). The intracellular Ca^{2+} concentration is computed in this model using a simplified description in which rises in intracellular Ca^{2+} are caused by influx through membrane Ca^{2+} channels, and Ca^{2+} removal is described by an exponential process. The resulting equation for the intracellular Ca^{2+} concentration, $[Ca^{2+}]$, is

$$\frac{d[Ca^{2+}]}{dt} = -\gamma i_{Ca} - \frac{[Ca^{2+}]}{\tau_{Ca}}. \qquad (6.7)$$

Here i_{Ca} is the total Ca^{2+} current per unit area of membrane, τ_{Ca} is the time constant determining the rate at which intracellular Ca^{2+} is removed, and γ is a factor that converts from the electric current due to Ca^{2+} ion flow

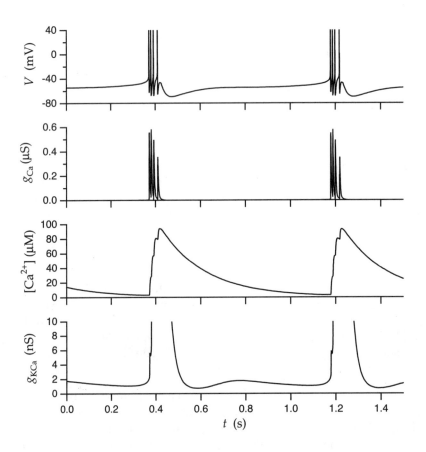

Figure 6.4 Periodic bursting in a model STG neuron. From the top, the panels show the membrane potential, the Ca^{2+} conductance, the intracellular Ca^{2+} concentration, and the Ca^{2+}-dependent K^+ conductance. The Ca^{2+}-dependent K^+ conductance is shown at an expanded scale so the reduction of the conductance due to the falling intracellular Ca^{2+} concentration during the interburst intervals can be seen. In this example, $\tau_{Ca} = 200$ ms. (Simulation by M. Goldman based on a variant of a model of Turrigiano et al., 1995, due to Z. Liu and M. Goldman.)

to the rate at which the Ca^{2+} ion concentration changes within the cell. Because the Ca^{2+} concentration is determined by dividing the number of Ca^{2+} ions in a cell by the total cellular volume and the Ca^{2+} influx is computed by multiplying i_{Ca} by the membrane surface area, γ is proportional to the surface-to-volume ratio for the cell. It also contains a factor that converts from coulombs per second of electrical current to moles per second of Ca^{2+} ions. This factor is $1/(zF)$, where z is the number of charges on the ion ($z = 2$ for Ca^{2+}) and F is the Faraday constant. If, as is normally the case, $[Ca^{2+}]$ is in moles/liter, γ should also contain a factor that converts the volume measure to liters, 10^6 mm^3/liter. Finally, γ is sometimes multiplied by an additional factor that reflects fast intracellular Ca^{2+} buffering. Most of the Ca^{2+} ions that enter a neuron are rapidly bound to intracellular buffers, so only a fraction of the Ca^{2+} current through membrane channels is actually available to change the concentration $[Ca^{2+}]$ of free Ca^{2+} ions in the cell. This factor is a few percent. The minus sign in front of the γ in equation 6.7 is due to the definition of membrane currents as positive in the outward direction.

Figure 6.4 shows the model STG neuron firing action potentials in bursts. As in the models of figures 6.2 and 6.3, the bursts are transient Ca^{2+} spikes with action potentials riding on top of them. The Ca^{2+} current during these bursts causes a dramatic increase in the intracellular Ca^{2+} concentration. This activates the Ca^{2+}-dependent K^+ current, which, along with the inactivation of the Ca^{2+} current, terminates the burst. The interburst interval is determined primarily by the time it takes for the intracellular Ca^{2+} concentration to return to a low value, which deactivates the Ca^{2+}-dependent K^+ current, allowing another burst to be generated. Although figure 6.4 shows that the conductance of the Ca^{2+}-dependent K^+ current reaches a low value immediately after each burst (due to its voltage dependence), this initial dip is too early for another burst to be generated at that point in the cycle.

6.3   The Cable Equation

Single-compartment models describe the membrane potential over an entire neuron with a single variable. Membrane potentials can vary considerably over the surface of the cell membrane, especially for neurons with long and narrow processes, or if we consider rapidly changing membrane potentials. Figure 6.5A shows the delay and attenuation of an action potential as it propagates from the soma out to the dendrites of a cortical pyramidal neuron. Figure 6.5B shows the delay and attenuation of an excitatory postsynaptic potential (EPSP) initiated in the dendrite by synaptic input as it spreads to the soma. Understanding these features is crucial for determining whether and when a given synaptic input will cause a neuron to fire an action potential.

The attenuation and delay within a neuron are most severe when electrical signals travel down the long, narrow, cablelike structures of dendritic or axonal branches. For this reason, the mathematical analysis of signal propagation within neurons is called cable theory. Dendritic and axonal cables *cable theory* are typically narrow enough that variations of the potential in the radial or axial directions are negligible compared to longitudinal variations. Therefore, the membrane potential along a neuronal cable is expressed as a function of a single longitudinal spatial coordinate x and time, $V(x, t)$, and the basic problem is to solve for this potential.

Current flows within a neuron due to voltage gradients. In chapter 5, we discussed how the potential difference across a segment of neuronal cable is related to the longitudinal current flowing down the cable. The longitudinal resistance of a cable segment of length Δx and radius a is given by multiplying the intracellular resistivity r_L by Δx and dividing by the cross-sectional area, πa^2, so that $R_L = r_L \Delta x / (\pi a^2)$. The voltage drop

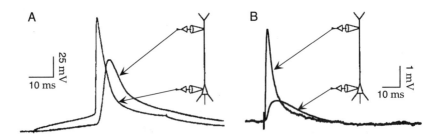

Figure 6.5 Simultaneous intracellular recordings from the soma and apical dendrite of cortical pyramidal neurons in slice preparations. (A) A pulse of current was injected into the soma of the neuron to produce the action potential seen in the somatic recording. The action potential appears delayed and with smaller amplitude in the dendritic recording. (B) A set of axon fibers was stimulated, producing an excitatory synaptic input. The excitatory postsynaptic potential (EPSP) is larger and peaks earlier in the dendrite than in the soma. Note that the scale for the potential is smaller than in A. (A adapted from Stuart and Sakmann, 1994; B adapted from Stuart and Spruston, 1998.)

across this length of cable, $\Delta V = V(x + \Delta x) - V(x)$, is then related to the amount of longitudinal current flow by Ohm's law. In chapter 5, we discussed the magnitude of this current flow, but for the present purposes, we also need to define a sign convention for its direction. We define currents flowing in the direction of increasing x as positive. By this convention, the relationship between ΔV and I_L given by Ohm's law is $\Delta V = -R_L I_L$ or $\Delta V = -r_L \Delta x I_L/(\pi a^2)$. Solving this for the longitudinal current, we find $I_L = -\pi a^2 \Delta V/(r_L \Delta x)$. It is useful to take the limit of this expression for infinitesimally short cable segments, that is, as $\Delta x \to 0$. In this limit, the ratio of ΔV to Δx becomes the derivative $\partial V/\partial x$. We use a partial derivative here because V can also depend on time. Thus, at any point along a cable of radius a and intracellular resistivity r_L, the longitudinal current flowing in the direction of increasing x is

$$I_L = -\frac{\pi a^2}{r_L} \frac{\partial V}{\partial x}. \tag{6.8}$$

The membrane potential $V(x, t)$ is determined by solving a partial differential equation, the cable equation, that describes how the currents entering, leaving, and flowing within a neuron affect the rate of change of the membrane potential. To derive the cable equation, we consider the currents within the small segment shown in figure 6.6. This segment has a radius a and a short length Δx. The rate of change of the membrane potential due to currents flowing into and out of this region is determined by its capacitance. Recall from chapter 5 that the capacitance of a membrane is determined by multiplying the specific membrane capacitance c_m by the area of the membrane. The cylinder of membrane shown in figure 6.6 has a surface area of $2\pi a \Delta x$, and hence a capacitance of $2\pi a \Delta x c_m$. The amount of current needed to change the membrane potential at a rate $\partial V/\partial t$ is thus $2\pi a \Delta x c_m \partial V/\partial t$.

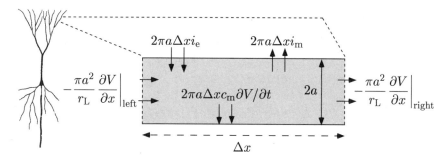

Figure 6.6 The segment of neuron used in the derivation of the cable equation. The longitudinal, membrane, and electrode currents that determine the rate of change of the membrane potential within this segment are denoted. The segment has length Δx and radius a. The expression involving the specific membrane capacitance refers to the rate at which charge builds up on the cell membrane, generating changes in the membrane potential. (The neuron diagram here and in figures 6.15 and 6.16 is from Haberly, 1990.)

All of the currents that can change the membrane potential of the segment being considered are shown in figure 6.6. Current can flow longitudinally into the segment from neighboring segments, and expression 6.8 has been used in figure 6.6 to specify the longitudinal currents at both ends of the segment. Current can flow across the membrane of the segment we are considering through ion and synaptic receptor channels, or through an electrode. The contribution from ion and synaptic channels is expressed as a current per unit area of membrane i_m times the surface area of the segment, $2\pi a \Delta x$. The electrode current is not normally expressed as a current per unit area, but for the present purposes it is convenient to define i_e to be the total electrode current flowing into a given region of the neuronal cable divided by the surface area of that region. The total amount of electrode current being injected into the cable segment of figure 6.6 is then $i_e 2\pi a \Delta x$. Because the electrode current is normally specified by I_e, not by a current per unit area, all the results we obtain will ultimately be re-expressed in terms of I_e. Following the standard convention, membrane and synaptic currents are defined as positive when they are outward, and electrode currents are defined as positive when they are inward.

The cable equation is derived by setting the sum of all the currents shown in figure 6.6 equal to the current needed to charge the membrane. The total longitudinal current entering the cylinder is the difference between the current flowing in on the left and that flowing out on the right. Thus,

$$2\pi a \Delta x c_m \frac{\partial V}{\partial t} = -\left(\frac{\pi a^2}{r_L}\frac{\partial V}{\partial x}\right)\bigg|_{\text{left}} + \left(\frac{\pi a^2}{r_L}\frac{\partial V}{\partial x}\right)\bigg|_{\text{right}} - 2\pi a \Delta x (i_m - i_e)\,.$$

$$(6.9)$$

Dividing both sides of this equation by $2\pi a \Delta x$, we note that the right side involves the term

$$\frac{1}{\Delta x}\left[\left(\frac{\pi a^2}{r_L}\frac{\partial V}{\partial x}\right)\bigg|_{\text{right}} - \left(\frac{\pi a^2}{r_L}\frac{\partial V}{\partial x}\right)\bigg|_{\text{left}}\right] \to \frac{\partial}{\partial x}\left(\frac{\pi a^2}{r_L}\frac{\partial V}{\partial x}\right)\,. \qquad (6.10)$$

The arrow refers to the limit $\Delta x \to 0$, which we now take. We can move r_{L} outside the derivative in this equation under the assumption that it is not a function of position. However, the factor of a^2 must remain inside the derivative unless it is independent of x. Substituting the result 6.10 into 6.9, we obtain the cable equation,

cable equation

$$c_{\mathrm{m}}\frac{\partial V}{\partial t} = \frac{1}{2ar_{\mathrm{L}}}\frac{\partial}{\partial x}\left(a^2\frac{\partial V}{\partial x}\right) - i_{\mathrm{m}} + i_{\mathrm{e}}\,. \qquad (6.11)$$

*boundary
conditions for the
cable equation*

To determine the membrane potential, equation (6.11) must be augmented by appropriate boundary conditions. The boundary conditions specify what happens to the membrane potential when the neuronal cable branches or terminates. The point at which a cable branches, or equivalently where multiple cable segments join, is called a node. At such a branching node, the potential must be continuous, that is, the functions $V(x, t)$ defined along each of the segments must yield the same result when evaluated at the x value corresponding to the node. In addition, charge must be conserved, which means that the sum of the longitudinal currents entering (or leaving) a node along all of its branches must be 0. According to equation 6.8, the longitudinal current entering a node is proportional to the square of the cable radius times the derivative of the potential evaluated at that point, $a^2\partial V/\partial x$. The sum of the longitudinal currents entering the node, computed by evaluating these derivatives along each cable segment at the point where they meet at the node, must be 0.

Several different boundary conditions can be imposed at the end of a terminating cable segment. One simple condition is that no current flows out of the end of the cable. By equation 6.8, this means that the spatial derivative of the potential must vanish at a termination point.

Due to the complexities of neuronal membrane currents and morphologies, the cable equation is most often solved numerically, using multicompartmental techniques described later in this chapter. However, it is useful to study analytic solutions of the cable equation in simple cases to get a feel for how different morphological features, such as long dendritic cables, branching nodes, changes in cable radii, and cable ends, affect the membrane potential.

Linear Cable Theory

Before we can solve the cable equation by any method, the membrane current i_{m} must be specified. We discussed models of various ion channel contributions to the membrane current in chapter 5 and earlier in this chapter. These models typically produce nonlinear expressions that are too complex to allow analytic solution of the cable equation. The analytic solutions we discuss use two rather drastic approximations: synaptic currents are ignored, and the membrane current is written as a linear function of the membrane potential. Eliminating synaptic currents requires us to examine how a neuron responds to the electrode current i_e. In some cases, electrode current can mimic the effects of a synaptic conductance, although the two are not equivalent. In any case, studying responses to electrode current allows us to investigate the effects of different morphologies on membrane potentials.

Typically, a linear approximation for the membrane current is valid only if the membrane potential stays within a limited range, for example, close to the resting potential of the cell. The resting potential is defined as the potential where no net current flows across the membrane. Near this potential, we approximate the membrane current per unit area as

$$i_m = (V - V_{rest})/r_m \, , \qquad (6.12)$$

where V_{rest} is the resting potential and r_m is the specific membrane resistance. It is convenient to define v as the membrane potential relative to the resting potential, $v = V - V_{rest}$, so that $i_m = v/r_m$.

$v = V - V_{rest}$

If the radii of the cable segments used to model a neuron are constant except at branches and abrupt junctions, the factor a^2 in equation 6.11 can be taken out of the derivative and combined with the prefactor $1/2ar_L$ to produce a factor $a/2r_L$ that multiplies the spatial second derivative. With this modification and use of the linear expression for the membrane current, the cable equation for v is

$$c_m \frac{\partial v}{\partial t} = \frac{a}{2r_L} \frac{\partial^2 v}{\partial x^2} - \frac{v}{r_m} + i_e \, . \qquad (6.13)$$

It is convenient to multiply this equation by r_m, turning the factor that multiplies the time derivative on the left side into the membrane time constant $\tau_m = r_m c_m$. This also changes the expression multiplying the spatial second derivative on the right side of equation 6.13 to $ar_m/2r_L$. This factor has the dimensions of length squared, and it defines a fundamental length constant for a segment of cable of radius a, the electrotonic length,

electrotonic length λ

$$\lambda = \sqrt{\frac{ar_m}{2r_L}} \, . \qquad (6.14)$$

Using the values $r_m = 1\,\text{M}\Omega\cdot\text{mm}^2$ and $r_L = 1\,\text{k}\Omega\cdot\text{mm}$, a cable of radius $a = 2\,\mu\text{m}$ has an electrotonic length of 1 mm. A segment of cable with radius a and length λ has a membrane resistance that is equal to its longitudinal resistance, as can be seen from equation 6.14,

R_λ

$$R_\lambda = \frac{r_m}{2\pi a\lambda} = \frac{r_L \lambda}{\pi a^2} \, . \qquad (6.15)$$

The resistance R_λ defined by this equation is a useful quantity that enters into a number of calculations.

linear cable equation

Expressed in terms of τ_m and λ, the cable equation becomes

$$\tau_m \frac{\partial v}{\partial t} = \lambda^2 \frac{\partial^2 v}{\partial x^2} - v + r_m i_e \, . \qquad (6.16)$$

Equation 6.16 is a linear equation for v similar to the diffusion equation, and it can be solved by standard methods of mathematical analysis. The constants τ_m and λ set the scale for temporal and spatial variations in the membrane potential. For example, the membrane potential requires a time of order τ_m to settle down after a transient, and deviations in the membrane potential due to localized electrode currents decay back to 0 over a length of order λ.

The membrane potential is affected both by the form of the cable equation and by the boundary conditions imposed at branching nodes and terminations. To isolate these two effects, we consider two idealized cases: an infinite cable that does not branch or terminate, and a single branching node that joins three semi-infinite cables. Of course, real neuronal cables are not infinitely long, but the solutions we find are applicable for long cables far from their ends. We determine the potential for both of these morphologies when current is injected at a single point. Because the equation we are studying is linear, the membrane potential for any other spatial distribution of electrode current can be determined by summing solutions corresponding to current injection at different points. The use of point injection to build more general solutions is a standard method of linear analysis. In this context, the solution for a point source of current injection *Green's function* is called a Green's function.

An Infinite Cable

In general, solutions to the linear cable equation are functions of both position and time. However, if the current being injected is held constant, the membrane potential settles to a steady-state solution that is independent of time. Solving for this time-independent solution is easier than solving the full time-dependent equation, because the cable equation reduces to an ordinary differential equation in the static case,

$$\lambda^2 \frac{d^2v}{dx^2} = v - r_m i_e . \tag{6.17}$$

For the localized current injection we wish to study, i_e is 0 everywhere except within a small region of size Δx around the injection site, which we take to be $x = 0$. Eventually we will let $\Delta x \to 0$. Away from the injection site, the linear cable equation is $\lambda^2 d^2v/dx^2 = v$, which has the general solution $v(x) = B_1 \exp(-x/\lambda) + B_2 \exp(x/\lambda)$ with as yet undetermined coefficients B_1 and B_2. These constant coefficients are determined by imposing boundary conditions appropriate to the particular morphology being considered. For an infinite cable, on physical grounds we simply require that the solution does not grow without bound when $x \to \pm\infty$. This means that we must choose the solution with $B_1 = 0$ for the region $x < 0$ and the solution with $B_2 = 0$ for $x > 0$. Because the solution must be continuous at $x = 0$, we must require $B_1 = B_2 = B$, and these two solutions can be combined into a single expression, $v(x) = B \exp(-|x|/\lambda)$. The remaining task

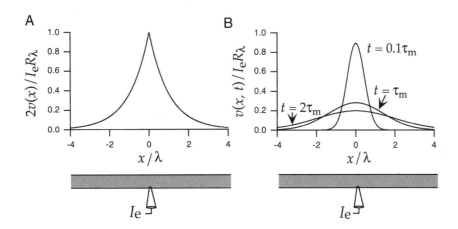

Figure 6.7 The potential for current injection at the point $x = 0$ along an infinite cable. (A) Static solution for a constant electrode current. The potential decays exponentially away from the site of current injection. (B) Time-dependent solution for a δ function pulse of current. The potential is described by a Gaussian function centered at the site of current injection that broadens and shrinks in amplitude over time.

is to determine B, which we do by balancing the current injected with the current that diffuses away from $x = 0$.

In the small region of size Δx around $x = 0$ where the current is injected, the full equation $\lambda^2 d^2 v/dx^2 = v - r_{\mathrm{m}} i_{\mathrm{e}}$ must be solved. If the total amount of current injected by the electrode is I_{e}, the current per unit area injected into this region is $I_{\mathrm{e}}/(2\pi a \Delta x)$. This grows without bound as $\Delta x \to 0$. The first derivative of the membrane potential $v(x) = B \exp(-|x|/\lambda)$ is discontinuous at the point $x = 0$. For small Δx, the derivative at one side of the region we are discussing (at $x = -\Delta x/2$) is approximately B/λ, while at the other side (at $x = +\Delta x/2$) it is $-B/\lambda$. In these expressions, we have used the fact that Δx is small to set $\exp(-|\Delta x|/2\lambda) \approx 1$. For small Δx, the second derivative is approximately the difference between these two first derivatives divided by Δx, which is $-2B/(\lambda \Delta x)$. We can ignore the term v in the cable equation within this small region, because it is not proportional to $1/\Delta x$. Substituting the expressions we have derived for the remaining terms in the equation, we find that $-2\lambda^2 B/(\lambda \Delta x) = -r_{\mathrm{m}} I_{\mathrm{e}}/(2\pi a \Delta x)$, which means that $B = I_{\mathrm{e}} R_\lambda/2$, using R_λ from equation 6.15. Thus, the membrane potential for static current injection at the point $x = 0$ along an infinite cable is

$$v(x) = \frac{I_{\mathrm{e}} R_\lambda}{2} \exp\left(-\frac{|x|}{\lambda}\right). \tag{6.18}$$

According to this result, the membrane potential away from the site of current injection ($x = 0$) decays exponentially with length constant λ (see figure 6.7A). The ratio of the membrane potential at the injection site to the magnitude of the injected current is called the input resistance of the cable. The value of the potential at $x = 0$ is $I_{\mathrm{e}} R_\lambda/2$, indicating that the

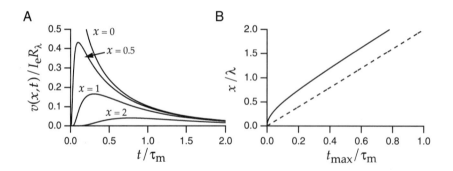

Figure 6.8 Time dependence of the potential on an infinite cable in response to a pulse of current injected at the point $x = 0$ at time $t = 0$. (A) The potential is always largest at the site of current injection. At any fixed point, it reaches its maximum value as a function of time later for measurement sites located farther away from the current source. (B) Movement of the temporal maximum of the potential. The solid line shows the relationship between the measurement location x and the time t_{max} when the potential reaches its maximum value at that location. The dashed line corresponds to a constant velocity $2\lambda/\tau_m$.

infinite cable has an input resistance of $R_\lambda/2$. Each direction of the cable acts like a resistance of R_λ, and these two act in parallel to produce a total resistance half as big. Note that each semi-infinite cable extending from the point $x = 0$ has a resistance equal to a finite cable of length λ.

We now consider the membrane potential produced by an instantaneous pulse of current injected at the point $x = 0$ at the time $t = 0$. Specifically, we consider $i_e = I_e \tau_m \delta(x)\delta(t)/2\pi a$, which means that the current pulse delivers a total charge of $I_e \tau_m$. We do not derive the solution for this case (see Tuckwell, 1988, for example), but simply state the answer,

$$v(x, t) = \frac{I_e R_\lambda}{\sqrt{4\pi t/\tau_m}} \exp\left(-\frac{\tau_m x^2}{4\lambda^2 t}\right) \exp\left(-\frac{t}{\tau_m}\right). \qquad (6.19)$$

In this case, the spatial dependence of the potential is determined by a Gaussian, rather than an exponential function. The Gaussian is always centered around the injection site, so the potential is always largest at $x = 0$. The width of the Gaussian curve around $x = 0$ is proportional to $\lambda\sqrt{t/\tau_m}$. As expected, λ sets the scale for this spatial variation, but the width also grows as the square root of the time measured in units of τ_m. The factor $(4\pi t/\tau_m)^{-1/2}$ in equation 6.19 preserves the total area under this Gaussian curve, but the additional exponential factor $\exp(-t/\tau_m)$ reduces the integrated amplitude over time. As a result, the spatial dependence of the membrane potential is described by a spreading Gaussian function with an integral that decays exponentially (figure 6.7B).

Figure 6.8 shows the solution of equation 6.19 plotted at various fixed positions as a function of time. Figure 6.8A shows that the membrane potential measured farther from the injection site reaches its maximum value at later times. It is important to keep in mind that the membrane potential spreads out from the region $x = 0$; it does not propagate like a wave. Nevertheless, we can define a type of "velocity" for this solution by computing the time t_{max} when the maximum of the potential occurs at a given spatial location. This is done by setting the time derivative of $v(x, t)$ in equation 6.19 to 0, giving

$$t_{max} = \frac{\tau_m}{4} \left(\sqrt{1 + 4(x/\lambda)^2} - 1 \right) . \tag{6.20}$$

For large x, $t_{max} \approx x\tau_m/2\lambda$, corresponding to a velocity of $2\lambda/\tau_m$. For smaller x values, the location of the maximum moves faster than this "velocity" would imply (figure 6.8B).

An Isolated Branching Node

To illustrate the effects of branching on the membrane potential in response to a point source of current injection, we consider a single isolated junction of three semi-infinite cables, as shown in the bottom panels of figure 6.9. For simplicity, we discuss the solution for static current injection at a point, but the results generalize directly to the case of time-dependent currents. We label the potentials along the three segments v_1, v_2, and v_3, and label the distance outward from the junction point along any given segment by the coordinate x (although in figure 6.9 a slightly different convention is used). The electrode injection site is located a distance y away from the junction along segment 2. The solution for the three segments is then

$$v_1(x) = p_1 I_e R_{\lambda_1} \exp(-x/\lambda_1 - y/\lambda_2)$$
$$v_2(x) = \frac{I_e R_{\lambda_2}}{2} \left[\exp(-|y - x|/\lambda_2) + (2p_2 - 1) \exp(-(y + x)/\lambda_2) \right]$$
$$v_3(x) = p_3 I_e R_{\lambda_3} \exp(-x/\lambda_3 - y/\lambda_2) , \tag{6.21}$$

where, for $i = 1, 2$, and 3,

$$p_i = \frac{a_i^{3/2}}{a_1^{3/2} + a_2^{3/2} + a_3^{3/2}} \quad , \quad \lambda_i = \sqrt{\frac{a_i r_m}{2r_L}} \quad , \text{and} \quad R_{\lambda_i} = \frac{r_L \lambda_i}{\pi a_i^2} . \tag{6.22}$$

Note that the distances x and y appearing in the exponential functions are divided by the electrotonic length of the segment along which the potential is measured or the current is injected. This solution satisfies the cable equation, because it is constructed by combining solutions of the form 6.18. The only term that has a discontinuous first derivative within the range being considered is the first term in the expression for v_2, and this solves the cable equation at the current injection site because it is identical to 6.18. We leave it to the reader to verify that this solution satisfies the boundary conditions $v_1(0) = v_2(0) = v_3(0)$ and $\sum a_i^2 \partial v_i / \partial x = 0$.

Figure 6.9 shows the potential near a junction where a cable of radius 2 μ breaks into two thinner cables of radius 1 μ. In figure 6.9A, current is injected along the thicker cable, and in figure 6.9B it is injected along one

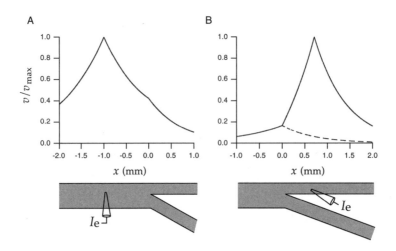

Figure 6.9 The potentials along the three branches of an isolated junction for a current injection site one electrotonic length constant away from the junction. The potential v is plotted relative to v_{max}, which is v at the site of the electrode. The thick branch has a radius of 2 μ and an electrotonic length constant $\lambda = 1$ mm, and the two thin branches have radii of 1 μ and $\lambda = 2^{-1/2}$ mm. (A) Current injection along the thick branch. The potentials along both of the thin branches, shown by the solid curve over the range $x > 0$, are identical. The solid curve over the range $x < 0$ shows the potential on the thick branch where current is being injected. (B) Current injection along one of the thin branches. The dashed line shows the potential along the thin branch where current injection does not occur. The solid line shows the potential along the thick branch for $x < 0$ and along the thin branch receiving the injected current for $x > 0$.

of the thinner branches. In both cases, the site of current injection is one electrotonic length constant away from the junction. The two daughter branches have little effect on the falloff of the potential away from the electrode site in figure 6.9A. This is because the thin branches do not represent a large current sink. The thick branch has a bigger effect on the attenuation of the potential along the thin branch receiving the electrode current in figure 6.9B. This can be seen as an asymmetry in the falloff of the potential on either side of the electrode. Loading by the thick cable segment contributes to a quite severe attenuation between the two thin branches in figure 6.9B. Comparison of figures 6.9A and B reveals a general feature of static attenuation in a passive cable: attenuation near the soma due to potentials arising in the periphery is typically greater than attenuation in the periphery due to potentials arising near the soma.

The Rall Model

The infinite and semi-infinite cables we have considered are clearly mathematical idealizations. We now turn to a model neuron introduced by Rall (1959, 1977) that, though still highly simplified, captures some of the important elements that affect the responses of real neurons. Most neurons receive their synaptic inputs over complex dendritic trees. The integrated

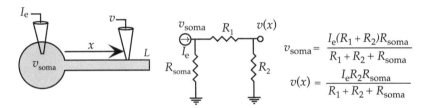

$$v_{soma} = \frac{I_e(R_1 + R_2)R_{soma}}{R_1 + R_2 + R_{soma}}$$

$$v(x) = \frac{I_e R_2 R_{soma}}{R_1 + R_2 + R_{soma}}$$

Figure 6.10 The Rall model with static current injected into the soma. The schematic at left shows the recording setup. The potential is measured at the soma and at a distance x along the equivalent cable. The central diagram is the equivalent circuit for this case, and the corresponding formulas for the somatic and dendritic voltages are given at the right. The symbols at the bottom of the resistances R_{soma} and R_2 indicate that $v = 0$ at these points. R_{soma} is the membrane resistance of the soma, and R_1 and R_2 are the resistances given in equations 6.23 and 6.24.

effect of these inputs is usually measured from the soma, and the spike-initiation region of the axon that determines whether the neuron fires an action potential is typically located near the soma. In Rall's model, a compact soma region (represented by one compartment) is connected to a single equivalent cylindrical cable that replaces the entire dendritic region of the neuron (see the schematics in figures 6.10 and 6.12). The critical feature of the model is the choice of the radius and length for the equivalent cable to best match the properties of the dendritic structure being approximated.

The radius a and length L of the equivalent cable are determined by matching two important elements of the full dendritic tree. These are its average length in electrotonic units, which determines the amount of attenuation, and the total surface area, which determines the total membrane resistance and capacitance. The average electrotonic length of a dendrite is determined by considering direct paths from the soma to the terminals of the dendrite. The electrotonic lengths for these paths are constructed by measuring the distance traveled along each of the cable segments traversed in units of the electrotonic length constant for that segment. In general, the total electrotonic length measured by summing these electrotonic segment lengths depends on which terminal of the tree is used as the end point. However, an average value can be used to define an electrotonic length for the full dendritic structure. The length L of the equivalent cable is then chosen so that L/λ is equal to this average electrotonic length, where λ is the length constant for the equivalent cable. The radius of the equivalent cable, which is needed to compute λ, is determined by setting the surface area of the equivalent cable, $2\pi a L$, equal to the surface area of the full dendritic tree.

Under some restrictive circumstances the equivalent cable reproduces the effects of a full tree exactly. Among these conditions is the requirement $a_1^{3/2} = a_2^{3/2} + a_3^{3/2}$ on the radii of any three segments being joined at a node within the tree. Note from equation 6.22 that this condition makes $p_1 = p_2 + p_3 = 1/2$. However, even when the so-called 3/2 law is not exact, the equivalent cable is an extremely useful and often reasonably accurate simplification.

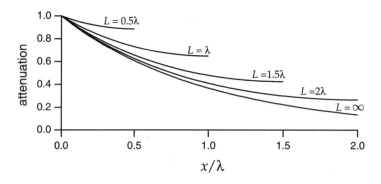

Figure 6.11 Voltage and current attenuation for the Rall model. The attenuation plotted is the ratio of the dendritic voltage to the somatic voltage for the recording setup of figure 6.10, or the ratio of the somatic current to the electrode current for the arrangement in figure 6.12. Attenuation is plotted as a function of x/λ for different equivalent cable lengths.

Figures 6.10 and 6.12 depict static solutions of the Rall model for two different recording configurations, expressed in the form of equivalent circuits. The equivalent circuits are an intuitive way of describing the solution of the cable equation. In figure 6.10, constant current is injected into the soma. The circuit diagram shows an arrangement of resistors that replicates the results of solving the time-independent cable equation (equation 6.17) for the purposes of voltage measurements at the soma, v_{soma}, and at a distance x along the equivalent cable, $v(x)$. The values for these resistances (and similarly the values of R_3 and R_4 given below) are set so that the equivalent circuit reconstructs the solution of the cable equation obtained using standard methods (see, for example, Tuckwell, 1988). R_{soma} is the membrane resistance of the soma, and

$$R_1 = \frac{R_\lambda \left(\cosh\left(L/\lambda\right) - \cosh\left((L - x)/\lambda\right)\right)}{\sinh\left(L/\lambda\right)} \tag{6.23}$$

$$R_2 = \frac{R_\lambda \cosh\left((L - x)/\lambda\right)}{\sinh\left(L/\lambda\right)}. \tag{6.24}$$

Expressions for v_{soma} and $v(x)$, arising directly from the equivalent circuit using standard rules of circuit analysis (see the Mathematical Appendix), are given at the right side of figure 6.10.

The input resistance of the Rall model neuron, as measured from the soma, is determined by the somatic resistance R_{soma} acting in parallel with the effective resistance of the cable, and is $(R_1 + R_2)R_{\text{soma}}/(R_1 + R_2 + R_{\text{soma}})$. The effective resistance of the cable, $R_1 + R_2 = R_\lambda/\tanh(L/\lambda)$, approaches the value R_λ when $L \gg \lambda$. The effect of lengthening a cable saturates when it gets much longer than its electrotonic length. The voltage attenuation caused by the cable is defined as the ratio of the dendritic potential to the somatic potential, and in this case it is given by

$$\frac{v(x)}{v_{\text{soma}}} = \frac{R_2}{R_1 + R_2} = \frac{\cosh\left((L - x)/\lambda\right)}{\cosh\left(L/\lambda\right)}. \tag{6.25}$$

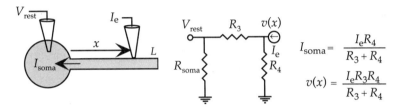

Figure 6.12 The Rall model with static current injected a distance x along the equivalent cable while the soma is clamped at its resting potential. The schematic at left shows the recording setup. The potential at the site of the current injection and the current entering the soma are measured. The central diagram is the equivalent circuit for this case, and the corresponding formulas for the somatic current and dendritic voltage are given at the right. R_{soma} is the membrane resistance of the soma, and R_3 and R_4 are the resistances given in equations 6.26 and 6.27.

This result is plotted in figure 6.11.

Figure 6.12 shows the equivalent circuit for the Rall model when current is injected at a location x along the cable, and the soma is clamped at $v_{\text{soma}} = 0$ (or equivalently $V_{\text{soma}} = V_{\text{rest}}$). The equivalent circuit can be used to determine the current entering the soma and the voltage at the site of current injection. In this case, the somatic resistance is irrelevant because the soma is clamped at its resting potential. The other resistances are

$$R_3 = R_\lambda \sinh(x/\lambda) \tag{6.26}$$

and

$$R_4 = \frac{R_\lambda \sinh(x/\lambda)\cosh((L-x)/\lambda)}{\cosh(L/\lambda) - \cosh((L-x)/\lambda)}. \tag{6.27}$$

The input resistance for this configuration, as measured from the dendrite, is determined by R_3 and R_4 acting in parallel, and is $R_3 R_4/(R_3 + R_4) = R_\lambda \sinh(x/\lambda)\cosh((L-x)/\lambda)/\cosh(L/\lambda)$. When L and x are both much larger than λ, this approaches the limiting value R_λ. The current attenuation is defined as the ratio of the somatic current to the electrode current, and is given by

$$\frac{I_{\text{soma}}}{I_e} = \frac{R_4}{R_3 + R_4} = \frac{\cosh((L-x)/\lambda)}{\cosh(L/\lambda)}. \tag{6.28}$$

The inward current attenuation (plotted in figure 6.11) for the recording configuration of figure 6.12 is identical to the outward voltage attenuation for figure 6.10 given by equation 6.25. Equality of the voltage attenuation measured in one direction and the current attenuation measured in the opposite direction is a general feature of linear cable theory.

The Morphoelectrotonic Transform

The membrane potential for a neuron of complex morphology is obviously much more difficult to compute than the simple cases we have

anatomy   attenuation (in)   delay (in)

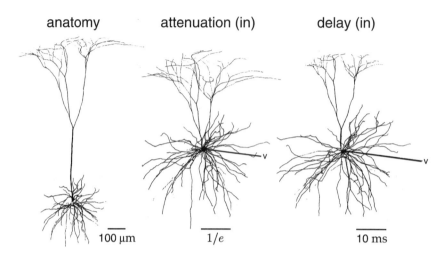

100 μm   1/e   10 ms

Figure 6.13 The morphoelectrotonic transform of a cortical neuron. The left panel is a normal drawing of the neuron. The central panel is a diagram in which the distance between any point and the soma is proportional to the logarithm of the steady-state attenuation between the soma and that point for static current injected at the terminals of the dendrites. The scale bar denotes the distance corresponding to an attenuation of $\exp(-1)$. In the right panel, the distance from the soma to a given point is proportional to the inward delay, which is the centroid of the soma potential minus the centroid at the periphery when a pulse of current is injected peripherally. The v labels in the diagrams indicate that the reference potential in these cases is the somatic potential. (Adapted from Zador et al, 1995.)

considered. Fortunately, efficient numerical schemes (discussed later in this chapter) exist for generating solutions for complex cable structures. However, even when the solution is known, it is still difficult to visualize the effects of a complex morphology on the potential. Zador et al. (1995; see also Tsai et al., 1994) devised a scheme for depicting the attenuation and delay of the membrane potential for complex morphologies. The voltage attenuation, as plotted in figure 6.11, is not an appropriate quantity to represent geometrically because it is not additive. Consider three points along a cable satisfying $x_1 > x_2 > x_3$. The attenuation between x_1 and x_3 is the product of the attenuation from x_1 to x_2 and from x_2 to x_3, $v(x_1)/v(x_3) = (v(x_1)/v(x_2))(v(x_2)/v(x_3))$. An additive quantity can be obtained by taking the logarithm of the attenuation, due to the identity $\ln(v(x_1)/v(x_3)) = \ln(v(x_1)/v(x_2)) + \ln(v(x_2)/v(x_3))$. The morphoelectro-

morphoelectrotonic tonic transform is a diagram of a neuron in which the distance between
transform any two points is determined by the logarithm of the ratio of the membrane potentials at these two locations, not by the actual size of the neuron.

Another morphoelectrotonic transform can be used to indicate the amount of delay in the voltage waveform produced by a transient input current. The morphoelectrotonic transform uses a definition of delay different from that used in Figure 6.8B. The delay between any two points is defined as the difference between the centroid, or center of "mass", of the voltage response at these points. Specifically, the centroid at point x is defined as $\int dt\, t v(x,t) / \int dt\, v(x,t)$. Like the log-attenuation, the delay between any two points on a neuron is represented in the morphoelectrotonic transform as a distance.

Morphoelectrotonic transforms of a pyramidal cell from layer 5 of cat visual cortex are shown in figures 6.13 and 6.14. The left panel of figure 6.13 is a normal drawing of the neuron being studied, the middle panel shows the steady-state attenuation, and the right panel shows the delay. The transformed diagrams correspond to current being injected peripherally, with somatic potentials being compared to dendritic potentials. These figures indicate that, for potentials generated in the periphery, the apical and basal dendrites are much more uniform than the morphology would suggest.

The small neuron diagram at the upper left of figure 6.14 shows attenuation for the reverse situation from figure 6.13, when constant current is injected into the soma and dendritic potentials are compared with the somatic potential. Note how much smaller this diagram is than the one in the central panel of figure 6.13. This illustrates the general feature, mentioned previously, that potentials are attenuated much less in the outward than in the inward direction. This is because the thin dendrites provide less of a current sink for potentials arising from the soma than the soma provides for potentials coming from the dendrites.

The capacitance of neuronal cables causes the voltage attenuation for time-dependent current injection to increase as a function of frequency. Figure 6.14 compares the attenuation of dendritic potentials relative to the somatic potential when constant or sinusoidal current of two different frequencies is injected into the soma. Clearly, attenuation increases dramatically as a function of frequency. Thus, a neuron that appears electrotonically compact for static or low frequency current injection may be not compact when higher frequencies are considered. For example, action potential waveforms, which correspond to frequencies around 500 Hz, are much more severely attenuated within neurons than slower varying potentials.

6.4   Multi-compartment Models

The cable equation can be solved analytically only in relatively simple cases. When the complexities of real membrane conductances are included, the membrane potential must be computed numerically. This is done by splitting the neuron being modeled into separate regions or compartments, and approximating the continuous membrane potential $V(x,t)$ by a discrete set of values representing the potentials within the different compartments. This assumes that each compartment is small enough so that there is negligible variation of the membrane potential across it. The precision of such a multi-compartmental description depends on the

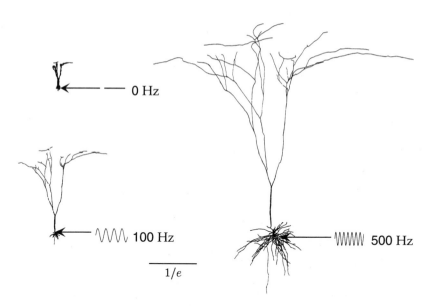

Figure 6.14 Morphoelectrotonic transforms of the same neuron as in figure 6.13 but showing the outward log-attenuation for constant and oscillating input currents. Distances in these diagrams are proportional to the logarithm of the amplitude of the voltage oscillations at a given point divided by the amplitude of the oscillations at the soma when a sinusoidal current is injected into the soma. The upper left panel corresponds to constant current injection, the lower left panel to sinusoidal current injection at a frequency of 100 Hz, and the right panel to an injection frequency of 500 Hz. The scale bar denotes the distance corresponding to an attenuation of $\exp(-1)$. (Adapted from Zador et al., 1995.)

number of compartments used and on their size relative to the length constants that characterize their electrotonic compactness. Figure 6.15 shows a schematic diagram of a cortical pyramidal neuron, along with a series of compartmental approximations of its structure. The number of compartments used can range from thousands, in some models, to one, for the description at the extreme right of figure 6.15.

In a multi-compartment model, each compartment has its own membrane potential V_μ (where μ labels compartments), and its own gating variables that determine the membrane current for compartment μ, i_m^μ. Each membrane potential V_μ satisfies an equation similar to 6.1 except that the compartments couple to their neighbors in the multi-compartment structure (figure 6.16). For a nonbranching cable, each compartment is coupled to two neighbors and the equations for the membrane potentials of the compartments are

$$c_m \frac{dV_\mu}{dt} = -i_m^\mu + \frac{I_e^\mu}{A_\mu} + g_{\mu,\mu+1}(V_{\mu+1} - V_\mu) + g_{\mu,\mu-1}(V_{\mu-1} - V_\mu). \quad (6.29)$$

Here I_e^μ is the total electrode current flowing into compartment μ, and A_μ is its surface area. Compartments at the ends of a cable have only one neighbor, and thus only a single term replacing the last two terms in equation 6.29. For a compartment where a cable branches in two, there are

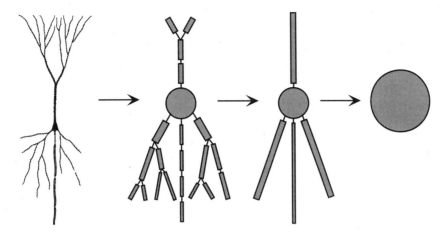

Figure 6.15 A sequence of approximations of the structure of a neuron. The neuron is represented by a variable number of discrete compartments, each representing a region that is described by a single membrane potential. The connectors between compartments represent resistive couplings. The simplest description is the single-compartment model furthest to the right.

three such terms, corresponding to coupling of the branching node to the first compartment in each of the daughter branches.

The constant $g_{\mu,\mu'}$ that determines the resistive coupling from neighboring compartment μ' to compartment μ is determined by computing the current that flows from one compartment to its neighbor due to Ohm's law. For simplicity, we begin by computing the coupling between two compartments that have the same length L and radius a. Using the results of chapter 5, the resistance between two such compartments, measured from their centers, is the intracellular resistivity, r_L times the distance between the compartment centers divided by the cross-sectional area, $r_L L/(\pi a^2)$. The total current flowing from compartment $\mu + 1$ to compartment μ is then $\pi a^2 (V_{\mu+1} - V_\mu)/r_L L$. Equation 6.29 for the potential within a compartment μ refers to currents per unit area of membrane. Thus, we must divide the total current from compartment μ' by the surface area of compartment μ, $2\pi a L$, and we find that $g_{\mu,\mu'} = a/(2r_L L^2)$.

The value of $g_{\mu,\mu'}$ is given by a more complex expression if the two neighboring compartments have different lengths or radii. This can occur when a tapering cable is approximated by a sequence of cylindrical compartments, or at a branch point where a single compartment connects with two other compartments, as in figure 6.16. In either case, suppose that compartment μ has length L_μ and radius a_μ, and compartment μ' has length $L_{\mu'}$ and radius $a_{\mu'}$. The resistance between these two compartments is the sum of the two resistances from the middle of each compartment to the junction between them, $r_L L_\mu/(2\pi a_\mu^2) + r_L L_{\mu'}/(2\pi a_{\mu'}^2)$. To compute $g_{\mu,\mu'}$ we invert this expression and divide the result by the total surface area of

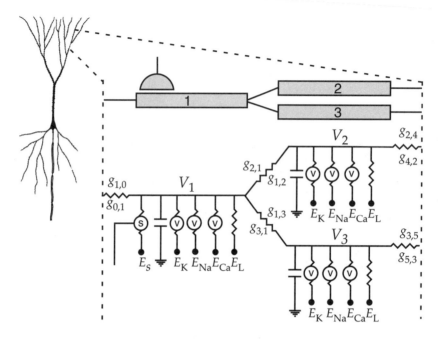

Figure 6.16 A multi-compartment model of a neuron. The expanded region shows three compartments at a branch point where a single cable splits into two. Each compartment has membrane and synaptic conductances, as indicated by the equivalent electrical circuit, and the compartments are coupled together by resistors. Although a single resistor symbol is drawn, note that $g_{\mu,\mu'}$ is not necessarily equal to $g_{\mu',\mu}$.

compartment μ, $2\pi a_\mu L_\mu$, which gives

$$g_{\mu,\mu'} = \frac{a_\mu a_{\mu'}^2}{r_L L_\mu (L_\mu a_{\mu'}^2 + L_{\mu'} a_\mu^2)}\,. \tag{6.30}$$

Equations 6.29 for all of the compartments of a model determine the membrane potential throughout the neuron with a spatial resolution given by the compartment size. An efficient method for integrating the coupled multi-compartment equations is discussed in appendix B. Using this scheme, models can be integrated numerically with excellent efficiency, even those involving large numbers of compartments. Such integration schemes are built into neuron simulation software packages such as Neuron and Genesis.

Action-Potential Propagation Along an Unmyelinated Axon

As an example of multi-compartment modeling, we simulate the propagation of an action potential along an unmyelinated axon. In this model, each compartment has the same membrane conductances as the single-compartment Hodgkin-Huxley model discussed in chapter 5. The different compartments are joined together in a single nonbranching cable

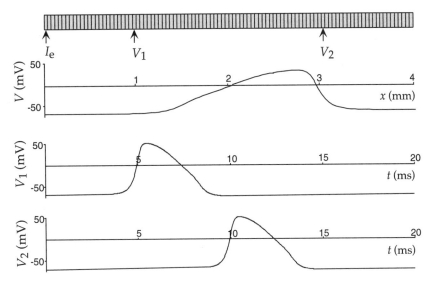

Figure 6.17 Propagation of an action potential along a multi-compartment model axon. The upper panel shows the multi-compartment representation of the axon with 100 compartments. The axon segment shown is 4 mm long and has a radius of 1 μm. An electrode current sufficient to initiate action potentials is injected at the point marked I_e. The panel beneath this shows the membrane potential as a function of position along the axon, at $t = 9.75$ ms. The spatial position in this panel is aligned with the axon depicted above it. The action potential is moving to the right. The bottom two panels show the membrane potential as a function of time at the two locations denoted by the arrows and symbols V_1 and V_2 in the upper panel.

representing a length of axon. Figure 6.17 shows an action-potential propagating along an axon modeled in this way. The action potential extends over more than 1 mm of axon and travels about 2 mm in 5 ms, for a speed of 0.4 m/s.

Although action potentials typically move along axons in a direction outward from the soma (called orthodromic propagation), the basic process of action-potential propagation does not favor one direction over the other. Propagation in the reverse direction, called antidromic propagation, is possible under certain stimulation conditions. For example, if an axon is stimulated in the middle of its length, action potentials will propagate in both directions away from the point of stimulation. Once an action potential starts moving along an axon, it does not generate a second action potential moving in the opposite direction because of refractory effects. The region in front of a moving action potential is ready to generate a spike as soon as enough current moves longitudinally down the axon from the region currently spiking to charge the next region up to spiking threshold. However, Na$^+$ conductances in the region just behind the moving action potential are still partially inactivated, so this region cannot generate another spike until after a recovery period. By the time the trailing region has recovered, the action potential has moved too far away to generate a second spike.

orthodromic and antidromic propagation

Refractoriness following spiking has a number of other consequences for action-potential propagation. Two action potentials moving in opposite directions that collide annihilate one another because they cannot pass through each other's trailing refractory regions. Refractoriness also keeps action potentials from reflecting off the ends of axon cables, which avoids the impedance matching needed to prevent reflection from the ends of ordinary electrical cables.

The propagation velocity for an action potential along an unmyelinated axon is proportional to the ratio of the electrotonic length constant to the membrane time constant, $\lambda/\tau_m = (a/(2c_m^2 r_L r_m))^{1/2}$. This is proportional to the square root of the axon radius. The square-root dependence of the propagation speed on the axon radius means that thick axons are required to achieve high action-potential propagation speeds, and the squid giant axon is an extreme example. Action-potential propagation can also be sped up by covering the axon with an insulating myelin wrapping, as we discuss next.

Action-Potential Propagation Along a Myelinated Axon

saltatory propagation

Many axons in vertebrates are covered with an insulating sheath of myelin except at gaps, called the nodes of Ranvier, where there is a high density of fast voltage-dependent Na^+ channels (see figure 6.18A). The myelin sheath consists of many layers of glial cell membrane wrapped around the axon. This gives the myelinated region of the axon a very high membrane resistance and a small membrane capacitance. This results in what is called saltatory propagation, in which membrane potential depolarization is transferred passively down the myelin-covered sections of the axon, and action potentials are actively regenerated at the nodes of Ranvier. Figure 6.18A shows an equivalent circuit for a multi-compartment model of a myelinated axon.

We can compute the capacitance of a myelin-covered axon by treating the myelin sheath as an extremely thick cell membrane. Consider the geometry shown in the cross-sectional diagram of figure 6.18B. The myelin sheath extends from the radius a_1 of the axon core to the outer radius a_2. For calculational purposes, we can think of the myelin sheath as being made of a series of thin, concentric cylindrical shells. The capacitances of these shells combine in series to make up the full capacitance of the myelinated axon. If a single layer of cell membrane has thickness d_m and capacitance per unit area c_m, the capacitance of a cylinder of membrane of radius a, thickness Δa, and length L is $c_m 2\pi d_m L/\Delta a$. According to the rule for capacitors in series, the inverse of the total capacitance is obtained by adding the inverses of the individual capacitances. The capacitance of a myelinated cylinder of length L and the dimensions in figure 6.18B is then obtained by taking the limit $\Delta a \to 0$ and integrating,

$$\frac{1}{C_m} = \frac{1}{c_m 2\pi d_m L} \int_{a_1}^{a_2} \frac{da}{a} = \frac{\ln(a_2/a_1)}{c_m 2\pi d_m L}. \qquad (6.31)$$

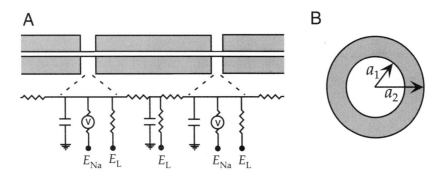

Figure 6.18 A myelinated axon. (A) The equivalent circuit for a multi-compartment representation of a myelinated axon. The myelinated segments are represented by a membrane capacitance, a longitudinal resistance, and a leakage conductance. The nodes of Ranvier also contain a voltage-dependent Na^+ conductance. (B) A cross section of a myelinated axon consisting of a central axon core of radius a_1 and a myelin sheath making the outside radius a_2.

A re-evaluation of the derivation of the linear cable equation earlier in this chapter indicates that the equation describing the membrane potential along the myelinated sections of an axon, in the limit of infinite resistance for the myelinated membrane and with $i_e = 0$, is

$$\frac{C_m}{L}\frac{\partial v}{\partial t} = \frac{\pi a_1^2}{r_L}\frac{\partial^2 v}{\partial x^2}. \tag{6.32}$$

This is equivalent to the diffusion equation, $\partial v/\partial t = D\partial^2 v/\partial x^2$, with diffusion constant $D = \pi a_1^2 L/(C_m r_L) = a_1^2 \ln(a_2/a_1)/(2c_m r_L d_m)$. It is interesting to compute the inner core radius, a_1, that maximizes this diffusion constant for a fixed outer radius a_2. Setting the derivative of D with respect to a_1 to 0 gives the optimal inner radius $a_1 = a_2 \exp(-1/2)$ or $a_1 \approx 0.6a_2$. An inner core fraction of 0.6 is typical for myelinated axons. This indicates that for a given outer radius, the thickness of myelin maximizes the diffusion constant along the myelinated axon segment.

At the optimal ratio of radii, $D = a_2^2/(4ec_m r_L d_m)$, which is proportional to the square of the axon radius. Because of the form of the diffusion equation it obeys with this value of D, v can be written as a function of x/a_2 and t. This scaling implies that the propagation velocity for a myelinated cable is proportional to a_2, that is, to the axon radius, not its square root (as in the case of an unmyelinated axon). Increasing the axon radius by a factor of 4, for example, increases the propagation speed of an unmyelinated cable only by a factor of 2, while it increases the speed for a myelinated cable fourfold.

6.5   Chapter Summary

We continued the discussion of neuron modeling that began in chapter 5 by considering models with more complete sets of conductances and techniques for incorporating neuronal morphology. We introduced A-type K^+, transient Ca^{2+}, and Ca^{2+}-dependent K^+ conductances, and noted their effect on neuronal activity. The cable equation and its linearized version were introduced to examine the effects of morphology on membrane potentials. Finally, multi-compartment models were presented and used to discuss propagation of action potentials along unmyelinated and myelinated axons.

6.6   Appendices

A: Gating Functions for Conductance-Based Models

Connor-Stevens Model

The rate functions used for the gating variables n, m, and h of the Connor-Stevens model, in units of $1/\text{ms}$ with V in units of mV, are

$$\alpha_m = \frac{0.38(V + 29.7)}{1 - \exp(-0.1(V + 29.7))} \quad \beta_m = 15.2\exp(-0.0556(V + 54.7))$$
$$\alpha_h = 0.266\exp(-0.05(V + 48)) \quad \beta_h = 3.8/(1 + \exp(-0.1(V + 18)))$$
$$\alpha_n = \frac{0.02(V + 45.7)}{1 - \exp(-0.1(V + 45.7))} \quad \beta_n = 0.25\exp(-0.0125(V + 55.7)).$$

$$(6.33)$$

The A-current is described directly in terms of the asymptotic values and τ functions for its gating variables (with τ_a and τ_b in units of ms and V in units of mV),

$$a_\infty = \left(\frac{0.0761\exp(0.0314(V + 94.22))}{1 + \exp(0.0346(V + 1.17))} \right)^{1/3} \quad (6.34)$$

$$\tau_a = 0.3632 + 1.158/(1 + \exp(0.0497(V + 55.96))) \quad (6.35)$$

$$b_\infty = \left(\frac{1}{1 + \exp(0.0688(V + 53.3))} \right)^4 \quad (6.36)$$

and

$$\tau_b = 1.24 + 2.678/(1 + \exp(0.0624(V + 50))). \quad (6.37)$$

Transient Ca^{2+} Conductance

The gating functions used for the variables M and H in the transient Ca^{2+} conductance model we discussed, with V in units of mV and τ_M and τ_H in ms, are

$$M_\infty = \frac{1}{1 + \exp\left(-(V+57)/6.2\right)} \tag{6.38}$$

$$H_\infty = \frac{1}{1 + \exp\left((V+81)/4\right)} \tag{6.39}$$

$$\tau_M = 0.612 + \left(\exp\left(-(V+132)/16.7\right) + \exp\left((V+16.8)/18.2\right)\right)^{-1} \tag{6.40}$$

and

$$\tau_H = \begin{cases} \exp\left((V+467)/66.6\right) & \text{if } V < -80 \text{ mV} \\ 28 + \exp\left(-(V+22)/10.5\right) & \text{if } V \ge -80 \text{ mV.} \end{cases} \tag{6.41}$$

Ca^{2+}-dependent K$^+$ Conductance

The gating functions used for the Ca^{2+}-dependent K$^+$ conductance we discussed, with V in units of mV and τ_c in ms, are

$$c_\infty = \left(\frac{[\text{Ca}^{2+}]}{[\text{Ca}^{2+}] + 3\mu\text{M}}\right) \frac{1}{1 + \exp(-(V+28.3)/12.6)} \tag{6.42}$$

and

$$\tau_c = 90.3 - \frac{75.1}{1 + \exp(-(V+46)/22.7)} \,. \tag{6.43}$$

B: Integrating Multi-compartment Models

Multi-compartment models are defined by a coupled set of differential equations (equation 6.29), one for each compartment. There are also gating variables for each compartment, but these involve only the membrane potential (and possibly Ca^{2+} concentration) within that compartment, and integrating their equations can be handled as in the single-compartment case using the approach discussed in appendix B of chapter 5. Integrating the membrane potentials for the different compartments is more complex because they are coupled to each other.

Equation 6.29, for the membrane potential within compartment μ, can be written in the form

$$\frac{dV_\mu}{dt} = B_\mu V_{\mu-1} + C_\mu V_\mu + D_\mu V_{\mu+1} + F_\mu \,, \tag{6.44}$$

where

$$B_\mu = c_m^{-1} g_{\mu,\mu-1}, \quad C_\mu = -c_m^{-1}(\sum_i g_i^\mu + g_{\mu,\mu+1} + g_{\mu,\mu-1}),$$

$$D_\mu = c_m^{-1} g_{\mu,\mu+1}, \quad F_\mu = c_m^{-1}(\sum_i g_i^\mu E_i + I_e^\mu / A_\mu). \qquad (6.45)$$

Note that the gating variables and other parameters have been absorbed into the values of the coefficients B_μ, C_μ, D_μ, and F_μ in this equation. Equation 6.44, with μ running over all of the compartments of the model, generates a set of coupled differential equations. Because of the coupling between compartments, we cannot use the method discussed in appendix A of chapter 5 to integrate these equations. Instead, we present another method that shares some of the positive features of that approach. The Runge-Kutta method, which is a standard numerical integrator, is poorly suited for this application and is likely to run orders of magnitude slower than the method described below.

Two of the most important features of an integration method are accuracy and stability. Accuracy refers to how closely numerical finite-difference methods reproduce the exact solution of a differential equation as a function of the integration step size Δt. Stability refers to what happens when Δt is chosen to be excessively large and the method starts to become inaccurate. A stable integration method will degrade smoothly as Δt is increased, producing results of steadily decreasing accuracy. An unstable method, on the other hand, will at some point display a sudden transition and generate wildly inaccurate results. Given the tendency of impatient modelers to push the limits on Δt, it is highly desirable to have a method that is stable.

Defining

$$V_\mu(t + \Delta t) = V_\mu(t) + \Delta V_\mu, \qquad (6.46)$$

the finite difference form of equation 6.44 gives the update rule

$$\Delta V_\mu = (B_\mu V_{\mu-1}(t) + C_\mu V_\mu(t) + D_\mu V_{\mu+1}(t) + F_\mu) \Delta t, \qquad (6.47)$$

which is how ΔV_μ is computed using the so-called Euler method. This method is both inaccurate and unstable. The stability of the method can be improved dramatically by evaluating the membrane potentials on the right side of equation 6.47 not at time t, but at a later time $t + z\Delta t$, so that

$$\Delta V_\mu = (B_\mu V_{\mu-1}(t + z\Delta t) + C_\mu V_\mu(t + z\Delta t) + D_\mu V_{\mu+1}(t + z\Delta t) + F_\mu) \Delta t. \qquad (6.48)$$

Two such methods are predominantly used, the reverse Euler method, for which $z = 1$, and the Crank-Nicholson method with $z = 0.5$. The reverse Euler method is the more stable of the two and the Crank-Nicholson is the more accurate. In either case, ΔV_μ is determined from equation 6.48. These methods are called implicit because equation 6.48 must be solved to determine ΔV_μ. To do this, we write $V_\mu(t + z\Delta t) \approx V_\mu(t) + z\Delta V_\mu$ and likewise for $V_{\mu \pm 1}$. Substituting this into equation 6.48 gives

$$\Delta V_\mu = b_\mu \Delta V_{\mu-1} + c_\mu \Delta V_\mu + d_\mu \Delta V_{\mu+1} + f_\mu, \qquad (6.49)$$

where

$$b_\mu = B_\mu z\Delta t, \quad c_\mu = C_\mu z\Delta t, \quad d_\mu = D_\mu z\Delta t,$$
$$f_\mu = (F_\mu + B_\mu V_{\mu-1}(t) + C_\mu V_\mu(t) + D_\mu V_{\mu+1}(t))\Delta t. \qquad (6.50)$$

Equation 6.49 for all μ values provides a set of coupled linear equations for the quantities ΔV_μ. An efficient method exists for solving these equations (Hines, 1984; Tuckwell, 1988). We illustrate the method for a single, nonbranching cable that begins at compartment $\mu = 1$, so that $b_1 = 0$, and ends at compartment $\mu = N$, so $d_N = 0$. The method consists of solving equation 6.49 for ΔV_μ in terms of $\Delta V_{\mu+1}$ sequentially, starting at one end of the cable and proceeding to the other end. For example, if we start the procedure at compartment 1, ΔV_1 can be expressed as

$$\Delta V_1 = \frac{d_1 \Delta V_2 + f_1}{1 - c_1}. \qquad (6.51)$$

Substituting this into the equation 6.49 for $\mu = 2$ gives

$$\Delta V_2 = c_2' \Delta V_2 + d_2 \Delta V_3 + f_2', \qquad (6.52)$$

where $c_2' = c_2 + b_2 d_1/(1 - c_1)$ and $f_2' = f_2 + b_2 f_1/(1 - c_1)$. We now repeat the procedure going down the cable. At each stage, we solve for $\Delta V_{\mu-1}$ in terms of ΔV_μ, finding

$$\Delta V_{\mu-1} = \frac{d_{\mu-1} \Delta V_\mu + f_{\mu-1}'}{1 - c_{\mu-1}'}, \qquad (6.53)$$

where

$$c_{\mu+1}' = c_{\mu+1} + \frac{b_{\mu+1} d_\mu}{1 - c_\mu'} \qquad (6.54)$$

and

$$f_{\mu+1}' = f_{\mu+1} + \frac{b_{\mu+1} f_\mu'}{1 - c_\mu'}. \qquad (6.55)$$

Finally, when we get to the end of the cable, we can solve for

$$\Delta V_N = \frac{f_N'}{1 - c_N'} \qquad (6.56)$$

because $d_N = 0$.

The procedure for computing all the ΔV_μ is the following. Define $c_1' = c_1$ and $f_1' = f_1$ and iterate equations 6.54 and 6.55 down the length of the cable to define all the c' and f' parameters. Then solve for ΔV_N from equation 6.56 and iterate back up the cable, solving for the ΔV's using 6.53. This process takes only $2N$ steps.

We leave the extension of this method to the case of a branched cable as an exercise for the reader. The general procedure is similar to the one we presented for a nonbranching cable. The equations are solved by starting at the ends of the branches and moving in toward their branching node, then continuing on as for a nonbranching cable, and finally reversing direction and completing the solution moving in the opposite direction along the cable and its branches.

6.7   Annotated Bibliography

Many of the references for chapter 5 apply to this chapter as well, including **Jack et al. (1975), Tuckwell (1988), Johnston & Wu (1995), Koch & Segev (1998), Koch (1998), Hille (1992)**, and **Mascagni & Sherman (1998)**. **Rall (1977)** describes cable theory, the equivalent cable model of dendritic trees, and the 3/2 law. The solution of equation 6.21 can be constructed using the set of rules for solving the linear cable equation on arbitrary trees found in Abbott (1992; see also Abbott et al., 1991). **Marder & Calabrese (1996)** reviews neuromodulation.

Two freely available software packages for detailed neuronal modeling are in wide use, Neuron (see Hines & Carnevale, 1997) and Genesis (see Bower & Beeman, 1998). These are available at http://www.neuron.yale.edu and http://genesis.bbb.caltech.edu/GENESIS/genesis.html.

7 Network Models

7.1 Introduction

Extensive synaptic connectivity is a hallmark of neural circuitry. For example, a typical neuron in the mammalian neocortex receives thousands of synaptic inputs. Network models allow us to explore the computational potential of such connectivity, using both analysis and simulations. As illustrations, we study in this chapter how networks can perform the following tasks: coordinate transformations needed in visually guided reaching, selective amplification leading to models of simple and complex cells in primary visual cortex, integration as a model of short-term memory, noise reduction, input selection, gain modulation, and associative memory. Networks that undergo oscillations are also analyzed, with application to the olfactory bulb. Finally, we discuss network models based on stochastic rather than deterministic dynamics, using the Boltzmann machine as an example.

Neocortical circuits are a major focus of our discussion. In the neocortex, which forms the convoluted outer surface of the (for example) human brain, neurons lie in six vertical layers highly coupled within cylindrical columns. Such columns have been suggested as basic functional units, *cortical columns* and stereotypical patterns of connections both within a column and between columns are repeated across cortex. There are three main classes of interconnections within cortex, and in other areas of the brain as well. Feedforward connections bring input to a given region from another re- *feedforward,* gion located at an earlier stage along a particular processing pathway. Re- *recurrent,* current synapses interconnect neurons within a particular region that are *and top-down* considered to be at the same stage along the processing pathway. These *connections* may include connections within a cortical column as well as connections between both nearby and distant cortical columns within a region. Top-down connections carry signals back from areas located at later stages. These definitions depend on how the region being studied is specified and on the hierarchical assignment of regions along a pathway. In general, neurons within a given region send top-down projections back to the areas from which they receive feedforward input, and receive top-down input from the areas to which they project feedforward output. The numbers, though not necessarily the strengths, of feedforward and top-down fibers between connected regions are typically comparable, and recurrent synapses typically outnumber feedforward or top-down inputs. We begin this chapter by studying networks with purely feedforward input and then study the effects of recurrent connections. The analysis of top-down connections, for which it is more difficult to establish clear computational roles, is left until chapter 10.

The most direct way to simulate neural networks is to use the methods discussed in chapters 5 and 6 to synaptically connect model spiking neurons. This is a worthwhile and instructive enterprise, but it presents significant computational, calculational, and interpretational challenges. In this chapter, we follow a simpler approach and construct networks of neuron-like units with outputs consisting of firing rates rather than action potentials. Spiking models involve dynamics over time scales ranging from channel openings that can take less than a millisecond, to collective network processes that may be several orders of magnitude slower. Firing-rate models avoid the short time scale dynamics required to simulate action potentials and thus are much easier to simulate on computers. Firing-rate models also allow us to present analytic calculations of some aspects of network dynamics that could not be treated in the case of spiking neurons. Finally, spiking models tend to have more free parameters than firing-rate models, and setting these appropriately can be difficult.

There are two additional arguments in favor of firing-rate models. The first concerns the apparent stochasticity of spiking. The models discussed in chapters 5 and 6 produce spike sequences deterministically in response to injected current or synaptic input. Deterministic models can predict spike sequences accurately only if all their inputs are known. This is unlikely to be the case for the neurons in a complex network, and network models typically include only a subset of the many different inputs to individual neurons. Therefore, the greater apparent precision of spiking models may not actually be realized in practice. If necessary, firing-rate models can be used to generate stochastic spike sequences from a deterministically computed rate, using the methods discussed in chapters 1 and 2.

The second argument involves a complication with spiking models that arises when they are used to construct simplified networks. Although cortical neurons receive many inputs, the probability of finding a synaptic connection between a randomly chosen pair of neurons is actually quite low. Capturing this feature, while retaining a high degree of connectivity through polysynaptic pathways, requires including a large number of neurons in a network model. A standard way of dealing with this problem is to use a single model unit to represent the average response of several neurons that have similar selectivities. These "averaging" units can then be interconnected more densely than the individual neurons of the actual network, so fewer of them are needed to build the model. If neural responses are characterized by firing rates, the output of the model unit is simply the average of the firing rates of the neurons it represents collectively. However, if the response is a spike, it is not clear how the spikes of the represented neurons can be averaged. The way spiking models are

typically constructed, an action potential fired by the model unit dupli-
cates the effect of all the neurons it represents firing synchronously. Not
surprisingly, such models tend to exhibit large-scale synchronization unlike anything seen in a healthy brain.

Firing-rate models also have their limitations. They cannot account for
aspects of spike timing and spike correlations that may be important for
understanding nervous system function. Firing-rate models are restricted
to cases where the firing of neurons in a network is uncorrelated, with little
synchronous firing, and where precise patterns of spike timing are unim-
portant. In such cases, comparisons of spiking network models with models that use firing-rate descriptions have shown that they produce similar
results. Nevertheless, the exploration of neural networks undoubtedly requires the use of both firing-rate and spiking models.

7.2   Firing-Rate Models

As discussed in chapter 1, the sequence of spikes generated by a neuron
is completely characterized by the neural response function $\rho(t)$, which
consists of δ function spikes located at times when the neuron fired action
potentials. In firing-rate models, the exact description of a spike sequence
provided by the neural response function $\rho(t)$ is replaced by the approximate description provided by the firing rate $r(t)$. Recall from chapter 1 that
$r(t)$ is defined as the probability density of firing and is obtained from $\rho(t)$
by averaging over trials. The validity of a firing-rate model depends on
how well the trial-averaged firing rate of network units approximates the
effect of actual spike sequences on the dynamic behavior of the network.

The replacement of the neural response function by the corresponding firing rate is typically justified by the fact that each network neuron has a
large number of inputs. Replacing $\rho(t)$, which describes an actual spike
train, with the trial-averaged firing rate $r(t)$ is justified if the quantities of
relevance for network dynamics are relatively insensitive to the trial-to-
trial fluctuations in the spike sequences represented by $\rho(t)$. In a network
model, the relevant quantities that must be modeled accurately are the
total inputs for the neurons within the network. For any single synaptic
input, the trial-to-trial variability is likely to be large. However, if we sum
the input over many synapses activated by uncorrelated presynaptic spike
trains, the mean of the total input typically grows linearly with the number
of synapses, while its standard deviation grows only as the square root of
the number of synapses. Thus, for uncorrelated presynaptic spike trains,
using presynaptic firing rates in place of the actual presynaptic spike trains
may not significantly modify the dynamics of the network. Conversely, a
firing-rate model will fail to describe a network adequately if the presynaptic inputs to a substantial fraction of its neurons are correlated. This
can occur, for example, if the presynaptic neurons fire synchronously.

The synaptic input arising from a presynaptic spike train is effectively filtered by the dynamics of the conductance changes that each presynaptic action potential evokes in the postsynaptic neuron (see chapter 5) and the dynamics of propagation of the current from the synapse to the soma. The temporal averaging provided by slow synaptic or membrane dynamics can reduce the effects of spike-train variability and help justify the approximation of using firing rates instead of presynaptic spike trains. Firing-rate models are more accurate if the network being modeled has a significant amount of synaptic transmission that is slow relative to typical presynaptic interspike intervals.

The construction of a firing-rate model proceeds in two steps. First, we determine how the total synaptic input to a neuron depends on the firing rates of its presynaptic afferents. This is where we use firing rates to approximate neural response functions. Second, we model how the firing rate of the postsynaptic neuron depends on its total synaptic input. Firing-rate response curves are typically measured by injecting current into the soma of a neuron. We therefore find it most convenient to define the total synaptic input as the total current delivered to the soma as a result of all the synaptic conductance changes resulting from presynaptic action po-
synaptic current I_s tentials. We denote this total synaptic current by I_s. We then determine the postsynaptic firing rate from I_s. In general, I_s depends on the spatially inhomogeneous membrane potential of the neuron, but we assume that, other than during action potentials or transient hyperpolarizations, the membrane potential remains close to, but slightly below, the threshold for action potential generation. An example of this type of behavior is seen in the upper panels of figure 7.2. I_s is then approximately equal to the synaptic current that would be measured from the soma in a voltage-clamp experiment, except for a reversal of sign. In the next section, we model how I_s depends on presynaptic firing rates.

In the network models we consider, both the output from, and the input to, a neuron are characterized by firing rates. To avoid a proliferation of sub- and superscripts on the quantity $r(t)$, we use the letter u to denote a
input rate u presynaptic firing rate, and v to denote a postsynaptic rate. Note that v is
output rate v used here to denote a firing rate, not a membrane potential. In addition, we use these two letters to distinguish input and output firing rates in network models, a convention we retain through the remaining chapters. When we consider multiple input or output neurons, we use vectors $\mathbf{u}$ and
input rate vector $\mathbf{u}$ $\mathbf{v}$ to represent their firing rates collectively, with the components of these
output rate vector $\mathbf{v}$ vectors representing the firing rates of the individual input and output units.

The Total Synaptic Current

Consider a neuron receiving N_u synaptic inputs labeled by $b = 1, 2, \ldots, N_u$ (figure 7.1). The firing rate of input b is denoted by u_b, and the input rates are represented collectively by the N_u-component vector $\mathbf{u}$. We model how the synaptic current I_s depends on presynaptic firing rates by first consid-

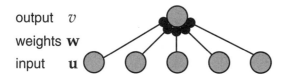

output v

weights **w**

input   **u**

Figure 7.1 Feedforward inputs to a single neuron. Input rates **u** drive a neuron at an output rate v through synaptic weights given by the vector **w**.

ering how it depends on presynaptic spikes. If an action potential arrives at input b at time 0, we write the synaptic current generated in the soma of the postsynaptic neuron at time t as $w_b K_s(t)$, where w_b is the synaptic weight and $K_s(t)$ is called the synaptic kernel. Collectively, the synaptic weights are represented by a synaptic weight vector **w**, which has N_u *synaptic weights* **w** components w_b. The amplitude and sign of the synaptic current generated by input b are determined by w_b. For excitatory synapses, $w_b > 0$, and for inhibitory synapses, $w_b < 0$. In this formulation of the effect of presynaptic spikes, the probability of transmitter release from a presynaptic terminal is absorbed into the synaptic weight factor w_b, and we do not include short-term plasticity in the model (although this can be done by making w_b a dynamic variable).

The synaptic kernel, $K_s(t) \geq 0$, describes the time course of the synaptic   *synaptic* current in response to a presynaptic spike arriving at time $t=0$. This time   *kernel $K_s(t)$* course depends on the dynamics of the synaptic conductance activated by the presynaptic spike, and also on both the passive and the active properties of the dendritic cables that carry the synaptic current to the soma. For example, long passive cables broaden the synaptic kernel and slow its rise from 0. Cable calculations or multi-compartment simulations, such as those discussed in chapter 6, can be used to compute $K_s(t)$ for a specific dendritic structure. To avoid ambiguity, we normalize $K_s(t)$ by requiring its integral over all positive times to be 1. At this point, for simplicity, we use the same function $K_s(t)$ to describe all synapses.

Assuming that the effects of the spikes at a single synapse sum linearly, the total synaptic current at time t arising from a sequence of presynaptic spikes occurring at input b at times t_i is given by

$$w_b \sum_{t_i < t} K_s(t - t_i) = w_b \int_{-\infty}^{t} d\tau \, K_s(t - \tau) \rho_b(\tau) . \qquad (7.1)$$

In the second expression, we have used the neural response function, $\rho_b(\tau) = \sum_i \delta(\tau - t_i)$, to describe the sequence of spikes fired by presynaptic neuron b. The equality follows from integrating over the sum of δ functions in the definition of $\rho_b(\tau)$. If there is no nonlinear interaction between different synaptic currents, the total synaptic current coming from all presynaptic inputs is obtained simply by summing,

$$I_s = \sum_{b=1}^{N_u} w_b \int_{-\infty}^{t} d\tau \, K_s(t - \tau) \rho_b(\tau) . \qquad (7.2)$$

As discussed previously, the critical step in the construction of a firing-rate model is the replacement of the neural response function $\rho_b(\tau)$ in equation 7.2 with the firing rate of neuron b, $u_b(\tau)$, so that we write

$$I_s = \sum_{b=1}^{N_u} w_b \int_{-\infty}^{t} d\tau\, K_s(t-\tau) u_b(\tau). \qquad (7.3)$$

The synaptic kernel most frequently used in firing-rate models is an exponential, $K_s(t) = \exp(-t/\tau_s)/\tau_s$. With this kernel, we can describe I_s by a differential equation if we take the derivative of equation 7.3 with respect to t,

$$\tau_s \frac{dI_s}{dt} = -I_s + \sum_{b=1}^{N_u} w_b u_b = -I_s + \mathbf{w} \cdot \mathbf{u}. \qquad (7.4)$$

dot product

In the second equality, we have expressed the sum $\sum w_b u_b$ as the dot product of the weight and input vectors, $\mathbf{w} \cdot \mathbf{u}$. In this and the following chapters, we primarily use the vector versions of equations such as 7.4, but when we first introduce an important new equation, we often write it in its subscripted form as well.

Recall that K describes the temporal evolution of the synaptic current due to both synaptic conductance and dendritic cable effects. For an electrotonically compact dendritic structure, τ_s will be close to the time constant that describes the decay of the synaptic conductance. For fast synaptic conductances such as those due to AMPA glutamate receptors, this may be as short as a few milliseconds. For a long, passive dendritic cable, τ_s may be larger than this, but its measured value is typically quite small.

The Firing Rate

Equation 7.4 determines the synaptic current entering the soma of a postsynaptic neuron in terms of the firing rates of the presynaptic neurons. To finish formulating a firing-rate model, we must determine the postsynaptic firing rate from our knowledge of I_s. For constant synaptic current, the firing rate of the postsynaptic neuron can be expressed as $v = F(I_s)$, where F is the steady-state firing rate as a function of somatic input current.

activation function $F(I_s)$

F is also called an activation function. F is sometimes taken to be a saturating function such as a sigmoid function. This is useful in cases where the derivative of F is needed in the analysis of network dynamics. It is also bounded from above, which can be important in stabilizing a network against excessively high firing rates. More often, we use a threshold

threshold γ

linear function $F(I_s) = [I_s - \gamma]_+$, where γ is the threshold and the notation $[\ ]_+$ denotes half-wave rectification, as in previous chapters. For convenience, we treat I_s in this expression as if it were measured in units of a firing rate (Hz), that is, as if I_s is multiplied by a constant that converts its units from nA to Hz. This makes the synaptic weights dimensionless. The threshold γ also has units of Hz.

For time-independent inputs, the relation $v = F(I_s)$ is all we need to know to complete the firing-rate model. The total steady-state synaptic current predicted by equation 7.4 for time-independent $\mathbf{u}$ is $I_s = \mathbf{w} \cdot \mathbf{u}$. This generates a steady-state output firing rate $v = v_\infty$ given by

$$v_\infty = F(\mathbf{w} \cdot \mathbf{u}). \tag{7.5}$$

The steady-state firing rate tells us how a neuron responds to constant current, but not to a current that changes with time. To model time-dependent inputs, we need to know the firing rate in response to a time-dependent synaptic current $I_s(t)$. The simplest assumption is that this is still given by the activation function, so $v = F(I_s(t))$ even when the total synaptic current varies with time. This leads to a firing-rate model in which the dynamics arises exclusively from equation 7.4,

firing-rate model
with current
dynamics

$$\tau_s \frac{dI_s}{dt} = -I_s + \mathbf{w} \cdot \mathbf{u} \quad \text{with} \quad v = F(I_s). \tag{7.6}$$

An alternative formulation of a firing-rate model can be constructed by assuming that the firing rate does not follow changes in the total synaptic current instantaneously, as was assumed for the model of equation 7.6. Action potentials are generated by the synaptic current through its effect on the membrane potential of the neuron. Due to the membrane capacitance and resistance, the membrane potential is, roughly speaking, a low-pass filtered version of I_s (see the Mathematical Appendix). For this reason, the time-dependent firing rate is often modeled as a low-pass filtered version of the steady-state firing rate,

$$\tau_r \frac{dv}{dt} = -v + F(I_s(t)). \tag{7.7}$$

The constant τ_r in this equation determines how rapidly the firing rate approaches its steady-state value for constant I_s, and how closely v can follow rapid fluctuations for a time-dependent $I_s(t)$. Equivalently, it measures the time scale over which v averages $F(I_s(t))$. The low-pass filtering effect of equation 7.7 is described in the Mathematical Appendix in the context of electrical circuit theory. The argument we have used to motivate equation 7.7 would suggest that τ_r should be approximately equal to the membrane time constant of the neuron. However, this argument really applies to the membrane potential, not the firing rate, and the dynamics of the two are not the same. Some network models use a value of τ_r that is considerably less than the membrane time constant. We re-examine this issue in the following section.

The second model that we have described involves the pair of equations 7.4 and 7.7. If one of these equations relaxes to its equilibrium point much more rapidly than the other, the pair can be reduced to a single equation. We discuss cases in which this occurs in the following section. For example, if $\tau_r \ll \tau_s$, we can make the approximation that equation 7.7 rapidly sets $v = F(I_s(t))$, and then the second model reduces to the first model that is defined by equation 7.6. If instead, $\tau_r \gg \tau_s$, we can make the approximation that equation 7.4 comes to equilibrium quickly compared with equation 7.7. Then we can make the replacement $I_s = \mathbf{w} \cdot \mathbf{u}$ in equation 7.7 and write

firing-rate equation

$$\tau_r \frac{dv}{dt} = -v + F(\mathbf{w} \cdot \mathbf{u}) . \tag{7.8}$$

For most of this chapter, we analyze network models described by the firing-rate dynamics of equation 7.8, although occasionally we consider networks based on equation 7.6.

Firing-Rate Dynamics

The firing-rate models described by equations 7.6 and 7.8 differ in their assumptions about how firing rates respond to and track changes in the input current to a neuron. In one case (equation 7.6), it is assumed that firing rates follow time-varying input currents instantaneously, without attenuation or delay. In the other case (equation 7.8), the firing rate is a low-pass filtered version of the input current. To study the relationship between input current and firing rate, it is useful to examine the firing rate of a spiking model neuron in response to a time-varying injected current, $I(t)$. The model used for this purpose in figure 7.2 is an integrate-and-fire neuron receiving balanced excitatory and inhibitory synaptic input along with a current injected into the soma that is the sum of constant and oscillating components. This model was discussed in chapter 5. The balanced synaptic input is used to represent background input not included in the computation of I_s, and it acts as a source of noise. The noise prevents effects, such as locking of the spiking to the oscillations of the injected current, that would invalidate a firing-rate description.

Figure 7.2 shows the firing rates of the model integrate-and-fire neuron in response to an input current $I(t) = I_0 + I_1 \cos(\omega t)$. The firing rate is plotted at different times during the cycle of the input current oscillations for ω corresponding to frequencies of 1, 50, and 100 Hz. For the panels on the left side, the constant component of the injected current (I_0) was adjusted so the neuron never stopped firing during the cycle. In this case, the relation $v(t) = F(I(t))$ (solid curves) provides an accurate description of the firing rate for all of the oscillation frequencies shown. As long as the neuron keeps firing fairly rapidly, the low-pass filtering properties of the membrane potential are not relevant for the dynamics of the firing rate. Low-pass filtering is irrelevant in this case, because the neuron is continually being shuttled between the threshold and reset values, so it never has a chance to settle exponentially anywhere near its steady-state value.

The right panels in figure 7.2 show that the situation is different if the input current is below the threshold for firing through a significant part

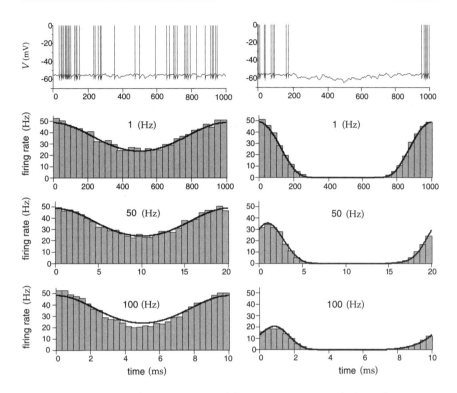

Figure 7.2 Firing rate of an integrate-and-fire neuron receiving balanced excitatory and inhibitory synaptic input and an injected current consisting of a constant and a sinusoidally varying term. For the left panels, the constant component of the injected current was adjusted so the firing never stopped during the oscillation of the varying part of the injected current. For the right panel, the constant component was lowered so the firing stopped during part of the cycle. The upper panels show two representative voltage traces of the model cell. The histograms beneath these traces were obtained by binning spikes generated over multiple cycles. They show the firing rate as a function of the time during each cycle of the injected current oscillations. The different rows correspond to 1, 50, and 100 Hz oscillation frequencies for the injected current. The solid curves show the fit of a firing-rate model that involves both instantaneous and low-pass filtered effects of the injected current. For the left panel, this reduces to the simple prediction $v = F(I(t))$. (Adapted from Chance, 2000.)

of the oscillation cycle. In this case, the firing is delayed and attenuated at high frequencies, as would be predicted by equation 7.7. In this case, the membrane potential stays below threshold for long enough periods of time that its dynamics become relevant for the firing of the neuron.

The essential message from figure 7.2 is that neither equation 7.6 nor equation 7.8 provides a completely accurate prediction of the dynamics of the firing rate at all frequencies and for all levels of injected current. A more complex model can be constructed that accurately describes the firing rate over the entire range of input current amplitudes and frequencies. The solid curves in figure 7.2 were generated by a model that expresses the firing rate as a function of both I from equation 7.6 and v from equation 7.8. In other words, it is a combination of the two models discussed in the pre-

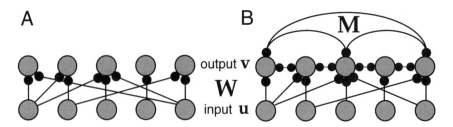

Figure 7.3 Feedforward and recurrent networks. (A) A feedforward network with input rates **u**, output rates **v**, and a feedforward synaptic weight matrix **W**. (B) A recurrent network with input rates **u**, output rates **v**, a feedforward synaptic weight matrix **W**, and a recurrent synaptic weight matrix **M**. Although we have drawn the connections between the output neurons as bidirectional, this does not necessarily imply connections of equal strength in both directions.

vious section with the firing rate given neither by $F(I)$ nor by v but by another function, $G(I, v)$. This compound model provides quite an accurate description of the firing rate of the integrate-and-fire model, but it is more complex than the models used in this chapter.

Feedforward and Recurrent Networks

Figure 7.3 shows examples of network models with feedforward and recurrent connectivity. The feedforward network of figure 7.3A has N_v output units with rates v_a ($a = 1, 2, \ldots, N_v$), denoted collectively by the vector **v**, driven by N_u input units with rates **u**. Equations 7.8 and 7.6 can easily be extended to cover this case by replacing the vector of synaptic weights **w** with a matrix **W**, with the matrix component W_{ab} representing the strength of the synapse from input unit b to output unit a. Using the formulation of equation 7.8, the output firing rates are then determined by

feedforward model

$$\tau_r \frac{d\mathbf{v}}{dt} = -\mathbf{v} + \mathbf{F}(\mathbf{W} \cdot \mathbf{u}) \quad \text{or} \quad \tau_r \frac{dv_a}{dt} = -v_a + F\left(\sum_{b=1}^{N_u} W_{ab} u_b\right). \quad (7.9)$$

We use the notation $\mathbf{W} \cdot \mathbf{u}$ to denote the vector with components $\sum_b W_{ab} u_b$. The use of the dot to represent a sum of a product of two quantities over a shared index is borrowed from the notation for the dot product of two vectors. The expression $\mathbf{F}(\mathbf{W} \cdot \mathbf{u})$ represents the vector with components $F(\sum W_{ab} u_b)$ for $a = 1, 2, \ldots, N_v$.

The recurrent network of figure 7.3B also has two layers of neurons with rates **u** and **v**, but in this case the neurons of the output layer are interconnected with synaptic weights described by a matrix **M**. Matrix element $M_{aa'}$ describes the strength of the synapse from output unit a' to output *recurrent model* unit a. The output rates in this case are determined by

$$\tau_r \frac{d\mathbf{v}}{dt} = -\mathbf{v} + \mathbf{F}(\mathbf{W} \cdot \mathbf{u} + \mathbf{M} \cdot \mathbf{v}). \quad (7.10)$$

It is often convenient to define the total feedforward input to each neuron in the network of figure 7.3B as $\mathbf{h} = \mathbf{W} \cdot \mathbf{u}$. Then, the output rates are determined by the equation

$$\tau_r \frac{d\mathbf{v}}{dt} = -\mathbf{v} + \mathbf{F}(\mathbf{h} + \mathbf{M} \cdot \mathbf{v}). \qquad (7.11)$$

Neurons are typically classified as either excitatory or inhibitory, meaning that they have either excitatory or inhibitory effects on all of their postsynaptic targets. This property is formalized in Dale's law, which states that   *Dale's law*
a neuron cannot excite some of its postsynaptic targets and inhibit others. In terms of the elements of $\mathbf{M}$, this means that for each presynaptic neuron a', $M_{aa'}$ must have the same sign for all postsynaptic neurons a. To impose this restriction, it is convenient to describe excitatory and inhibitory neurons separately. The firing-rate vectors $\mathbf{v}_E$ and $\mathbf{v}_I$ for the excitatory and inhibitory neurons are then described by a coupled set of equations identical in form to equation 7.11,   *excitatory-inhibitory network*

$$\tau_E \frac{d\mathbf{v}_E}{dt} = -\mathbf{v}_E + \mathbf{F}_E\left(\mathbf{h}_E + \mathbf{M}_{EE} \cdot \mathbf{v}_E + \mathbf{M}_{EI} \cdot \mathbf{v}_I\right) \qquad (7.12)$$

and

$$\tau_I \frac{d\mathbf{v}_I}{dt} = -\mathbf{v}_I + \mathbf{F}_I\left(\mathbf{h}_I + \mathbf{M}_{IE} \cdot \mathbf{v}_E + \mathbf{M}_{II} \cdot \mathbf{v}_I\right). \qquad (7.13)$$

There are now four synaptic weight matrices describing the four possible types of neuronal interactions. The elements of $\mathbf{M}_{EE}$ and $\mathbf{M}_{IE}$ are greater than or equal to 0, and those of $\mathbf{M}_{EI}$ and $\mathbf{M}_{II}$ are less than or equal to 0. These equations allow the excitatory and inhibitory neurons to have different time constants, activation functions, and feedforward inputs.

In this chapter, we consider several recurrent network models described by equation 7.11 with a symmetric weight matrix, $M_{aa'} = M_{a'a}$ for all a and a'. Requiring $\mathbf{M}$ to be symmetric simplifies the mathematical analysis, but   *symmetric coupling*
it violates Dale's law. Suppose, for example, that neuron a, which is excitatory, and neuron a', which is inhibitory, are mutually connected. Then, $M_{aa'}$ should be negative and $M_{a'a}$ positive, so they cannot be equal. Equation 7.11 with symmetric $\mathbf{M}$ can be interpreted as a special case of equations 7.12 and 7.13 in which the inhibitory dynamics are instantaneous ($\tau_I \to 0$) and the inhibitory rates are given by $\mathbf{v}_I = \mathbf{M}_{IE}\mathbf{v}_E$. This produces an effective recurrent weight matrix $\mathbf{M} = \mathbf{M}_{EE} + \mathbf{M}_{EI} \cdot \mathbf{M}_{IE}$, which can be made symmetric by the appropriate choice of the dimension and form of the matrices $\mathbf{M}_{EI}$ and $\mathbf{M}_{IE}$. The dynamic behavior of equation 7.11 is restricted by requiring the matrix $\mathbf{M}$ to be symmetric. For example symmetric coupling typically does not allow for network oscillations. In the latter part of this chapter, we consider the richer dynamics of models described by equations 7.12 and 7.13.

Continuously Labeled Networks

It is often convenient to identify each neuron in a network by using a parameter that describes some aspect of its selectivity rather than the integer label a or b. For example, neurons in primary visual cortex can be characterized by their preferred orientation angles, preferred spatial phases and frequencies, or other stimulus-related parameters (see chapter 2). In many of the examples in this chapter, we consider stimuli characterized by a single angle Θ, which represents, for example, the orientation of a visual stimulus. Individual neurons are identified by their preferred stimulus angles, which are typically the values of Θ for which they fire at maximum rates. Thus, neuron a is identified by an angle θ_a. The weight of the synapse from neuron b or neuron a' to neuron a is then expressed as a function of the preferred stimulus angles θ_b, $\theta_{a'}$ and θ_a of the pre- and post-synaptic neurons, $W_{ab} = W(\theta_a, \theta_b)$ or $M_{aa'} = M(\theta_a, \theta_{a'})$. We often consider cases in which these synaptic weight functions depend only on the difference between the pre- and postsynaptic angles, so that $W_{ab} = W(\theta_a - \theta_b)$ or $M_{aa'} = M(\theta_a - \theta_{a'})$.

density of coverage ρ_θ

In large networks, the preferred stimulus parameters for different neurons will typically take a wide range of values. In the models we consider, the number of neurons is large and the angles θ_a, for different values of a, cover the range from 0 to 2π densely. For simplicity, we assume that this coverage is uniform, so that the density of coverage, the number of neurons with preferred angles falling within a unit range, which we denote by ρ_θ, is constant. For mathematical convenience in these cases, we allow the preferred angles to take continuous values rather than restricting them to the actual discrete values θ_a for $a = 1, 2, \ldots, N$. Thus, we label the neurons by a continuous angle θ and express the firing rate as a function of θ, so that $u(\theta)$ and $v(\theta)$ describe the firing rates of neurons with preferred angles θ. Similarly, the synaptic weight matrices **W** and **M** are replaced by functions $W(\theta, \theta')$ and $M(\theta, \theta')$ that characterizes the strength of synapses from a presynaptic neuron with preferred angle θ' to a post-synaptic neuron with preferred angle θ in the feedforward and recurrent cases, respectively.

continuous model

If the number of neurons in a network is large and the density of coverage of preferred stimulus values is high, we can approximate the sums in equation 7.10 by integrals over θ'. The number of postsynaptic neurons with preferred angles within a range $\Delta\theta'$ is $\rho_\theta\Delta\theta'$, so, when we take the limit $\Delta\theta' \to 0$, the integral over θ' is multiplied by the density factor ρ_θ. Thus, in the case of continuous labeling of neurons, equation 7.10 becomes (for constant ρ_θ)

$$\tau_r \frac{dv(\theta)}{dt} = -v(\theta) + F\left(\rho_\theta \int_{-\pi}^{\pi} d\theta' \, W(\theta, \theta')u(\theta') + M(\theta, \theta')v(\theta')\right). \quad (7.14)$$

As we did in equation 7.11, we can write the first term inside the integral of this expression as an input function $h(\theta)$. We make frequent use of continuous labeling for network models, and we often approximate sums over neurons by integrals over their preferred stimulus parameters.

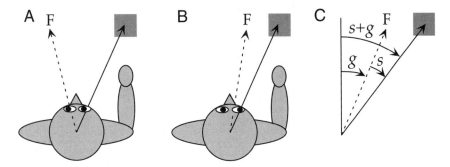

Figure 7.4 Coordinate transformations during a reaching task. (A, B) The location of the target (the gray square) relative to the body is the same in A and B, and thus the movements required to reach toward it are identical. However, the image of the object falls on different parts of the retina in A and B due to a shift in the gaze direction produced by an eye rotation that shifts the fixation point F. (C) The angles used in the analysis: s is the angle describing the location of the stimulus (the target) in retinal coordinates, that is, relative to a line directed to the fixation point; g is the gaze angle, indicating the direction of gaze relative to an axis straight out from the body. The direction of the target relative to the body is $s + g$.

7.3   Feedforward Networks

Substantial computations can be performed by feedforward networks in the absence of recurrent connections. Much of the work done on feed-forward networks centers on plasticity and learning, as discussed in the following chapters. Here, we present an example of the computational power of feedforward circuits, the calculation of the coordinate transformations needed in visually guided reaching tasks.

Neural Coordinate Transformations

Reaching for a viewed object requires a number of coordinate transformations that turn information about where the image of the object falls on the retina into movement commands in shoulder-, arm-, or hand-based coordinates. To perform a transformation from retinal to body-based co-ordinates, information about the retinal location of an image and about the direction of gaze relative to the body must be combined. Figure 7.4A and B illustrate, in a one-dimensional example, how a rotation of the eyes affects the relationship between gaze direction, retinal location, and location relative to the body. Figure 7.4C introduces the notation we use. The angle g describes the orientation of a line extending from the head to the point of visual fixation. The visual stimulus in retinal coordinates is given by the angle s between this line and a line extending out to the target. The angle describing the reach direction, the direction to the target relative to the body, is the sum $s + g$.

Visual neurons have receptive fields fixed to specific locations on the retina. Neurons in motor areas can display visually evoked responses that

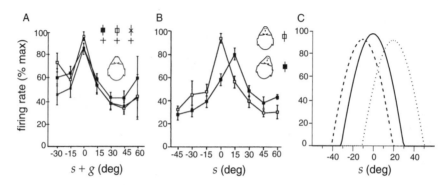

Figure 7.5 Tuning curves of a visually responsive neuron in the premotor cortex of a monkey. Incoming objects approaching at various angles provided the visual stimulation. (A) When the monkey fixated on the three points denoted by the cross symbols, the response tuning curve did not shift with the eyes. In this panel, unlike B and C, the horizontal axis refers to the stimulus location in body-based, not retinal, coordinates ($s + g$, not s). (B) Turning the monkey's head by 15° produced a 15° shift in the response tuning curve as a function of retinal location, indicating that this neuron encoded the stimulus direction in a body-based system. (C) Model tuning curves based on equation 7.15 shift their retinal tuning to remain constant in body-based coordinates. The solid, heavy dashed, and light dashed curves refer to $g = 0°$, 10°, and −20° respectively. The small changes in amplitude arise from the limited range of preferred retinal location and gaze angles in the model. (A, B adapted from Graziano et al., 1997; C adapted from Salinas and Abbott, 1995.)

are not tied to specific retinal locations but, rather, depend on the relationship of a visual image to various parts of the body. Figures 7.5A and B show tuning curves of a neuron in the premotor cortex of a monkey that responded to visual images of approaching objects. Surprisingly, when the head of the monkey was held stationary during fixation on three different targets, the tuning curves did not shift as the eyes rotated (figure 7.5A). Although the recorded neurons respond to visual stimuli, the responses do not depend directly on the location of the image on the retina. When the head of the monkey is rotated but the fixation point remains the same, the tuning curves shift by precisely the amount of the head rotation (figure 7.5B). Thus, these neurons encode the location of the image in a body-based, not a retinal, coordinate system.

To account for these data, we need to construct a model neuron that is driven by visual input, but that nonetheless has a tuning curve for image location that is not a function of s, the retinal location of the image, but of $s + g$, the location of the object in body-based coordinates. A possible basis for this construction is provided by a combined representation of s and g by neurons in area 7a in the posterior parietal cortex of the monkey. Recordings made in area 7a reveal neurons that fire at rates that depend on both the location of the stimulating image on the retina and the direction of gaze (figure 7.6A). The response tuning curves, expressed as functions of the retinal location of the stimulus, do not shift when the direction of gaze is varied. Instead, shifts of gaze direction affect the magnitude of the visual response. Thus, responses in area 7a exhibit gaze-dependent gain

gain modulation   modulation of a retinotopic visual receptive field.

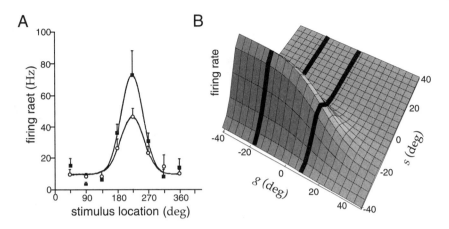

Figure 7.6 Gaze-dependent gain modulation of visual responses of neurons in posterior parietal cortex. (A) Average firing-rate tuning curves of an area 7a neuron as a function of the location of the spot of light used to evoke the response. Stimulus location is measured as an angle around a circle of possible locations on the screen and is related to, but not equal to, our stimulus variable s. The two curves correspond to the same visual images but with two different gaze directions. (B) A three-dimensional plot of the activity of a model neuron as a function of both retinal position and gaze direction. The striped bands correspond to tuning curves with different gains similar to those shown in A. (A adapted from Brotchie et al., 1995; B adapted from Pouget and Sejnowski, 1995.)

Figure 7.6B shows a mathematical description of a gain-modulated tuning curve. The response tuning curve is expressed as a product of a Gaussian function of $s-\xi$, where ξ is the preferred retinal location ($\xi=-20°$ in figure 7.6B), and a sigmoidal function of $g - \gamma$, where γ is the gaze direction producing half of the maximum gain ($\gamma=20°$ in figure 7.6B). Although it does not correspond to the maximum neural response, we refer to γ as the "preferred" gaze direction.

To model a neuron with a body-centered response tuning curve, we construct a feedforward network with a single output unit representing, for example, the premotor cortex neuron shown in figure 7.5. The input layer of the network consists of a population of area 7a neurons with gain-modulated responses similar to those shown in figure 7.6B. Neurons with gains that both increase and decrease as a function of g are included in the model. The average firing rates of the input layer neurons are described by tuning curves $u = f_u(s-\xi, g-\gamma)$, with the different neurons taking different ξ and γ values.

We use continuous labeling of neurons, and replace the sum over presynaptic neurons by an integral over their ξ and γ values, inserting the appropriate density factors ρ_ξ and ρ_γ, which we assume are constant. The steady-state response of the single output neuron is determined by the continuous analog of equation 7.5. The synaptic weight from a presynaptic neuron with preferred stimulus location ξ and preferred gaze direction γ is denoted by $w(\xi, \gamma)$, so the steady-state response of the output neuron is given by

$$v_\infty = F\left(\rho_\xi \rho_\gamma \int d\xi d\gamma \, w(\xi, \gamma) f_u(s - \xi, g - \gamma) \right). \qquad (7.15)$$

For the output neuron to respond to the stimulus location in body-based coordinates, its firing rate must be a function of $s+g$. To see if this is possible, we shift the integration variables in 7.15 by $\xi \to \xi - g$ and $\gamma \to \gamma + g$. Ignoring effects from the end points of the integration (which is valid if s and g are not too close to these limits), we find

$$v_\infty = F\left(\rho_\xi \rho_\gamma \int d\xi d\gamma \, w(\xi - g, \gamma + g) f_u(s + g - \xi, -\gamma) \right). \qquad (7.16)$$

This is a function of $s+g$ provided that $w(\xi - g, \gamma + g) = w(\xi, \gamma)$, which holds if $w(\xi, \gamma)$ is a function of the sum $\xi + \gamma$. Thus, the coordinate transformation can be accomplished if the synaptic weight from a given neuron depends only on the sum of its preferred retinal and gaze angles. It has been suggested that weights of this form can arise naturally from random hand and gaze movements through correlation-based synaptic modification of the type discussed in chapter 8.

Figure 7.5C shows responses predicted by equation 7.15 when the synaptic weights are given by a function $w(\xi + \gamma)$. The retinal location of the tuning curve shifts as a function of gaze direction, but would remain stationary if it were plotted instead as a function of $s + g$. This can be seen by noting that the peaks of all three curves in figure 7.5C occur at $s + g = 0$.

Gain-modulated neurons provide a general basis for combining two different input signals in a nonlinear way. In the network we studied, it is possible to find appropriate synaptic weights $w(\xi, \gamma)$ to generate output neuron responses with a wide range of different dependencies on s and g. The mechanism by which sensory and modulatory inputs combine in a multiplicative way in gain-modulated neurons is not known. Later in this chapter, we discuss a recurrent network model for generating gain-modulated responses.

7.4 Recurrent Networks

Recurrent networks have richer dynamics than feedforward networks, but they are more difficult to analyze. To get a feel for recurrent circuitry, we begin by analyzing a linear model, that is, a model for which the relationship between firing rate and synaptic current is linear, $\mathbf{F(h + M \cdot v)} = \mathbf{h + M \cdot v}$. The linear approximation is a drastic one that allows, among other things, the components of $\mathbf{v}$ to become negative, which is impossible for real firing rates. Furthermore, some of the features we discuss in connection with linear, as opposed to nonlinear, recurrent networks can also be achieved by a feedforward architecture. Nevertheless, the linear model is extremely useful for exploring properties of recurrent circuits, and this approach will be used both here and in the following chapters. In addition, the analysis of linear networks forms the basis for studying the stability properties of nonlinear networks. We augment the discussion of linear networks with results from simulations of nonlinear networks.

Linear Recurrent Networks

Under the linear approximation, the recurrent model of equation 7.11 takes the form   *linear recurrent*
                                                                             model

$$\tau_r \frac{d\mathbf{v}}{dt} = -\mathbf{v} + \mathbf{h} + \mathbf{M} \cdot \mathbf{v} \,. \tag{7.17}$$

Because the model is linear, we can solve analytically for the vector of output rates $\mathbf{v}$ in terms of the feedforward inputs $\mathbf{h}$ and the initial values $\mathbf{v}(0)$. The analysis is simplest when the recurrent synaptic weight matrix is symmetric, and we assume this to be the case. Equation 7.17 can be solved by expressing $\mathbf{v}$ in terms of the eigenvectors of $\mathbf{M}$. The eigenvectors $\mathbf{e}_\mu$ for   *eigenvector* $\mathbf{e}$
$\mu = 1, 2, \dots, N_v$ satisfy

$$\mathbf{M} \cdot \mathbf{e}_\mu = \lambda_\mu \mathbf{e}_\mu \tag{7.18}$$

for some value of the constant λ_μ, which is called the eigenvalue. For a   *eigenvalue* λ
symmetric matrix, the eigenvectors are orthogonal, and they can be normalized to unit length so that $\mathbf{e}_\mu \cdot \mathbf{e}_\nu = \delta_{\mu\nu}$. Such eigenvectors define an orthogonal coordinate system or basis that can be used to represent any N_v-dimensional vector. In particular, we can write   *eigenvector*
                                                                             expansion

$$\mathbf{v}(t) = \sum_{\mu=1}^{N_v} c_\mu(t) \mathbf{e}_\mu \,, \tag{7.19}$$

where $c_\mu(t)$ for $\mu = 1, 2, \dots, N_v$ are a set of time-dependent coefficients describing $\mathbf{v}(t)$.

It is easier to solve equation 7.17 for the coefficients c_μ than for $\mathbf{v}$ directly. Substituting the expansion 7.19 into equation 7.17 and using property 7.18, we find that

$$\tau_r \sum_{\mu=1}^{N_v} \frac{dc_\mu}{dt} \mathbf{e}_\mu = -\sum_{\mu=1}^{N_v} (1 - \lambda_\mu) c_\mu(t) \mathbf{e}_\mu + \mathbf{h} \,. \tag{7.20}$$

The sum over μ can be eliminated by taking the dot product of each side of this equation with one of the eigenvectors, $\mathbf{e}_\nu$, and using the orthogonality property $\mathbf{e}_\mu \cdot \mathbf{e}_\nu = \delta_{\mu\nu}$ to obtain

$$\tau_r \frac{dc_\nu}{dt} = -(1 - \lambda_\nu) c_\nu(t) + \mathbf{e}_\nu \cdot \mathbf{h} \,. \tag{7.21}$$

The critical feature of this equation is that it involves only one of the co-efficients, c_ν. For time-independent inputs $\mathbf{h}$, the solution of equation 7.21 is

$$c_\nu(t) = \frac{\mathbf{e}_\nu \cdot \mathbf{h}}{1 - \lambda_\nu} \left(1 - \exp\left(-\frac{t(1 - \lambda_\nu)}{\tau_r}\right)\right) + c_\nu(0) \exp\left(-\frac{t(1 - \lambda_\nu)}{\tau_r}\right),$$

(7.22)

where $c_\nu(0)$ is the value of c_ν at time 0, which is given in terms of the initial firing-rate vector $\mathbf{v}(0)$ by $c_\nu(0) = \mathbf{e}_\nu \cdot \mathbf{v}(0)$.

Equation 7.22 has several important characteristics. If $\lambda_\nu > 1$, the exponential functions grow without bound as time increases, reflecting a fundamental instability of the network. If $\lambda_\nu < 1$, c_ν approaches the steady-state value $\mathbf{e}_\nu \cdot \mathbf{h}/(1 - \lambda_\nu)$ exponentially with time constant $\tau_r/(1 - \lambda_\nu)$. This steady-state value is proportional to $\mathbf{e}_\nu \cdot \mathbf{h}$, which is the projection of the input vector onto the relevant eigenvector. For $0 < \lambda_\nu < 1$, the steady-state value is amplified relative to this projection by the factor $1/(1 - \lambda_\nu)$, which is greater than 1. The approach to equilibrium is slowed relative to the basic time constant τ_r by an identical factor. The steady-state value of $\mathbf{v}(t)$, *steady state* $\mathbf{v}_\infty$ which we call $\mathbf{v}_\infty$, can be derived from equation 7.19 as

$$\mathbf{v}_\infty = \sum_{\nu=1}^{N_v} \frac{(\mathbf{e}_\nu \cdot \mathbf{h})}{1 - \lambda_\nu} \mathbf{e}_\nu.$$

(7.23)

This steady-state response can also arise from a purely feedforward scheme if the feedforward weight matrix is chosen appropriately, as we invite the reader to verify as an exercise.

Selective Amplification

Suppose that one of the eigenvalues of a recurrent weight matrix, denoted by λ_1, is very close to 1, and all the others are significantly smaller than 1. In this case, the denominator of the $\nu = 1$ term on the right side of equation 7.23 is near 0, and, unless $\mathbf{e}_1 \cdot \mathbf{h}$ is extremely small, this single term will dominate the sum. As a result, we can write

$$\mathbf{v}_\infty \approx \frac{(\mathbf{e}_1 \cdot \mathbf{h})\mathbf{e}_1}{1 - \lambda_1}.$$

(7.24)

Such a network performs selective amplification. The response is dominated by the projection of the input vector along the axis defined by $\mathbf{e}_1$, and the amplitude of the response is amplified by the factor $1/(1 - \lambda_1)$, which may be quite large if λ_1 is near 1. The steady-state response of such a network, which is proportional to $\mathbf{e}_1$, therefore encodes an amplified projection of the input vector onto $\mathbf{e}_1$.

Further information can be encoded if more eigenvalues are close to 1. Suppose, for example, that two eigenvectors, $\mathbf{e}_1$ and $\mathbf{e}_2$, have the same eigenvalue, $\lambda_1 = \lambda_2$, close to but less than 1. Then, equation 7.24 is replaced by

$$\mathbf{v}_\infty \approx \frac{(\mathbf{e}_1 \cdot \mathbf{h})\mathbf{e}_1 + (\mathbf{e}_2 \cdot \mathbf{h})\mathbf{e}_2}{1 - \lambda_1}, \tag{7.25}$$

which shows that the network now amplifies and encodes the projection of the input vector onto the plane defined by $\mathbf{e}_1$ and $\mathbf{e}_2$. In this case, the activity pattern of the network is not simply scaled when the input changes. Instead, changes in the input shift both the magnitude and the pattern of network activity. Eigenvectors that share the same eigenvalue are termed degenerate, and degeneracy is often the result of a symmetry. Degeneracy is not limited to just two eigenvectors. A recurrent network with n degenerate eigenvalues near 1 can amplify and encode a projection of the input vector from the N-dimensional space in which it is defined onto the n-dimensional subspace spanned by the degenerate eigenvectors.

Input Integration

If the recurrent weight matrix has an eigenvalue exactly equal to 1, $\lambda_1 = 1$, and all the other eigenvalues satisfy $\lambda_\nu < 1$, a linear recurrent network can act as an integrator of its input. In this case, c_1 satisfies the equation

$$\tau_r \frac{dc_1}{dt} = \mathbf{e}_1 \cdot \mathbf{h}, \tag{7.26}$$

obtained by setting $\lambda_1 = 1$ in equation 7.21. For arbitrary time-dependent inputs, the solution of this equation is

$$c_1(t) = c_1(0) + \frac{1}{\tau_r} \int_0^t dt'\, \mathbf{e}_1 \cdot \mathbf{h}(t'). \tag{7.27}$$

If $\mathbf{h}(t)$ is constant, $c_1(t)$ grows linearly with t. This explains why equation 7.24 diverges as $\lambda_1 \to 1$. Suppose, instead, that $\mathbf{h}(t)$ is nonzero for a while, and then is set to 0 for an extended period of time. When $\mathbf{h} = 0$, equation 7.22 shows that $c_\nu \to 0$ for all $\nu \neq 1$, because for these eigenvectors $\lambda_\nu < 1$. Assuming that $c_1(0) = 0$, this means that after such a period, the firing-rate vector is given, from equations 7.27 and 7.19, by

network integration

$$\mathbf{v}(t) \approx \frac{\mathbf{e}_1}{\tau_r} \int_0^t dt'\, \mathbf{e}_1 \cdot \mathbf{h}(t'). \tag{7.28}$$

This shows that the network activity provides a measure of the running integral of the projection of the input vector onto $\mathbf{e}_1$. One consequence of this is that the activity of the network does not cease if $\mathbf{h} = 0$, provided that the integral up to that point in time is nonzero. The network thus exhibits sustained activity in the absence of input, which provides a memory of the integral of prior input.

Networks in the brain stem of vertebrates responsible for maintaining eye position appear to act as integrators, and networks similar to the one we

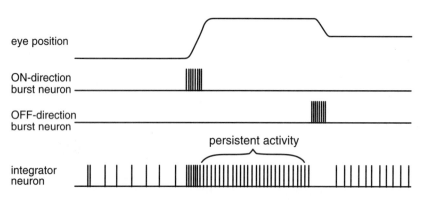

Figure 7.7 Cartoon of burst and integrator neurons involved in horizontal eye positioning. The upper trace represents horizontal eye position during two saccadic eye movements. Motion of the eye is driven by burst neurons that move the eyes in opposite directions (second and third traces from top). The steady-state firing rate (labeled persistent activity) of the integrator neuron is proportional to the time integral of the burst rates, integrated positively for the ON-direction burst neuron and negatively for the OFF-direction burst neuron, and thus provides a memory trace of the maintained eye position. (Adapted from Seung et al., 2000.)

have been discussing have been suggested as models of this system. As outlined in figure 7.7, eye position changes in response to bursts of activity in ocular motor neurons located in the brain stem. Neurons in the medial vestibular nucleus and prepositus hypoglossi appear to integrate these motor signals to provide a persistent memory of eye position. The sustained firing rates of these neurons are approximately proportional to the angular orientation of the eyes in the horizontal direction, and activity persists at an approximately constant rate when the eyes are held fixed (bottom trace in figure 7.7).

The ability of a linear recurrent network to integrate and display persistent activity relies on one of the eigenvalues of the recurrent weight matrix being exactly 1. Any deviation from this value will cause the persistent activity to change over time. Eye position does indeed drift, but matching the performance of the ocular positioning system requires fine-tuning of the eigenvalue to a value extremely close to 1. Including nonlinear interactions does not alleviate the need for a precisely tuned weight matrix. Synaptic modification rules can be used to establish the necessary synaptic weights, but it is not clear how such precise tuning is accomplished in the biological system.

Continuous Linear Recurrent Networks

For a linear recurrent network with continuous labeling, the equation for the firing rate $v(\theta)$ of a neuron with preferred stimulus angle θ is a linear version of equation 7.14,

$$\tau_r \frac{dv(\theta)}{dt} = -v(\theta) + h(\theta) + \rho_\theta \int_{-\pi}^{\pi} d\theta'\, M(\theta - \theta')v(\theta')\,, \qquad (7.29)$$

where $h(\theta)$ is the feedforward input to a neuron with preferred stimulus angle θ, and we have assumed a constant density ρ_θ. Because θ is an angle, h, M, and v must all be periodic functions with period 2π. By making M a function of $\theta - \theta'$, we are imposing a symmetry with respect to translations or shifts of the angle variables. In addition, we assume that M is an even function, $M(\theta - \theta') = M(\theta' - \theta)$. This is the analog, in a continuously labeled model, of a symmetric synaptic weight matrix.

Equation 7.29 can be solved by methods similar to those used for discrete networks. We introduce eigenfunctions that satisfy

$$\rho_\theta \int_{-\pi}^{\pi} d\theta'\, M(\theta - \theta') e_\mu(\theta') = \lambda_\mu e_\mu(\theta)\,. \tag{7.30}$$

We leave it as an exercise to show that the eigenfunctions (normalized so that ρ_θ times the integral from $-\pi$ to π of their square is 1) are $1/(2\pi\rho_\theta)^{1/2}$, corresponding to $\mu = 0$, and $\cos(\mu\theta)/(\pi\rho_\theta)^{1/2}$ and $\sin(\mu\theta)/(\pi\rho_\theta)^{1/2}$ for $\mu = 1, 2, \ldots$. The eigenvalues are identical for the sine and cosine eigenfunctions and are given (including the case $\mu = 0$) by

$$\lambda_\mu = \rho_\theta \int_{-\pi}^{\pi} d\theta'\, M(\theta') \cos(\mu\theta')\,. \tag{7.31}$$

The steady-state firing rates for a constant input are given by the continuous analog of equation 7.23,

$$\begin{aligned}
v_\infty(\theta) = \ & \frac{1}{1 - \lambda_0} \int_{-\pi}^{\pi} \frac{d\theta'}{2\pi}\, h(\theta') \\
& + \sum_{\mu=1}^{\infty} \frac{\cos(\mu\theta)}{1 - \lambda_\mu} \int_{-\pi}^{\pi} \frac{d\theta'}{\pi}\, h(\theta')\cos(\mu\theta') \\
& + \sum_{\mu=1}^{\infty} \frac{\sin(\mu\theta)}{1 - \lambda_\mu} \int_{-\pi}^{\pi} \frac{d\theta'}{\pi}\, h(\theta')\sin(\mu\theta')\,.
\end{aligned} \tag{7.32}$$

The integrals in this expression are the coefficients in a Fourier series for the function h and are known as cosine and sine Fourier integrals (see the Mathematical Appendix).

Fourier series

Figure 7.8 shows an example of selective amplification by a linear recurrent network. The input to the network, shown in panel A of figure 7.8, is a cosine function that peaks at $0°$ to which random noise has been added. Figure 7.8C shows Fourier amplitudes for this input. The Fourier amplitude is the square root of the sum of the squares of the cosine and sine Fourier integrals. No particular μ value is overwhelmingly dominant. In this and the following examples, the recurrent connections of the network are given by

$$M(\theta - \theta') = \frac{\lambda_1}{\pi\rho_\theta} \cos(\theta - \theta')\,, \tag{7.33}$$

which has all eigenvalues except λ_1 equal to 0. The network model shown in figure 7.8 has $\lambda_1 = 0.9$, so that $1/(1 - \lambda_1) = 10$. Input amplification can

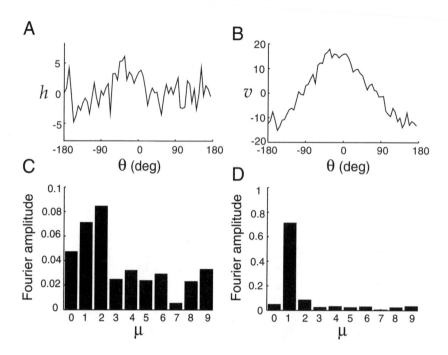

Figure 7.8 Selective amplification in a linear network. (A) The input to the neurons of the network as a function of their preferred stimulus angle. (B) The activity of the network neurons plotted as a function of their preferred stimulus angle in response to the input of panel A. (C) The Fourier transform amplitudes of the input shown in panel A. (D) The Fourier transform amplitudes of the output shown in panel B. The recurrent coupling of this network model took the form of equation 7.33 with $\lambda_1 = 0.9$. (This figure, and figures 7.9, 7.12, 7.13, and 7.14, were generated using software from Carandini and Ringach, 1997.)

be quantified by comparing the Fourier amplitude of v_∞, for a given μ value, with the analogous amplitude for the input h. According to equation 7.32, the ratio of these quantities is $1/(1 - \lambda_\mu)$, so, in this case, the $\mu = 1$ amplitude should be amplified by a factor of 10 while all other amplitudes are unamplified. This factor of 10 amplification can be seen by comparing the $\mu = 1$ Fourier amplitudes in figures 7.8C and D (note the different scales for the vertical axes). All the other components are unamplified. As a result, the output of the network is primarily in the form of a cosine function with $\mu = 1$, as seen in figure 7.8B.

Nonlinear Recurrent Networks

A linear model does not provide an adequate description of the firing rates of a biological neural network. The most significant problem is that the firing rates in a linear network can take negative values. This problem can

rectification   be fixed by introducing rectification into equation 7.11 by choosing

$$\mathbf{F}(\mathbf{h} + \mathbf{M} \cdot \mathbf{r}) = [\mathbf{h} + \mathbf{M} \cdot \mathbf{r} - \boldsymbol{\gamma}]_+ \,, \qquad (7.34)$$

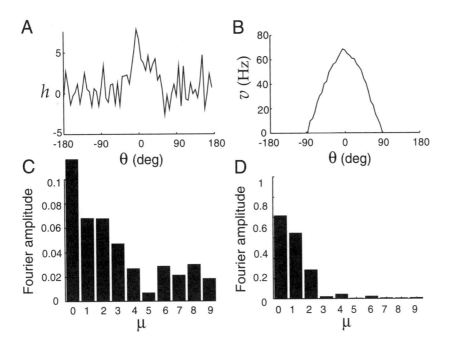

Figure 7.9 Selective amplification in a recurrent network with rectification. (A) The input $h(\theta)$ of the network plotted as a function of preferred angle. (B) The steady-state output $v(\theta)$ as a function of preferred angle. (C) Fourier transform amplitudes of the input $h(\theta)$. (D) Fourier transform amplitudes of the output $v(\theta)$. The recurrent coupling took the form of equation 7.33 with $\lambda_1 = 1.9$.

where $\boldsymbol{\gamma}$ is a vector of threshold values that we often take to be $\mathbf{0}$ (we use the notation $\mathbf{0}$ to denote a vector with all its components equal to zero). In this section, we show some examples illustrating the effect of including such a rectifying nonlinearity. Some of the features of linear recurrent networks remain when rectification is included, but several new features also appear.

vector of zeros $\mathbf{0}$

In the examples given below, we consider a continuous model, similar to that of equation 7.29, with recurrent couplings given by equation 7.33 but now including a rectification nonlinearity, so that

$$\tau_r \frac{dv(\theta)}{dt} = -v(\theta) + \left[h(\theta) + \frac{\lambda_1}{\pi} \int_{-\pi}^{\pi} d\theta' \, \cos(\theta - \theta')v(\theta') \right]_+ . \qquad (7.35)$$

If λ_1 is not too large, this network converges to a steady state for any constant input (we consider conditions for steady-state convergence in a later section), and therefore we often limit the discussion to the steady-state activity of the network.

Nonlinear Amplification

Figure 7.9 shows the nonlinear analog of the selective amplification shown for a linear network in figure 7.8. Once again, a noisy input (figure 7.9A) generates a much smoother output response profile (figure 7.9B). The output response of the rectified network corresponds roughly to the positive part of the sinusoidal response profile of the linear network (figure 7.8B). The negative output has been eliminated by the rectification. Because fewer neurons in the network have nonzero responses than in the linear case, the value of the parameter λ_1 in equation 7.33 has been increased to 1.9. This value, being larger than 1, would lead to an unstable network in the linear case. While nonlinear networks can also be unstable, the restriction to eigenvalues less than 1 is no longer the relevant condition.

In a nonlinear network, the Fourier analysis of the input and output responses is no longer as informative as it is for a linear network. Due to the rectification, the $\nu = 0, 1$, and 2 Fourier components are all amplified (figure 7.9D) compared to their input values (figure 7.9C). Nevertheless, except for rectification, the nonlinear recurrent network amplifies the input signal selectively in a manner similar to the linear network.

A Recurrent Model of Simple Cells in Primary Visual Cortex

In chapter 2, we discussed a feedforward model in which the elongated receptive fields of simple cells in primary visual cortex were formed by summing the inputs from neurons of the lateral geniculate nucleus (LGN) with their receptive fields arranged in alternating rows of ON and OFF cells. While this model quite successfully accounts for a number of features of simple cells, such as orientation tuning, it is difficult to reconcile with the anatomy and circuitry of the cerebral cortex. By far the majority of the synapses onto any cortical neuron arise from other cortical neurons, not from thalamic afferents. Therefore, feedforward models account for the response properties of cortical neurons while ignoring the inputs that are numerically most prominent. The large number of intracortical connections suggests, instead, that recurrent circuitry might play an important role in shaping the responses of neurons in primary visual cortex.

Ben-Yishai, Bar-Or, and Sompolinsky (1995) developed a model in which orientation tuning is generated primarily by recurrent rather than feedforward connections. The model is similar in structure to the model of equations 7.35 and 7.33, except that it includes a global inhibitory interaction. In addition, because orientation angles are defined over the range from $-\pi/2$ to $\pi/2$, rather than over the full 2π range, the cosine functions in the model have extra factors of 2 in them. The basic equation of the model, as we implement it, is

$$\tau_r \frac{dv(\theta)}{dt} = -v(\theta) + \left[h(\theta) + \int_{-\pi/2}^{\pi/2} \frac{d\theta'}{\pi} \left(-\lambda_0 + \lambda_1 \cos(2(\theta - \theta')) \right) v(\theta') \right]_+ ,$$
(7.36)

where $v(\theta)$ is the firing rate of a neuron with preferred orientation θ.

The input to the model represents the orientation-tuned feedforward input arising from ON-center and OFF-center LGN cells responding to an oriented image. As a function of preferred orientation, the input for an image with orientation angle $\Theta = 0$ is

$$h(\theta) = Ac\left(1 - \epsilon + \epsilon \cos(2\theta)\right),$$
(7.37)

where A sets the overall amplitude and c is equal to the image contrast. The factor ϵ controls how strongly the input is modulated by the orientation angle. For $\epsilon = 0$, all neurons receive the same input, while $\epsilon = 0.5$ produces the maximum modulation consistent with a positive input. We study this model in the case when ϵ is small, which means that the input is only weakly tuned for orientation and any strong orientation selectivity must arise through recurrent interactions.

To study orientation selectivity, we want to examine the tuning curves of individual neurons in response to stimuli with different orientation angles Θ. The plots of network responses that we have been using show the firing rates $v(\theta)$ of all the neurons in the network as a function of their preferred stimulus angles θ when the input stimulus has a fixed value, typically $\Theta = 0$. As a consequence of the translation invariance of the network model, the response for other values of Θ can be obtained simply by shifting this curve so that it plots $v(\theta - \Theta)$. Furthermore, except for the asymmetric effects of noise on the input, $v(\theta - \Theta)$ is a symmetric function. These features follow from the fact that the network we are studying is invariant with respect to translations and sign changes of the angle variables that characterize the stimulus and response selectivities. An important consequence of this result is that the curve $v(\theta)$, showing the response of the entire population, can also be interpreted as the tuning curve of a single neuron. If the response of the population to a stimulus angle Θ is $v(\theta - \Theta)$, the response of a single neuron with preferred angle $\theta = 0$ is $v(-\Theta) = v(\Theta)$ from the symmetry of v. Because $v(\Theta)$ is the tuning curve of a single neuron with $\theta = 0$ to a stimulus angle Θ, the plots we show of $v(\theta)$ can be interpreted as both population responses and individual neuronal tuning curves.

Figure 7.10A shows the feedforward input to the model network for four different levels of contrast. Because the parameter ϵ was chosen to be 0.1, the modulation of the input as a function of orientation angle is small. Due to network amplification, the response of the network is much more strongly tuned to orientation (figure 7.10B). This is the result of the selective amplification of the tuned part of the input by the recurrent network. The modulation and overall height of the input curve in figure 7.10A increase linearly with contrast. The response shown in figure 7.10B, interpreted as a tuning curve, increases in amplitude for higher contrast but does not broaden. This can be seen by noting that all four curves in figure 7.10B go to 0 at the same two points. This effect, which occurs because the shape and width of the response tuning curve are determined primarily by the recurrent interactions within the network, is a feature of orientation curves of real simple cells, as seen in figure 7.10C. The width of the

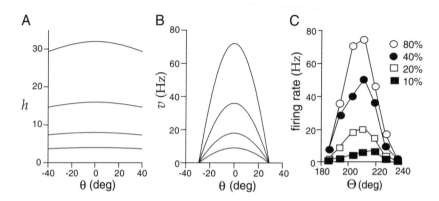

Figure 7.10 The effect of contrast on orientation tuning. (A) The feedforward input as a function of preferred orientation. The four curves, from top to bottom, correspond to contrasts of 80%, 40%, 20%, and 10%. (B) The output firing rates in response to different levels of contrast as a function of orientation preference. These are also the response tuning curves of a single neuron with preferred orientation 0. As in A, the four curves, from top to bottom, correspond to contrasts of 80%, 40%, 20%, and 10%. The recurrent model had $\lambda_0 = 7.3$, $\lambda_1 = 11$, $A = 40$ Hz, and $\epsilon = 0.1$. (C) Tuning curves measured experimentally at four contrast levels, as indicated in the legend. (C adapted from Sompolinsky and Shapley, 1997 based on data from Sclar and Freeman, 1982.)

tuning curve can be reduced by including a positive threshold in the response function of equation 7.34, or by changing the amount of inhibition, but it stays roughly constant as a function of stimulus strength.

A Recurrent Model of Complex Cells in Primary Visual Cortex

In the model of orientation tuning discussed in the previous section, recurrent amplification enhances selectivity. If the pattern of network connectivity amplifies nonselective rather than selective responses, recurrent interactions can also decrease selectivity. Recall from chapter 2 that neurons in the primary visual cortex are classified as simple or complex, depending on their sensitivity to the spatial phase of a grating stimulus. Simple cells respond maximally when the spatial positioning of the light and dark regions of a grating matches the locations of the ON and OFF regions of their receptive fields. Complex cells do not have distinct ON and OFF regions in their receptive fields, and respond to gratings of the appropriate orientation and spatial frequency relatively independently of where their light and dark stripes fall. In other words, complex cells are insensitive to spatial phase.

Chance, Nelson, and Abbott (1999) showed that complex cell responses could be generated from simple cell responses by a recurrent network. As in chapter 2, we label spatial phase preferences by the angle ϕ. The feedforward input $h(\phi)$ in the model is set equal to the rectified response of a simple cell with preferred spatial phase ϕ (figure 7.11A). Each neuron in the network is labeled by the spatial phase preference of its feedfor-

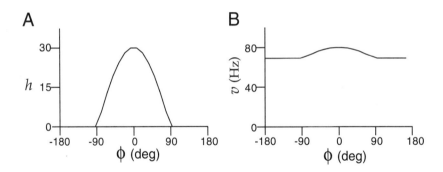

Figure 7.11 A recurrent model of complex cells. (A) The input to the network as a function of spatial phase preference. The input $h(\phi)$ is equivalent to that of a simple cell with spatial phase preference ϕ responding to a grating of 0 spatial phase. (B) Network response, which can also be interpreted as the spatial phase tuning curve of a network neuron. The network was given by equation 7.38 with $\lambda_1 = 0.95$. (Adapted from Chance et al., 1999.)

ward input. The network neurons also receive recurrent input given by the weight function $M(\phi - \phi') = \lambda_1/(2\pi\rho_\phi)$, which is the same for all connected neuron pairs. As a result, their firing rates are determined by

$$\tau_r \frac{dv(\phi)}{dt} = -v(\phi) + \left[h(\phi) + \frac{\lambda_1}{2\pi} \int_{-\pi}^{\pi} d\phi' \, v(\phi') \right]_+ . \qquad (7.38)$$

In the absence of recurrent connections ($\lambda_1 = 0$), the response of a neuron labeled by ϕ is $v(\phi) = h(\phi)$, which is equal to the response of a simple cell with preferred spatial phase ϕ. However, for λ_1 sufficiently close to 1, the recurrent model produces responses that resemble those of complex cells. Figure 7.11B shows the population response, or equivalently the single-cell response tuning curve, of the model in response to the tuned input shown in Figure 7.11A. The input, being the response of a simple cell, shows strong tuning for spatial phase. The output tuning curve, however, is almost constant as a function of spatial phase, like that of a complex cell. The spatial-phase insensitivity of the network response is due to the fact that the network amplifies the component of the input that is independent of spatial phase, because the eigenfunction of M with the largest eigenvalue is spatial-phase invariant. This changes simple cell inputs into complex cell outputs.

Winner-Takes-All Input Selection

For a linear network, the response to two superimposed inputs is simply the sum of the responses to each input separately. Figure 7.12 shows one way in which a rectifying nonlinearity modifies this superposition property. In this case, the input to the recurrent network consists of activity centered around two preferred stimulus angles, $\pm 90°$. The output of the nonlinear network shown in figure 7.12B is not of this form, but instead

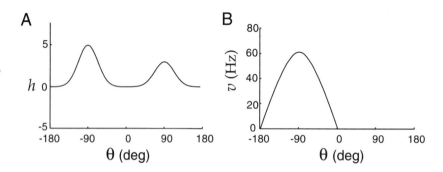

Figure 7.12 Winner-takes-all input selection by a nonlinear recurrent network. (A) The input to the network consisting of two peaks. (B) The output of the network has a single peak at the location of the higher of the two peaks of the input. The model is the same as that used in figure 7.9.

has a single peak at the location of the input bump with the larger amplitude (the one at $-90°$). This occurs because the nonlinear recurrent network supports the stereotyped unimodal activity pattern seen in figure 7.12B, so a multimodal input tends to generate a unimodal output. The height of the input peak has a large effect in determining where the single peak of the network output is located, but it is not the only feature that determines the response. For example, the network output can favor a broader, lower peak over a narrower, higher one.

Gain Modulation

A nonlinear recurrent network can generate an output that resembles the gain-modulated responses of posterior parietal neurons shown in figure 7.6, as noted by Salinas and Abbott (1996). To obtain this result, we interpret the angle θ as a preferred direction in the visual field in retinal coordinates (the variable we called s earlier in the chapter). The signal corresponding to gaze direction (what we called g before) is represented as a constant input to all neurons irrespective of their preferred stimulus angle. Figure 7.13 shows the effect of adding such a constant term to the input of the nonlinear network. The input shown in figure 7.13A corresponds to a visual target located at a retinal position of $0°$. The different lines show different values of the constant input, representing three different gaze directions.

The responses shown in figure 7.13B all have localized activity centered around $\theta = 0°$, indicating that the individual neurons have fixed tuning curves expressed in retinal coordinates. The effect of the constant input, representing gaze direction, is to scale up or gain-modulate these tuning curves, producing a result similar to that shown in figure 7.6. The additive constant in the input shown in figure 7.13A has a multiplicative effect on the output activity shown in 7.13B. This is primarily due to the fact that the width of the activity profiles is fixed by the recurrent network interaction,

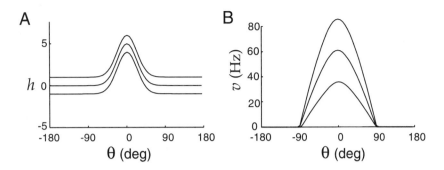

Figure 7.13 Effect of adding a constant to the input of a nonlinear recurrent network. (A) The input to the network consists of a single peak to which a constant factor has been added. (B) The gain-modulated output of the nonlinear network. The three curves correspond to the three input curves in panel A, in the same order. The model is the same as that used in figures 7.9 and 7.12.

so a constant positive input raises (and a negative input lowers) the peak of the response curve without broadening the base of the curve.

Sustained Activity

The effects illustrated in figures 7.12 and 7.13 arise because the nonlinear recurrent network has a stereotyped pattern of activity that is largely determined by interactions with other neurons in the network rather than by the feedforward input. If the recurrent connections are strong enough, the pattern of population activity, once established, can become independent of the structure of the input. For example, the recurrent network we have been studying can support a pattern of activity localized around a given preferred stimulus value, even when the input is uniform. This is seen in figure 7.14. The neurons of the network initially receive inputs that depend on their preferred angles, as seen in figure 7.14A. This produces a localized pattern of network activity (figure 7.14B). When the input is switched to the same constant value for all neurons (figure 7.14C), the network activity does not become uniform. Instead, it stays localized around the value $\theta = 0$ (figure 7.14D). This means that constant input can maintain a state that provides a memory of previous localized input activity. Networks similar to this have been proposed as models of sustained activity in the head-direction system of the rat and in prefrontal cortex during tasks involving working memory.

This memory mechanism is related to the integration seen in the linear model of eye position maintenance discussed previously. The linear network has an eigenvector $\mathbf{e}_1$ with eigenvalue $\lambda_1 = 1$. This allows $\mathbf{v} = c_1 \mathbf{e}_1$ to be a static solution of the equations of the network (7.17) in the absence of input for any value of c_1. As a result, the network can preserve any initial value of c_1 as a memory. In the case of figure 7.14, the steady-state activity in the absence of tuned input is a function of $\theta - \Theta$, for any value of the angle Θ. As a result, the network can preserve any initial value of

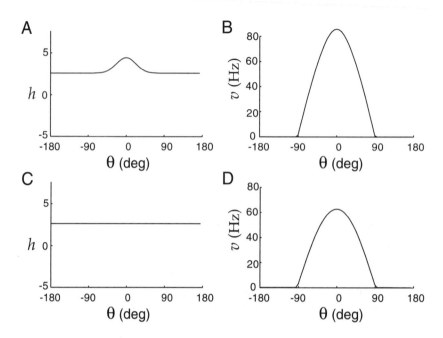

Figure 7.14 Sustained activity in a recurrent network. (A) Input to the neurons of the network consisting of localized excitation and a constant background. (B) The activity of the network neurons in response to the input of panel A. (C) Constant network input. (D) Response to the constant input of panel C when it immediately followed the input in A. The model is the same as that used in figures 7.9, 7.12, and 7.13.

Θ as a memory ($\Theta = 0°$ in the figure). The activities of the units $v(\theta)$ depend on Θ in an essentially nonlinear manner, but if we consider linear perturbations around this nonlinear solution, there is an eigenvector with eigenvalue $\lambda_1 = 1$ associated with shifts in the value of Θ. In this case, it can be shown that $\lambda_1 = 1$ because the network was constructed to be translationally invariant.

Maximum Likelihood and Network Recoding

Recurrent networks can generate characteristic patterns of activity even when they receive complex inputs (figure 7.9), and can maintain these patterns while receiving constant input (figure 7.14). Pouget et al. (1998) suggested that the location of the characteristic pattern (i.e., the value of Θ associated with the peak of the population activity profile) could be interpreted as a match of a fixed template curve to the input activity profile. This curve-fitting operation is at the heart of the maximum likelihood decoding method we described in chapter 3 for estimating a stimulus variable such as Θ. In the maximum likelihood method, the fitting curve is determined by the tuning functions of the neurons, and the curve-fitting procedure is defined by the characteristics of the noise perturbing the input activities. If the properties of the recurrent network match these op-

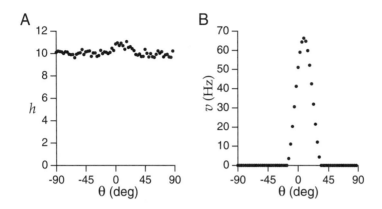

Figure 7.15 Recoding by a network model. (A) The noisy initial inputs $h(\theta)$ to 64 network neurons are shown as dots. The standard deviation of the noise is 0.25 Hz. After a short settling time, the input is set to a constant value of $h(\theta) = 10$. (B) The smooth activity profile that results from the recurrent interactions. The network model was similar to that used in figure 7.9, except that the recurrent synaptic weights were in the form of a Gabor-like function rather than a cosine, and the recurrent connections had short-range excitation and long-range inhibition. (see Pouget et al., 1998.)

timal characteristics, the network can approximate maximum likelihood decoding. Once the activity of the population of neurons has stabilized to its stereotyped shape, a simple decoding method such as vector decoding (see chapter 3) can be applied to extract the estimated value of Θ. This allows the accuracy of a vector decoding method to approach that of more complex optimal methods, because the computational work of curve fitting has been performed by the nonlinear recurrent interactions.

Figure 7.15 shows how this idea works in a network of 64 neurons receiving inputs that have Gaussian (rather than cosine) tuning curves as a function of Θ. Vector decoding applied to the reconstruction of Θ from the activity of the network or its inputs turns out to be almost unbiased. The way to judge decoding accuracy is therefore to compute the standard deviation of the decoded Θ values (chapter 3). The noisy input activity in figure 7.15A shows a slight bump around the value $\theta = 10°$. Vector decoding applied to input activities with this level of noise gives a standard deviation in the decoded angle of 4.5°. Figure 7.15B shows the output of the network obtained by starting with initial activities $v(\theta) = 0$ and input $h(\theta)$ as in figure 7.15A, and then setting $h(\theta)$ to a constant (θ-independent) value to maintain sustained activity. This generates a smooth pattern of sustained population activity. Vector decoding applied to the output activities generated in this way gives a standard deviation in the decoded angle of 1.7°. This is not too far from the Cramér-Rao bound, which gives the maximum possible accuracy for any unbiased decoding scheme applied to this system (see chapter 3), which is 0.88°.

Network Stability

fixed-point

When a network responds to a constant input by relaxing to a steady state with $d\mathbf{v}/dt=\mathbf{0}$, it is said to exhibit fixed-point behavior. Almost all the network activity we have discussed thus far involves such fixed points. This is by no means the only type of long-term activity that a network model can display. In a later section of this chapter, we discuss networks that oscillate, and chaotic behavior is also possible. But if certain conditions are met, a network will inevitably reach a fixed point in response to constant input. The theory of Lyapunov functions, to which we give an informal introduction, can be used to prove when this occurs.

recurrent model with current dynamics

It is easier to discuss the Lyapunov function for a network if we use the firing-rate dynamics of equation 7.6 rather than equation 7.8. For a network model, this means expressing the vector of network firing rates as $\mathbf{v} = \mathbf{F}(\mathbf{I})$, where $\mathbf{I}$ is the total synaptic current vector (i.e., I_a represents the total synaptic current for unit a). We assume that $F'(I) > 0$ for all I, where F' is the derivative of F. $\mathbf{I}$ obeys the dynamic equation derived from generalizing equation 7.6 to a network situation,

$$\tau_s \frac{d\mathbf{I}}{dt} = -\mathbf{I} + \mathbf{h} + \mathbf{M} \cdot \mathbf{F}(\mathbf{I}) . \tag{7.39}$$

Note that we have made the substitution $\mathbf{v} = \mathbf{F}(\mathbf{I})$ in the last term of the right side of this equation. Equation 7.39 can be used instead of equation 7.11 to provide a firing-rate model of a recurrent network.

Lyapunov function L

For the firing-rate model of equation 7.39 with a symmetric recurrent weight matrix, Cohen and Grossberg (1983) showed that the function

$$L(\mathbf{I}) = \sum_{a=1}^{N_v} \left(\int_0^{I_a} dz_a \, z_a F'(z_a) - h_a F(I_a) - \frac{1}{2} \sum_{a'=1}^{N_v} F(I_a) M_{aa'} F(I_{a'}) \right) \tag{7.40}$$

has $dL/dt < 0$ whenever $d\mathbf{I}/dt \neq \mathbf{0}$. To see this, take the time derivative of equation 7.40 and use 7.39 to obtain

$$\frac{dL(\mathbf{I})}{dt} = -\tau_s \sum_{a=1}^{N_v} F'(I_a) \left(\frac{dI_a}{dt} \right)^2 . \tag{7.41}$$

Because $F' > 0$, L decreases unless $d\mathbf{I}/dt = \mathbf{0}$. If L is bounded from below, it cannot decrease indefinitely, so $\mathbf{I} = \mathbf{h} + \mathbf{M} \cdot \mathbf{v}$ must converge to a fixed point. This implies that $\mathbf{v}$ must converge to a fixed point as well.

We have required that $F'(I) > 0$ for all values of its argument I. However, with some technical complications, it can be shown that the Lyapunov function we have presented also applies to the case of the rectifying activation function $F(I) = [I]_+$, even though it is not differentiable at $I = 0$ and $F'(I) = 0$ for $I < 0$. Convergence to a fixed point, or one of a set of fixed points, requires the Lyapunov function to be bounded from below. One way to ensure this is to use a saturating activation function, so that $F(I)$ is bounded as $I \to \infty$. Another way is to keep the eigenvalues of $\mathbf{M}$ sufficiently small.

Associative Memory

The models of memory discussed previously in this chapter store information by means of persistent activity. This is called working or short-term memory. In biological systems, persistent activity appears to play a role in retaining information over periods of seconds to minutes. Retention of long-term memories, over periods of hours to years, is thought to involve storage by means of synaptic strengths rather than persistent activity. One general idea is that synaptic weights in a recurrently connected network are set when a memory is stored so that the network can, at a later time, internally recreate the pattern of activity that represents the stored memory. In such networks, persistent activity is used to signal memory recall and to register the identity of the retrieved item, but the synaptic weights provide the long-term storage of the possible memory patterns. The pattern of activity of the units in the network at the start of memory retrieval determines which memory is recalled through its relationship to, or association with, the pattern of activity representing that memory. Such associative networks have been used to model regions of the mammalian brain implicated in various forms of memory, including area CA3 of the hippocampus and parts of the prefrontal cortex.

In an associative (or more strictly, autoassociative) memory, a partial or approximate representation of a stored item is used to recall the full item. Unlike a standard computer memory, recall in an associative memory is based on content rather than on an address. An example would be recalling every digit of a familiar phone number, given a few of its digits as an initial clue. In a network associative memory, recurrent weights are adjusted so that the network has a set of discrete fixed points identical (or very similar) to the patterns of activity that represent the stored memories. In many cases, the dynamics of the network are governed by a Lyapunov function (equation 7.40), ensuring the existence of fixed points. Provided that not too many memories are stored, these fixed points can perfectly, or at least closely, match the memory patterns. During recall, an associative memory network performs the computational operation of pattern matching by finding the fixed-point that most closely matches the initial state of the network. Each memory pattern has a basin of attraction, defined as the set of initial activities from which the network evolves to that particular fixed point. These basins of attraction define the pattern-matching properties of the network.

Associative memory networks can be constructed from units with either continuous-valued or binary (typically on or off) activities. We consider a network of continuous-valued units described by equation 7.11 with $\mathbf{h} = \mathbf{0}$. To use this model for memory storage, we define a set of memory patterns, denoted by $\mathbf{v}^m$ with $m = 1, 2, \ldots, N_{\mathrm{mem}}$, that we wish to store and recall. Note that $\mathbf{v}^m$ does not signify a component of a vector, but rather an entire vector identified by the superscript m. Associative recall is achieved by starting the network in an initial state that is similar to one of the memory patterns. That is, $\mathbf{v}(0) \approx \mathbf{v}^m$ for one of the m values, where "approximately

memory patterns $\mathbf{v}^m$

number of memories N_{mem}

equal" means that a significant number, but not necessarily all, of the elements of $\mathbf{v}(0)$ are close to the corresponding elements of $\mathbf{v}^m$. The network then evolves according to equation 7.11. If recall is successful, the dynamics converge to a fixed point equal (or at least significantly more similar than $\mathbf{v}(0)$) to the memory pattern associated with the initial state (i.e., $\mathbf{v}(t) \to \mathbf{v}^m$ for large t). Failure of recall occurs if the fixed point reached by the network is not similar to $\mathbf{v}^m$, or if a fixed point is not reached at all.

For exact recall to occur, $\mathbf{v}^m$ must be a fixed point of the network dynamics, which means it must satisfy the equation

$$\mathbf{v}^m = \mathbf{F}(\mathbf{M} \cdot \mathbf{v}^m) . \tag{7.42}$$

memory capacity

Therefore, we examine conditions under which such solutions exist for all the memory patterns. The capacity of a network is determined in part by the number of different pre-specified vectors that can simultaneously satisfy equation 7.42 for an appropriate choice of $\mathbf{M}$. In the limit of large N_v, the capacity is typically proportional to N_v. Capacity is not the only relevant measure of the performance of an associative memory. Memory function can be degraded if there are spurious fixed points of the network dynamics in addition to the fixed points that represent the memory patterns. Finally, useful pattern matching requires each fixed point to have a sufficiently large basin of attraction. Analyzing spurious fixed points and the sizes of basins of attraction is beyond the scope of this text.

Although the units in the network have continuous-valued activities, we consider the simple case in which the units are either inactive or active in the memory patterns themselves. Inactive units correspond to components of $\mathbf{v}^m$ that are equal to 0, and active units, to components that are equal to some constant value c. To simplify the discussion, we assume that each of the memory patterns has exactly αN_v active and $(1 - \alpha)N_v$ inactive units. The choice of which units are active in each pattern is random, and independent of the other patterns. The parameter α is known

sparseness α

as the sparseness of the memory patterns. As α decreases, making the patterns more sparse, more of them can be stored but each contains less information.

To build an associative memory network, we need to construct a matrix that allows all the memory patterns to satisfy equation 7.42. To begin, suppose that we knew of a matrix $\mathbf{K}$ for which all the memory patterns were degenerate eigenvectors with eigenvalue λ,

$$\mathbf{K} \cdot \mathbf{v}^m = \lambda \mathbf{v}^m \tag{7.43}$$

for all m. Then, consider the matrix

$$\mathbf{M} = \mathbf{K} - \frac{\mathbf{nn}}{\alpha N_v} \quad \text{or} \quad M_{aa'} = K_{aa'} - \frac{1}{\alpha N_v} . \tag{7.44}$$

vector of ones $\mathbf{n}$

Here $\mathbf{n}$ is a vector that has each of its N_v components equal to 1. The term $\mathbf{nn}$ in the matrix represents uniform inhibition between network units. $\mathbf{M}$ satisfies

$$\mathbf{M} \cdot \mathbf{v}^m = \lambda \mathbf{v}^m - c\mathbf{n} \tag{7.45}$$

for any memory pattern $\mathbf{v}^m$. The second term on the right side follows from the fact that $\mathbf{n} \cdot \mathbf{v}^m = c\alpha N_v$. Treated component by component, equation 7.42 for this matrix separates into two conditions: one for the components of $\mathbf{v}^m$ that are 0 and another for the components of $\mathbf{v}^m$ equal to c,

$$F(-c) = 0 \quad \text{and} \quad c = F(c(\lambda - 1)). \tag{7.46}$$

It is relatively easy to find conditions for which these equations have a solution. For positive c, the first condition is automatically satisfied for a rectifying activation function, $F(I) = 0$ for $I \leq 0$. For such a function satisfying $F'(I) > 0$ for all positive I, the second equation will be satisfied and equation 7.11 will have a stable fixed-point solution with $c > 0$ if, for example, $\lambda > 1$, $(\lambda - 1)F'(0) > 1$, and $F(c(\lambda - 1))$ grows more slowly than c for large c.

The existence of spurious fixed points decreases the usefulness of a network associative memory. This might seem to be a problem in the example we are discussing because the degeneracy of the eigenvalues means that any linear combination of memory patterns also satisfies equation 7.43. However, the nonlinearity in the network can prevent linear combinations of memory patterns from satisfying equation 7.42, even if they satisfy equation 7.43, thereby eliminating at least some of the spurious fixed points.

The problem of constructing an associative memory network thus reduces to finding the matrix $\mathbf{K}$ of equation 7.43, or at least constructing a matrix with similar properties. Because the choice of active units in each memory pattern is independent, the probability that a given unit is active in two different memory patterns is α^2. Thus, $\mathbf{v}^n \cdot \mathbf{v}^m \approx \alpha^2 c^2 N_v$ if $m \neq n$. Consider the dot product of one of the memory patterns, $\mathbf{v}^m$, with the vector $\mathbf{v}^n - \alpha c\mathbf{n}$, for some value of n. If $m = n$, $(\mathbf{v}^n - \alpha c\mathbf{n}) \cdot \mathbf{v}^m = c^2 \alpha N_v (1 - \alpha)$, whereas if $m \neq n$, $(\mathbf{v}^n - \alpha c\mathbf{n}) \cdot \mathbf{v}^m \approx c^2 N_v(\alpha^2 - \alpha^2) = 0$. It follows from these results that the matrix

$$\mathbf{K} = \frac{\lambda}{c^2 \alpha N_v (1 - \alpha)} \sum_{n=1}^{N_{\text{mem}}} \mathbf{v}^n (\mathbf{v}^n - \alpha c\mathbf{n}) \tag{7.47}$$

has properties similar to those of the matrix in equation 7.43, that is, $\mathbf{K} \cdot \mathbf{v}^m \approx \lambda \mathbf{v}^m$ for all m.

Recall that the Lyapunov function in equation 7.40 guarantees that the network has fixed points only if it is bounded from below and the matrix $\mathbf{M}$ is symmetric. Bounding of the Lyapunov function can be achieved if the activation function saturates. However, the recurrent weight matrix obtained by substituting expression 7.47 into equation 7.44 is not likely to be symmetric. A symmetric form of the recurrent weight matrix can be constructed by writing

$$\mathbf{M} = \frac{\lambda}{c^2 \alpha N_v (1 - \alpha)} \sum_{n=1}^{N_{\text{mem}}} (\mathbf{v}^n - \alpha c\mathbf{n})(\mathbf{v}^n - \alpha c\mathbf{n}) - \frac{\mathbf{n}\mathbf{n}}{\alpha N_v}. \tag{7.48}$$

The reader is urged to verify that, due to the additional terms in the sum over memory patterns, the conditions that must be satisfied when using 7.48 are slightly modified from 7.46 to

$$F(-c(1+\alpha\lambda)) = 0 \quad \text{and} \quad c = F(c(\lambda - 1 - \alpha\lambda)). \quad (7.49)$$

One way of looking at the recurrent weights in equation 7.48 is in terms of a learning rule used to construct the matrix. In this learning rule, an excitatory contribution to the coupling between two units is added whenever both of them are either active or inactive for a particular memory pattern. An inhibitory term is added whenever one unit is active and the other is not. The learning rule associated with equation 7.48 is called a covariance rule because of its relationship to the covariance matrix of the memory patterns. Learning rules for constructing networks that perform associative memory and other tasks are discussed in chapter 8.

Figure 7.16 shows an associative memory network of $N_v = 50$ units that stores four patterns, using the matrix from equation 7.48. Two of these patterns were generated randomly as discussed above. The other two patterns were assigned nonrandomly to make them easy to identify in the figure. Recall of these two nonrandom patterns is shown in figures 7.16B and 7.16C. From an initial pattern of activity only vaguely resembling one of the stored patterns, the network attains a fixed point very similar to the best matching memory pattern. The same results apply for the other two memory patterns stored by the network, but they are more difficult to identify in a figure because they are random.

The matrix 7.48 that we use as a basis for constructing an associative memory network satisfies the conditions required for exact storage and recall of the memory patterns only approximately. This introduces some errors in recall. As the number of memory patterns increases, the approximation becomes worse and the performance of the associative memory deteriorates, which limits the number of memories that can be stored. The simple covariance prescription for the weights in equation 7.48 is far from optimal. Other prescriptions for constructing $\mathbf{M}$ can achieve significantly higher storage capacities.

The basic conclusions from studies of associative memory models is that large networks can store large numbers of patterns, particularly if they are sparse (α is small) and if a few errors in recall can be tolerated. The capacity of certain associative memory networks can be calculated analytically. The number of memory patterns that can be stored is on the order of the number of neurons in the network, N_v, and depends on the sparseness, α, as $1/(\alpha \log(1/\alpha))$. The amount of information that can be stored is proportional to N_v^2, which is roughly the number of synapses in the network, but the information stored per synapse (i.e., the constant of proportionality) is typically quite small.

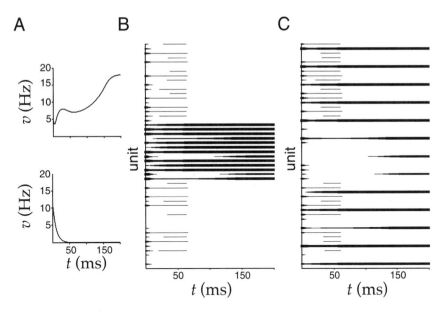

Figure 7.16 Associative recall of memory patterns in a network model. Panel A shows two representative units, and panels B and C show the firing rates of all 50 units plotted against time. The thickness of the horizontal lines in these plots is proportional to the firing rate of the corresponding neuron. (A) Firing rates of representative neurons. The upper panel shows the firing rate of one of the neurons corresponding to a nonzero component of the recalled memory pattern. The firing rate achieves a nonzero (and nonsaturated) steady-state value. The lower panel shows the firing rate of a neuron corresponding to a zero component of the recalled memory pattern. This goes to 0. (B) Recall of one of the stored memory patterns. The stored pattern had nonzero values for units 18 through 31. The initial state of the network was random, but with a bias toward this particular pattern. The final state is similar to the memory pattern. (C) Recall of another of the stored memory patterns. The stored pattern had nonzero values for every fourth unit. The initial state of the network was again random, but biased toward this pattern. The final state is similar to the memory pattern. This model uses the matrix of equation 7.48 with $\alpha = 0.25$ and $\lambda = 1.25$, and the activation function $F(I) = 150 \text{ Hz}[\tanh((I + 20 \text{ Hz})/(150 \text{ Hz}))]_+$.

7.5   Excitatory-Inhibitory Networks

In this section, we discuss models in which excitatory and inhibitory neurons are described separately by equations 7.12 and 7.13. These models exhibit richer dynamics than the single population models with symmetric coupling matrices we have analyzed up to this point. In models with excitatory and inhibitory subpopulations, the full synaptic weight matrix is not symmetric, and network oscillations can arise. We begin by analyzing a model of homogeneous coupled excitatory and inhibitory populations. We introduce methods for determining whether this model exhibits constant or oscillatory activity. We then present two network models in which oscillations appear. The first is a model of the olfactory bulb, and the second displays selective amplification in an oscillatory mode.

Homogeneous Excitatory and Inhibitory Populations

As an illustration of the dynamics of excitatory-inhibitory network models, we analyze a simple model in which all of the excitatory neurons are described by a single firing rate, v_E, and all of the inhibitory neurons are described by a second rate, v_I. Although we think of this example as a model of interacting neuronal populations, it is constructed as if it consists of just two neurons. Equations 7.12 and 7.13, with threshold linear response functions, are used to describe the two firing rates, so that

$$\tau_E \frac{dv_E}{dt} = -v_E + [M_{EE}v_E + M_{EI}v_I - \gamma_E]_+ \qquad (7.50)$$

and

$$\tau_I \frac{dv_I}{dt} = -v_I + [M_{II}v_I + M_{IE}v_E - \gamma_I]_+ \,. \qquad (7.51)$$

The synaptic weights M_{EE}, M_{IE}, M_{EI}, and M_{II} are numbers rather than matrices in this model. In the example we consider, we set $M_{EE} = 1.25$, $M_{IE} = 1$, $M_{II} = 0$, $M_{EI} = -1$, $\gamma_E = -10$ Hz, $\gamma_I = 10$ Hz, $\tau_E = 10$ ms, and we vary the value of τ_I. The negative value of γ_E means that this parameter serves as a source of constant background activity rather than as a threshold.

Phase-Plane Methods and Stability Analysis

The model of interacting excitatory and inhibitory populations given by equations 7.50 and 7.51 provides an opportunity for us to illustrate some of the techniques used to study the dynamics of nonlinear systems. This model exhibits both fixed-point (constant v_E and v_I) and oscillatory activity, depending on the values of its parameters. Stability analysis can be used to determine the parameter values where transitions between these two types of activity take place.

The firing rates $v_E(t)$ and $v_I(t)$ arising from equations 7.50 and 7.51 can be displayed by plotting them as functions of time, as in figures 7.18A and 7.19A. Another useful way of depicting these results, illustrated in figures 7.18B and 7.19B, is to plot pairs of points $(v_E(t), v_I(t))$ for a range of t values. As the firing rates change, these points trace out a curve or
phase plane   trajectory in the v_E-v_I plane, which is called the phase plane of the model. Phase-plane plots can be used to give a geometric picture of the dynamics of a model.

Values of v_E and v_I for which the right side of either equation 7.50 or equation 7.51 vanishes are of particular interest in phase-plane analysis. Sets of such values form two curves in the phase plane known as nullclines.
nullcline   The nullclines for equations 7.50 and 7.51 are the straight lines drawn in figure 7.17A. The nullclines are important because they divide the phase plane into regions with opposite flow patterns. This is because dv_E/dt and

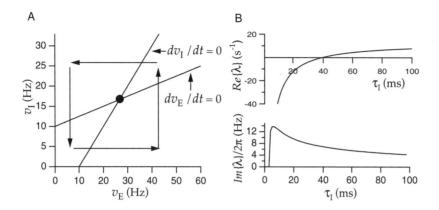

Figure 7.17 (A) Nullclines, flow directions, and fixed point for the firing-rate model of interacting excitatory and inhibitory neurons. The two straight lines are the nullclines along which $dv_E/dt = 0$ or $dv_I/dt = 0$. The filled circle is the fixed point of the model. The horizontal and vertical arrows indicate the directions that v_E (horizontal arrows) and v_I (vertical arrows) flow in different regions of the phase plane relative to the nullclines. (B) Real (upper panel) and imaginary (lower panel) parts of the eigenvalue determining the stability of the fixed point. To the left of the point where the imaginary part of the eigenvalue goes to 0, both eigenvalues are real. The imaginary part has been divided by 2π to give the frequency of oscillations near the fixed point.

dv_I/dt are positive on one side of their nullclines and negative on the other, as the reader can verify from equations 7.50 and 7.51 Above the nullcline along which $dv_E/dt = 0$, $dv_E/dt < 0$, and below it $dv_E/dt > 0$. Similarly, $dv_I/dt > 0$ to the right of the nullcline where $dv_I/dt = 0$, and $dv_I/dt < 0$ to the left of it. This determines the direction of flow in the phase plane, as denoted by the horizontal and vertical arrows in figure 7.17A. Furthermore, the rate of flow typically slows if the phase-plane trajectory approaches a nullcline.

At a fixed point of a dynamic system, the dynamic variables remain at constant values. In the model being considered, a fixed point occurs when the firing rates v_E and v_I take values that make $dv_E/dt = dv_I/dt = 0$. Because a fixed point requires both derivatives to vanish, it can occur only at an intersection of nullclines. The model we are considering has a single fixed point (at $v_E = 26.67$, $v_I = 16.67$) denoted by the filled circle in figure 7.17A. A fixed point provides a potential static configuration for the system, but it is critically important whether the fixed point is stable or unstable. If a fixed point is stable, initial values of v_E and v_I near the fixed point will be drawn toward it over time. If the fixed point is unstable, nearby configurations are pushed away from the fixed point, and the system will remain at the fixed point indefinitely only if the rates are set initially to the fixed-point values with infinite precision.

Linear stability analysis can be used to determine whether a fixed point is stable or unstable. To do this we take derivatives of the expressions for dv_E/dt and dv_I/dt obtained by dividing the right sides of equations 7.50

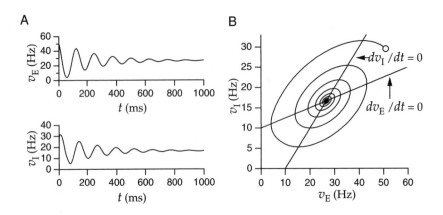

Figure 7.18 Activity of the excitatory-inhibitory firing-rate model when the fixed point is stable. (A) The excitatory and inhibitory firing rates settle to the fixed point over time. (B) The phase-plane trajectory is a counterclockwise spiral collapsing to the fixed point. The open circle marks the initial values $v_E(0)$ and $v_I(0)$. For this example, $\tau_I = 30$ ms.

and 7.51 by τ_E and τ_I respectively. We then evaluate these derivatives at the values of v_E and v_I that correspond to the fixed point. The four combinations of derivatives computed in this way can be arranged into a matrix

stability matrix

$$\begin{pmatrix} (M_{EE} - 1)/\tau_E & M_{EI}/\tau_E \\ M_{IE}/\tau_I & (M_{II} - 1)/\tau_I \end{pmatrix}. \qquad (7.52)$$

As discussed in the Mathematical Appendix, the stability of the fixed point is determined by the real parts of the eigenvalues of this matrix. The eigenvalues are given by

$$\lambda = \frac{1}{2} \left(\frac{M_{EE} - 1}{\tau_E} + \frac{M_{II} - 1}{\tau_I} \pm \sqrt{\left(\frac{M_{EE} - 1}{\tau_E} - \frac{M_{II} - 1}{\tau_I} \right)^2 + \frac{4 M_{EI} M_{IE}}{\tau_E \tau_I}} \right).$$

$$(7.53)$$

If the real parts of both eigenvalues are less than 0, the fixed point is stable, whereas if either is greater than 0, the fixed point is unstable. If the factor under the radical sign in equation 7.53 is positive, both eigenvalues are real, and the behavior near the fixed point is exponential. This means that there is exponential movement toward the fixed point if both eigenvalues are negative, or away from the fixed point if either eigenvalue is positive. We focus on the case when the factor under the radical sign is negative, so that the square root is imaginary and the eigenvalues form a complex conjugate pair. In this case, the behavior near the fixed point is oscillatory and the trajectory either spirals into the fixed point, if the real part of the eigenvalues is negative, or out from the fixed point if the real part of the eigenvalues is positive. The imaginary part of the eigenvalue determines the frequency of oscillations near the fixed point. The real and imaginary parts of one of these eigenvalues are plotted as a function of τ_I

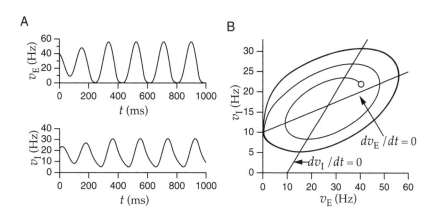

Figure 7.19 Activity of the excitatory-inhibitory firing-rate model when the fixed point is unstable. (A) The excitatory and inhibitory firing rates settle into periodic oscillations. (B) The phase-plane trajectory is a counterclockwise spiral that joins the limit cycle, which is the closed orbit. The open circle marks the initial values $v_E(0)$ and $v_I(0)$. For this example, $\tau_I = 50$ ms.

in figure 7.17B. This figure indicates that the fixed point is stable if $\tau_I < 40$ ms and unstable for larger values of τ_I.

Figures 7.18 and 7.19 show examples in which the fixed point is stable and unstable, respectively. In figure 7.18A, the oscillations in v_E and v_I are damped, and the firing rates settle down to the stable fixed point. The corresponding phase-plane trajectory is a collapsing spiral (figure 7.18B). In figure 7.19A the oscillations grow, and in figure 7.19B the trajectory is a spiral that expands outward until the system enters a limit cycle. A limit cycle is a closed orbit in the phase plane indicating periodic behavior. The fixed point is unstable in this case, but the limit cycle is stable. Without rectification, the phase-plane trajectory would spiral out from the unstable fixed point indefinitely. The rectification nonlinearity prevents the spiral trajectory from expanding past 0 and thereby stabilizes the limit cycle.

limit cycle

There are a number of ways that a nonlinear system can make a transition from a stable fixed point to a limit cycle. Such transitions are called bifurcations. The transition seen between figures 7.18 and 7.19 is a Hopf bifurcation. In this case, a fixed point becomes unstable as a parameter is changed (in this case τ_I) when the real part of a complex eigenvalue changes sign. In a Hopf bifurcation, the limit cycle emerges at a finite frequency, which is similar to the behavior of a type II neuron when it starts firing action potentials, as discussed in chapter 6. Other types of bifurcations produce type I behavior with oscillations emerging at 0 frequency (chapter 6). One example of this is a saddle-node bifurcation, which occurs when parameters are changed such that two fixed points, one stable and one unstable, meet at the same point in the phase plane.

Hopf bifurcation

saddle-node bifurcation

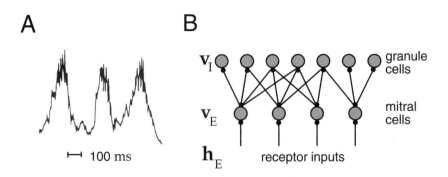

Figure 7.20 (A) Extracellular field potential recorded in the olfactory bulb during respiratory waves representing three successive sniffs. (B) Schematic diagram of the olfactory bulb model. (A adapted from Freeman and Schneider, 1982; B adapted from Li, 1995.)

The Olfactory Bulb

The olfactory bulb, and analogous olfactory areas in insects, provide examples of sensory processing involving oscillatory activity. The olfactory bulb represents the first stage of processing beyond the olfactory receptors in the vertebrate olfactory system. Olfactory receptor neurons respond to odor molecules and send their axons to the olfactory bulb. These axons *mitral cells* terminate in glomeruli where they synapse onto mitral and tufted cells, as *tufted cells* well as local interneurons. The mitral and tufted cells provide the output of the olfactory bulb by sending projections to the primary olfactory cortex. They also synapse onto the larger population of inhibitory granule *granule cells* cells. The granule cells in turn inhibit the mitral and tufted cells.

The activity in the olfactory bulb of many vertebrates is strongly influenced by a sniff cycle in which a few quick sniffs bring odors past the olfactory receptors. Figure 7.20A shows an extracellular potential recorded during three successive sniffs. The three large oscillations in the figure are due to the sniffs. The oscillations we discuss in this section are the smaller, higher-frequency oscillations seen around the peak of each sniff cycle. These arise from oscillatory neural activity. Individual mitral cells have quite low firing rates, and do not fire on each cycle of the oscillations. The oscillations are phase-locked across the bulb, in that different neurons fire at fixed phase lags from each other, but different odors induce oscillations of different amplitudes and phases.

Li and Hopfield (1989) modeled the mitral and granule cells of the olfactory bulb as a nonlinear input-driven network oscillator. Figure 7.20B shows the architecture of the model, which uses equations 7.12 and 7.13 with $\mathbf{M}_{EE} = \mathbf{M}_{II} = 0$. The absence of these couplings in the model is in accord with the anatomy of the bulb. The rates $\mathbf{v}_E$ and $\mathbf{v}_I$ refer to the mitral and granule cells, respectively. Figure 7.21A shows the activation functions of the model. The time constants for the two populations of cells are the same, $\tau_E = \tau_I = 6.7$ ms. $\mathbf{h}_E$ is the input from the receptors to the mitral

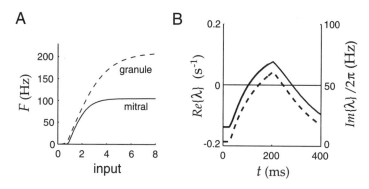

Figure 7.21 Activation functions and eigenvalues for the olfactory bulb model. (A) The activation functions F_E (solid curve) for the mitral cells, and F_I (dashed curve) for the granule cells. (B) The real (solid line, left axis) and imaginary (dashed line, right axis) parts of the eigenvalue that determines whether the network model exhibits fixed-point or oscillatory behavior. These are plotted as a function of time during a sniff cycle. When the real part of the eigenvalue becomes greater than 0, it determines the growth rate away from the fixed point, and the imaginary part divided by 2π determines the initial frequency of the resulting oscillations. (Adapted from Li, 1995.)

cells, and $\mathbf{h_I}$ is a constant representing top-down input that exists from the olfactory cortex to the granule cells.

The field potential in figure 7.20A shows oscillations during each sniff, but not between sniffs. For the model to match this pattern of activity, the input from the olfactory receptors, $\mathbf{h_E}$, must induce a transition between fixed-point and oscillatory activity. Before a sniff, the network must have a stable fixed point with low activities. As $\mathbf{h_E}$ increases during a sniff, this steady-state configuration must become unstable, leading to oscillatory activity. The analysis of the stability of the fixed point and the onset of oscillations is closely related to our previous stability analysis of the model of homogeneous populations of coupled excitatory and inhibitory neurons. It is based on properties of the eigenvalues of the linear stability matrix (see the Mathematical Appendix). In this case, the stability matrix includes contributions from the derivatives of the activation functions evaluated at the fixed point. For the fixed point to become unstable, the real part of at least one of the eigenvalues that arise in this analysis must become larger than 0. To ensure oscillations, at least one of these destabilizing eigenvalues should have a nonzero imaginary part. These requirements impose constraints on the connections between the mitral and granule cells and on the inputs.

Figure 7.21B shows the real and imaginary parts of the relevant eigenvalue, labeled λ, during one sniff cycle. About 100 ms into the cycle the real part gets bigger than 0. Reading off the imaginary part at this point, we see that this sets off roughly 40 Hz oscillations in the network. These oscillations stop about 300 ms into the sniff cycle when the real part of λ drops below 0. The input $\mathbf{h_E}$ from the receptors plays two critical roles in this process. First, it makes the real part of the eigenvalue greater than 0 by

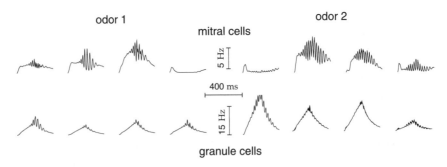

Figure 7.22 Activities of four of ten mitral (upper) and granule (lower) cells during a single sniff cycle for two different odors. (Adapted from Li and Hopfield, 1989.)

modifying where the fixed point lies on the activation function curves in figure 7.21A. Second, it affects which particular neurons are destabilized, and thus which begin to oscillate. The ultimate pattern of oscillatory activity is determined both by the input $\mathbf{h}_E$ and by the recurrent couplings of the network.

Figure 7.22 shows the behavior of the network during a single sniff cycle in the presence of two different odors, represented by two different values of $\mathbf{h}_E$. The top rows show the activity of four mitral cells, and the bottom rows of four granule cells. The amplitudes and phases of the oscillations seen in these traces, along with the identities of the mitral cells taking part in them, provide a signature of the identity of the odor that was presented.

Oscillatory Amplification

As a final example of network oscillations, we return to amplification of input signals by a recurrently connected network. Two factors control the amount of selective amplification that is viable in networks such as that shown in figure 7.9. The most important constraint on the recurrent weights is that the network must be stable, so the activity does not increase without bound. Another possible constraint is suggested by figure 7.14D, where the output shows a tuned response even though the input to the network is constant as a function of θ. Tuned output in the absence of tuned input can serve as a memory mechanism, but it will produce persistent perceptions if it occurs in a primary sensory area, for example. Avoiding this in the network limits the recurrent weights and the amount of amplification that can be supported.

Li and Dayan (1999) showed that this restriction can be significantly eased using the richer dynamics of networks of coupled inhibitory and excitatory neurons. Figure 7.23 shows an example with continuous neuron labeling based on a continuous version of equations 7.12 and 7.13. The input is $h_E(\theta) = 8 + 5\cos(2\theta)$ in the modulated case (figure 7.23B) or $h_E(\theta) = 8$ in the unmodulated case (figure 7.23C). Noise with standard deviation 0.4 corrupts this input. The input to the network is constant in time.

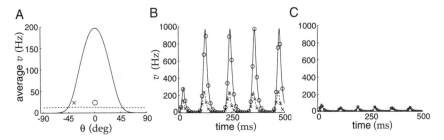

Figure 7.23 Selective amplification in an excitatory-inhibitory network. (A) Time-averaged response of the network to a tuned input with $\Theta = 0°$ (solid curve) and to an untuned input (dashed curve). Symbols "o" and "x" mark the $0°$ and $-37°$ points seen in B and C. (B) Activities over time of neurons with preferred angles of $\theta = 0°$ (solid curve) and $\theta = -37°$ (dashed curve) in response to a modulated input with $\Theta = 0°$. (C) Activities of the same units shown in B to a constant input. The lines lie on top of each other, showing that the two units respond identically. The parameters are $\tau_E = \tau_I = 10$ ms, $h_I = 0$, $M_{EI} = -\delta(\theta - \theta')/\rho_\theta$, $M_{EE} = (1/\pi\rho_\theta)[5.9 + 7.8\cos(2(\theta - \theta'))]_+$, $M_{IE} = 13.3/\pi\rho_\theta$, and $M_{II} = 0$. (After Li and Dayan, 1999.)

The network oscillates in response to either constant or tuned input. Figure 7.23A shows the time average of the oscillating activities of the neurons in the network as a function of their preferred angles for noisy tuned (solid curve) and untuned (dashed curve) inputs. Neurons respond to the tuned input in a highly tuned and amplified manner. Despite the high degree of amplification, the average response of the neurons to untuned input is almost independent of θ. Figures 7.23B and 7.23C show the activities of individual neurons with $\theta = 0°$ ("o") and $\theta = -37°$ ("x") over time for the tuned and untuned inputs respectively. The network does not produce persistent perception, because the output to an untuned input is itself untuned. In contrast, a nonoscillatory version of this network, with $\tau_I = 0$, exhibits tuned sustained activity in response to an untuned input for recurrent weights this strong. The oscillatory network can thus operate in a regime of high selective amplification without generating spurious tuned activity.

7.6   Stochastic Networks

Up to this point, we have considered models in which the output of a cell is a deterministic function of its input. In this section, we introduce a network model called the Boltzmann machine, for which the input-output relationship is stochastic. Boltzmann machines are interesting from the perspective of learning, and also because they offer an alternative interpretation of the dynamics of network models.

Boltzmann machine

In the simplest form of Boltzmann machine, the neurons are treated as binary, so $v_a(t) = 1$ if unit a is active at time t (e.g., it fires a spike between times t and $t + \Delta t$ for some small value of Δt), and $v_a(t) = 0$ if it is inactive.

The state of unit a is determined by its total input current,

$$I_a(t) = h_a(t) + \sum_{a'=1}^{N_v} M_{aa'} v_{a'}(t), \qquad (7.54)$$

where $M_{aa'} = M_{a'a}$ and $M_{aa} = 0$ for all a and a', and h_a is the total feedforward input into unit a. In the model, units are permitted to change state only at integral multiples of Δt. At each time step, a single unit is selected, usually at random, to be updated. This update is based on a probabilistic rather than a deterministic rule. If unit a is selected, its state at the next time step is set stochastically to 1 with probability

$$P[v_a(t + \Delta t) = 1] = F(I_a(t)), \quad \text{with} \quad F(I_a) = \frac{1}{1 + \exp(-I_a)}. \qquad (7.55)$$

It follows that $P[v_a(t + \Delta t) = 0] = 1 - F(I_a(t))$. F is a sigmoidal function, which has the property that the larger the value of I_a, the more likely unit a is to take the value 1.

Markov chain

Under equation 7.55, the state of activity of the network evolves as a Markov chain. This means that the components of $\mathbf{v}$ at different times are sequences of random variables with the property that $\mathbf{v}(t + \Delta t)$ depends only on $\mathbf{v}(t)$, and not on the previous history of the network. Equation 7.55

Glauber dynamics

implements what is known as Glauber dynamics.

An advantage of using Glauber dynamics to define the evolution of a network model is that general results from statistical mechanics can be used to determine the equilibrium distribution of activities. Under Glauber dynamics, $\mathbf{v}$ does not converge to a fixed point, but can be described by a

energy function

probability distribution associated with an energy function

$$E(\mathbf{v}) = -\mathbf{h} \cdot \mathbf{v} - \frac{1}{2} \mathbf{v} \cdot \mathbf{M} \cdot \mathbf{v}. \qquad (7.56)$$

The probability distribution characterizing $\mathbf{v}$, once the network has converged to an equilibrium state, is

$$P[\mathbf{v}] = \frac{\exp(-E(\mathbf{v}))}{Z} \quad \text{where} \quad Z = \sum_{\mathbf{v}} \exp(-E(\mathbf{v})). \qquad (7.57)$$

partition function

The notion of convergence as $t \to \infty$ can be formalized precisely, but informally it means that after repeated updating according to equation 7.55, the states of the network are described statistically by equation 7.57. Z is called the partition function and $P[\mathbf{v}]$, the Boltzmann distribution. Un-

Boltzmann distribution

der the Boltzmann distribution, states with lower energies are more likely. In this case, Glauber dynamics implements a statistical operation called Gibbs sampling for the distribution given in equation 7.57. From now on, we refer to the update procedure described by equation 7.55 as Gibbs sam-

Gibbs sampling

pling.

mean-field approximation

The Boltzmann machine is inherently stochastic. However, an approximation to the Boltzmann machine, known as the mean-field approximation, can be constructed on the basis of the deterministic synaptic current dynamics of a firing-rate model. In this case, $\mathbf{I}$ is determined by the dynamic equation 7.39 rather than by equation 7.54, with the function F in equation 7.39 set to the same sigmoidal function as in equation 7.55. The output v_a is determined from I_a at discrete times (integer multiples of Δt). The rule used for this is not the deterministic relationship $v_a = F(I_a)$ used in the firing-rate version of the model. Instead, v_a is determined from I_a stochastically, being set to either 1 or 0 with probability $F(I_a)$ or $1 - F(I_a)$ respectively. Thus, although the mean-field formulation for $\mathbf{I}$ is deterministic, $\mathbf{I}$ is used to generate a probability distribution over a binary output vector $\mathbf{v}$. Because $v_a = 1$ has probability $F(I_a)$ and $v_a = 0$ has probability $1 - F(I_a)$, and the units are independent, the probability distribution for the entire vector $\mathbf{v}$ is

$$Q[\mathbf{v}] = \prod_{a=1}^{N_v} F(I_a)^{v_a} (1 - F(I_a))^{1-v_a} . \qquad (7.58)$$

This is called the mean-field distribution for the Boltzmann machine. Note that this distribution (and indeed $\mathbf{v}$ itself) plays no role in the dynamics of the mean-field formulation of the Boltzmann machine. It is, rather, a way of interpreting the outputs.

mean-field distribution

We have presented two formulations of the Boltzmann machine, Gibbs sampling and the mean-field approach, that lead to the two distributions $P[\mathbf{v}]$ and $Q[\mathbf{v}]$ (equations 7.57 and 7.58). The Lyapunov function of equation 7.40, which decreases steadily under the dynamics of equation 7.39 until a fixed point is reached, provides a key insight into the relationship between these two distributions. In the appendix to this chapter, we show that this Lyapunov function can be expressed as

$$L(\mathbf{I}) = D_{\mathrm{KL}}(Q, P) + K , \qquad (7.59)$$

where K is a constant, and D_{KL} is the Kullback-Leibler divergence (see chapter 4). $D_{\mathrm{KL}}(Q, P)$ is a measure of how different the two distributions Q and P are from each other. The fact that the dynamics of equation 7.39 reduces the Lyapunov function to a minimum value means that it also reduces the difference between Q and P, as measured by the Kullback-Leibler divergence. This offers an interesting interpretation of the mean-field dynamics; it modifies the current value of the vector $\mathbf{I}$ until the distribution of binary output values generated by the mean-field formulation of the Boltzmann machine matches as closely as possible (finding at least a local minimum of $D_{\mathrm{KL}}(Q, P)$) the distribution generated by Gibbs sampling. In this way, the mean-field procedure can be viewed as an approximation of Gibbs sampling.

The power of the Boltzmann machine lies in the relationship between the distribution of output values, equation 7.57, and the quadratic energy function of equation 7.56. This makes it possible to determine how changing the weights $\mathbf{M}$ affects the distribution of output states. In chapter 8, we present a learning rule for the weights of the Boltzmann machine that allows $P[\mathbf{v}]$ to approximate a probability distribution extracted from a set of inputs. In chapter 10, we study other models that construct output distributions in this way.

Note that the mean-field distribution $Q[\mathbf{v}]$ is simpler than the full Boltzmann distribution $P[\mathbf{v}]$ because the units are statistically independent. This prevents $Q[\mathbf{v}]$ from providing a good approximation in some cases, particularly if there are negative weights between units that tend to make their activities mutually exclusive. The mean-field analysis of the Boltzmann machine illustrates the limitations of rate-based descriptions in capturing the full extent of the correlations that can exist between spiking neurons.

7.7 Chapter Summary

The models in this chapter mark the start of our discussion of computation, as opposed to coding. Using a description of the firing rates of network neurons, we showed how to construct linear and nonlinear feedforward and recurrent networks that transform information from one coordinate system to another, selectively amplify input signals, integrate inputs over extended periods of time, select between competing inputs, sustain activity in the absence of input, exhibit gain modulation, allow simple decoding with performance near the Cramér-Rao bound, and act as content-addressable memories. We used network responses to a continuous stimulus variable as an extended example. This led to models of simple and complex cells in primary visual cortex. We described a model of the olfactory bulb as an example of a system for which computation involves oscillations arising from asymmetric couplings between excitatory and inhibitory neurons. Linear stability analysis was applied to a simplified version of this model. We also considered a stochastic network model called the Boltzmann machine.

7.8 Appendix

Lyapunov Function for the Boltzmann Machine

Here, we show that the Lyapunov function of equation 7.40 can be reduced to equation 7.59 when applied to the mean-field version of the Boltzmann machine. Recall, from equation 7.40, that

$$L(\mathbf{I}) = \sum_{a=1}^{N_v} \left(\int_0^{I_a} dz_a\, z_a F'(z_a) - h_a F(I_a) - \frac{1}{2} \sum_{a'=1}^{N_v} F(I_a) M_{aa'} F(I_{a'}) \right). \quad (7.60)$$

When F is given by the sigmoidal function of equation 7.55,

$$\int_0^{I_a} dz_a\, z_a F'(z_a) = F(I_a) \ln F(I_a) + (1 - F(I_a)) \ln(1 - F(I_a)) + k, \quad (7.61)$$

where k is a constant, as can be verified by differentiating the right side. The nonconstant part of the right side of this equation is just (minus) the entropy associated with the binary variable v_a. In fact,

$$\sum_{a=1}^{N_v} \int_0^{I_a} dz_a\, z_a F'(z_a) = \langle \ln Q[\mathbf{v}] \rangle_Q + N_v k, \tag{7.62}$$

where the average is over all values of $\mathbf{v}$ weighted by their probabilities $Q[\mathbf{v}]$.

To evaluate the remaining terms in equation 7.60, we note that because the components of $\mathbf{v}$ are binary and independent for the Boltzmann machine, relations such as $\langle v_a \rangle_Q = F(I_a)$ and $\langle v_a v_b \rangle_Q = F(I_a)F(I_b)$ (for $a \neq b$) are valid. Then, using equation 7.56, we find

$$\sum_{a=1}^{N_v} \left(-h_a F(I_a) - \frac{1}{2}\sum_{a'=1}^{N_v} F(I_a) M_{aa'} F(I_{a'}) \right) = \langle E(\mathbf{v}) \rangle_Q. \tag{7.63}$$

Similarly, from equation 7.57, we can show that

$$\langle \ln P[\mathbf{v}] \rangle_Q = \langle -E(\mathbf{v}) \rangle_Q - \ln Z. \tag{7.64}$$

Combining the results of equations 7.62, 7.63, and 7.64, we obtain

$$L(\mathbf{I}) = \langle \ln Q[\mathbf{v}] - \ln P[\mathbf{v}] \rangle_Q + N_v k - \ln Z. \tag{7.65}$$

which gives equation 7.59 with $K = N_v k - \log Z$ because $\langle \ln Q[\mathbf{v}] - \ln P[\mathbf{v}] \rangle_Q$ is, by definition, the Kullback-Leibler divergence $D_{\mathrm{KL}}(Q, P)$. Note that in this (and subsequent) chapters, we define the Kullback-Leibler divergence using a natural logarithm, rather than the base 2 logarithm used in chapter 4. The two definitions differ only by an overall multiplicative constant.

7.9 Annotated Bibliography

Wilson & Cowan (1972, 1973) provides pioneering analyses of firing-rate models. Subsequent treatments related to the discussion in this chapter are presented in **Abbott (1994), Ermentrout (1998), Gerstner (1998)**, Amit & Tsodyks (1991a, 1991b), and Bressloff & Coombes (2000). The notion of a regular repeating unit of cortical computation dates back to the earliest investigations of cortex and is discussed by Douglas & Martin (1998).

Our discussion of the feedforward coordinate transformation model is based on Pouget & Sejnowski (1995, 1997) and Salinas & Abbott (1995), which built on theoretical work by Zipser & Andersen (1988) concerning parietal gain fields (see Andersen, 1989). Amplification by recurrent circuits is discussed in Douglas et al. (1995) and **Abbott (1994)**. We followed

Seung's (1996) analysis of neural integration for eye position (see also Seung et al., 2001), which builds on Robinson (1989). In general, we followed Seung (1996) and Zhang (1996) in adopting the theoretical context of continuous line or surface attractors.

Sompolinsky & Shapley 1997 (see also Somers, Nelson & Sur, 1995; Carandini & Ringach, 1997) reviews the debate about the relative roles of feedforward and recurrent input as the source of orientation selectivity in primary visual cortex. We presented a model of a hypercolumn; an extension to multiple hypercolumns is used by Li (1998, 1999) to link psychophysical and physiological data on contour integration and texture segmentation. Persistent activity in prefrontal cortex during short-term memory tasks is reviewed by Goldman-Rakic (1994) and Fuster (1995) and is modeled by Compte et al. (2000).

Network associative memories were described and analyzed in Hopfield (1982; 1984) and Cohen & Grossberg (1983), where a general Lyapunov function is introduced. Grossberg (1988), **Amit (1989)**, and **Hertz, et al. (1991)** present theoretical results concerning associative networks, in particular their capacity to store information. Associative memory in non-binary recurrent networks has been studied in particular by Treves and collaborators (see Rolls & Treves, 1998) and, in the context of line attractor networks, in Samsonovich & McNaughton (1997) and Battaglia & Treves (1998).

Rinzel and Ermentrout (1998) discusses phase-plane methods, and XPP (see http://www.pitt.edu/~phase) provides a computer environment for performing phase-plane and other forms of mathematical analysis on neuron and network models. We followed Li's (1995) presentation of Li & Hopfield's (1989) oscillatory model of the olfactory bulb.

The Boltzmann machine was introduced by Hinton & Sejnowski (1986) as a stochastic generalization of the Hopfield network (Hopfield, 1982). The mean-field model is due to Hopfield (1984), and we followed the probabilistic discussion given in Jordan et al. (1998). Markov chain methods for performing probabilistic inference are discussed in Neal (1993).

III Adaptation and Learning

8 Plasticity and Learning

8.1 Introduction

Activity-dependent synaptic plasticity is widely believed to be the basic phenomenon underlying learning and memory, and it is also thought to play a crucial role in the development of neural circuits. To understand the functional and behavioral significance of synaptic plasticity, we must study how experience and training modify synapses, and how these modifications change patterns of neuronal firing to affect behavior. Experimental work has revealed ways in which neuronal activity can affect synaptic strength, and experimentally inspired synaptic plasticity rules have been applied to a wide variety of tasks including auto- and heteroassociative memory, pattern recognition, storage and recall of temporal sequences, and function approximation.

In 1949, Donald Hebb conjectured that if input from neuron A often contributes to the firing of neuron B, then the synapse from A to B should be strengthened. Hebb suggested that such synaptic modification could produce neuronal assemblies that reflect the relationships experienced during training. The Hebb rule forms the basis of much of the research done on the role of synaptic plasticity in learning and memory. For example, consider applying this rule to neurons that fire together during training due to an association between a stimulus and a response. These neurons would develop strong interconnections, and subsequent activation of some of them by the stimulus could produce the synaptic drive needed to activate the remaining neurons and generate the associated response. Hebb's original suggestion concerned increases in synaptic strength, but it has been generalized to include decreases in strength arising from the repeated failure of neuron A to be involved in the activation of neuron B. General forms of the Hebb rule state that synapses change in proportion to the correlation or covariance of the activities of the pre- and postsynaptic neurons.

Hebb rule

Experimental work in a number of brain regions, including hippocampus, neocortex, and cerebellum, has revealed activity-dependent processes that can produce changes in the efficacies of synapses that persist for varying amounts of time. Figure 8.1 shows an example in which the data points

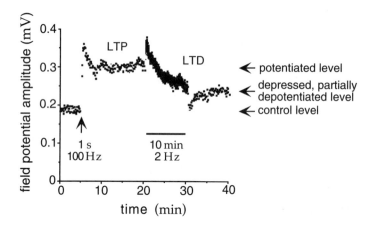

Figure 8.1 LTP and LTD at the Schaffer collateral inputs to the CA1 region of a rat hippocampal slice. The points show the amplitudes of field potentials evoked by constant amplitude stimulation. At the time marked by the arrow (at time 5 minutes), stimulation at 100 Hz for 1 s caused a significant increase in the response amplitude. Some of this increase decayed away following the stimulation, but most of it remained over the following 15 min test period, indicating LTP. Next, stimulation at 2 Hz was applied for 10 min (between times 20 and 30 minutes). This reduced the amplitude of the response. After a transient dip, the response amplitude remained at a reduced level approximately midway between the original and post-LTP values, indicating LTD. The arrows at the right show the levels initially (control), after LTP (potentiated), and after LTD (depressed, partially depotentiated). (Unpublished data of J. Fitzpatrick and J. Lisman.)

indicate amplitudes of field potentials evoked in the CA1 region of a slice of rat hippocampus by stimulation of the Schaffer collateral afferents. In experiments such as this, field potential amplitudes (or more often slopes) are used as a measure of synaptic strength. In figure 8.1, high-frequency *potentiation* stimulation induced synaptic potentiation (an increase in strength), and then long-lasting, low-frequency stimulation resulted in synaptic depres- *depression* sion (a decrease in strength) that partially removed the effects of the previous potentiation. This is in accord with a generalized Hebb rule because high-frequency presynaptic stimulation evokes a postsynaptic response, whereas low-frequency stimulation does not. Changes in synaptic strength involve both transient and long-lasting effects, as seen in figure 8.1. Changes that persist for tens of minutes or longer are generally *LTP and LTD* called long-term potentiation (LTP) and long-term depression (LTD). The longest-lasting forms appear to require protein synthesis.

A wealth of data is available on the underlying cellular basis of activity-dependent synaptic plasticity. For instance, the postsynaptic concentration of calcium ions appears to play a critical role in the induction of both LTP and LTD. However, we will not consider mechanistic models. Rather, we study synaptic plasticity at a functional level, attempting to relate the impact of synaptic plasticity on neurons and networks to the basic rules governing its induction.

Studies of plasticity and learning involve analyzing how synapses are affected by activity over the course of a training period. In this and the following chapters, we consider three types of training procedures. In unsupervised (also called self-supervised) learning, a network responds to a series of inputs during training solely on the basis of its intrinsic connections and dynamics. The network then self-organizes in a manner that depends on the synaptic plasticity rule being applied and on the nature of the inputs presented during training. We consider unsupervised learning in a more general setting called density estimation in chapter 10.

unsupervised learning

In supervised learning, which we consider in the last section of this chapter, a desired set of input-output relationships is imposed on the network by a "teacher" during training. Networks that perform particular tasks can be constructed in this way by letting a modification rule adjust their synapses until the desired computation emerges as a consequence of the training process. This is an alternative to explicitly specifying the synaptic weights, as was done in chapter 7. In this case, finding a biologically plausible teaching mechanism may not be a concern if the question being addressed is whether any weights can be found that allow a network to implement a particular function. In more biologically plausible examples of supervised learning, one network acts as the teacher for another network.

supervised learning

In chapter 9, we discuss a third form of learning, reinforcement learning, that is intermediate between these cases. In reinforcement learning, the network output is not constrained by a teacher, but evaluative feedback about network performance is provided in the form of reward or punishment. This can be used to control the synaptic modification process.

reinforcement learning

In this chapter we largely focus on activity-dependent synaptic plasticity of the Hebbian type, meaning plasticity based on correlations of pre- and postsynaptic firing. To ensure stability and to obtain interesting results, we often must augment Hebbian plasticity with more global forms of synaptic modification that, for example, scale the strengths of all the synapses onto a given neuron. These can have a major impact on the outcome of development or learning. Non-Hebbian forms of synaptic plasticity, such as those that modify synaptic strengths solely on the basis of pre- or postsynaptic firing, are likely to play important roles in homeostatic, developmental, and learning processes. Activity can also modify the intrinsic excitability and response properties of neurons. Models of such intrinsic plasticity show that neurons can be remarkably robust to external perturbations if they adjust their conductances to maintain specified functional characteristics. Intrinsic and synaptic plasticity can interact in interesting ways. For example, shifts in intrinsic excitability can compensate for changes in the level of input to a neuron caused by synaptic plasticity. It is likely that all of these forms of plasticity, and many others, are important elements of both the stability and the adaptability of nervous systems.

non-Hebbian plasticity

In this chapter, we describe and analyze basic correlation- and covariance-based synaptic plasticity rules in the context of unsupervised learning, and discuss their extension to supervised learning. As an example, we discuss the development of ocular dominance in cells of the primary visual cortex, and the formation of the map of ocular dominance preferences across the cortical surface. In the models we discuss, synaptic strengths are characterized by synaptic weights, defined as in chapter 7.

Stability and Competition

Increasing synaptic strength in response to activity is a positive feedback process. The activity that modifies synapses is reinforced by Hebbian plasticity, which leads to more activity and further modification. Without appropriate adjustments of the synaptic plasticity rules or the imposition of constraints, Hebbian modification tends to produce uncontrolled growth of synaptic strengths.

synaptic saturation

The easiest way to control synaptic strengthening is to impose an upper limit on the value that a synaptic weight can take. Such an upper limit is supported by LTP experiments. It also makes sense to prevent weights from changing sign, because the plasticity processes we are modeling cannot change an excitatory synapse into an inhibitory synapse or vice versa. We therefore impose the constraint, which we call a saturation constraint, that all excitatory synaptic weights must lie between 0 and a maximum value w_{max}, which is a constant. The simplest implementation of saturation is to set any weight that would cross a saturation bound due to application of a plasticity rule to the limiting value.

synaptic competition

Uncontrolled growth is not the only problem associated with Hebbian plasticity. Synapses are modified independently under a Hebbian rule, which can have deleterious consequences. For example, all of the synaptic weights may be driven to their maximum allowed values w_{max}, causing the postsynaptic neuron to lose selectivity to different patterns of input. The development of input selectivity typically requires competition between different synapses, so that some are forced to weaken when others become strong. We discuss a variety of synaptic plasticity rules that introduce competition between synapses. In some cases, the same mechanism that leads to competition also stabilizes growth of the synaptic weights. In other cases, it does not, and saturation constraints must also be imposed.

8.2   Synaptic Plasticity Rules

Rules for synaptic modification take the form of differential equations describing the rate of change of synaptic weights as a function of the pre- and postsynaptic activity and other possible factors. In this section, we give examples of such rules. In later sections, we discuss their computational implications.

In the models of plasticity that we study, the activity of each neuron is described by a continuous variable, not by a spike train. As in chapter 7, we use the letter u to denote the presynaptic level of activity and v to denote the postsynaptic activity. Normally, u and v represent the firing rates of the pre- and postsynaptic neurons, in which case they should be restricted to nonnegative values. Sometimes, to simplify the analysis, we ignore this constraint. An activity variable that takes both positive and negative values can be interpreted as the difference between a firing rate and a fixed background rate, or between the firing rates of two neurons being treated as a single unit. Finally, to avoid extraneous conversion factors in our equations, we take u and v to be dimensionless measures of the corresponding neuronal firing rates or activities. For example, u and v could be the firing rates of the pre- and postsynaptic neurons divided by their maximum or average values.

In the first part of this chapter, we consider unsupervised learning as applied to a single postsynaptic neuron driven by N_u presynaptic inputs with activities represented by u_b for $b = 1, 2, \ldots, N_u$, or collectively by the vector $\mathbf{u}$. In unsupervised learning, the postsynaptic activity v is evoked directly by the presynaptic activity $\mathbf{u}$. We describe v using a linear version of the firing-rate model discussed in chapter 7,

$$\tau_r \frac{dv}{dt} = -v + \mathbf{w} \cdot \mathbf{u} = -v + \sum_{b=1}^{N_u} w_b u_b \,, \tag{8.1}$$

where τ_r is a time constant that controls the firing-rate response dynamics. Recall that w_b is the synaptic weight that describes the strength of the synapse from presynaptic neuron b to the postsynaptic neuron, and $\mathbf{w}$ is the vector formed by all N_u synaptic weights. The individual synaptic weights can be either positive, representing excitation, or negative, representing inhibition. Equation 8.1 does not include any nonlinear dependence of the firing rate on the total synaptic input, not even rectification. Using such a linear firing-rate model considerably simplifies the analysis of synaptic plasticity. The restriction to nonnegative v either will be imposed by hand or, sometimes, will be ignored to simplify the analysis.

weight vector $\mathbf{w}$

The processes of synaptic plasticity are typically much slower than the dynamics characterized by equation 8.1. If, in addition, the stimuli are presented slowly enough to allow the network to attain its steady-state activity during training, we can replace the dynamic equation 8.1 by

$$v = \mathbf{w} \cdot \mathbf{u} \,, \tag{8.2}$$

which instantaneously sets v to the asymptotic steady-state value determined by equation 8.1. This is the equation we primarily use in our analysis of synaptic plasticity in unsupervised learning. Synaptic modification is included in the model by specifying how the vector $\mathbf{w}$ changes as a function of the pre- and postsynaptic levels of activity. The complex time course of plasticity seen in figure 8.1 is simplified by modeling only the longer-lasting changes.

The Basic Hebb Rule

The simplest plasticity rule that follows the spirit of Hebb's conjecture takes the form

$$\tau_w \frac{d\mathbf{w}}{dt} = v\mathbf{u}\,, \tag{8.3}$$

basic Hebb rule where τ_w is a time constant that controls the rate at which the weights change. This equation, which we call the basic Hebb rule, implies that simultaneous pre- and postsynaptic activity increases synaptic strength. If the activity variables represent firing rates, the right side of this equation can be interpreted as a measure of the probability that the pre- and postsynaptic neurons both fire spikes during a small time interval.

Synaptic plasticity is generally modeled as a slow process that gradually modifies synaptic weights over a time period during which the components of $\mathbf{u}$ take a variety of different values. Each different set of $\mathbf{u}$ values is called an input pattern. The direct way to compute the weight changes induced by a series of input patterns is to sum the small changes caused by each of them separately. A convenient alternative is to average over all of the different input patterns and compute the weight changes induced by this average. As long as the synaptic weights change slowly enough, the averaging method provides a good approximation of the weight changes produced by the set of input patterns.

In this chapter, we use angle brackets $\langle\,\rangle$ to denote averages over the ensemble of input patterns presented during training (which is a slightly different usage from earlier chapters). The Hebb rule of equation 8.3, when *averaged Hebb rule* averaged over the inputs used during training, becomes

$$\tau_w \frac{d\mathbf{w}}{dt} = \langle v\mathbf{u}\rangle\,. \tag{8.4}$$

In unsupervised learning, v is determined by equation 8.2, and if we replace v with $\mathbf{w}\cdot\mathbf{u}$, we can rewrite the averaged plasticity rule (equa-
correlation-based tion 8.4) as
rule

$$\tau_w \frac{d\mathbf{w}}{dt} = \mathbf{Q}\cdot\mathbf{w} \quad \text{or} \quad \tau_w \frac{dw_b}{dt} = \sum_{b'=1}^{N_u} Q_{bb'} w_{b'}\,, \tag{8.5}$$

input correlation where $\mathbf{Q}$ is the input correlation matrix given by
matrix $\mathbf{Q}$

$$\mathbf{Q} = \langle\mathbf{u}\mathbf{u}\rangle \quad \text{or} \quad Q_{bb'} = \langle u_b u_{b'}\rangle\,. \tag{8.6}$$

Equation 8.5 is called a correlation-based plasticity rule because of the presence of the input correlation matrix.

Whether or not the pre- and postsynaptic activity variables are restricted to nonnegative values, the basic Hebb rule is unstable. To show this, we consider the square of the length of the weight vector, $|\mathbf{w}|^2 = \mathbf{w}\cdot\mathbf{w} = \sum_b w_b^2$. Taking the dot product of equation 8.3 with $\mathbf{w}$, and noting that $d|\mathbf{w}|^2/dt =$

$2\mathbf{w} \cdot d\mathbf{w}/dt$ and that $\mathbf{w} \cdot \mathbf{u} = v$, we find that $\tau_w d|\mathbf{w}|^2/dt = 2v^2$, which is always positive (except in the trivial case $v = 0$). Thus, the length of the weight vector grows continuously when the rule 8.3 is applied. To avoid unbounded growth, we must impose an upper saturation constraint. A lower limit is also required if the activity variables are allowed to be negative. Even with saturation, the basic Hebb rule fails to induce competition between different synapses.

Sometimes, synaptic modification is modeled as a discrete rather than continuous process, particularly if the learning procedure involves the sequential presentation of inputs. In this case, equation 8.5 is replaced by a discrete updating rule

$$\mathbf{w} \to \mathbf{w} + \epsilon \mathbf{Q} \cdot \mathbf{w} \qquad (8.7)$$

where ϵ is a parameter, analogous to the learning rate $1/\tau_w$ in the continuous rule, that determines the amount of modification per application of the rule.

The Covariance Rule

If, as in Hebb's original conjecture, u and v are interpreted as representing firing rates (which must be positive), the basic Hebb rule describes only LTP. Experiments, such as the one shown in figure 8.1, indicate that synapses can depress in strength if presynaptic activity is accompanied by a low level of postsynaptic activity. High levels of postsynaptic activity, on the other hand, produce potentiation. These results can be modeled by a synaptic plasticity rule of the form

$$\tau_w \frac{d\mathbf{w}}{dt} = (v - \theta_v)\mathbf{u}\,, \qquad (8.8)$$

where θ_v is a threshold that determines the level of postsynaptic activity above which LTD switches to LTP. As an alternative to equation 8.8, we can impose the threshold on the input rather than output activity, and write

postsynaptic threshold θ_v

$$\tau_w \frac{d\mathbf{w}}{dt} = v(\mathbf{u} - \boldsymbol{\theta}_u)\,. \qquad (8.9)$$

Here, $\boldsymbol{\theta}_u$ is a vector of thresholds that determines the levels of presynaptic activities above which LTD switches to LTP. It is also possible to combine these two rules by subtracting thresholds from both the $\mathbf{u}$ and v terms, but this has the undesirable feature of predicting LTP when pre- and postsynaptic activity levels are both low.

presynaptic threshold $\boldsymbol{\theta}_u$

A convenient choice for the thresholds is the average value of the corresponding variable over the training period. In other words, we set the threshold in equation 8.8 to the average postsynaptic activity, $\theta_v = \langle v \rangle$, or the threshold vector in equation 8.9 to the average presynaptic activity vector, $\boldsymbol{\theta}_u = \langle \mathbf{u} \rangle$. As we did for equation 8.5, we use the relation $v = \mathbf{w} \cdot \mathbf{u}$ and *input covariance matrix* **C**

average over training inputs to obtain an averaged form of the plasticity rule. When the thresholds are set to their corresponding activity averages, equations 8.8 and 8.9 both produce the same averaged rule,

$$\tau_w \frac{d\mathbf{w}}{dt} = \mathbf{C} \cdot \mathbf{w} \,, \tag{8.10}$$

where **C** is the input covariance matrix,

$$\mathbf{C} = \langle (\mathbf{u} - \langle \mathbf{u} \rangle)(\mathbf{u} - \langle \mathbf{u} \rangle) \rangle = \langle \mathbf{uu} \rangle - \langle \mathbf{u} \rangle^2 = \langle (\mathbf{u} - \langle \mathbf{u} \rangle)\mathbf{u} \rangle \,. \tag{8.11}$$

covariance rules

Because of the presence of the covariance matrix in equation 8.10, equations 8.8 and 8.9 are known as covariance rules.

homosynaptic and heterosynaptic depression

Although they both average to give equation 8.10, the rules in equations 8.8 and 8.9 have their differences. Equation 8.8 modifies synapses only if they have nonzero presynaptic activities. When $v < \theta_v$, this produces an effect called homosynaptic depression. In contrast, equation 8.9 reduces the strengths of inactive synapses if $v > 0$. This is called heterosynaptic depression. Note that maintaining $\theta_v = \langle v \rangle$ in equation 8.8 requires changing θ_v as the weights are modified. In contrast, the threshold in equation 8.9 is independent of the weights and does not need to be changed during the training period to keep $\theta_u = \langle \mathbf{u} \rangle$.

Even though covariance rules include LTD and thus allow weights to decrease, they are unstable because of the same positive feedback that makes the basic Hebb rule unstable. For either rule 8.8 with $\theta_v = \langle v \rangle$ or rule 8.9 with $\theta_u = \langle \mathbf{u} \rangle$, $\tau_w d|\mathbf{w}|^2/dt = 2v(v - \langle v \rangle)$. The time average of the right side of this equation is proportional to the variance of the output, $\langle v^2 \rangle - \langle v \rangle^2$, which is positive except in the trivial case when v is constant. Also similar to the case of the Hebb rule is the fact that the covariance rules are noncompetitive, but competition can be introduced by allowing the thresholds to slide, as described below.

The BCM Rule

BCM rule

The covariance-based rule of equation 8.8 does not require any postsynaptic activity to produce LTD, and rule 8.9 can produce LTD without presynaptic activity. Bienenstock, Cooper, and Munro (1982) suggested an alternative plasticity rule, for which there is experimental evidence, that requires both pre- and postsynaptic activity to change a synaptic weight. This rule, which is called the BCM rule, takes the form

$$\tau_w \frac{d\mathbf{w}}{dt} = v\mathbf{u} \left(v - \theta_v \right) \,. \tag{8.12}$$

As in equation 8.8, θ_v acts as a threshold on the postsynaptic activity that determines whether synapses are strengthened or weakened.

If the threshold θ_v is held fixed, the BCM rule, like the basic Hebbian rule, is unstable. Synaptic modification can be stabilized against unbounded growth by allowing the threshold to vary. The critical condition for stability is that θ_v must grow more rapidly than v as the output activity grows large. In one instantiation of the BCM rule with a sliding threshold, θ_v acts as a low-pass filtered version of v^2, as determined by the equation

$$\tau_\theta \frac{d\theta_v}{dt} = v^2 - \theta_v \,. \tag{8.13}$$

sliding threshold

Here τ_θ sets the time scale for modification of the threshold. This is usually slower than the presentation of individual presynaptic patterns, but faster than the rate at which the weights change, which is determined by τ_w. With a sliding threshold, the BCM rule implements competition between synapses because strengthening some synapses increases the postsynaptic firing rate, which raises the threshold and makes it more difficult for other synapses to be strengthened or even to remain at their current strengths.

Synaptic Normalization

The BCM rule stabilizes Hebbian plasticity by means of a sliding threshold that reduces synaptic weights if the postsynaptic neuron becomes too active. This amounts to using the postsynaptic activity as an indicator of the strengths of synaptic weights. A more direct way to stabilize a Hebbian plasticity rule is to add terms that depend explicitly on the weights. This typically leads to some form of weight normalization, which corresponds to the idea that postsynaptic neurons can support only a fixed total synaptic weight, so increases in some weights must be accompanied by decreases in others.

Normalization of synaptic weights involves imposing some sort of global constraint. Two types of constraints are typically used. If the synaptic weights are nonnegative, their growth can be limited by holding the sum of all the weights of the synapses onto a given postsynaptic neuron to a constant value. An alternative, which also works for weights that can be either positive or negative, is to constrain the sum of the squares of the weights instead of their linear sum. In either case, the constraint can be imposed either rigidly, requiring that it be satisfied at all times during the training process, or dynamically, requiring only that it be satisfied asymptotically at the end of training. We discuss one example of each type: a rigid scheme for imposing a constraint on the sum of synaptic weights, and a dynamic scheme for constraining the sum over their squares. Dynamic constraints can be applied in the former case and rigid constraints in the latter, but we restrict our discussion to two widely used schemes. We discuss synaptic normalization in connection with the basic Hebb rule, but the results we present can be applied to covariance rules as well. Weight normalization can drastically alter the outcome of a training procedure, and different normalization methods may lead to different outcomes.

Subtractive Normalization

The sum over synaptic weights that is constrained by subtractive normalization can be written as $\sum w_b = \mathbf{n} \cdot \mathbf{w}$ where $\mathbf{n}$ is an N_u-dimensional vector with all its components equal to 1 (as introduced in chapter 7). This sum can be constrained by replacing equation 8.3 with

Hebb rule with subtractive normalization

$$\tau_w \frac{d\mathbf{w}}{dt} = v\mathbf{u} - \frac{v(\mathbf{n} \cdot \mathbf{u})\mathbf{n}}{N_u}.$$

(8.14)

Note that $\mathbf{n} \cdot \mathbf{u}$ is simply the sum of all the inputs. This rule imposes what is called subtractive normalization because the same quantity is subtracted from the change to each weight, whether that weight is large or small. Subtractive normalization imposes the constraint on the sum of weights rigidly because it does not allow the Hebbian term to change $\mathbf{n} \cdot \mathbf{w}$. To see this, we take the dot product of equation 8.14 with $\mathbf{n}$ to obtain

$$\tau_w \frac{d\mathbf{n} \cdot \mathbf{w}}{dt} = v\mathbf{n} \cdot \mathbf{u} \left(1 - \frac{\mathbf{n} \cdot \mathbf{n}}{N_u} \right) = 0.$$

(8.15)

The last equality follows because $\mathbf{n} \cdot \mathbf{n} = N_u$. Hebbian modification with subtractive normalization is nonlocal in that it requires the sum of all the input activities, $\mathbf{n} \cdot \mathbf{u}$, to be available to the mechanism that modifies each synapse. It is not obvious how such a rule could be implemented biophysically.

Subtractive normalization must be augmented by a saturation constraint that prevents weights from becoming negative. If the rule 8.14 attempts to drive any of the weights below 0, the saturation constraint prevents this change. At this point, the rule is not applied to any saturated weights, and its effect on the other weights is modified. Both modifications can be achieved by setting the components of the vector $\mathbf{n}$ corresponding to any saturated weights to 0 and the factor of N_u in equation 8.14 equal to the sum of the components of this modified $\mathbf{n}$ vector (i.e., the number of unsaturated components). Without any upper saturation limit, this procedure often results in a final outcome in which all the weights but one have been set to 0. To avoid this, an upper saturation limit is also typically imposed. Hebbian plasticity with subtractive normalization is highly competitive because small weights are reduced by a larger proportion of their sizes than are large weights.

Multiplicative Normalization and the Oja Rule

A constraint on the sum of the squares of the synaptic weights can be imposed dynamically by using a modification of the basic Hebb rule known as the Oja rule (Oja, 1982),

Oja rule

$$\tau_w \frac{d\mathbf{w}}{dt} = v\mathbf{u} - \alpha v^2 \mathbf{w},$$

(8.16)

where α is a positive constant. This rule involves only information that is local to the synapse being modified (namely, the pre- and postsynaptic activities and the local synaptic weight), but its form is based more on theoretical arguments than on experimental data. The normalization it imposes is called multiplicative because the amount of modification induced by the second term in equation 8.16 is proportional to $\mathbf{w}$.

The stability of the Oja rule can be established by taking the dot product of equation 8.16 with the weight vector $\mathbf{w}$ to obtain

$$\tau_w \frac{d|\mathbf{w}|^2}{dt} = 2v^2(1 - \alpha|\mathbf{w}|^2).$$

(8.17)

This indicates that $|\mathbf{w}|^2$ will relax over time to the value $1/\alpha$, which obviously prevents the weights from growing without bound, proving stability. It also induces competition between the different weights, because when one weight increases, the maintenance of a constant length for the weight vector forces other weights to decrease.

Timing-Based Rules

Experiments have shown that the relative timing of pre- and postsynaptic action potentials plays a critical role in determining the sign and amplitude of the changes in synaptic efficacy produced by activity. Figure 8.2 shows examples from an intracellular recording of a pair of cortical pyramidal cells in a slice experiment, and from an in vivo experiment on retinotectal synapses in a *Xenopus* tadpole. Both experiments involve repeated pairing of pre- and postsynaptic action potentials, and both show that the relative timing of these spikes is critical in determining the amount and type of synaptic modification that takes place. Synaptic plasticity occurs only if the difference in the pre- and postsynaptic spike times falls within a window of roughly ± 50 ms. Within this window, the sign of the synaptic modification depends on the order of stimulation. Presynaptic spikes that precede postsynaptic action potentials produce LTP. Presynaptic spikes that follow postsynaptic action potentials produce LTD. This is in accord with Hebb's original conjecture, because a synapse is strengthened only when a presynaptic action potential precedes a postsynaptic action potential and therefore can be interpreted as having contributed to it. When the order is reversed and the presynaptic action potential could not have contributed to the postsynaptic response, the synapse is weakened. The maximum amount of synaptic modification occurs when the paired spikes are separated by only a few milliseconds, and the evoked plasticity decreases to 0 as this separation increases.

Simulating the spike-timing dependence of synaptic plasticity requires a spiking model. However, an approximate model can be constructed on the basis of firing rates. The effect of pre- and postsynaptic timing can be included in a synaptic modification rule by including a temporal difference

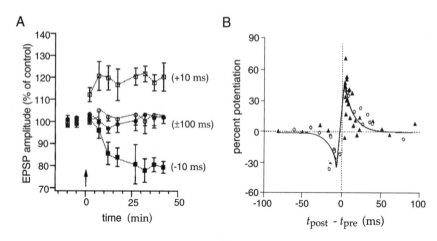

Figure 8.2 LTP and LTD produced by 50 to 75 pairs of pre- and postsynaptic action potential with various timings. (A) The amplitude of the excitatory postsynaptic potential (EPSP) evoked by stimulation of the presynaptic neuron plotted at various times as a percentage of the amplitude prior to paired stimulation. At the time indicated by the arrow, paired stimulations of the presynaptic and postsynaptic neurons were performed. For the traces marked by open symbols, the presynaptic spike occurred either 10 or 100 ms before the postsynaptic neuron fired an action potential. The traces marked by solid symbols denote the reverse ordering in which the presynaptic spike occurred either 10 or 100 ms after the postsynaptic spike. Separations of 100 ms had no long-lasting effect. In contrast, the 10 ms delays produced effects that built up to a maximum over a 5-to-10-minute period and lasted for the duration of the experiment. Pairing a presynaptic action potential with a postsynaptic action potential 10 ms later produced LTP, whereas the reverse ordering generated LTD. (B) LTP and LTD of retinotectal synapses recorded in vivo in *Xenopus* tadpoles. The percent change in synaptic strength produced by multiple pairs of action potentials is plotted as a function of their time difference. The filled symbols correspond to extracellular stimulation of the postsynaptic neuron, and the open symbols, to intracellular stimulation. The H function in equation 8.18 is proportional to the solid curve. (A adapted from Markram et al., 1997; B adapted from Zhang et al., 1998.)

τ between the times when the firing rates of the pre- and postsynaptic neurons are evaluated. A function $H(\tau)$ determines the rate of synaptic modification that occurs due to postsynaptic activity separated in time from presynaptic activity by an interval τ. The total rate of synaptic modification is determined by integrating over all time differences τ. If we assume that the rate of synaptic modification is proportional to the product of the pre- and postsynaptic rates, as it is for a Hebbian rule, the rate of change

timing-based rule   of the synaptic weights at time t is given by

$$\tau_w \frac{d\mathbf{w}}{dt} = \int_0^\infty d\tau \, (H(\tau)v(t)\mathbf{u}(t-\tau) + H(-\tau)v(t-\tau)\mathbf{u}(t)) \,. \qquad (8.18)$$

If $H(\tau)$ is positive for positive τ and negative for negative τ, the first term on the right side of this equation represents LTP, and the second, LTD. The solid curve in figure 8.2B is an example of such an H function. The temporal asymmetry of H has a dramatic effect on synaptic weight changes because it causes them to depend on the temporal order of the pre- and postsynaptic activity during training. Among other things, this allows synaptic weights to store information about temporal sequences.

Rules in which synaptic plasticity is based on the relative timing of pre- and postsynaptic action potentials still require saturation constraints for stability, but, in spiking models, they can generate competition between synapses without further constraints or modifications. This is because different synapses compete to control the timing of postsynaptic spikes. Synapses that are able to evoke postsynaptic spikes rapidly get strengthened. These synapses then exert a more powerful influence on the timing of postsynaptic spikes, and they tend to generate spikes at times that lead to the weakening of other synapses that are less capable of controlling postsynaptic spike timing.

8.3 Unsupervised Learning

We now consider the computational properties of the different synaptic modification rules we have introduced in the context of unsupervised learning. Unsupervised learning provides a model for the effects of activity on developing neural circuits and the effects of experience on mature networks. We separate the discussion of unsupervised learning into cases involving a single postsynaptic neuron and cases in which there are multiple postsynaptic neurons.

A major area of research in unsupervised learning concerns the development of neuronal selectivity and the formation of cortical maps. Chapters 1 and 2 provided examples of the selectivities of neurons in various cortical areas to features of the stimuli to which they respond. In many cases, neuronal selectivities are arranged across the cortical surface in an orderly and regular pattern known as a cortical map. The patterns of connection that give rise to neuronal selectivities and cortical maps are established during development by both activity-independent and activity-dependent processes. A conventional view is that activity-independent mechanisms control the initial targeting of axons, determine the appropriate layer for them to innervate, and establish a coarse order or patterning in their projections. Other activity-independent and activity-dependent mechanisms then refine this order and help to create and preserve neuronal selectivities and cortical maps.

cortical map

Although the relative roles of activity-independent and activity-dependent processes in cortical development are the subject of extensive debate, developmental models based on activity-dependent plasticity have played an important role in suggesting key experiments and successfully predicting their outcomes. A detailed analysis of the more complex pattern-forming models that have been proposed is beyond the scope of this book. Instead, in this and later sections, we give a brief overview of some different approaches and the results that have been obtained. In this section, we consider the case of ocular dominance.

ocular dominance

Ocular dominance refers to the tendency of input to neurons in the adult primary visual cortex of many mammalian species to favor one eye over the other. This is especially true for neurons in layer 4, which receive extensive innervation from the LGN. Neurons dominated by one eye or the other occupy different patches of cortex, and areas with left- or right-eye ocular dominance alternate across the cortex in fairly regular bands known

ocular dominance stripes

as ocular dominance stripes. In a later section, we discuss how this cortical map can arise from Hebbian plasticity.

Single Postsynaptic Neuron

Equation 8.5, which shows the consequence of averaging the basic Hebb rule over all the presynaptic training patterns, is a linear equation for $\mathbf{w}$. Provided that we ignore any constraints on $\mathbf{w}$, it can be analyzed using standard techniques for solving differential equations (see chapter 7 and the Mathematical Appendix). In particular, we use the method of matrix diagonalization, which involves expressing $\mathbf{w}$ in terms of the eigenvectors of $\mathbf{Q}$. These are denoted by $\mathbf{e}_\mu$ with $\mu = 1, 2, \cdots, N_u$, and they satisfy $\mathbf{Q} \cdot \mathbf{e}_\mu = \lambda_\mu \mathbf{e}_\mu$. For correlation or covariance matrices, all the eigenvalues, λ_μ for all μ, are real and nonnegative, and for convenience we order them so that $\lambda_1 \geq \lambda_2 \geq \ldots \geq \lambda_{N_u}$.

Any N_u-dimensional vector can be represented using the eigenvectors as a basis, so we can write

$$\mathbf{w}(t) = \sum_{\mu=1}^{N_u} c_\mu(t)\mathbf{e}_\mu \,, \tag{8.19}$$

where the coefficients are equal to the dot products of $\mathbf{w}$ with the corresponding eigenvectors. For example, at time 0, $c_\mu(0) = \mathbf{w}(0) \cdot \mathbf{e}_\mu$. Writing $\mathbf{w}$ as a sum of eigenvectors turns matrix multiplication into ordinary multiplication, making calculations easier. Substituting the expansion 8.19 into 8.5 and following the procedure presented in chapter 7 for isolating uncoupled equations for the coefficients, we find that $c_\mu(t) = c_\mu(0) \exp(\lambda_\mu t / \tau_w)$. Going back to equation 8.19, this means that

$$\mathbf{w}(t) = \sum_{\mu=1}^{N_u} \exp\left(\frac{\lambda_\mu t}{\tau_w}\right)\left(\mathbf{w}(0) \cdot \mathbf{e}_\mu\right)\mathbf{e}_\mu \,. \tag{8.20}$$

The exponential factors in 8.20 all grow over time, because the eigenvalues λ_μ are positive for all μ values. For large t, the term with the largest value of λ_μ (assuming it is unique) becomes much larger than any of the other terms and dominates the sum for $\mathbf{w}$. This largest eigenvalue has the label $\mu = 1$, and its corresponding eigenvector $\mathbf{e}_1$ is called the principal eigen-

principal eigenvector

vector. Thus, for large t, $\mathbf{w} \propto \mathbf{e}_1$ to a good approximation, provided that $\mathbf{w}(0) \cdot \mathbf{e}_1 \neq 0$. After training, the response to an arbitrary input vector $\mathbf{u}$ is well approximated by

$$v \propto \mathbf{e}_1 \cdot \mathbf{u} \,. \tag{8.21}$$

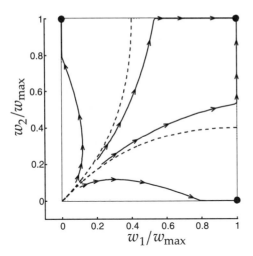

Figure 8.3 Hebbian weight dynamics with saturation. The correlation matrix of the input vectors had diagonal elements equal to 1 and off-diagonal elements of -0.4. The principal eigenvector, $\mathbf{e}_1 = (1, -1)/2^{1/2}$, dominates the dynamics if the initial values of the weights are small enough (below or to the left of the dashed lines). This makes the weight vector move to the corners $(w_{max}, 0)$ or $(0, w_{max})$. However, starting the weights with larger values (between the dashed lines) allows saturation to occur at the corner (w_{max}, w_{max}). (Adapted from MacKay & Miller, 1990.)

Because the dot product corresponds to a projection of one vector onto another, Hebbian plasticity can be interpreted as producing an output proportional to the projection of the input vector onto the principal eigenvector of the correlation matrix of the inputs used during training. We discuss the significance of this result in the next section.

The proportionality sign in equation 8.21 hides the large multiplicative factor $\exp(\lambda_1 t/\tau_w)$ that is a consequence of the positive feedback inherent in Hebbian plasticity. One way to limit growth of the weight vector in equation 8.5 is to impose a saturation constraint. This can have significant effects on the outcome of Hebbian modification, including, in some cases, preventing the weight vector from ending up proportional to the principal eigenvector. Figure 8.3 shows examples of the Hebbian development of the weights in a case with just two inputs. For the correlation matrix used in this example, the principal eigenvector is $\mathbf{e}_1 = (1, -1)/2^{1/2}$, so an analysis that ignored saturation would predict that one weight would increase while the other decreased. Which weight moves in which direction is controlled by the initial conditions. Given the constraints, this would suggest that $(w_{max}, 0)$ and $(0, w_{max})$ are the most likely final configurations. This analysis gives the correct answer only for the regions in figure 8.3 below or to the left of the dashed lines. Between the dashed lines, the final state is $\mathbf{w} = (w_{max}, w_{max})$ because the weights hit the saturation boundary before the exponential growth is large enough to allow the principal eigenvector to dominate.

Another way to eliminate the large exponential factor in the weights is to use the Oja rule, 8.16, instead of the basic Hebb rule. The weight vector generated by the Oja rule, in the example we have discussed, approaches $\mathbf{w} = \mathbf{e}_1/\alpha^{1/2}$ as $t \to \infty$. In other words, the Oja rule gives a weight vector that is parallel to the principal eigenvector, but normalized to a length of $1/\alpha^{1/2}$ rather than growing without bound.

averaged Hebb rule with subtractive constraint

Finally, we examine the effect of including a subtractive constraint, as in equation 8.14. Averaging equation 8.14 over the training inputs, we find

$$\tau_w \frac{d\mathbf{w}}{dt} = \mathbf{Q} \cdot \mathbf{w} - \frac{(\mathbf{w} \cdot \mathbf{Q} \cdot \mathbf{n})\mathbf{n}}{N_u}. \qquad (8.22)$$

If we once again express $\mathbf{w}$ as a sum of eigenvectors of $\mathbf{Q}$, we find that the growth of each coefficient in this sum is unaffected by the extra term in equation 8.22, provided that $\mathbf{e}_\mu \cdot \mathbf{n} = 0$. However, if $\mathbf{e}_\mu \cdot \mathbf{n} \neq 0$, the extra term modifies the growth. In our discussion of ocular dominance, we consider a case in which the principal eigenvector of the correlation matrix is proportional to $\mathbf{n}$. In this case, $\mathbf{Q} \cdot \mathbf{e}_1 - (\mathbf{e}_1 \cdot \mathbf{Q} \cdot \mathbf{n})\mathbf{n}/N = 0$, so the projection in the direction of the principal eigenvector is unaffected by the synaptic plasticity rule. Furthermore, $\mathbf{e}_\mu \cdot \mathbf{n} = 0$ for $\mu \geq 2$ because the eigenvectors of the correlation matrix are mutually orthogonal, which implies that the evolution of the remaining eigenvectors is unaffected by the constraint. As a result,

$$\mathbf{w}(t) = (\mathbf{w}(0) \cdot \mathbf{e}_1)\,\mathbf{e}_1 + \sum_{\mu=2}^{N_u} \exp\left(\frac{\lambda_\mu t}{\tau_w}\right) (\mathbf{w}(0) \cdot \mathbf{e}_\mu)\,\mathbf{e}_\mu. \qquad (8.23)$$

Thus, ignoring the effects of any possible saturation constraints, the synaptic weight matrix tends to become parallel to the eigenvector with the second largest eigenvalue as $t \to \infty$.

In summary, if weight growth is limited by some form of multiplicative normalization, as in the Oja rule, the configuration of synaptic weights produced by Hebbian modification typically will be proportional to the principal eigenvector of the input correlation matrix. When subtractive normalization is used and the principal eigenvector is proportional to $\mathbf{n}$, the eigenvector with the next largest eigenvalue provides an estimate of the configuration of final weights, again up to a proportionality factor. If, however, saturation constraints are used, as they must be in a subtractive scheme, this can invalidate the results of a simplified analysis based solely on these eigenvectors (as in figure 8.3). Nevertheless, we base our discussion of the Hebbian development of ocular dominance and cortical maps on an analysis of the eigenvectors of the input correlation matrix. We present simulation results to verify that this analysis is not invalidated by the constraints imposed in the full models.

Principal Component Projection

If applied for a long enough time, both the basic Hebb rule (equation 8.3) and the Oja rule (equation 8.16) generate weight vectors that are parallel

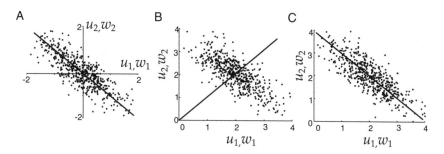

Figure 8.4 Unsupervised Hebbian learning and principal component analysis. The axes in these figures are used to represent the components of both **u** and **w**. (A) The filled circles show the inputs $\mathbf{u} = (u_1, u_2)$ used during a training period while a Hebbian plasticity rule was applied. After training, the vector of synaptic weights was aligned parallel to the solid line. (B) Correlation-based modification with nonzero mean input. Input vectors were generated as in A except that the distribution was shifted to produce an average value $\langle \mathbf{u} \rangle = (2, 2)$. After a training period during which a Hebbian plasticity rule was applied, the synaptic weight vector was aligned parallel to the solid line. (C) Covariance-based modification. Points from the same distribution as in B were used while a covariance-based Hebbian rule was applied. The weight vector becomes aligned parallel to the solid line.

to the principal eigenvector of the correlation matrix of the inputs used during training. Figure 8.4A provides a geometric picture of the significance of this result. In this example, the basic Hebb rule was applied to a unit described by equation 8.2 with two inputs ($N_u = 2$). This model is not constrained to have positive **u** and v. The inputs used during the training period were chosen from a two-dimensional Gaussian distribution with unequal variances, resulting in the elliptical distribution of points seen in the figure. The initial weight vector $\mathbf{w}(0)$ was chosen randomly. The two-dimensional weight vector produced by a Hebbian rule is proportional to the principal eigenvector of the input correlation matrix. The line in figure 8.4A indicates the direction along which the final **w** lies, with the u_1 and u_2 axes used to represent w_1 and w_2 as well. The weight vector points in the direction along which the cloud of input points has the largest variance, a result with interesting implications.

Any unit that obeys equation 8.2 characterizes the state of its N_u inputs by a single number v, which is equal to the projection of **u** onto the weight vector **w**. Intuition suggests, and a technique known as principal component analysis (PCA) formalizes, that setting the projection direction **w** proportional to the principal eigenvector $\mathbf{e}_1$ is often the optimal choice if a set of vectors is to be represented by, and reconstructed from, a set of single numbers through a linear relation. For example, if $\mathbf{e}_1$ is normalized so that $|\mathbf{e}_1| = 1$, the vectors $v\mathbf{w}$ with $v = \mathbf{w} \cdot \mathbf{u}$ and $\mathbf{w} = \mathbf{e}_1$ provide the best estimates that can be generated from single numbers (v) of the set of input vectors **u** used to construct **Q**. Furthermore, the fact that projection along this direction maximizes the variance of the resulting outputs can be interpreted using information theory. The entropy of a Gaussian distributed random variable with variance σ^2 grows with increasing variance as $\log_2 \sigma$. If the input statistics are Gaussian, and the output is corrupted

principal components analysis PCA

by noise that is also Gaussian, maximizing the variance of v by a Hebbian rule maximizes not only the output entropy but also the amount of information v carries about $\mathbf{u}$. In chapter 10, we further consider the computational significance of finding the direction of maximum variance in the input data set, and we discuss the relationship between this and general techniques for extracting structure from input statistics.

Figure 8.4B shows the consequence of applying correlation-based Hebbian plasticity when the average activities of the inputs are not 0, as is inevitable if real firing rates are employed. In this example, correlation-based Hebbian modification aligns the weight vector parallel to a line from the origin to the point $\langle \mathbf{u} \rangle$. This clearly fails to capture the essence of the distribution of inputs. Figure 8.4C shows the result of applying covariance-based Hebbian modification instead. Now the weight vector is aligned with the cloud of data points because this rule finds the direction of the principal eigenvector of the covariance matrix $\mathbf{C}$ of equation 8.11 rather the correlation matrix $\mathbf{Q}$.

Hebbian Development of Ocular Dominance

As an example of a developmental model of ocular dominance at the single neuron level, we consider the highly simplified case of a single layer 4 cell that receives input from just two LGN afferents. One afferent is associated with the right eye and has activity u_R, and the other is from the left eye and has activity u_L. Two synaptic weights $\mathbf{w} = (w_R, w_L)$ describe the strengths of these projections, and the output activity v is determined by equation 8.2,

$$v = w_R u_R + w_L u_L .\tag{8.24}$$

The weights in this model are constrained to nonnegative values. Initially, the weights for the right- and left-eye inputs are taken to be approximately equal. Ocular dominance arises when one of the weights is pushed to 0 while the other remains positive.

We can estimate the results of a Hebbian developmental process by considering the input correlation matrix. We assume that the two eyes are statistically equivalent, so the correlation matrix of the right- and left-eye inputs takes the form

$$\mathbf{Q} = \langle \mathbf{uu} \rangle = \begin{pmatrix} \langle u_R u_R \rangle & \langle u_R u_L \rangle \\ \langle u_L u_R \rangle & \langle u_L u_L \rangle \end{pmatrix} = \begin{pmatrix} q_S & q_D \\ q_D & q_S \end{pmatrix} .\tag{8.25}$$

The subscripts S and D denote same- and different-eye correlations. The eigenvectors are $\mathbf{e}_1 = (1, 1)/2^{1/2}$ and $\mathbf{e}_2 = (1, -1)/2^{1/2}$ for this correlation matrix, and their eigenvalues are $\lambda_1 = q_S + q_D$ and $\lambda_2 = q_S - q_D$.

If the right- and left-eye weights evolve according to equation 8.5, it is straightforward to show that the eigenvector combinations $w_+ = w_R + w_L$

and $w_- = w_{\mathrm{R}} - w_{\mathrm{L}}$ obey the uncoupled equations

$$\tau_w \frac{dw_+}{dt} = (q_{\mathrm{S}} + q_{\mathrm{D}})w_+ \quad \text{and} \quad \tau_w \frac{dw_-}{dt} = (q_{\mathrm{S}} - q_{\mathrm{D}})w_- . \tag{8.26}$$

Positive correlations between the two eyes are likely to exist ($q_{\mathrm{D}} > 0$) after eye opening has occurred. This means that $q_{\mathrm{S}} + q_{\mathrm{D}} > q_{\mathrm{S}} - q_{\mathrm{D}}$, so, according to equations 8.26, w_+ grows more rapidly than w_-. Equivalently, $\mathbf{e}_1 = (1, 1)/\sqrt{2}$ is the principal eigenvector. The basic Hebbian rule thus predicts a final weight vector proportional to $\mathbf{e}_1$, which implies equal innervation from both eyes. This is not the observed outcome of cortical development.

Figure 8.3 shows, in a different example, that in the presence of constraints certain initial conditions can lead to final configurations that are not proportional to the principal eigenvector. However, in the present case initial conditions with $w_{\mathrm{R}} > 0$ and $w_{\mathrm{L}} > 0$ imply that $w_+ > w_-$, and this coupled with the faster growth of w_+ means that w_+ will always dominate over w_- at saturation. Multiplicative normalization does not change this situation.

To obtain ocular dominance in a Hebbian model, we must impose a constraint that prevents w_+ from dominating the final configuration. One way of doing this is to use equation 8.14, the Hebbian rule with subtractive normalization. This completely eliminates the growth of the weight vector in the direction of $\mathbf{e}_1$ (i.e., the increase of w_+) because, in this case, $\mathbf{e}_1$ is proportional to $\mathbf{n}$. On the other hand, it has no effect on growth in the direction $\mathbf{e}_2$ (i.e., the growth of w_-) because $\mathbf{e}_2 \cdot \mathbf{n} = 0$. Thus, with subtractive normalization, the weight vector grows parallel (or anti-parallel) to the direction $\mathbf{e}_2 = (1, -1)/\sqrt{2}$. The direction of this growth depends on initial conditions through the value of $\mathbf{w}(0) \cdot \mathbf{e}_2 = (w_{\mathrm{R}}(0) - w_{\mathrm{L}}(0))/\sqrt{2}$. If this is positive, w_{R} increases and w_{L} decreases; if it is negative, w_{L} increases and w_{R} decreases. Eventually, the decreasing weight will hit the saturation limit of 0, and the other weight will stop increasing due to the normalization constraint. At this point, total dominance by one eye or the other has been achieved. This simple model shows that ocular dominance can arise from Hebbian plasticity if there is sufficient competition between the growth of the left- and right-eye weights.

Hebbian Development of Orientation Selectivity

Hebbian models can also account for the development of the orientation selectivity displayed by neurons in primary visual cortex. The model of Hubel and Wiesel for generating an orientation-selective simple cell response by summing linear arrays of alternating ON and OFF LGN inputs was presented in chapter 2. The necessary pattern of LGN inputs can arise from Hebbian plasticity on the basis of correlations between the responses of different LGN cells and competition between ON and OFF units. Such a model can be constructed by considering a simple cell receiving input from ON-center and OFF-center cells of the LGN, and applying Hebbian

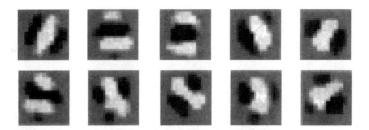

Figure 8.5 Different cortical receptive fields arising from a correlation-based developmental model. White and black regions correspond to areas in the visual field where ON-center (white regions) or OFF-center (black regions) LGN cells excite the cortical neuron being modeled. (Adapted from Miller, 1994.)

plasticity, subject to appropriate constraints, to the feedforward weights of the model.

As in the case of ocular dominance, the development of orientation selectivity can be partially analyzed on the basis of properties of the correlation matrix driving Hebbian development. The critical feature required to produce orientation-selective receptive fields is the growth of components proportional to eigenvectors that are not spatially uniform or rotationally invariant. Application of a Hebbian rule without constraints leads to growth of a uniform component resulting in an unstructured receptive field. Appropriate constraints can eliminate growth of this component, producing spatial structured receptive fields. The development of receptive fields that are not rotationally invariant, and that therefore exhibit orientation selectivity, relies on nonlinear aspects of the model and is therefore difficult to study analytically. For this reason, we simply present simulation results.

arbor function

Neurons in primary visual cortex receive afferents from LGN cells centered over a finite region of the visual field. This anatomical constraint can be included in developmental models by introducing what is called an arbor function. The arbor function, which is often taken to be Gaussian, characterizes the density of innervation from different visual locations to the cell being modeled. As a simplification, this density is not altered during the Hebbian developmental process, but the strengths of synapses within the arbor are modified by the Hebbian rule. The outcome is oriented receptive fields of a limited spatial extent. Figure 8.5 shows the weights resulting from a simulation of receptive-field development using a large array of ON- and OFF-center LGN afferents. This illustrates the variety of oriented receptive field structures that can arise through a Hebbian developmental rule.

Temporal Hebbian Rules and Trace Learning

Unlike correlation- or covariance-based rules, temporal Hebbian rules allow changes of synaptic strength to depend on the temporal sequence of activity across the synapse. This allows temporal Hebbian rules to exhibit a phenomenon called trace learning, where the term trace refers to the history of synaptic activity.

We can approximate the final result of applying a temporal plasticity rule by integrating equation 8.18 from $t = 0$ to a large final time $t = T$, assuming that $w = 0$ initially, and shifting the integration variable to obtain

$$\mathbf{w} = \frac{1}{\tau_w} \int_0^T dt \, v(t) \int_{-\infty}^\infty d\tau \, H(\tau) \mathbf{u}(t - \tau) \,. \tag{8.27}$$

The approximation comes from ignoring both small contributions associated with the end points of the integral and the change in $\mathbf{v}$ produced during training by the modification of $\mathbf{w}$. Equation 8.27 shows that temporally dependent Hebbian plasticity depends on the correlation between the postsynaptic activity and the presynaptic activity temporally filtered by the function H.

Equation 8.27 (with a suitably chosen H) can be used to model the development of invariant responses. Neurons in inferotemporal cortex, for example, can respond selectively to particular objects independent of their location within a wide receptive field. The idea underlying the application of equation 8.27 is that objects persist in the visual environment for characteristic lengths of time as their images move across the retina. If the plasticity rule in equation 8.27 filters presynaptic activity over this persistence time, it strengthens synapses from all the presynaptic units that are activated by the image of the object while it persists and moves. As a result, the response of the postsynaptic cell comes to be independent of the position of the object, and position-invariant responses can be generated.

Multiple Postsynaptic Neurons

To study the effect of plasticity on multiple neurons, we introduce the network of figure 8.6, in which N_v output neurons receive input through N_u feedforward connections and from recurrent interconnections. A vector $\mathbf{v}$ represents the activities of the multiple output units, and the feedforward synaptic connections are described by a matrix $\mathbf{W}$, with the element W_{ab} giving the strength and sign of the synapse from input unit b to output unit a. The strength of the recurrent connection from output unit a' to output unit a is described by element $M_{aa'}$ of the recurrent weight matrix $\mathbf{M}$.

The activity vector for the output units of the network in figure 8.6 is determined by a linear version of the recurrent model of chapter 7,

$$\tau_r \frac{d\mathbf{v}}{dt} = -\mathbf{v} + \mathbf{W} \cdot \mathbf{u} + \mathbf{M} \cdot \mathbf{v} \,. \tag{8.28}$$

Provided that the real parts of the eigenvalues of $\mathbf{M}$ are less than 1, this equation has a stable fixed point with a steady-state output activity vector

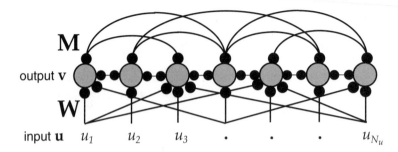

Figure 8.6 A network with multiple output units driven by feedforward synapses with weights **W**, and interconnected by recurrent synapses with weights **M**.

determined by

$$\mathbf{v} = \mathbf{W} \cdot \mathbf{u} + \mathbf{M} \cdot \mathbf{v}.$$ (8.29)

$\mathbf{K} = (\mathbf{I} - \mathbf{M})^{-1}$ Equation 8.29 can be solved by defining the matrix inverse $\mathbf{K} = (\mathbf{I} - \mathbf{M})^{-1}$, where **I** is the identity matrix, to obtain

$$\mathbf{v} = \mathbf{K} \cdot \mathbf{W} \cdot \mathbf{u} \quad \text{or} \quad v_a = \sum_{a'=1}^{N_v} \sum_{b=1}^{N_u} K_{aa'} W_{a'b} u_b.$$ (8.30)

It is important for different output neurons in a multi-unit network to be selective for different aspects of the input, or else their responses will be redundant. For the case of a single cell, competition between different synapses could be used to ensure that synapse-specific plasticity rules do not modify all of the synapses onto a postsynaptic neuron in the same way. For multiple output networks, fixed or plastic linear or nonlinear recurrent interactions can be used to ensure that the units do not all develop the same selectivity. In the following sections, we consider three different patterns of plasticity: plastic feedforward and fixed recurrent synapses, plastic feedforward and recurrent synapses, and, finally, fixed feedforward and plastic recurrent synapses.

Hebbian Development of Ocular Dominance Stripes

Ocular dominance stripes can arise in a Hebbian model with plastic feedforward but fixed recurrent synapses. In the single-cell model of ocular dominance considered previously, the ultimate ocular preference of the output cell depends on the initial conditions of its synaptic weights. A multiple-output version of the model without any recurrent connections would therefore generate a random pattern of selectivities across the cortex if it started with random weights. Figure 8.7B shows that ocular dominance is actually organized in a highly structured cortical map. Dominance by the left and right eye alternates across the cortex, forming a striped pattern. Such a map can arise in the context of Hebbian development of feedforward weights if we include a fixed intracortical connection matrix **M**.

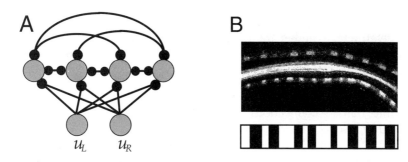

Figure 8.7 The development of ocular dominance in a Hebbian model. (A) The simplified model in which right- and left- eye inputs from a single retinal location drive an array of cortical neurons. (B) Ocular dominance maps. The upper panel shows an area of cat primary visual cortex radioactively labeled to distinguish regions activated by one eye or the other. The light and dark areas along the cortical regions at the top and bottom indicate alternating right- and left-eye innervation. The central region is white matter where fibers are not segregated by ocular dominance. The lower panel shows the pattern of innervation for a 512 unit model after Hebbian development. White and black regions denote units dominated by right- and left-eye projections respectively. (B data of S. LeVay, adapted from Nicholls et al., 1992.)

With fixed recurrent weights $\mathbf{M}$ and plastic feedforward weights $\mathbf{W}$, the effect of averaging Hebbian modifications over the training inputs is

$$\tau_w \frac{d\mathbf{W}}{dt} = \langle \mathbf{v}\mathbf{u} \rangle = \mathbf{K} \cdot \mathbf{W} \cdot \mathbf{Q}\,, \qquad (8.31)$$

where $\mathbf{Q} = \langle \mathbf{u}\mathbf{u} \rangle$ is the input autocorrelation matrix. Equation 8.31 has the same form as the single unit equation 8.5, except that both $\mathbf{K}$ and $\mathbf{Q}$ affect the growth of $\mathbf{W}$.

We consider a highly simplified model of the development of ocular dominance maps that considers only a single direction across the cortex and a single point in the visual field. Figure 8.7A shows the simplified model, which has only two input activities, u_R and u_L, that have the correlation matrix of equation 8.25. These are connected to multiple output units through weight vectors $\mathbf{w}_R$ and $\mathbf{w}_L$. The output units are connected to each other through weights $\mathbf{M}$, so $\mathbf{v} = \mathbf{w}_R u_R + \mathbf{w}_L u_L + \mathbf{M} \cdot \mathbf{v}$. The index a denoting the identity of a given output unit also represents the location of that unit on the cortex. This linking of a to locations on the cortical surface allows us to interpret the results of the model in terms of a cortical map.

Writing $\mathbf{w}_+ = \mathbf{w}_R + \mathbf{w}_L$ and $\mathbf{w}_- = \mathbf{w}_R - \mathbf{w}_L$, the equivalent of equation 8.26 for these sum and difference vectors is

$$\tau_w \frac{d\mathbf{w}_+}{dt} = (q_S + q_D)\mathbf{K} \cdot \mathbf{w}_+ \qquad \tau_w \frac{d\mathbf{w}_-}{dt} = (q_S - q_D)\mathbf{K} \cdot \mathbf{w}_-\,. \qquad (8.32)$$

As in the single-cell model of ocular dominance, subtractive normalization, which holds the value of $\mathbf{w}_+$ fixed while leaving the growth of $\mathbf{w}_-$ unaffected, eliminates the tendency for the cortical cells to become binocular. Then only the equation for $\mathbf{w}_-$ is relevant, and the growth of $\mathbf{w}_-$ is dominated by the principal eigenvector of $\mathbf{K}$. The sign of the component $[\mathbf{w}_-]_a$ determines whether the neuron in the region of the cortex corresponding to the label a is dominated by the right eye (if it is positive) or the left eye (if it is negative). Oscillations in the signs of the components of $\mathbf{w}_-$ translate into ocular dominance stripes.

We assume that the connections between the output neurons are translation invariant, so that $\mathbf{K}_{aa'} = K(|a - a'|)$ depends only on the separation between the cortical cells a and a'. Note that it is more convenient to consider the form of the function K than to discuss the connections between output neurons ($\mathbf{M}$) directly. We use a convenient trick to remove edge effects, which is to impose periodic boundary conditions that require the activities of the units with $a = 0$ and $a = N_v$ to be identical. This provides a reasonable model of a patch of the cortex away from regional boundaries. In some cases, edge effects can impose important constraints on the overall structure of maps, but we do not analyze this here.

In the case of periodic boundary conditions, the eigenvectors of $\mathbf{K}$ have the form

$$e_a^\mu = \cos\left(\frac{2\pi\mu a}{N_v} - \phi\right) \tag{8.33}$$

for $\mu = 0, 1, 2, \ldots, N_v/2$, and with a phase parameter ϕ that can take any value from 0 to 2π. The eigenvalues are given by the discrete Fourier transform $\tilde{K}(\mu)$ of $K(|a - a'|)$, which is real in the case we consider (see the Mathematical Appendix). The principal eigenvector is the eigenfunction from equation 8.33 with μ value corresponding to the maximum of $\tilde{K}(\mu)$. The functions K and $\tilde{K}$ in figure 8.8 are each the difference of two Gaussian functions. $\tilde{K}$ has been plotted as a function of the spatial frequency $k = 2\pi\mu/(N_v d)$, where d is the cortical distance between locations a and $a + 1$. The value of μ corresponding to the principal eigenvector is determined by the k value corresponding to the maximum of the curve in figure 8.8B.

The oscillations in sign of the principal eigenvector, which is indicated by the dotted line in figure 8.8A, generate an alternating pattern of left- and right-eye innervation resembling the ocular dominance stripes seen in primary visual cortex (upper panel figure 8.7B). The lower panel of figure 8.7B shows the result of a simulation of Hebbian development of an ocular dominance map for a one-dimensional line across cortex consisting of 512 units. In this simulation, constraints that limit the growth of synaptic weights have been included, but these do not dramatically alter the conclusions of our analysis.

Competitive Hebb Rule

In this section and the next, we consider models that deal with Hebbian learning and development on a more abstract level. Although inspired by features of neural circuitry, these models are less directly linked to neurons and synapses. For example, the model discussed in this section considers

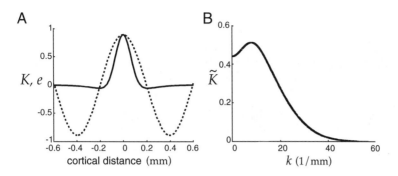

Figure 8.8 Hypothetical K function. (A) The solid line is $K(|a - a'|)$ given by the difference of two Gaussian functions. We have plotted this as a function of the distance between the cortical locations corresponding to the indices a and a', rather than of $|a - a'|$. The dotted line is the principal eigenvector plotted on the same scale. (B) $\tilde{K}$, the Fourier transform of K. This is also given by the difference of two Gaussians. As in A, we use cortical distance units and plot $\tilde{K}$ in terms of the spatial frequency k rather than the integer index μ.

the effects of excitation and inhibition by modeling competitive and cooperative aspects of how activity is generated in, and spread across, the cortex. The advantage of this approach is that it allows us to deal with complex, nonlinear features of Hebbian models in a more controlled and manageable way. As we have seen, competition between neurons is an important element in the development of realistic patterns of selectivity. Linear recurrent connections can produce only a limited amount of differentiation among network neurons, because they induce fairly weak competition between output units. As detailed in chapter 7, recurrent connections can lead to much stronger competition if the interactions are nonlinear. The model discussed in this section allows strongly nonlinear competition to arise in a Hebbian setting.

nonlinear competition

As mentioned in the previous paragraph, the model we discuss represents the effect of cortical processing in two somewhat abstract stages. One stage, modeling the effects of long-range inhibition, involves competition among all the cortical cells for feedforward input in a scheme related to that used in chapter 2 for contrast saturation. The second stage, modeling shorter-range excitation, involves cooperation in which neurons that receive feedforward input excite their neighbors.

In the first stage, the feedforward input for unit a, and that for all the other units, is fed through a nonlinear function to generate a competitive measure of the local excitation generated at location a,

$$z_a = \frac{\left(\sum_b W_{ab} u_b\right)^{\delta}}{\sum_{a'} \left(\sum_b W_{a'b} u_b\right)^{\delta}} \, . \tag{8.34}$$

The activities and weights are all assumed to be positive. The parameter δ controls the degree of competition. For large δ, only the largest feedforward input survives. The case $\delta = 1$ is quite similar to the linear recurrent connections of the previous section.

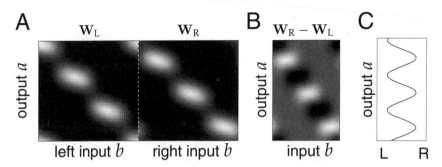

Figure 8.9 Ocular dominance patterns from a competitive Hebb rule. (A) Final stable weights W_{ab} plotted as a function of a and b, showing the relative strengths of the connections from left- and right-eye inputs centered at various retinal locations. White represents a large positive value and black represents 0. (B) The difference in the connections between right- and left-eye inputs. White indicates a positive value and black, a negative value. (C) Difference in the connections summed across all the inputs b to each cortical cell a, showing the net ocularity for each cell. The model used here has 100 input units for each eye and for the output layer, and a coarse initial topography was assumed. Circular (toroidal) boundary conditions were imposed to avoid edge effects. The input activity patterns during training represented single Gaussian illuminations in both eyes centered on a randomly chosen input unit b, with a larger magnitude for one eye (chosen randomly) than for the other. The recurrent weights $\mathbf{M}$ took the form of a Gaussian.

In the cooperative stage, the local excitation of equation 8.34 is distributed across the cortex by recurrent connections, producing a level of activity in unit a given by

$$v_a = \sum_{a'} M_{aa'} z_{a'} . \tag{8.35}$$

This ensures that the excitation characterized by z_a is spread across a local neighborhood of the cortex rather than being concentrated entirely at location a. In this scheme, the recurrent connections described by the matrix $\mathbf{M}$ are usually purely excitatory and of fairly short range, because the effect of longer-range inhibition has been modeled by the competition.

competitive Hebbian learning

Using the outputs of equation 8.35 in conjunction with a Hebbian rule for the feedforward weights is called competitive Hebbian learning. The competition between neurons implemented by this scheme does not ensure competition among the synapses onto a given neuron, so some mechanism such as a normalization constraint is still required. For these models, the outcome of training cannot be analyzed simply by considering eigenvectors of the covariance or correlation matrix because the activation process is nonlinear. Rather, higher-order statistics of the input distribution are important. Nonlinear competition can lead to strong differentiation of output units.

An example of the use of competitive Hebbian learning is shown in figure 8.9, in the form of a one-dimensional cortical map of ocular dominance with inputs arising from LGN neurons with receptive fields covering an extended region of the visual field (rather than the single location of our simpler model). This example uses competitive Hebbian plasticity with nondynamic multiplicative weight normalization. Two weight matrices, $\mathbf{W}_R$ and $\mathbf{W}_L$, corresponding to right- and left-eye inputs, characterize the connectivity of the model. These are shown separately in figure 8.9A, which illustrates that the cortical cells develop retinotopically ordered receptive fields, and they segregate into alternating patches dominated by one eye or the other. The index a indicates the identity and cortical location of the ouput unit, whereas b indicates the retinal location of the center of the corresponding input unit. The ocular dominance pattern is easier to see in figure 8.9B, which shows the difference between the right- and left-eye weights, $\mathbf{W}_R - \mathbf{W}_L$, and figure 8.9C, which shows the net ocularity of the total input to each output neuron of the model ($\sum_b [W_R - W_L]_{ab}$ for each a). It is possible to analyze the structure shown in figure 8.9 and reveal the precise effect of the competition (i.e., the effect of changing the competition parameter δ in equation 8.34). Such an analysis shows, for example, that subtractive normalization of the synaptic weight is not necessary to ensure the robust development of ocular dominance, as it is in the noncompetitive case.

Feature-Based Models

Models of cortical map formation can get extremely complex when multiple neuronal selectivities, such as retinotopic location, ocular dominance, and orientation preference, are considered simultaneously. To deal with this, a class of more abstract models, called competitive feature-based models, has been developed. These use a general approach similar to the competitive Hebbian models discussed in the previous section. Feature-based models are not directly related to the biophysical reality of neuronal firing rates and synaptic strengths, but they provide a compact description of map development.

In a feedforward model, the selectivity of neuron a is determined by the feedforward weights W_{ab} for all values of b, describing how this neuron is connected to the input units of the network. The input units are driven by the stimulus and their responses reflect various stimulus features. Thus, selectivity in these models is determined by how the synaptic weights transfer the selectivities of the input units to the output units. The idea of a feature-based model is to simplify this by directly relating the output unit selectivities to the corresponding features of the stimulus. In feature-based models, the index b is not used to label different input units, but rather to label different features of the stimulus. N_u is thus equal to the number of parameters being used to characterize the stimulus. Also, the input variable u_b is set to the parameter used to characterize feature b of the stimulus. Similarly, W_{ab} does not describe the coupling between input unit b and output unit a, but instead it represents the selectivity of output unit a to stimulus feature b. For example, suppose $b = 1$ represents the location of a visual stimulus, and $b = 2$ represents its ocularity, the difference in strength between the left- and right-eye inputs. Then, u_1 and u_2 would be the location coordinate and the ocularity of the stimulus, and W_{a1} and W_{a2} would be the preferred stimulus location (the center of the

receptive field) and the preferred eye for neuron a. Map development in
such a model is studied by noting how the appearance of various stimulus features and neuronal selectivities affects the matrix $\mathbf{W}$. By associating
the index a with cortical location, the structure of the final matrix $\mathbf{W}$ that
arises from a plasticity rule predicts how selectivities are mapped across
the cortex.

The activity of a particular output unit in a feature-based model is determined by how closely the stimulus being presented matches its preferred
stimulus. The weights W_{ab} for all b values determine the preferred stimulus features for neuron a, and we assume that the activation of neuron
a is high if the components of the input u_b match the components of W_{ab}.
A convenient way to achieve this is to express the activation for unit a as
$\exp(-\sum_b (u_b - W_{ab})^2/(2\sigma_b^2))$, which has its maximum at $u_b = W_{ab}$ for all b,
and falls off as a Gaussian function for less perfect matches of the stimulus
to the selectivity of the cell. The parameter σ_b determines how selective
the neuron is for characteristic b of the stimulus.

The Gaussian expression for the activation of neuron a is not used directly
to determine its level of activity. Rather, as in the case of competitive Hebbian learning, we introduce a competitive activity variable for cortical site
a,

$$z_a = \frac{\exp\left(-\sum_b (u_b - W_{ab})^2/(2\sigma_b^2)\right)}{\sum_{a'} \exp\left(-\sum_b (u_b - W_{a'b})^2/(2\sigma_b^2)\right)}. \tag{8.36}$$

In addition, some cooperative mechanism must be included to keep the
maps smooth, which means that nearby neurons should, as far as possible, have similar selectivities. The two algorithms we discuss, the self-
organizing map and the elastic net, differ in how they introduce this second element.

self-organizing map The self-organizing map spreads the activity defined by equation 8.36 to
nearby cortical sites through equation 8.35, $v_a = \sum_{a'} M_{aa'} z_{a'}$. This gives
cortical cells a and a' similar selectivities if they are nearby, because v_a and
$v_{a'}$ affect one another through local recurrent excitation. The elastic net
elastic net sets the activity of unit a to the result of equation 8.36, $v_a = z_a$, which generates competition. Smoothness of the map is ensured not by spreading
this activity, as in the self-organizing map, but by including an additional
term in the plasticity rule that tends to make nearby selectivities the same
(see below).

Hebbian development of the selectivities characterized by $\mathbf{W}$ is generated
by an activity-dependent rule. In general, Hebbian plasticity adjusts the
weights of activated units so that they become more responsive to, and selective for, input patterns that excite them. Feature-based models achieve
feature-based the same thing by modifying the selectivities W_{ab} so they more closely
learning rule match the input parameters u_b when output unit a is activated by $\mathbf{u}$. In
the case of the self-organized map, this is achieved through the averaged

rule

$$\tau_w \frac{dW_{ab}}{dt} = \langle v_a(u_b - W_{ab}) \rangle .\qquad(8.37)$$

The elastic net modification rule is similar, but an additional term is included to make the maps smooth, because smoothing is not included in the rule that generates the activity in this case. The elastic net plasticity rule is   *elastic net rule*

$$\tau_w \frac{dW_{ab}}{dt} = \langle v_a(u_b - W_{ab}) \rangle + \beta \sum_{a' \text{ neighbor of } a} (W_{a'b} - W_{ab}) ,\qquad(8.38)$$

where the sum is over all points a' that are neighbors of a, and β is a parameter that controls the degree of smoothness in the map. The elastic net makes W_{ab} similar to $W_{a'b}$, if a and a' are nearby on the cortex, by reducing $(W_{a'b} - W_{ab})^2$. Both the feature-based and elastic net rules make $W_{ab} \to u_b$ when v_a is positive.

Figure 8.10A shows the results of an optical imaging experiment that reveals how ocularity and orientation selectivity are arranged across a region of the primary visual cortex of a macaque monkey. The dark lines show the boundaries of the ocular dominance stripes. The lighter lines show iso-orientation contours, which are locations where the preferred orientations are roughly the same. They indicate, by the regions they enclose, neighborhoods (called domains) of cells that favor similar orientations. They also show how these neighborhoods are arranged with respect to each other and to the ocular dominance stripes. There are singularities, called pinwheels, in the orientation map, where regions with different orientation preferences meet at a point. These tend to occur near the centers of the ocular dominance stripes. There are also linear zones where the iso-orientation domains are parallel. These tend to occur at, and run perpendicular to, the boundaries of the ocular dominance stripes.

Figure 8.10B shows the result of an elastic net model plotted in the same form as the macaque map of figure 8.10A. The similarity is evident and striking. Here $\mathbf{u} = (x, y, o, e\cos\theta, e\sin\theta)$ includes 5 stimulus features ($N_u = 5$): two (x, y) for retinal location, one (o) for ocularity, and two (θ, e) for the direction and strength of orientation. The self-organizing map can produce almost identical results, and noncompetitive and competitive Hebbian developmental algorithms can also lead to similar structures.

Anti-Hebbian Modification

We previously alluded to the problem of redundancy among multiple output neurons that can arise from feedforward Hebbian modification. The Oja rule of equation 8.16 for multiple output units, which takes the form

$$\tau_w \frac{dW_{ab}}{dt} = v_a u_b - \alpha v_a^2 W_{ab} ,\qquad(8.39)$$

A   B

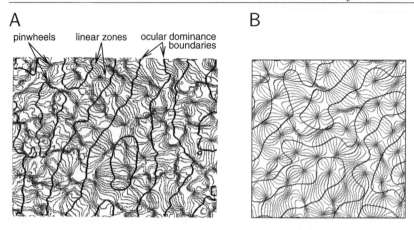

Figure 8.10 Orientation domains and ocular dominance. (A) Contour map showing iso-orientation contours (gray lines) and the boundaries of ocular dominance
stripes (black lines) in a 1.7×1.7 mm patch of macaque primary visual cortex.
Iso-orientation contours are drawn at intervals of $11.25°$. Pinwheels are singularities in the orientation map where all the orientations meet, and linear zones are
extended patches over which the iso-orientation contours are parallel. (B) Ocular
dominance and orientation map produced by the elastic net model. The significance of the lines is the same as in (A), except that the darker gray lines show
orientation preferences of $0°$. (A adapted from Obermayer & Blasdel, 1993; B from
Erwin et al., 1995.)

provides a good illustration of this problem. In the absence of recurrent
connections, this rule sets each row of the feedforward weight matrix to
the principal eigenvector of the input correlation matrix, making each output unit respond identically.

One way to reduce redundancy in a linear model is to make the linear
recurrent interactions of equation 8.29 plastic rather than fixed, using an
anti-Hebbian modification rule. As the name implies, anti-Hebbian plasticity causes synapses to decrease (rather than increase) in strength when
there is simultaneous pre- and postsynaptic activity. The recurrent interactions arising from an anti-Hebb rule can prevent different output units
from representing the same eigenvector. This occurs because the recurrent interactions tend to make output units less correlated by canceling
the effects of common feedforward input. Anti-Hebbian modification is
believed to be the predominant form of plasticity at synapses from parallel fibers to Purkinje cells in the cerebellum, although this may be a special
case because Purkinje cells inhibit rather than excite their targets. A basic
anti-Hebb rule for $M_{aa'}$ can be created simply by changing the sign on the
right side of equation 8.3. However, just as Hebbian plasticity tends to
make weights increase without bound, anti-Hebbian modification tends
to make them decrease to 0, and to avoid this it is necessary to use

$$\tau_M \frac{d\mathbf{M}}{dt} = -\mathbf{v}\mathbf{v} + \beta\mathbf{M} \quad or \quad \tau_M \frac{dM_{aa'}}{dt} = -v_a v_{a'} + \beta M_{aa'} \qquad (8.40)$$

to modify the off-diagonal components of $\mathbf{M}$ (the diagonal components are
defined to be 0). Here, β is a positive constant. For suitably chosen β and
τ_M, the combination of rules 8.39 and 8.40 produces a stable configuration

in which the rows of the weight matrix $\mathbf{W}$ are different eigenvectors of the correlation matrix $\mathbf{Q}$, and all the elements of the recurrent weight matrix $\mathbf{M}$ are ultimately set to 0.

Goodall (1960) proposed an alternative scheme for decorrelating different output units. In his model, the feedforward weights $\mathbf{W}$ are kept constant, whereas the recurrent weights adapt according to the anti-Hebb rule   *Goodall rule*

$$\tau_M \frac{d\mathbf{M}}{dt} = -(\mathbf{W} \cdot \mathbf{u})\mathbf{v} + \mathbf{I} - \mathbf{M}. \qquad (8.41)$$

The minus sign in the term $-(\mathbf{W} \cdot \mathbf{u})\mathbf{v}$ embodies the anti-Hebbian modification. This term is nonlocal because the change in the weight of a given synapse depends on the total feedforward input to the postsynaptic neuron, not merely on the input at that particular synapse (recall that $\mathbf{v} \neq \mathbf{W} \cdot \mathbf{u}$ in this case because of the recurrent connections). The term $\mathbf{I} - \mathbf{M}$ prevents the weights from going to 0 by pushing them toward the identity matrix $\mathbf{I}$. Unlike 8.40, this rule requires the existence of autapses, synapses that a neuron makes onto itself (i.e., the diagonal elements of $\mathbf{M}$ are not 0).

If the Goodall plasticity rule converges and stops changing $\mathbf{M}$, the right side of equation 8.41 must vanish on average, which requires (using the definition of $\mathbf{K}$)

$$\langle(\mathbf{W} \cdot \mathbf{u})\mathbf{v}\rangle = \mathbf{I} - \mathbf{M} = \mathbf{K}^{-1}. \qquad (8.42)$$

Multiplying both sides by $\mathbf{K}$ and using equation 8.30, we find

$$\langle \mathbf{v}\mathbf{v} \rangle = \langle (\mathbf{K} \cdot \mathbf{W} \cdot \mathbf{u})\mathbf{v}\rangle = \mathbf{I}. \qquad (8.43)$$

This means that the outputs are decorrelated and also indicates histogram equalization in the sense, discussed in chapter 4, that all the elements of $\mathbf{v}$ have the same variance. Indeed, the Goodall algorithm can be used to implement the decorrelation and whitening discussed in chapter 4. Because the anti-Hebb and Goodall rules are based on linear models, they are capable of removing only second-order redundancy, meaning redundancy characterized by the covariance matrix. In chapter 10, we consider models that are based on eliminating higher orders of redundancy as well.

Timing-Based Plasticity and Prediction

Temporal Hebbian rules have been used in the context of multi-unit networks to store information about temporal sequences. To illustrate this, we consider a network with the architecture of figure 8.6. We study the effect of time-dependent synaptic plasticity, as given by equation 8.18, on the recurrent synapses of the model, leaving the feedforward synapses constant.

Suppose that before training the average response of output unit a to a stimulus characterized by a parameter s is given by the tuning curve $f_a(s)$,

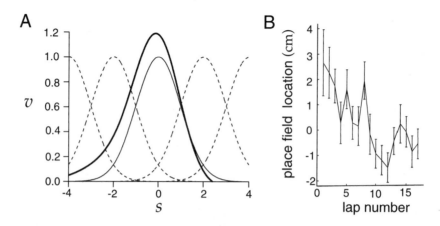

Figure 8.11 Predicted and experimental shifts of place fields. (A) Shift in a neuronal firing-rate tuning curve caused by repeated exposure to a time-dependent stimulus during training. The dashed curves and thin solid curve indicate the initial response tuning curves of a network of interconnected neurons. The thick solid curve is the response tuning curve of the neuron that initially had the thin solid tuning curve after a training period involving a time-dependent stimulus. The tuning curve increases in amplitude, asymmetrically broadens, and shifts as a result of temporally asymmetric Hebbian plasticity. The shift shown corresponds to a training stimulus with a positive rate of change, that is, one that moves rightward on this plot as a function of time. The corresponding shift in the tuning curve is to the left. The shift has been calculated using more neurons and tuning curves than are shown in this plot. (B) Location of place field centers while a rat traversed laps around a closed track (0 is defined as the average center location across the whole experiment). Over sequential laps, the place fields shifted backward relative to the direction the rat moved. (A adapted and modified from Abbott & Blum, 1996; B adapted from Mehta et al., 1997.)

which reaches a maximum value for the optimal stimulus $s = s_a$. Different neurons have different optimal stimulus values, as depicted by the dashed and thin solid curves in figure 8.11A. We now examine what happens when the plasticity rule 8.18 is applied throughout a training period during which the stimulus being presented is an increasing function of time, i.e., moves to the right in figure 8.11A. Such a stimulus excites the different neurons in the network sequentially. For example, the neuron with $s_a = -2$ is active before the neuron with $s_a = 0$, which in turn is active before the neuron with $s_a = 2$. If the stimulus changes rapidly enough, the interval between the firing of the neuron with $s_a = -2$ and that with $s_a = 0$ will fall within the window for LTP depicted in figure 8.2B. This means that a synapse from the neuron with $s_a = -2$ to the $s_a = 0$ neuron will be strengthened. On the other hand, because the neuron with $s_a = 2$ fires after the $s_a = 0$ neuron, falling within the window for LTD, a synapse from it to the $s_a = 0$ neuron will be weakened.

The effect of this type of modification on the tuning curve in the middle of the array (the thin solid curve in figure 8.11A centered at $s = 0$) is shown by the thick solid curve in figure 8.11A. After the training period, the neuron with $s_a = 0$ receives strengthened input from neurons with $s_a < 0$ and weakened input from neurons with $s_a > 0$. This asymmetrically broadens and shifts the tuning curve of the neuron with $s_a = 0$ to lower stimulus values. The leftward shift seen in figure 8.11A is a result of the temporal character of the plasticity rule and the temporal evolution of the stimulus during training. Note that the shift is in the direction opposite to the motion of the stimulus during training. This backward shift has an interesting interpretation. If the same time-dependent stimulus is presented again after training, the neuron with $s_a = 0$ will respond earlier than it did prior to training. Thus, the training experience causes neurons to develop responses that predict the behavior of the stimulus. Although we chose to discuss the neuron with $s_a = 0$ as a representative example, the responses of other neurons shift in a similar manner.

Asymmetric enlargements and backward shifts of neural response tuning curves similar to those predicted from temporally asymmetric LTP and LTD induction have been seen in recordings of hippocampal place cells in rats (see chapter 1) made by Mehta et al. (1997, 2000). Figure 8.11B shows the average location of place fields (the place-cell analog of receptive fields) recorded while a rat ran repeated laps around a closed track. Over time, the place field shifted backward along the track relative to the direction the rat moved.

8.4 Supervised Learning

In unsupervised learning, inputs are imposed during a training period and the output is determined by the network dynamics, using the current values of the weights. This means that the network and plasticity rule must uncover patterns and regularities in the input data (such as the direction of maximal variance) by themselves. In supervised learning, both a set of inputs and the corresponding desired outputs are imposed during training, so the network is essentially given the answer.

Two basic problems addressed in supervised learning are storage, which means learning the relationship between the input and output patterns provided during training, and generalization, which means being able to provide appropriate outputs for inputs that were not presented during training but are similar to those that were. The main tasks we consider within the context of supervised learning are classification of inputs into two categories and function approximation (or regression), in which the output of a network unit is trained to approximate a specified function of the input. Understanding generalization in such settings has been a major focus of theoretical investigations in statistics and computer science but lies outside the scope of our discussion.

Supervised Hebbian Learning

In supervised learning, paired input and output samples are presented during training. We label these pairs by $\mathbf{u}^m$ and v^m for $m = 1 \ldots N_S$, where the superscript is a label and does not signify either a component of **u** or a power of v. For a feedforward network, an averaged Hebbian plasticity rule for supervised learning can be obtained from equation 8.4 by averaging across all the input-output pairs,

$$\tau_w \frac{d\mathbf{w}}{dt} = \langle v\mathbf{u} \rangle = \frac{1}{N_S} \sum_{m=1}^{N_S} v^m \mathbf{u}^m \,. \tag{8.44}$$

cross-correlation   As in the unsupervised Hebbian learning case, the synaptic modification process depends on the input-output cross-correlation $\langle v\mathbf{u} \rangle$. However, for supervised learning, the output $v = v^m$ is imposed on the network rather than being determined by it.

Unless the cross-correlation is 0, equation 8.44 never stops changing the synaptic weights. The methods introduced to stabilize Hebbian modification in the case of unsupervised learning can be applied to supervised learning as well. However, stabilization is easier in the supervised case, because the right side of equation 8.44 does not depend on **w**. Therefore, the growth is only linear rather than exponential in time, making a simple multiplicative synaptic weight decay term sufficient for stability. This is

supervised learning   introduced by writing the supervised learning rule as
with decay

$$\tau_w \frac{d\mathbf{w}}{dt} = \langle v\mathbf{u} \rangle - \alpha\mathbf{w} \,, \tag{8.45}$$

for some positive constant α. Asymptotically, equation 8.45 makes $\mathbf{w} = \langle v\mathbf{u} \rangle / \alpha$, that is, the weights become proportional to the input-output cross-correlation.

We discuss supervised Hebbian learning in the case of a single output unit, but the results can be generalized to multiple outputs.

Classification and the Perceptron

perceptron   The perceptron is a nonlinear map that classifies inputs into one of two
binary classifier   categories. It thus acts as a binary classifier. To make the model consistent when units are connected in a network, we also require the inputs to be binary. We can think of the two possible states as representing units that are either active or inactive. As such, we would naturally assign them the values 1 and 0. However, the analysis is simpler (while producing similar results) if, instead, we require the inputs u_a and output v to take the two values $+1$ and -1.

The output of the perceptron is based on a modification of the linear rule of equation 8.2 to

$$v = \begin{cases} +1 & \text{if } \mathbf{w} \cdot \mathbf{u} - \gamma \geq 0 \\ -1 & \text{if } \mathbf{w} \cdot \mathbf{u} - \gamma < 0 \,. \end{cases} \tag{8.46}$$

The threshold γ determines the dividing line between values of $\mathbf{w} \cdot \mathbf{u}$ that generate $+1$ and -1 outputs. The supervised learning task for the perceptron is to place each of N_S input patterns $\mathbf{u}^m$ into one of two classes designated by the desired binary output v^m. How well the perceptron performs this task depends on the nature of the classification. The weight vector and threshold define a subspace (a hyperplane) of dimension $N_u - 1$ (the subspace of points satisfying $\mathbf{w} \cdot \mathbf{u} = \gamma$, which is perpendicular to $\mathbf{w}$) that cuts the N_u-dimensional space of input vectors into two regions. It is possible for a perceptron to classify inputs perfectly only if a hyperplane exists that divides the input space into one half-space containing all the inputs corresponding to $v = +1$, and another half-space containing all those for $v = -1$. This condition is called linear separability. An instructive case to consider is when each component of each input vector and the associated output values are chosen randomly and independently, with equal probabilities of being $+1$ and -1. For large N_u, the maximum number of random associations that can be described by a perceptron for typical examples of this type is $2N_u$.

linear separability

For linearly separable inputs, a set of weights exists that allows the perceptron to perform perfectly. However, this does not mean that a Hebbian modification rule can construct such weights. A Hebbian rule based on equation 8.45 with $\alpha = N_u/N_S$ constructs the weight vector

$$\mathbf{w} = \frac{1}{N_u} \sum_{m=1}^{N_S} v^m \mathbf{u}^m \,. \tag{8.47}$$

To see how well such weights allow the perceptron to perform, we compute the output generated by one input vector, $\mathbf{u}^n$, chosen from the training set. For this example, we set $\gamma = 0$. Nonzero threshold values are considered later in the chapter.

With $\gamma = 0$, the value of v for input $\mathbf{u}^n$ is determined solely by the sign of $\mathbf{w} \cdot \mathbf{u}^n$. Using the weights of equation 8.47, we find

$$\mathbf{w} \cdot \mathbf{u}^n = \frac{1}{N_u} \left(v^n \mathbf{u}^n \cdot \mathbf{u}^n + \sum_{m \neq n} v^m \mathbf{u}^m \cdot \mathbf{u}^n \right) \,. \tag{8.48}$$

If we set $\sum_{m \neq n} v^m \mathbf{u}^m \cdot \mathbf{u}^n / N_u = \eta^n$ (where the superscript on η, as on v, is a label, not a power), then $v^n \mathbf{u}^n \cdot \mathbf{u}^n / N_u = v^n$ because $1^2 = (-1)^2 = 1$, so we can write

$$\mathbf{w} \cdot \mathbf{u}^n = v^n + \eta^n \,. \tag{8.49}$$

Substituting this expression into equation 8.46 to determine the output of the perceptron for the input $\mathbf{u}^n$, we see that the term η^n acts as a source of noise, interfering with the ability of the perceptron to generate the correct answer $v = v^n$.

We can think of η^n as a sample drawn from a probability distribution of η values. Consider the case when all the components of $\mathbf{u}^m$ and v^m for all m

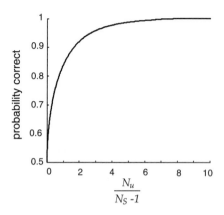

Figure 8.12 Percentage of correct responses for a perceptron with a Hebbian weight vector for a random binary input-output map. As the ratio of the number of inputs, N_u, to one less than the number of input vectors being learned, $N_S - 1$, grows, the percentage of correct responses goes to 1. When this ratio is small, the percentage of correct responses approaches the chance level of $1/2$.

are chosen randomly with equal probabilities of being $+1$ or -1. Including the dot product, the right side of the expression $N_u \eta^n = \sum_{m \neq n} v^m \mathbf{u}^m \cdot \mathbf{u}^n$ that defines η^n is the sum of $(N_S - 1)N_u$ terms, each of which is equally likely to be either $+1$ or -1. For large N_u and N_S, the central limit theorem (see the Mathematical Appendix) tells us that the distribution of η values is Gaussian with mean 0 and variance equal to $(N_S - 1)/N_u$. This suggests that the perceptron with Hebbian weights should work well if the number of input patterns being learned is significantly less than the number of input vector components. We can make this more precise by noting from equations 8.46 with $\gamma = 0$ and equation 8.49 that, for $v^n = +1$, the perceptron will give the correct answer if $-1 < \eta^n < \infty$. Similarly, for $v^n = -1$, the perceptron will give the correct answer if $-\infty < \eta^n < 1$. If v^n has probability $1/2$ of taking either value, the probability of the perceptron giving the correct answer is $1/2$ times the integral of the Gaussian distribution from -1 to ∞ plus $1/2$ times its integral from $-\infty$ to 1. Combining these two terms, we find

$$P[\text{correct}] = \sqrt{\frac{N_u}{2\pi(N_S - 1)}} \int_{-\infty}^{1} d\eta \, \exp\left(-\frac{N_u \eta^2}{2(N_S - 1)}\right). \qquad (8.50)$$

This result is plotted in figure 8.12, which shows that the Hebbian perceptron performs quite well if $N_S - 1$ is less than about $0.2 N_u$. It is possible for the perceptron to perform considerably better than this if a non-Hebbian weight vector is used. We return to this in a later section.

Function Approximation

In chapter 1, we studied examples in which the firing rate of a neuron was given by a function of a stimulus parameter, namely, the response tuning curve. When such a relationship exists, we can think of the neuronal

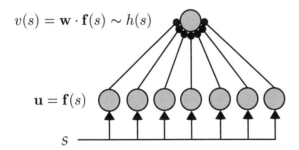

$v(s) = \mathbf{w} \cdot \mathbf{f}(s) \sim h(s)$

$\mathbf{u} = \mathbf{f}(s)$

s

Figure 8.13 A network for representing functions. The value of an input variable s is encoded by the activity of a population of neurons with tuning curves $\mathbf{f}(s)$. This activity drives an output neuron through a vector of weights $\mathbf{w}$ to create an output activity v that approximates the function $h(s)$.

firing rate as representing the function. Populations of neurons (labeled by an index $b = 1, 2, \ldots, N_u$) that respond to a stimulus value s, by firing at average rates $f_b(s)$, can similarly represent an entire set of functions. However, a function $h(s)$ that is not equal to any of the single neuron tuning curves must be represented by combining the responses of a number of units. This can be done using the network shown in figure 8.13. The average steady-state activity level of the output unit in this network, in response to stimulus value s, is given by equation 8.2,

function approximation

$$v(s) = \mathbf{w} \cdot \mathbf{u} = \mathbf{w} \cdot \mathbf{f}(s) = \sum_{b=1}^{N} w_b f_b(s) \,. \qquad (8.51)$$

Note that we have replaced $\mathbf{u}$ with $\mathbf{f}(s)$ where $\mathbf{f}(s)$ is the vector with components $f_b(s)$. The network presented in chapter 7 that performs coordinate transformation is an example of this type of function approximator.

In equation 8.51, the input tuning curves $\mathbf{f}(s)$ act as a basis for representing the output function $h(s)$, and for this reason they are called basis functions. Different sets of basis functions can be used to represent a given set of output functions. A set of basis functions that can represent any member of a class of functions using a linear sum, as in equation 8.51, is called complete for this class. For the basis sets typically used in mathematics, such as the sines and cosines in a Fourier series, the weights in equation 8.51 are unique. When neural tuning curves are used to expand a function, the weights tend not to be unique, and the set of input functions is called overcomplete. In this chapter, we assume that the basis functions are held fixed, and only the weights are adjusted to improve output performance. It is also interesting to consider methods for learning the best basis functions for a particular application. One way of doing this is by applying an algorithm called backpropagation, which develops the basis functions guided by the output errors of the network. Other methods, which we consider in chapter 10, involve unsupervised learning.

basis functions

completeness

overcomplete

Suppose that the function-representation network of figure 8.13 is provided with a sequence of N_S sample stimuli, s^m for $m = 1, 2, \ldots, N_S$, and the corresponding function values $h(s^m)$, during a training period. To make $v(s^m)$ match $h(s^m)$ as closely as possible for all m, we minimize the error

$$E = \frac{1}{2} \sum_{m=1}^{N_S} \left(h(s^m) - v(s^m) \right)^2 = \frac{N_S}{2} \left\langle (h(s) - \mathbf{w} \cdot \mathbf{f}(s))^2 \right\rangle . \qquad (8.52)$$

normal equations

We have made the replacement $v(s) = \mathbf{w} \cdot \mathbf{f}(s)$ in this equation and have used the bracket notation for the average over the training samples. Equations for the weights that minimize this error, called the normal equations, are obtained by setting its derivative with respect to the weights to 0, yielding the condition

$$\langle \mathbf{f}(s)\mathbf{f}(s) \rangle \cdot \mathbf{w} = \langle \mathbf{f}(s)h(s) \rangle . \qquad (8.53)$$

The supervised Hebbian rule of equation 8.45, applied in this case, ultimately sets the weight vector to $\mathbf{w} = \langle \mathbf{f}(s)h(s) \rangle / \alpha$. These weights must satisfy the normal equations 8.53 if they are to optimize function approximation. There are two circumstances under which this occurs. The obvious one is when the input units are orthogonal across the training stimuli, $\langle \mathbf{f}(s)\mathbf{f}(s) \rangle = \mathbf{I}$. In this case, the normal equations are satisfied with $\alpha = 1$. However, this condition is unlikely to hold for most sets of input tuning curves. An alternative possibility is that for all pairs of stimuli s^m and s^n in the training set,

$$\mathbf{f}(s^m) \cdot \mathbf{f}(s^n) = c\delta_{mn} \qquad (8.54)$$

tight frame

for some constant c. This is called a tight frame condition. If it is satisfied, the weights given by supervised Hebbian learning with decay can satisfy the normal equations. To see this, we insert the weights $\mathbf{w} = \langle \mathbf{f}(s)h(s) \rangle / \alpha$ into equation 8.53 and use 8.54 to obtain

$$\langle \mathbf{f}(s)\mathbf{f}(s) \rangle \cdot \mathbf{w} = \frac{\langle \mathbf{f}(s)\mathbf{f}(s) \rangle \cdot \langle \mathbf{f}(s)h(s) \rangle}{\alpha} = \frac{1}{\alpha N_S^2} \sum_{mn} \mathbf{f}(s^m)\mathbf{f}(s^m) \cdot \mathbf{f}(s^n)h(s^n)$$

$$= \frac{c}{\alpha N_S^2} \sum_m \mathbf{f}(s^m)h(s^m) = \frac{c}{\alpha N_S} \langle \mathbf{f}(s)h(s) \rangle . \qquad (8.55)$$

This shows that the normal equations are satisfied for $\alpha = c/N_S$. Thus, we have shown two ways that supervised Hebbian learning can solve the function approximation problem, but both require special conditions on the basis functions $\mathbf{f}(s)$. A more general scheme, discussed below, involves using an error-correcting rule.

Supervised Error-Correcting Rules

An essential limitation of supervised Hebbian rules is that synaptic modification does not depend on the actual performance of the network. An alternative learning strategy is to start with an initial guess for the weights, compare the output $v(\mathbf{u}^m)$ in response to input $\mathbf{u}^m$ with the desired output v^m, and change the weights to improve the performance. Two important error-correcting modification rules are the perceptron rule, which applies to binary classification, and the delta rule, which can be applied to function approximation and many other problems.

The Perceptron Learning Rule

Suppose that the perceptron of equation 8.46 (with nonzero γ) incorrectly classifies an input pattern $\mathbf{u}^m$. If the output is $v(\mathbf{u}^m) = -1$ when $v^m = 1$, the weight vector should be modified to make $\mathbf{w} \cdot \mathbf{u}^m - \gamma$ larger. Similarly, if $v(\mathbf{u}^m) = 1$ when $v^m = -1$, $\mathbf{w} \cdot \mathbf{u}^m - \gamma$ should be decreased. A plasticity rule that performs such an adjustment is the perceptron learning rule,

perceptron learning rule

$$\mathbf{w} \to \mathbf{w} + \frac{\epsilon_w}{2} \left(v^m - v(\mathbf{u}^m) \right) \mathbf{u}^m \quad \text{and} \quad \gamma \to \gamma - \frac{\epsilon_w}{2} (v^m - v(\mathbf{u}^m)). \quad (8.56)$$

Here, and in subsequent sections in this chapter, we use discrete updates for the weights (indicated by the $\to$) rather than the differential equations used up to this point. This is due to the discrete nature of the presentation of the training patterns. In equation 8.56, we have assumed that the threshold γ is also plastic. The learning rule for γ is inverted compared with that for the weights, because γ enters equation 8.46 with a minus sign.

To verify that the perceptron learning rule makes appropriate weight adjustments, we note that it implies that

$$\left(\mathbf{w} \cdot \mathbf{u}^m - \gamma \right) \to \left(\mathbf{w} \cdot \mathbf{u}^m - \gamma \right) + \frac{\epsilon_w}{2} (v^m - v(\mathbf{u}^m)) \left(|\mathbf{u}^m|^2 + 1 \right). \quad (8.57)$$

This result shows that if $v^m = 1$ and $v(\mathbf{u}^m) = -1$, the weight change increases $\mathbf{w} \cdot \mathbf{u}^m - \gamma$. If $v^m = -1$ and $v(\mathbf{u}^m) = 1$, $\mathbf{w} \cdot \mathbf{u}^m - \gamma$ is decreased. This is exactly what is needed to compensate for the error. Note that the perceptron learning rule does not modify the weights if the output is correct.

To learn a set of input pattern classifications, the perceptron learning rule is applied to each one repeatedly, either sequentially or in a random order. For fixed ϵ_w, the perceptron learning rule of equation 8.56 is guaranteed to find a set of weights $\mathbf{w}$ and threshold γ that solve any linearly separable problem. This is proved in the appendix.

The Delta Rule

The function approximation task with the error function E of equation 8.52 can be solved using an error-correcting scheme similar in spirit to the perceptron learning rule, but designed for continuous rather than binary outputs. A simple but extremely useful version of this is the gradient descent procedure, which modifies $\mathbf{w}$ according to

gradient descent

$$\mathbf{w} \to \mathbf{w} - \epsilon_w \nabla_\mathbf{w} E \quad \text{or} \quad w_b \to w_b - \epsilon_w \frac{\partial E}{\partial w_b}, \tag{8.58}$$

where $\nabla_\mathbf{w} E$ is the vector with components $\partial E/\partial w_b$. This rule is sensible because $-\nabla_\mathbf{w} E$ points in the direction (in the space of synaptic weights) along which E decreases most rapidly. This process tends to reduce E because, for small ϵ_w,

$$E(\mathbf{w} - \epsilon_w \nabla_\mathbf{w} E) \approx E(\mathbf{w}) - \epsilon_w |\nabla_\mathbf{w} E|^2 \leq E(\mathbf{w}). \tag{8.59}$$

If ϵ_w is too large or $\mathbf{w}$ is very near to a point where $\nabla_\mathbf{w} E(\mathbf{w}) = \mathbf{0}$, E can increase instead. We assume that ϵ_w is small enough so that E decreases at least until $\mathbf{w}$ is very close to a minimum. If E has many minima, gradient descent will lead to only one of them (a local minimum), and not necessarily the one with the lowest value of E (the global minimum). In the case of function approximation using basis functions as in equation 8.51, gradient descent finds a value of $\mathbf{w}$ that satisfies the normal equations, and therefore constructs an optimal function approximator, because the error function of equation 8.52 has only one minimum.

For function approximation, the error E in equation 8.52 is a sum over the set of examples. As a result, $\nabla_\mathbf{w} E$ also involves a sum,

$$\nabla_\mathbf{w} E = -\sum_{m=1}^{N_S} (h(s^m) - v(s^m))\mathbf{f}(s^m), \tag{8.60}$$

where we have used the fact that $\nabla_\mathbf{w} v(s^m) = \mathbf{f}(s^m)$. The presence of the sum means that the learning rule of equation 8.58 cannot be applied until all the sample patterns have been presented, because all of them are needed to compute the amount by which the weight vector should be changed. It is much more convenient if updating of the weights takes place continuously while sample inputs are presented. This can be done using *stochastic gradient* a procedure known as stochastic gradient descent. This alternative pro-*decent* cedure involves presenting randomly chosen input-output pairs s^m and $h(s^m)$, and change $\mathbf{w}$ according to

$$\mathbf{w} \to \mathbf{w} + \epsilon_w (h(s^m) - v(s^m))\mathbf{f}(s^m). \tag{8.61}$$

delta rule   This rule, called the delta rule, allows learning to take place one sample at a time. Use of this rule is based on the fact that summing the changes proportional to $(h(s^m) - v(s^m))\mathbf{f}(s^m)$ over the random choices of m is, on average, equivalent to doing the sum in equation 8.60. The effect of using equation 8.61 instead of equation 8.58 is the introduction of noise that causes the weights to fluctuate about average values that satisfy the normal equations. Replacing a full sum by an appropriately weighted sum of randomly chosen terms is an example of a so-called Monte Carlo method. There are more efficient methods of searching for minima of functions than stochastic gradient descent, but many of them are complicated to implement.

Figure 8.14 shows the result of modifying an initially random set of weights using the delta rule. Ultimately, an array of input neurons with

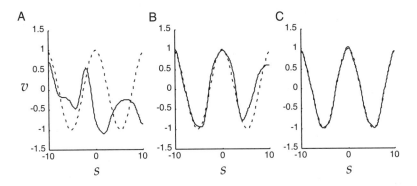

Figure 8.14 Eleven input neurons with Gaussian tuning curves drive an output neuron to approximate a sine function. The input tuning curves are $f_b(s) = \exp[-0.5(s - s_b)^2]$ with $s_b = -10, -8, -6, \ldots, 8, 10$. The delta rule was used to adjust the weights. Sample points were chosen randomly with s in the range between -10 and 10. The firing rate of the output neuron is plotted as a solid curve, and the sinusoidal target function as a dashed curve. (A) The firing rate of the output neuron when random weights in the range between -1 and 1 were used. (B) The output firing rate after weight modification using the delta rule with 20 sample points. (C) The output firing rate after weight modification using the delta rule with 100 sample points.

Gaussian tuning curves drives an output neuron so that it quite accurately represents a sine function. Figures 8.14B and C illustrate the difference between storage and generalization. The output $v(s)$ in figure 8.14B matches the sine function well for values of s that were in the training set, and, in this sense, has stored that information. However, it does not generalize well, in that it does not match the sine function for other values of s not in the training set. The output $v(s)$ in figure 8.14C has good storage and generalization properties, at least within the range of values of s used. The ability of the network to approximate the function $h(s)$ for stimulus values not presented during training depends in a complicated way on the smoothness of the target function, the number and smoothness of the basis functions $\mathbf{f}(s)$, and the size of the training set.

It is not immediately obvious how the delta rule of equation 8.61 could be implemented biophysically, because the network has to compute the difference $h(s^m)\mathbf{f}(s^m) - v(s^m)\mathbf{f}(s^m)$. One possibility is that the two terms $h(s^m)\mathbf{f}(s^m)$ and $v(s^m)\mathbf{f}(s^m)$ are computed in separate phases. First, the output of the network is clamped to the desired value $h(s^m)$ and Hebbian plasticity is applied. Then, the network runs freely to generate $v(s^m)$ and anti-Hebbian modifications are made. In the next section, we discuss a particular example of this in the case of the Boltzmann machine, and we show how learning rules intended for supervised learning can sometimes be used for unsupervised learning as well.

Contrastive Hebbian Learning in Stochastic Networks

In chapter 7, we presented the Boltzmann machine, a stochastic network with binary units. One of the key innovations associated with the Boltzmann machine is a synaptic modification rule that has a sound foundation in probability theory. We start by describing the case of supervised learning, although the underlying theory is similar for both supervised and unsupervised cases with the Boltzmann machine.

Recall from chapter 7 that the Boltzmann machine produces a stochastic output $\mathbf{v}$ from an input $\mathbf{u}$ through a process called Gibbs sampling. This has two main consequences. First, rather than being described by an equation such as 8.2, the input-output relationship of a stochastic network is described by a distribution $P[\mathbf{v}|\mathbf{u}; \mathbf{W}]$, which is the probability that input $\mathbf{u}$ generates output $\mathbf{v}$ when the weight matrix of the network is $\mathbf{W}$. Second, supervised learning in a deterministic network involves the development of an input-output relationship that matches, as closely as possible, a set of samples $(\mathbf{u}^m, \mathbf{v}^m)$ for $m = 1, 2, \ldots, N_S$. Such a task does not make sense for a model that has a stochastic rather than deterministic relationship between its input and output activities. Instead, stochastic networks are appropriate for representing statistical aspects of the relationship between two sets of variables. For example, suppose that we drew sample pairs from a joint probability distribution $P[\mathbf{u}, \mathbf{v}] = P[\mathbf{v}|\mathbf{u}]P[\mathbf{u}]$. In this case, a given value of $\mathbf{u}^m$, chosen from the distribution $P[\mathbf{u}]$, is associated with another vector $\mathbf{v}^m$ stochastically rather than deterministically. The probability that a particular output $\mathbf{v}^m$ is associated with $\mathbf{u}^m$ is given by $P[\mathbf{v}^m|\mathbf{u}^m]$. In other words, $P[\mathbf{u}]$ describes the probability of various $\mathbf{u}$ vectors appearing, and $P[\mathbf{v}|\mathbf{u}]$ is the probability that a particular vector $\mathbf{u}$ is associated with another vector $\mathbf{v}$.

A natural supervised learning task for a stochastic network is to make the input-output distribution of the network, $P[\mathbf{v}|\mathbf{u}; \mathbf{W}]$, match as closely as possible, the probability distribution $P[\mathbf{v}|\mathbf{u}]$ associated with the samples $(\mathbf{u}^m, \mathbf{v}^m)$. This is done by adjusting the feedforward weight matrix $\mathbf{W}$. Note that we are using the argument $\mathbf{W}$ to distinguish between two different distributions, $P[\mathbf{u}|\mathbf{v}]$, which is provided externally and generates the sample data, and $P[\mathbf{u}|\mathbf{v}; \mathbf{W}]$, which is the distribution generated by the Boltzmann machine with weight matrix $\mathbf{W}$. The idea of constructing networks that reproduce probability distributions inferred from sample data *density estimation* is central to the problem of density estimation, which is covered more fully in chapter 10.

We first consider a Boltzmann machine with only feedforward weights $\mathbf{W}$ connecting $\mathbf{u}$ to $\mathbf{v}$, and no recurrent weights among the $\mathbf{v}$ units. Given an input $\mathbf{u}$, an output $\mathbf{v}$ is computed by setting each component v_a to 1 with probability $F(\sum_b W_{ab} u_b)$ (and 0 otherwise) where $F(I) = 1/(1 + \exp(-I))$. This is the Gibbs sampling procedure discussed in chapter 7 applied to the feedforward Boltzmann machine. Because there are no recurrent connections, the states of the output units are independent, and they can all be sampled simultaneously. Analogous to the discussion in chapter 7, this procedure gives rise to a conditional probability distribution $P[\mathbf{v}|\mathbf{u};\mathbf{W}]$ for $\mathbf{v}$ given $\mathbf{u}$ that can be written as

$$P[\mathbf{v}|\mathbf{u};\mathbf{W}] = \frac{\exp(-E(\mathbf{u},\mathbf{v}))}{Z(\mathbf{u})} \quad \text{with} \quad Z(\mathbf{u}) = \sum_{\mathbf{v}} \exp(-E(\mathbf{u},\mathbf{v})), \quad (8.62)$$

where $E(\mathbf{u},\mathbf{v}) = -\mathbf{v}\cdot\mathbf{W}\cdot\mathbf{u}$. In this case, the partition function can be written as $Z(\mathbf{u}) = \prod_a(1+\exp(\sum_b W_{ab}u_b))$. However, for the Boltzmann machines with recurrent connections that we consider below, there is no simple closed form expression for the partition function.

The natural measure for determining how well the distribution generated by the network, $P[\mathbf{v}|\mathbf{u};\mathbf{W}]$, matches the sampled distribution, $P[\mathbf{v}|\mathbf{u}]$, for a particular input $\mathbf{u}$ is the Kullback-Leibler divergence,

$$D_{\mathrm{KL}}(P[\mathbf{v}|\mathbf{u}], P[\mathbf{v}|\mathbf{u};\mathbf{W}]) = \sum_{\mathbf{v}} P[\mathbf{v}|\mathbf{u}]\ln\left(\frac{P[\mathbf{v}|\mathbf{u}]}{P[\mathbf{v}|\mathbf{u};\mathbf{W}]}\right)$$

$$= -\sum_{\mathbf{v}} P[\mathbf{v}|\mathbf{u}]\ln\left(P[\mathbf{v}|\mathbf{u};\mathbf{W}]\right) + K, \quad (8.63)$$

where K is a term that is proportional to the entropy of the distribution $P[\mathbf{v}|\mathbf{u}]$ (see chapter 4). We do not write out this term explicitly because it does not depend on the feedforward weight matrix, so it does not affect the learning rule used to modify $\mathbf{W}$. As in chapter 7, we have, for convenience, used natural logarithms (rather than base 2 as in chapter 4) in the definition of the Kullback-Leibler divergence.

To estimate, from the samples, how well $P[\mathbf{v}|\mathbf{u};\mathbf{W}]$ matches $P[\mathbf{v}|\mathbf{u}]$ across the different values of $\mathbf{u}$, we average the Kullback-Leibler divergence over all the input samples $\mathbf{u}^m$. Furthermore, the sum over all $\mathbf{v}$ with weighting factor $P[\mathbf{v}|\mathbf{u}]$ in equation 8.63 can be replaced with an average over the sample pairs $(\mathbf{u}^m,\mathbf{v}^m)$ because $\mathbf{v}^m$ is chosen from the probability distribution $P[\mathbf{v}|\mathbf{u}]$. Using brackets to denote the average over samples, this results in the measure

$$\langle D_{\mathrm{KL}}(P[\mathbf{v}|\mathbf{u}], P[\mathbf{v}|\mathbf{u};\mathbf{W}])\rangle = -\frac{1}{N_{\mathrm{S}}}\sum_{m=1}^{N_{\mathrm{S}}} \ln\left(P[\mathbf{v}^m|\mathbf{u}^m;\mathbf{W}]\right) + \langle K\rangle \quad (8.64)$$

for comparing $P[\mathbf{v}|\mathbf{u};\mathbf{W}]$ and $P[\mathbf{v}|\mathbf{u}]$. Each logarithmic term in the sum on the right side of this equation is the logarithm of the probability that a sample output $\mathbf{v}^m$ would have been drawn from the distribution $P[\mathbf{v}|\mathbf{u}^m;\mathbf{W}]$, when in fact it was drawn from $P[\mathbf{v}|\mathbf{u}^m]$. This makes the sum in equation 8.64 equal to the logarithm of the likelihood that the sample data could have been produced from $P[\mathbf{v}|\mathbf{u}^m;\mathbf{W}]$. As a result, finding the network distribution $P[\mathbf{v}|\mathbf{u}^m;\mathbf{W}]$ that best matches $P[\mathbf{v}|\mathbf{u}^m]$ (in the sense of minimizing the Kullback-Leibler divergence) is equivalent to maximizing the likelihood that the sample $\mathbf{v}^m$ could have been drawn from $P[\mathbf{v}|\mathbf{u}^m;\mathbf{W}]$.

log likelihood

likelihood maximization

A learning rule that performs gradient ascent of the log likelihood can be derived by changing the weights by an amount proportional to the derivative of the logarithmic term in equation 8.64 with respect to the weights.

As in the discussion of the delta rule, it is more convenient to use a stochastic gradient ascent rule, choosing an index m at random to provide a Monte Carlo sample from the average of equation 8.64, and changing W_{ab} according to the derivative with respect to this sample,

$$\frac{\partial \ln P[\mathbf{v}^m | \mathbf{u}^m; \mathbf{W}]}{\partial W_{ab}} = \frac{\partial}{\partial W_{ab}} \left(-E(\mathbf{u}^m, \mathbf{v}^m) - \ln Z(\mathbf{u}^m) \right)$$

$$= v_a^m u_b^m - \sum_{\mathbf{v}} P[\mathbf{v} | \mathbf{u}^m; \mathbf{W}] v_a u_b^m . \qquad (8.65)$$

This derivative has a simple form for the Boltzmann machine because of equation 8.62.

Before we derive the stochastic gradient ascent learning rule, we need to evaluate the sum over $\mathbf{v}$ in the last term of the bottom line of equation 8.65. For Boltzmann machines with recurrent connections, like the ones we discuss below, this average cannot be calculated tractably. However, it can be estimated by stochastic sampling. In other words, we approximate the average over $\mathbf{v}$ by a single instance of a particular output $\mathbf{v}(\mathbf{u}^m)$ generated by the Boltzmann machine in response to the input $\mathbf{u}^m$. Making this replacement and setting the change in the weight matrix proportional to the derivative in equation 8.65, we obtain the learning rule

supervised learning for $\mathbf{W}$

$$W_{ab} \rightarrow W_{ab} + \epsilon_w \left(v_a^m u_b^m - v_a(\mathbf{u}^m) u_b^m \right) . \qquad (8.66)$$

Equation 8.66 is identical in form to the perceptron learning rule of equation 8.56, except that $\mathbf{v}(\mathbf{u}^m)$ is computed from the input $\mathbf{u}^m$ by Gibbs sampling rather than by a deterministic rule. As discussed at the end of the previous section, equation 8.66 can also be interpreted as the difference of Hebbian and anti-Hebbian terms. The Hebbian term $v_a^m u_b^m$ is based on the sample input $\mathbf{u}^m$ and output $\mathbf{v}^m$. The anti-Hebbian term $-v_a(\mathbf{u}^m) u_b^m$ involves the sample input $\mathbf{u}^m$ and an output $\mathbf{v}(\mathbf{u}^m)$ generated by the Boltzmann machine in response to this input, rather than the sample output $\mathbf{v}^m$. In other words, whereas $\mathbf{v}^m$ is provided externally, $\mathbf{v}(\mathbf{u}^m)$ is obtained by Gibbs sampling, using the input $\mathbf{u}^m$ and the current values of the network weights. The overall learning rule is called a contrastive Hebb rule

contrastive Hebb rule

because it depends on the difference between Hebbian and anti-Hebbian terms. W_{ab} stops changing when the average of $v_a(\mathbf{u}^m) u_b^m$ over the input samples and network outputs equals the average of $v_a^m u_b^m$ over the input and output samples.

Supervised learning for the Boltzmann machine is run in two phases, both of which use a sample input $\mathbf{u}^m$. The first phase, sometimes called the

wake phase

wake phase, involves Hebbian plasticity between sample inputs and outputs. The dynamics of the Boltzmann machine play no role during this phase. The second phase, called the sleep phase, consists of the network

sleep phase

"dreaming" (i.e., internally generating) $\mathbf{v}(\mathbf{u}^m)$ in response to $\mathbf{u}^m$ based on the current weights $\mathbf{W}$. Then, anti-Hebbian learning based on $\mathbf{u}^m$ and $\mathbf{v}(\mathbf{u}^m)$ is applied to the weight matrix. Gibbs sampling is typically used to generate $\mathbf{v}(\mathbf{u}^m)$ from $\mathbf{u}^m$. It is also possible to use the mean-field method we discussed in chapter 7 to approximate the average over the distribution $P[\mathbf{v}|\mathbf{u}^m; \mathbf{W}]$ in equation 8.65.

Supervised learning can also be implemented in a Boltzmann machine with recurrent connections. When the output units are connected by a symmetric recurrent weight matrix $\mathbf{M}$ (with $M_{aa} = 0$), the energy function is

$$E(\mathbf{u}, \mathbf{v}) = -\mathbf{v} \cdot \mathbf{W} \cdot \mathbf{u} - \frac{1}{2}\mathbf{v} \cdot \mathbf{M} \cdot \mathbf{v}. \quad (8.67)$$

Everything that has been described thus far applies to this case, except that the output $\mathbf{v}(\mathbf{u}^m)$ for the sample input $\mathbf{u}^m$ must now be computed by repeated Gibbs sampling, using $F(\sum_b W_{ab}u_b^m + \sum_{a'} M_{aa'}v_{a'})$ for the probability that $v_a = 1$ (see chapter 7). Repeated sampling is required to assure that the network relaxes to the equilibrium distribution of equation 8.62. Modification of the feedforward weight W_{ab} then proceeds as in equation 8.66. The contrastive Hebb rule for recurrent weight $M_{aa'}$ is similarly given by

supervised learning for $\mathbf{M}$

$$M_{aa'} \rightarrow M_{aa'} + \epsilon_m \left(v_a^m v_{a'}^m - v_a(\mathbf{u}^m)v_{a'}(\mathbf{u}^m) \right). \quad (8.68)$$

The Boltzmann machine was originally introduced in the context of unsupervised rather than supervised learning. In the supervised case, we try to make the distribution $P[\mathbf{v}|\mathbf{u}; \mathbf{W}]$ match the probability distribution $P[\mathbf{v}|\mathbf{u}]$ that generates the samples pairs $(\mathbf{u}^m, \mathbf{v}^m)$. In the unsupervised case, no output sample $\mathbf{v}^m$ is provided, and instead we try to make the network generate a probability distribution over $\mathbf{u}$ that matches the distribution $P[\mathbf{u}]$ from which the samples $\mathbf{u}^m$ were drawn. As we discuss in chapter 10, a frequent goal of probabilistic unsupervised learning is to generate network distributions that match the distributions of input data.

We consider the unsupervised Boltzmann machine without recurrent connections. In addition to the distribution of equation 8.62 for $\mathbf{v}$, given a specific input $\mathbf{u}$, the energy function of the Boltzmann machine can be used to define a distribution over both $\mathbf{u}$ and $\mathbf{v}$ defined by

$$P[\mathbf{u}, \mathbf{v}; \mathbf{W}] = \frac{\exp(-E(\mathbf{u}, \mathbf{v}))}{Z} \quad \text{with} \quad Z = \sum_{\mathbf{u}, \mathbf{v}} \exp(-E(\mathbf{u}, \mathbf{v})). \quad (8.69)$$

This can be used to construct a distribution for $\mathbf{u}$ alone by summing over the possible values of $\mathbf{v}$,

$$P[\mathbf{u}; \mathbf{W}] = \sum_{\mathbf{v}} P[\mathbf{u}, \mathbf{v}; \mathbf{W}] = \frac{1}{Z} \sum_{\mathbf{v}} \exp(-E(\mathbf{u}, \mathbf{v})). \quad (8.70)$$

The goal of unsupervised learning for the Boltzmann machine is to make this distribution match, as closely as possible, the distribution of inputs $P[\mathbf{u}]$.

The derivation of an unsupervised learning rule for this Boltzmann machine proceeds very much like the derivation we presented for the supervised case. The equivalent of equation 8.65 is

$$\frac{\partial \ln P[\mathbf{u}^m; \mathbf{W}]}{\partial W_{ab}} = \sum_{\mathbf{v}} P[\mathbf{v}|\mathbf{u}^m; \mathbf{W}]v_a u_b^m - \sum_{\mathbf{u},\mathbf{v}} P[\mathbf{u}, \mathbf{v}; \mathbf{W}]v_a u_b . \qquad (8.71)$$

In this case, both terms must be evaluated by Gibbs sampling. The wake phase Hebbian term requires a stochastic output $\mathbf{v}(\mathbf{u}^m)$, which is calculated from the sample input $\mathbf{u}^m$, just as it was for the anti-Hebbian term in equation 8.66. However, the sleep phase anti-Hebbian term in this case requires both an input $\mathbf{u}$ and an output $\mathbf{v}$ generated by the network. These are computed using a Gibbs sampling procedure in which both input and output states are stochastically generated through repeated Gibbs sampling. A randomly chosen component v_a is set to 1 with probability $F(\sum_b W_{ab} u_b)$ (or 0 otherwise), and a random component u_b is set to 1 with probability $F(\sum_a v_a W_{ab})$ (or 0 otherwise). Note that this corresponds to having the input units drive the output units in a feedforward manner through the weights $\mathbf{W}$, and having the output units drive the input units in a reversed manner using feedback weights with the same values. Once the network has settled to equilibrium through repeated Gibbs sampling of this sort, and the stochastic inputs and outputs have been generated, the full learning rule is

unsupervised learning for **W**

$$W_{ab} \to W_{ab} + \epsilon_w \left(v_a(\mathbf{u}^m) u_b^m - v_a u_b \right) . \qquad (8.72)$$

The unsupervised learning rule can be extended to include recurrent connections by following the same general procedure.

8.5   Chapter Summary

We presented a variety of forms of Hebbian synaptic plasticity, ranging from the basic Hebb rule to rules that involve multiplicative and subtractive normalization, constant or sliding thresholds, and spike-timing effects. Two important features, stability and competition, were emphasized. We showed how the effects of unsupervised Hebbian learning could be estimated by computing the principal eigenvector of the correlation matrix of the inputs used during training. Unsupervised Hebbian learning can be interpreted as a process that produces weights that project the input vector onto the direction of maximal variance in the training data set. In some cases, this requires an extension from correlation-based to covariance-based rules. We used the principal eigenvector approach to analyze Hebbian models of the development of ocular dominance and its associated map in primary visual cortex. Plasticity rules based on the dependence of synaptic modification on spike timing were shown to implement temporal sequence and trace learning.

Forcing multiple outputs to have different selectivities requires them to be connected, either through fixed weights or by weights that are themselves plastic. In the latter case, anti-Hebbian plasticity can ensure decorrelation of multiple output units. We also considered the role of competition and cooperation in models of activity-dependent development, and described two examples of feature-based models, the self-organizing map and the elastic net.

Finally, we considered supervised learning applied to binary classification and function approximation, using supervised Hebbian learning, the perceptron learning rule, and gradient descent learning through the delta rule. We also treated contrastive Hebbian learning for the Boltzmann machine, involving Hebbian and anti-Hebbian updates in different phases.

8.6 Appendix

Convergence of the Perceptron Learning Rule

For convenience, we take $\epsilon_w = 1$ and start the perceptron learning rule (equation 8.56) with $\mathbf{w} = 0$ and $\gamma = 0$. Then, under presentation of the sample m, the changes in the weights and threshold are given by

$$\Delta \mathbf{w} = \frac{1}{2}(v^m - v(\mathbf{u}^m))\mathbf{u}^m \quad \text{and} \quad \Delta \gamma = -\frac{1}{2}(v^m - v(\mathbf{u}^m)) . \tag{8.73}$$

Given a finite, linearly separable problem, there must be a set of weights $\mathbf{w}^*$ and a threshold γ^* that are normalized ($|\mathbf{w}^*|^2 + (\gamma^*)^2 = 1$) and allow the perceptron to categorize correctly, for which we require the condition $(\mathbf{w}^* \cdot \mathbf{u}^m - \gamma^*)v^m > \delta$ for some $\delta > 0$ and for all m.

Consider the cosine of the angle between the current weights and threshold $\mathbf{w}, \gamma$ and the solution $\mathbf{w}^*, \gamma^*$

$$\Phi(\mathbf{w}, \gamma) = \frac{\mathbf{w} \cdot \mathbf{w}^* + \gamma\gamma^*}{\sqrt{|\mathbf{w}|^2 + (\gamma)^2}} = \frac{\psi(\mathbf{w}, \gamma)}{|\mathbf{w}, \gamma|} , \tag{8.74}$$

to introduce some compact notation. The perceptron convergence theorem proves that the perceptron learning rule solves any solvable categorization problem, because assuming otherwise would imply that Φ would eventually grow larger than 1. This is impossible for a cosine function, which must lie between -1 and 1.

To show this, we consider the change in ψ due to one step of perceptron learning during which $\mathbf{w}$ and γ are modified because the current weights generated the wrong response. When an incorrect response is generated, $v(\mathbf{u}^m) = -v^m$, so $(v^m - v(\mathbf{u}^m))/2 = v^m$, and thus

$$\Delta \psi = (\mathbf{w}^* \cdot \mathbf{u}^m - \gamma^*)v^m > \delta . \tag{8.75}$$

The inequality follows from the condition imposed on $\mathbf{w}^*$ and γ^* as providing a solution of the categorization problem. Assuming that ψ is initially positive and iterating this result over n steps in which the weights change, we find that

$$\psi(\mathbf{w}, \gamma) \geq n\delta. \tag{8.76}$$

Similarly, over one learning step in which some change is made,

$$\Delta |\mathbf{w}, \gamma|^2 = 2(\mathbf{w} \cdot \mathbf{u}^m - \gamma)v^m + |\mathbf{u}^m|^2 + 1. \tag{8.77}$$

The first term on the right side is always negative when an error is made, and if we define D to be the maximum value of $|\mathbf{u}^m|^2$ over all the training samples, we find

$$\Delta |\mathbf{w}, \gamma|^2 < D + 1. \tag{8.78}$$

After n nontrivial learning iterations (iterations in which the weights and threshold are modified) starting from $|\mathbf{w}, \gamma|^2 = 0$, we therefore have

$$|\mathbf{w}, \gamma|^2 < n(D + 1) \tag{8.79}$$

Putting together equations 8.76 and 8.79, we find, after n nontrivial training steps,

$$\Phi(\mathbf{w}, \gamma) > \frac{n\delta}{\sqrt{n(D + 1)}}. \tag{8.80}$$

To ensure that $\Phi(\mathbf{w}, \gamma) \leq 1$, we must have $n \leq (D+1)/\delta^2$. Therefore, after a finite number of weight changes, the perceptron learning rule must stop changing the weights, and the perceptron must classify all the patterns correctly (although the weights and threshold that result are not necessarily equal to $\mathbf{w}^*$ and γ^*).

8.7 Annotated Bibliography

Hebb's (1949) original proposal about learning set the stage for many of the subsequent investigations. We followed the treatments of Hebbian, BCM, anti-Hebbian, and trace learning found in Goodall (1960), Sejnowski (1977), Bienenstock et al. (1982), Oja (1982), Földiák (1989; 1991), Leen (1991), Atick & Redlich (1993), and Wallis & Baddeley (1997). Extensive coverage of these topics and related analyses can be found in **Hertz et al. (1991)**. We followed Miller & MacKay (1994) and Miller (1996b) in the analysis of weight constraints and normalization. Jolliffe (1986) treats principal components analysis theoretically (see also chapter 10). Intrator & Cooper (1992) considers the BCM rule from the statistical perspective of projection pursuit (Huber, 1985).

Sejnowski (1999) comments on the relationship between Hebb's suggestions and recent experimental data and theoretical studies on temporal sensitivity in Hebbian plasticity (see Minai & Levy, 1993; Blum & Abbott,

1996; Kempter et al., 1999; Song et al., 2000). Plasticity of intrinsic conductance properties of neurons, as opposed to synaptic plasticity, is considered in LeMasson et al. (1993), Liu et al. (1999), and Stemmler & Koch (1999).

Descriptions of relevant data on the patterns of responsivity across cortical areas and the development of these patterns include Hubener et al. (1997), **Yuste & Sur (1999)**, and Weliky (2000). **Price & Willshaw (2000)** offers a broad-based, theoretically informed review. Recent experimental challenges to plasticity-based models include Crair et al. (1998) and Crowley & Katz (1999). Neural pattern formation mechanisms involving chemical matching, which are likely important at least for establishing coarse maps, are reviewed from a theoretical perspective in **Goodhill & Richards (1999)**. The use of learning algorithms to account for cortical maps is reviewed in **Erwin et al. (1995)**, **Miller (1996a)**, and **Swindale (1996)**. The underlying mathematical basis of some rules is closely related to the reaction diffusion theory of morphogenesis of **Turing (1952)**. Other rules are motivated on the basis of minimizing quantities such as wire length in cortex.

We described Hebbian models for the development of ocular dominance and orientation selectivity due to Linsker (1986), Miller et al. (1989), and Miller (1994); a competitive Hebbian model closely related to that of Goodhill (1993) and Piepenbrock & Obermayer (1999); a self-organizing map model related to that of Obermayer et al. (1992); and the elastic net (Durbin & Willshaw, 1987) model of Durbin & Mitchison (1990), Goodhill & Willshaw (1990), and **Erwin et al. (1995)**. The first feature-based models were called noise models (see **Swindale, 1996**).

The capacity of a perceptron for random associations is computed in Cover (1965) and Venkatesh (1986). The perceptron learning rule is due to Rosenblatt (1958; see also **Minsky & Papert, 1969**). The delta rule was introduced by Widrow & Hoff (1960; see also Widrow & Stearns, 1985) and independently arose in other fields. The widely used backpropagation algorithm (see Chauvin & Rumelhart, 1995) is a form of delta rule learning that works in a larger class of networks. O'Reilly (1996) suggests a more biologically plausible implementation.

Supervised learning for classification and function approximation, and its ties to Bayesian and frequentist statistical theory, are reviewed in **Duda & Hart (1973)**, **Duda at al. (2000)**, **Kearns & Vazirani (1994)**, and **Bishop (1995)**. Poggio and colleagues have explored basis function models of various representational and learning phenomena (see Poggio, 1990). Tight frames are discussed in Daubechies et al. (1986) and are applied to visual receptive fields by Salinas & Abbott (2000).

Contrastive Hebbian learning is due to Hinton & Sejnowski (1986). See Hinton (2000) for discussion of the Boltzmann machine without recurrent connections, and for an alternative learning rule.

9 Classical Conditioning and Reinforcement Learning

9.1 Introduction

The ability of animals to learn appropriate actions in response to particular stimuli on the basis of associated rewards or punishments is a focus of behavioral psychology. The field is traditionally separated into classical (or Pavlovian) and instrumental (or operant) conditioning. In classical conditioning, the reinforcers (i.e., the rewards or punishments) are delivered independently of any actions taken by the animal. In instrumental conditioning, the actions of the animal determine what reinforcement is provided. Learning about stimuli or actions solely on the basis of the rewards and punishments associated with them is called reinforcement learning. Reinforcement learning is minimally supervised because animals are not told explicitly what actions to take in particular situations, but must work this out for themselves on the basis of the reinforcement they receive.

classical and instrumental conditioning

reinforcement learning

We begin this chapter with a discussion of aspects of classical conditioning and the models that have been developed to account for them. We first discuss various pairings of one or more stimuli with presentation or denial of a reward, and present a simple learning algorithm that summarizes the results. We then present an algorithm, called temporal difference learning, that leads to predictions of both the presence and the timing of rewards delivered after a delay following stimulus presentation. Two neural systems, the cerebellum and the midbrain dopamine system, have been particularly well studied from the perspective of conditioning. The cerebellum has been studied in association with eyeblink conditioning, a paradigm in which animals learn to shut their eyes just in advance of disturbances, such as puffs of air, that are signaled by cues. The midbrain dopaminergic system has been studied in association with reward learning. We focus on the latter, together with a small fraction of the extensive behavioral data on conditioning.

There are two broad classes of instrumental conditioning tasks. In the first class, which we illustrate with an example of foraging by bees, the reinforcement is delivered immediately after the action is taken. This makes learning relatively easy. In the second class, the reward or punishment depends on an entire sequence of actions and is partly or wholly delayed

delayed rewards   until the sequence is completed. Thus, learning the appropriate action at each step in the sequence must be based on future expectation, rather than immediate receipt, of reward. This makes learning more difficult. Despite the differences between classical and instrumental conditioning, we show how to use the temporal difference model we discuss for classical conditioning as the heart of a model of instrumental conditioning when rewards are delayed.

For consistency with the literature on reinforcement learning, throughout this chapter, the letter r is used to represent a reward rather than a firing rate. Also, for convenience, we consider discrete actions such as a choice between two alternatives, rather than a continuous range of actions. We also consider trials that consist of a number of discrete events and use an integer time variable $t = 0, 1, 2, \ldots$ to indicate steps during a trial. We therefore also use discrete weight update rules (like those we discussed for supervised learning in chapter 8) rather than learning rules described by differential equations.

9.2 Classical Conditioning

Classical conditioning involves a wide range of different training and testing procedures and a rich set of behavioral phenomena. The basic procedures and results we discuss are summarized in table 9.1. Rather than going through the entries in the table at this point, we introduce a learning algorithm that serves to summarize and structure these results.

unconditioned stimulus and response   In the classic Pavlovian experiment, dogs are repeatedly fed just after a bell is rung. Subsequently, the dogs salivate whenever the bell sounds, as if they expect food to arrive. The food is called the unconditioned stimulus. Dogs naturally salivate when they receive food, and salivation is thus called the unconditioned response. The bell is called the conditioned stimulus because it elicits salivation only under the condition that there has been prior learning. The learned salivary response to the bell is called the conditioned response. We do not use this terminology in the following discussion. Instead, we treat those aspects of the conditioned responses that mark the animal's expectation of the delivery of reward, and build models of how these expectations are learned. We therefore refer to stimuli, rewards, and expectation of reward.

conditioned stimulus and response

Predicting Reward: The Rescorla-Wagner Rule

The Rescorla-Wagner rule (Rescorla and Wagner, 1972), which is a version of the delta rule of chapter 8, provides a concise account of certain aspects of classical conditioning. The rule is based on a simple linear prediction of the reward associated with a stimulus. We use a binary variable u to represent the presence or absence of the stimulus ($u = 1$ if the stimulus

Paradigm	Pre-Train	Train		Result	
Pavlovian		$s \to r$		$s \to {'r'}$	
Extinction	$s \to r$	$s \to \cdot$		$s \to {'\cdot'}$	
Partial		$s \to r$	$s \to \cdot$	$s \to \alpha\,{'r'}$	
Blocking	$s_1 \to r$	$s_1 + s_2 \to r$		$s_1 \to {'r'}$	$s_2 \to {'\cdot'}$
Inhibitory		$s_1 + s_2 \to \cdot$	$s_1 \to r$	$s_1 \to {'r'}$	$s_2 \to -{'r'}$
Overshadow		$s_1 + s_2 \to r$		$s_1 \to \alpha_1{'r'}$	$s_2 \to \alpha_2{'r'}$
Secondary	$s_1 \to r$	$s_2 \to s_1$		$s_2 \to {'r'}$	

Table 9.1 Classical conditioning paradigms. The columns indicate the training procedures and results, with some paradigms requiring a pre-training as well as a training period. Both training and pre-training periods consist of a moderate number of training trials. The arrows represent an association between one or two stimuli (s, or s_1 and s_2) and either a reward (r) or the absence of a reward ($\cdot$). In Partial and Inhibitory conditioning, the two types of training trials that are indicated are alternated. In the Result column, the arrows represent an association between a stimulus and the expectation of a reward (${'r'}$) or no reward (${'\cdot'}$). The factors of α denote a partial or weakened expectation, and the minus sign indicates the suppression of an expectation of reward.

is present, $u = 0$ if it is absent). The expected reward, denoted by v, is expressed as this stimulus variable multiplied by a weight w,

stimulus u
expected reward v

$$v = wu . \tag{9.1}$$

weight w

The value of the weight, w, is established by a learning rule designed to minimize the expected squared error between the actual reward r and the prediction v, $\langle (r - v)^2 \rangle$. The angle brackets indicate an average over the presentations of the stimulus and reward, either or both of which may be stochastic. As we saw in chapter 8, stochastic gradient descent in the form of the delta rule is one way of minimizing this error. This results in the trial-by-trial learning rule known as the Rescorla-Wagner rule,

Rescorla-Wagner rule

$$w \to w + \epsilon \delta u \quad \text{with} \quad \delta = r - v . \tag{9.2}$$

Here ϵ is the learning rate, which can be interpreted in psychological terms as the associability of the stimulus with the reward. The crucial term in this learning rule is the prediction error, δ. In a later section, we interpret the activity of dopaminergic cells in the ventral tegmental area (VTA) as encoding a form of this prediction error. If ϵ is sufficiently small and $u = 1$ on every trial (the stimulus is always presented), the rule ultimately makes w fluctuate about the equilibrium value $w = \langle r \rangle$, at which point the average value of δ is 0.

associability

The filled circles in figure 9.1 show how learning progresses according to the Rescorla-Wagner rule during the acquisition and extinction phases of Pavlovian conditioning. In this example, the stimulus and reward were both initially presented on each trial, but later the reward was removed. The weight approaches the asymptotic limit $w = r$ exponentially during the rewarded phase of training (conditioning), and exponentially decays to $w = 0$ during the unrewarded phase (extinction). Experimental learning

acquisition

extinction

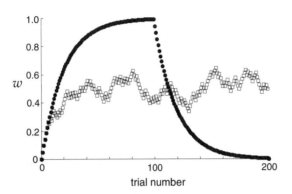

Figure 9.1 Acquisition and extinction curves for Pavlovian conditioning and partial reinforcement as predicted by the Rescorla-Wagner model. The filled circles show the time evolution of the weight w over 200 trials. In the first 100 trials, a reward of $r = 1$ was paired with the stimulus, while in trials 100-200 no reward was paired ($r = 0$). Open squares show the evolution of the weights when a reward of $r = 1$ was paired with the stimulus randomly on 50% of the trials. In both cases, $\epsilon = 0.05$.

curves are generally more sigmoidal in shape. There are various ways to account for this discrepancy, the simplest of which is to assume a nonlinear relationship between the expectation v and the behavior of the animal.

partial reinforcement The Rescorla-Wagner rule also accounts for aspects of the phenomenon of partial reinforcement, in which a reward is associated with a stimulus only on a random fraction of trials (table 9.1). Behavioral measures of the ultimate association of the reward with the stimulus in these cases indicate that it is weaker than when the reward is always presented. This is expected from the delta rule, because the ultimate steady-state average value of $w = \langle r \rangle$ is smaller than r in this case. The open squares in figure 9.1 show what happens to the weight when the reward is associated with the stimulus 50% of the time. After an initial rise from 0, the weight varies randomly around an average value of 0.5.

stimulus vector **u** To account for experiments in which more than one stimulus is used in association with a reward, the Rescorla-Wagner rule must be extended to include multiple stimuli. This is done by introducing a vector of binary variables **u**, with each of its components representing the presence or absence of a given stimulus, together with a vector of weights **w**. The expected reward is then the sum of each stimulus parameter multiplied by its corresponding weight, written compactly as a dot product,

$$v = \mathbf{w} \cdot \mathbf{u} \ . \tag{9.3}$$

delta rule Minimizing the prediction error by stochastic gradient descent in this case gives the delta learning rule,

$$\mathbf{w} \to \mathbf{w} + \epsilon \delta \mathbf{u} \quad \text{with} \quad \delta = r - v \ . \tag{9.4}$$

blocking Various classical conditioning experiments probe the way that predictions are shared between multiple stimuli (see table 9.1). Blocking is the paradigm that first led to the suggestion of the delta rule in connection with classical conditioning. In blocking, two stimuli are presented together with the reward, but only after an association has already developed for one stimulus by itself. In other words, during the pre-training period, a stimulus is associated with a reward, as in Pavlovian conditioning. Then, during the training period, a second stimulus is present along with the first, in association with the same reward. In this case, the pre-existing association of the first stimulus with the reward blocks an association from forming between the second stimulus and the reward. Thus, after training, a conditioned response is evoked only by the first stimulus, not by the second. This follows from the vector form of the delta rule, because training with the first stimulus makes $w_1 = r$. When the second stimulus is presented along with the first, its weight starts out at $w_2 = 0$, but the prediction of reward $v = w_1 u_1 + w_2 u_2$ is still equal to r. This makes $\delta = 0$, so no further weight modification occurs.

A standard way to induce inhibitory conditioning is to use trials in which one stimulus is shown in conjunction with the reward in alternation with trials in which that stimulus and an additional stimulus are presented in the absence of reward. In this case, the second stimulus becomes a conditioned inhibitor, predicting the absence of reward. This can be demonstrated by presenting a third stimulus that also predicts reward, in conjunction with the inhibitory stimulus, and showing that the net prediction of reward is reduced. It can also be demonstrated by showing that subsequent learning of a positive association between the inhibitory stimulus and reward is slowed. Inhibition emerges naturally from the delta rule. Trials in which the first stimulus is associated with a reward result in a positive value of w_1. Over trials in which both stimuli are presented together, the net prediction $v = w_1 + w_2$ comes to be 0, so w_2 is forced to be negative.

inhibitory conditioning

A further example of the interaction between stimuli is overshadowing. If two stimuli are presented together during training, the prediction of reward is shared between them. After application of the delta rule, $v = w_1 + w_2 = r$. However, the prediction is often shared unequally, as if one stimulus is more salient than the other. Overshadowing can be encompassed by generalizing the delta rule so that the two stimuli have different learning rates (different values of ϵ), reflecting unequal associabilities. Weight modification stops when $\langle \delta \rangle = 0$, at which point the faster growing weight will be larger than the slower growing weight. Various, more subtle effects come from having not only different but also modifiable learning rates, but these lie beyond the scope of our account.

overshadowing

The Rescorla-Wagner rule, binary stimulus parameters, and linear reward prediction are obviously gross simplifications of animal learning behavior. Yet they summarize and unify an impressive amount of classical conditioning data and are useful, provided their shortcomings are fully appreciated. As a reminder of this, we point out one experiment, secondary conditioning, that cannot be encompassed within this scheme.

secondary conditioning Secondary conditioning involves the association of one stimulus with a reward, followed by an association of a second stimulus with the first stimulus (table 9.1). This causes the second stimulus to evoke expectation of a reward with which it has never been paired (although if pairings of the two stimuli without the reward are repeated too many times, the result is extinction of the association of both stimuli with the reward). The delta rule cannot account for the positive expectation associated with the second stimulus. Indeed, because the reward does not appear when the second stimulus is presented, the delta rule would cause w_2 to become negative. In other words, in this case, the delta rule would predict inhibitory, not excitatory, secondary conditioning. Secondary conditioning is related to the problem of delayed rewards in instrumental conditioning that we discuss later in this chapter.

Secondary conditioning raises the important issue of keeping track of the time within a trial in which stimuli and rewards are present. This is evident because a positive association with the second stimulus is reliably established only if it precedes the first stimulus in the trials in which they are paired. If the two stimuli are presented simultaneously, the result may be inhibitory rather than secondary conditioning.

Predicting Future Reward: Temporal Difference Learning

We measure time within a trial using a discrete time variable t, which falls in the range $0 \leq t \leq T$. The stimulus $u(t)$, the prediction $v(t)$, and the reward $r(t)$ are all expressed as functions of t.

In addition to associating stimuli with rewards and punishments, animals can learn to predict the future time within a trial at which reinforcement will be delivered. We might therefore be tempted to interpret $v(t)$ as the reward predicted to be delivered at time step t. However, Sutton and Barto (1990) suggested an alternative interpretation of $v(t)$ that provides a better match to psychological and neurobiological data, and suggests how animals might use their predictions to optimize behavior when rewards are delayed. The suggestion is that the variable $v(t)$ should be interpreted as a prediction of the total future reward expected from time t onward to the *total future reward* end of the trial, namely

$$\left\langle \sum_{\tau=0}^{T-t} r(t+\tau) \right\rangle . \tag{9.5}$$

The brackets denote an average over trials. This quantity is useful for optimization, because it summarizes the total expected worth of the current state. To approximate $v(t)$, we generalize the linear relationship used for classical conditioning, equation 9.3. For the case of a single time-dependent stimulus $u(t)$, we write

$$v(t) = \sum_{\tau=0}^{t} w(\tau)u(t-\tau) . \tag{9.6}$$

This is just a discrete time version of the sort of linear filter used in chapters 1 and 2.

Arranging for $v(t)$ to predict the total future reward would appear to require a simple modification of the delta rule we have discussed previously,

$$w(\tau) \to w(\tau) + \epsilon\delta(t)u(t-\tau),   \quad (9.7)$$

with $\delta(t)$ being the difference between the actual and predicted total future reward, $\delta(t) = \sum_\tau r(t+\tau) - v(t)$. However, there is a problem with applying this rule in a stochastic gradient descent algorithm. Computation of $\delta(t)$ requires knowledge of the total future reward on a given trial. Although $r(t)$ is known at time t, the succeeding $r(t+1), r(t+2)\ldots$ have yet to be experienced, making it impossible to calculate $\delta(t)$. A possible solution is suggested by the recursive formula

$$\sum_{\tau=0}^{T-t} r(t+\tau) = r(t) + \sum_{\tau=0}^{T-t-1} r(t+1+\tau).   \quad (9.8)$$

The temporal difference model of prediction is based on the observation that $v(t+1)$ provides an approximation of the average value (across trials) of the last term in equation 9.8, so we can write

$$\sum_{\tau=0}^{T-t} r(t+\tau) \approx r(t) + v(t+1).   \quad (9.9)$$

Replacing the sum in the equation $\delta(t) = \sum_\tau r(t+\tau) - v(t)$ by this approximation gives the temporal difference learning rule,

temporal difference rule

$$w(\tau) \to w(\tau) + \epsilon\delta(t)u(t-\tau)   \quad \text{with} \quad \delta(t) = r(t) + v(t+1) - v(t).   \quad (9.10)$$

The name of the rule comes from the term $v(t+1) - v(t)$, which is the difference between two successive estimates. $\delta(t)$ is usually called the temporal difference error. Under a variety of circumstances, this rule is likely to converge to make the correct predictions.

Figure 9.2 shows what happens when the temporal difference rule is applied during a training period in which a stimulus appears at time $t = 100$, and a reward is given for a short interval around $t = 200$. Initially, $w(\tau) = 0$ for all τ. Figure 9.2A shows that the temporal difference error starts off being nonzero only at the time of the reward, $t = 200$, and then, over trials, moves backward in time, eventually stabilizing around the time of the stimulus, where it takes the value 2. This is equal to the (integrated) total reward provided over the course of each trial. Figure 9.2B shows the behavior during a trial of a number of variables before and after learning. After learning, the prediction $v(t)$ is 2 from the time the stimulus is first presented ($t = 100$) until the time the reward starts to be delivered. Thus, the temporal difference prediction error (δ) has a spike at $t = 99$. This spike persists, because $u(t) = 0$ for $t < 100$. The temporal difference term (Δv) is negative around $t = 200$, exactly compensating for the delivery of reward, and thus making $\delta = 0$.

A

B

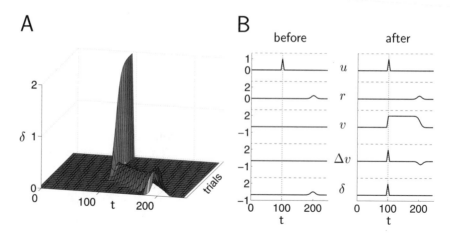

Figure 9.2 Learning to predict a reward. (A) The surface plot shows the prediction error $\delta(t)$ as a function of time within a trial, across trials. In the early trials, the peak error occurs at the time of the reward ($t = 200$), while in later trials it occurs at the time of the stimulus ($t = 100$). (B) The rows show the stimulus $u(t)$, the reward $r(t)$, the prediction $v(t)$, the temporal difference between predictions $\Delta v(t - 1) = v(t) - v(t-1)$, and the full temporal difference error $\delta(t - 1) = r(t - 1) + \Delta v(t - 1)$. The reward is presented over a short interval, and the prediction v sums the total reward. The left column shows the behavior before training, and the right column, after training. $\Delta v(t - 1)$ and $\delta(t - 1)$ are plotted instead of $\Delta v(t)$ and $\delta(t)$ because the latter quantities cannot be computed until time $t + 1$, when $v(t + 1)$ is available.

As the peak in δ moves backward from the time of the reward to the time of the stimulus, weights $w(\tau)$ for $\tau = 100, 99, \ldots$ successively grow. This gradually extends the prediction of future reward, $v(t)$, from an initial transient at the time of the reward to a broad plateau extending from the time of the stimulus to the time of the reward. Eventually, v predicts the correct total future reward from the time of the stimulus onward, and predicts the time of the reward delivery by dropping to 0 at the time when the reward is delivered. The exact shape of the ridge of activity that moves from $t = 200$ to $t = 100$ over the course of trials in figure 9.2A is sensitive to a number of factors, including the learning rate, and the form of the linear filter of equation 9.6.

Unlike the delta rule, the temporal difference rule provides an account of secondary conditioning. Suppose an association between stimulus s_1 and a future reward has been established, as in figure 9.2. When, as indicated in table 9.1, a second stimulus, s_2, is introduced before the first stimulus, the positive spike in $\delta(t)$ at the time that s_1 is presented drives an increase in the value of the weight associated with s_2, and thus establishes a positive association between the second stimulus and the reward. This exactly mirrors the primary learning process for s_1 described above. Of course, because the reward is not presented in these trials, there is a negative spike in $\delta(t)$ at the time of the reward itself, and ultimately the association between both s_1 and s_2 and the reward extinguishes.

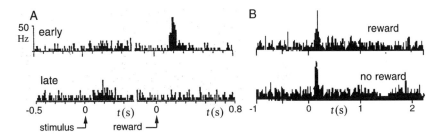

Figure 9.3 Activity of dopaminergic neurons in the VTA for a monkey performing reaction time tasks. (A) Activity of a dopamine cell accumulated over 20 trials showing the spikes time-locked to a stimulus (left panels) or to the reward (right panels) at the times marked 0. The top row is for early trials before the behavior is fully established. The bottom row is for late trials, when the monkey expects the reward on the basis of the stimulus. (B) Activity of a dopamine neuron with and without an expected reward delivery in a similar task. The top row shows the normal behavior of the cell when the reward is delivered. The bottom row shows the result of not delivering an expected reward. The basal firing rate of dopamine cells is rather low, but the inhibition at the time the reward would have been given is evident. (A adapted from Mirenowicz & Schultz, 1994; B adapted from Schultz, 1998.)

Dopamine and Predictions of Reward

The prediction error δ plays an essential role in both the Rescorla-Wagner and temporal difference learning rules, and we might hope to find a neural signal that represents this quantity. One suggestion is that the activity of dopaminergic neurons in the ventral tegmental area (VTA) in the midbrain plays this role.

ventral tegmental area VTA

There is substantial evidence that dopamine is involved in reward learning. Drugs of addiction, such as cocaine and amphetamines, act partly by increasing the longevity of the dopamine that is released onto target structures such as the nucleus accumbens. Other drugs, such as morphine and heroin, also affect the dopamine system. Further, dopamine delivery is important in self-stimulation experiments. Rats will compulsively press levers that cause current to be delivered through electrodes into various areas of their brains. One of the most effective self-stimulation sites is the medial forebrain ascending bundle, which is an axonal pathway. Stimulating this pathway is likely to cause increased delivery of dopamine to the nucleus accumbens and other areas of the brain, because the bundle contains many fibers from dopaminergic cells in the VTA projecting to the nucleus accumbens.

dopamine

In a series of studies by Schultz and his colleagues (Schultz, 1998), monkeys were trained through instrumental conditioning to respond to stimuli such as lights and sounds in order to obtain food and drink rewards. The activities of cells in the VTA were recorded while the monkeys learned these tasks. Figure 9.3A shows the activity of a dopamine cell at two times during learning. The figure is based on a reaction time task in which the monkey keeps a finger resting on a key until a sound is presented. The monkey then has to release the resting key and press another one to get a fruit juice reward. The reward is delivered a short time after the second key is pressed. The upper plot of figure 9.3A shows the response of a cell in early trials. The cell responds vigorously to the reward, but only fires a little above baseline in response to the sound. The lower plot shows the response after a moderate amount of training. Now the cell responds to the sound, but not to the reward. The responses show a distinct similarity to the plots of $\delta(t)$ in figure 9.2.

The similarity between the responses of the dopaminergic neurons and the quantity $\delta(t)$ suggests that their activity provides a prediction error for reward, i.e. an ongoing difference between the amount of reward that is delivered and the amount that is expected. Figure 9.3B provides further evidence for this interpretation. It shows the activity of a dopamine cell in a task similar to that of figure 9.3A. The top row of this figure shows normal performance, and is just like the bottom row of figure 9.3A. The bottom row shows what happens when the monkey is expecting reward but it is not delivered. In this case, the cell's activity is inhibited below baseline at just the time it would have been activated by the reward in the original trials. This is in agreement with the prediction error interpretation of this activity.

Something similar to the temporal difference learning rule could be realized in a neural system if the dopamine signal representing δ acts to gate and regulate the plasticity associated with learning. We discuss this possibility further in a later section.

9.3   Static Action Choice

policy

In classical conditioning experiments, rewards are directly associated with stimuli. In more natural settings, rewards and punishments are associated with the actions an animal takes. Animals develop policies, or plans of action, that increase reward. In studying how this might be done, we consider two different cases. In static action choice, the reward or punishment immediately follows the action taken. In sequential action choice, rewards may be delayed until several actions are completed.

foraging

As an example of static action choice, we consider bees foraging among flowers in search of nectar. We model an experiment in which single bees forage under controlled conditions among blue and yellow artificial flowers (small dishes of sugar water sitting on colored cards). In actual experiments, the bees learn within a single session (involving visits to 40 artificial flowers) about the reward characteristics of the blue and yellow flowers. All else being equal, they preferentially land on the color of flower that delivers more reward. This preference is maintained over multiple sessions. However, if the reward characteristics of the flowers are interchanged, the bees quickly swap their preferences.

We treat a simplified version of the problem, ignoring the spatial aspects of sampling, and assuming that a model bee is faced with repeated choices between two different flowers. If the bee chooses the blue flower on a trial, it receives a quantity of nectar r_b drawn from a probability density $p[r_b]$. If it chooses the yellow flower, it receives a quantity r_y, drawn from a probability density $p[r_y]$. The task of choosing between the flowers is a form of stochastic two-armed bandit problem (named after slot machines), *two-armed bandit* and is formally equivalent to many instrumental conditioning tasks..

The model bee has a stochastic policy, which means that it chooses blue *stochastic policy* and yellow flowers with probabilities that we write as $P[b]$ and $P[y]$, respectively. A convenient way to parameterize these probabilities is to use the softmax distribution *softmax*

$$P[b] = \frac{\exp(\beta m_b)}{\exp(\beta m_b) + \exp(\beta m_y)} \quad P[y] = \frac{\exp(\beta m_y)}{\exp(\beta m_b) + \exp(\beta m_y)}. \quad (9.11)$$

Here, m_b and m_y are parameters, known as action values, that are adjusted *action values m* by one of the learning processes described below. Note that $P[b] + P[y] = 1$, corresponding to the fact that the model bee invariably makes one of the two choices. Also note that $P[b] = \sigma(\beta(m_b - m_y))$, where $\sigma(m) = 1/(1 + \exp(-m))$ is the standard sigmoid function, which grows monotonically from 0 to 1 as m varies from $-\infty$ to ∞. $P[y]$ is similarly a sigmoid function of $\beta(m_y - m_b)$. The parameters m_b and m_y determine the frequency at which blue and yellow flowers are visited. Their values must be adjusted during the learning process on the basis of the reward provided.

The parameter β determines the variability of the bee's actions and exerts a strong influence on exploration. For large β, the probability of an action rises rapidly to 1, or falls rapidly to 0, as the difference between the action values increases or decreases. This makes the bee's action choice almost a deterministic function of the m variables. If β is small, the softmax probability approaches 1 or 0 more slowly, and the bee's actions are more variable and random. Thus, β controls the balance between exploration (small β) and exploitation (large β). The choice of whether to explore to determine if the current policy can be improved, or to exploit the available resources on the basis of the current policy, is known as the exploration- exploitation dilemma. Exploration is clearly critical, because the bee must *exploration-* sample from the two colors of flowers to determine which is better, and *exploitation* keep sampling to make sure that the reward conditions have not changed. *dilemma* But exploration is costly, because the bee has to sample flowers it believes to be less beneficial, to check if this is really the case. Some algorithms adjust β over trials, but we do not consider this possibility.

There are only two possible actions in the example we study, but the extension to multiple actions, $a = 1, 2, \ldots, N_a$, is straightforward. In this case, a vector **m** of parameters controls the decision process, and the probability *action value* $P[a]$ of choosing action a is *vector* **m**

$$P[a] = \frac{\exp(\beta m_a)}{\sum_{a'=1}^{N_a} \exp(\beta m_{a'})}. \quad (9.12)$$

We consider two simple methods of solving the bee foraging task. In the first method, called the indirect actor, the bee learns to estimate the expected nectar volumes provided by each flower by using a delta rule. It then bases its action choice on these estimates. In the second method, called the direct actor, the choice of actions is based directly on maximizing the expected average reward.

The Indirect Actor

indirect actor   One course for the bee to follow is to learn the average nectar volumes provided by each type of flower and base its action choice on these. This is called an indirect actor scheme, because the policy is mediated indirectly by the expected volumes. Here, this means setting the action values to

$$m_b = \langle r_b \rangle \quad \text{and} \quad m_y = \langle r_y \rangle . \tag{9.13}$$

In our discussion of classical conditioning, we saw that the Rescorla-Wagner or delta rule develops weights that approximate the average value of a reward, just as required for equation 9.13. Thus, if the bee chooses a blue flower on a trial and receives nectar volume r_b, it should update m_b according to the prediction error by

$$m_b \rightarrow m_b + \epsilon \delta \quad \text{with} \quad \delta = r_b - m_b , \tag{9.14}$$

and leave m_y unchanged. If it lands on a yellow flower, m_y is changed to $m_y + \epsilon \delta$ with $\delta = r_y - m_y$, and m_b is unchanged. If the probability densities of reward $p[r_b]$ and $p[r_y]$ change slowly relative to the learning rate, m_b and m_y will track $\langle r_b \rangle$ and $\langle r_y \rangle$, respectively.

Figure 9.4 shows the performance of the indirect actor on the two-flower foraging task. Figure 9.4A shows the course of weight change due to the delta rule in one example run. Figures 9.4B-D indicate the quality of the action choice by showing cumulative sums of the number of visits to blue and yellow flowers in three different runs. For ideal performance in this task, the dashed line would have slope 1 until trial 100 and 0 thereafter, and the solid line would show the reverse behavior, close to what is seen in figure 9.4C. This reflects the consistent choice of the optimal flower in both halves of the trial. A value of $\beta = 1$ (figure 9.4B) allows for continuous exploration, but at the cost of slow learning. When $\beta = 50$ (figure 9.4C & D), the tendency to exploit sometimes leads to good performance (figure 9.4C), but other times, the associated reluctance to explore causes the policy to perform poorly (figure 9.4D).

Figure 9.5A shows action choices of real bumblebees in a foraging experiment. This experiment was designed to test risk aversion in the bees, so the blue and yellow flowers differed in the reliability rather than the quantity of their nectar delivery. For the first 15 trials (each involving 40 visits to flowers), blue flowers always provided 2 μl of nectar, whereas $\frac{1}{3}$ of the yellow flowers provided 6 μl, and $\frac{2}{3}$ provided nothing (note that the mean

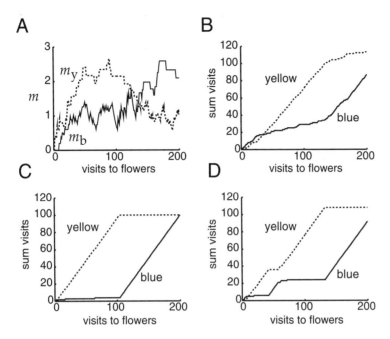

Figure 9.4 The indirect actor. Rewards were $\langle r_b \rangle = 1$, $\langle r_y \rangle = 2$ for the first 100 flower visits, and $\langle r_b \rangle = 2$, $\langle r_y \rangle = 1$ for the second 100 flower visits. Nectar was delivered stochastically on half the flowers of each type. (A) Values of m_b (solid) and m_y (dashed) as a function of visits for $\beta = 1$. Because a fixed value of $\epsilon = 0.1$ was used, the weights do not converge perfectly to the corresponding average reward, but they fluctuate around these values. (B-D) Cumulative visits to blue (solid) and yellow (dashed) flowers. (B) When $\beta = 1$, learning is slow, but ultimately the change to the optimal flower color is made reliably. (C, D) When $\beta = 50$, sometimes the bee performs well (C), and other times it performs poorly (D).

reward is the same for the two flower types). Between trials 15 and 16, the delivery characteristics of the flowers were swapped. Figure 9.5A shows the average performance of five bees on this task in terms of their percentage visits to the blue flowers across trials. They exhibit a strong preference for the constant flower type and switch this preference within only a few visits to the flowers when the contingencies change.

To apply the foraging model we have been discussing to the experiment shown in figure 9.5A, we need to model the risk avoidance exhibited by the bees, that is, their reluctance to choose the unreliable flower. One way to do this is to assume that the bees base their policy on the subjective utility function of the nectar volume shown in figure 9.5B, rather than on the nectar volume itself. Because the function is concave, the mean utility of the unreliable flowers is less than that of the reliable flowers. Figure 9.5C shows that the choices of the model bee match those of the real bees quite well. The model bee is less variable than the actual bees (even more than it appears, because the curve in 9.5A is averaged over five bees), perhaps because the model bees are not sampling from a two-dimensional array of flowers.

subjective utility

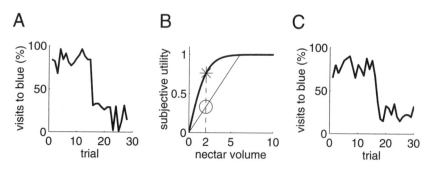

Figure 9.5 Foraging in bumblebees. (A) The mean preference of five real bumblebees for blue flowers over 30 trials involving 40 flower visits. There is a rapid switch of flower preference following the interchange of characteristics after trial 15. (B) Concave subjective utility function mapping nectar volume (in μl) to the subjective utility. The circle shows the average utility of the variable flowers, and the star shows the utility of the constant flowers. (C) The preference of a single model bee on the same task as the bumblebees. Here, $\epsilon = 3/10$ and $\beta = 23/8$. (Data in A from Real, 1991; B and C adapted from Montague et al., 1995.)

The Direct Actor

direct actor   An alternative to basing action choice on average rewards is to choose action values directly to maximize the average expected reward. The expected reward per trial is given in terms of the action probabilities and average rewards per flower by

$$\langle r \rangle = P[\text{b}]\langle r_{\text{b}} \rangle + P[\text{y}]\langle r_{\text{y}} \rangle . \qquad (9.15)$$

This can be maximized by stochastic gradient ascent. To see how this is done, we take the derivative of $\langle r \rangle$ with respect to m_{b},

$$\frac{\partial \langle r \rangle}{\partial m_{\text{b}}} = \beta \left(P[\text{b}]P[\text{y}]\langle r_{\text{b}} \rangle - P[\text{y}]P[\text{b}]\langle r_{\text{y}} \rangle \right) . \qquad (9.16)$$

In deriving this result, we have used the fact that, for the softmax distribution of equation 9.11,

$$\frac{\partial P[\text{b}]}{\partial m_{\text{b}}} = \beta P[\text{b}]P[\text{y}] \quad \text{and} \quad \frac{\partial P[\text{y}]}{\partial m_{\text{b}}} = -\beta P[\text{y}]P[\text{b}]. \qquad (9.17)$$

Using the relation $P[\text{y}] = 1 - P[\text{b}]$, we can rewrite equation 9.16 in a form convenient for later use,

$$\frac{\partial \langle r \rangle}{\partial m_{\text{b}}} = \beta P[\text{b}](1 - P[\text{b}])\langle r_{\text{b}} \rangle - \beta P[\text{y}]P[\text{b}]\langle r_{\text{y}} \rangle . \qquad (9.18)$$

Furthermore, we can include an arbitrary parameter $\bar{r}$ in both these terms, because it cancels out. Thus,

$$\frac{\partial \langle r \rangle}{\partial m_{\text{b}}} = \beta P[\text{b}](1 - P[\text{b}]) \left(\langle r_{\text{b}} \rangle - \bar{r} \right) - \beta P[\text{y}]P[\text{b}] \left(\langle r_{\text{y}} \rangle - \bar{r} \right) . \qquad (9.19)$$

A similar expression applies to $\partial\langle r\rangle/\partial m_y$, except that the blue and yellow labels are interchanged.

In stochastic gradient ascent, the changes in the parameter m_b are determined such that, averaged over trials, they end up proportional to $\partial\langle r\rangle/\partial m_b$. We can derive a stochastic gradient ascent rule for m_b from equation 9.19 in two steps. First, we interpret the two terms on the right side as changes associated with the choice of blue and yellow flowers. This accounts for the factors $P[b]$ and $P[y]$, respectively. Second, we note that over trials in which blue is selected, $r_b - \bar{r}$ averages to $\langle r_b\rangle - \bar{r}$, and over trials in which yellow is selected, $r_y - \bar{r}$ averages to $\langle r_y\rangle - \bar{r}$. Thus, if we change m_b according to

$$m_b \to m_b + \epsilon(1 - P[b])(r_b - \bar{r}) \quad \text{if b is selected}$$

$$m_b \to m_b - \epsilon P[b]\,(r_y - \bar{r}) \qquad \text{if y is selected,}$$

the average change in m_b is proportional to $\partial\langle r\rangle/\partial m_b$. Note that m_b is changed even when the bee chooses the yellow flower. We can summarize this learning rule as

$$m_b \to m_b + \epsilon(\delta_{ab} - P[b])(r_a - \bar{r})\,, \tag{9.20}$$

where a is the action selected (either b or y) and δ_{ab} is the Kronecker delta, $\delta_{ab} = 1$ if $a = b$ and $\delta_{ab} = 0$ if $a = y$. Similarly, the rule for m_y is

$$m_y \to m_y + \epsilon(\delta_{ay} - P[y])(r_a - \bar{r})\,. \tag{9.21}$$

The learning rule of equations 9.20 and 9.21 performs stochastic gradient ascent on the average reward, whatever the value of $\bar{r}$. Different values of $\bar{r}$ lead to different variances of the stochastic gradient terms, and thus different speeds of learning. A reasonable value for $\bar{r}$ is the mean reward under the specified policy or some estimate of this quantity.

Figure 9.6 shows the consequences of using the direct actor in the same stochastic foraging task as in figure 9.4. Two sample sessions are shown with widely differing levels of performance. Compared to the indirect actor, initial learning is quite slow, and the behavior after the reward characteristics of the flowers are interchanged can be poor. Explicit control of the trade-off between exploration and exploitation is difficult, because the action values can scale up to compensate for different values of β. Despite its comparatively poor performance in this task, the direct actor will be useful later as a model for how action choice can be separated from action evaluation.

The direct actor learning rule can be extended to multiple actions, $a = 1, 2, \ldots, N_a$, by using the multidimensional form of the softmax distribution (equation 9.12). In this case, when action a is taken, $m_{a'}$ for all values of a' is updated according to

$$m_{a'} \to m_{a'} + \epsilon\left(\delta_{aa'} - P[a']\right)(r_a - \bar{r})\,. \tag{9.22}$$

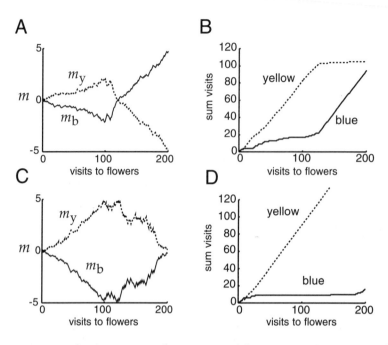

Figure 9.6 The direct actor. The statistics of the delivery of reward are the same as in figure 9.4, and $\epsilon = 0.1$, $\bar{r} = 1.5$, $\beta = 1$. The evolution of the weights and cumulative choices of flower type (with yellow dashed and blue solid) are shown for two sample sessions, one with good performance (A and B) and one with poor performance (C and D).

9.4   Sequential Action Choice

In the previous section, we considered ways that animals might learn to choose actions on the basis of immediate information about the consequences of those actions. A significant complication that arises when reward is based on a sequence of actions is illustrated by the maze task shown in figure 9.7. In this example, a hungry rat has to move through a maze, starting from point A, without retracing its steps. When it reaches one of the shaded boxes, it receives the associated number of food pellets and is removed from the maze. The rat then starts again at A. The task is to optimize the total reward, which in this case entails moving left at A and right at B. It is assumed that the animal starts knowing nothing about the structure of the maze or about the rewards.

If the rat started from point B or point C, it could learn to move right or left (respectively), using the methods of the previous section, because it experiences an immediate consequence of its actions in the delivery or nondelivery of food. The difficulty arises because neither action at the actual starting point, A, leads directly to a reward. For example, if the rat goes left at A and also goes left at B, it has to figure out that the former choice was good but the latter was bad. This is a typical problem in tasks that involve delayed rewards. The reward for going left at A is delayed

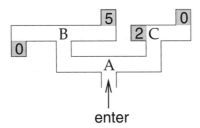

Figure 9.7 The maze task. The rat enters the maze from the bottom and has to move forward. Upon reaching one of the end points (the shaded boxes), it receives the number of food pellets indicated and the trial ends. Decision points are A, B, and C.

until after the rat also goes right at B.

There is an extensive body of theory in engineering, called dynamic programming, as to how systems of any sort can come to select appropriate actions in optimizing control problems similar to (and substantially more complicated than) the maze task. An important method on which we focus is called policy iteration. Our reinforcement learning version of policy iteration maintains and improves a stochastic policy, which determines the actions at each decision point (i.e., left or right turns at A, B, or C) through action values and the softmax distribution of equation 9.12. Policy iteration involves two elements. One, called the critic, uses temporal difference learning to estimate the total future reward that is expected when starting from A, B, or C, when the current policy is followed. The other element, called the actor, maintains and improves the policy. Adjustment of the action values at point A is based on predictions of the expected future rewards associated with points B and C that are provided by the critic. In effect, the rat learns the appropriate action at A, using the same methods of static action choice that allow it to learn the appropriate actions at B and C. However, rather than using an immediate reward as the reinforcement signal, it uses the expectations about future reward that are provided by the critic.

dynamic programming

policy iteration

critic

actor

The Maze Task

As we mentioned when discussing the direct actor, a stochastic policy is a way of assigning a probability distribution over actions (in this case choosing to turn either left or right) to each location (A, B, or C). The location is specified by a variable u that takes the values A, B, or C, and a two-component action value vector $\mathbf{m}(u)$ is associated with each location. The components of the action vector $\mathbf{m}(u)$ control the probability of taking a left or a right turn at u through the softmax distribution of equation 9.12.

The immediate reward provided when action a is taken at location u is written as $r_a(u)$. For the maze of figure 9.7, this takes the values 0, 2, or 5, depending on the values of u and a. The predicted future reward expected at location u is given by $v(u) = w(u)$. This is an estimate of the total award that the rat expects to receive, on average, if it starts at the point u and follows its current policy through to the end of the maze. The average is taken over the stochastic choices of actions specified by the policy. In this case, the expected reward is simply equal to the weight. The learning procedure consists of two separate steps: policy evaluation, in which $w(u)$ is adjusted to improve the predictions of future reward, and policy improvement, in which $\mathbf{m}(u)$ is adjusted to increase the total reward.

Policy Evaluation

In policy evaluation, the rat keeps its policy fixed (i.e., keeps all the $\mathbf{m}(u)$ fixed) and uses temporal difference learning to determine the expected total future reward starting from each location. Suppose that, initially, the rat has no preference for turning left or right, that is, $\mathbf{m}(u) = 0$ for all u, so the probability of left and right turns is $1/2$ at all locations. By inspection of the possible places the rat can go, we find that the values of the states are

$$v(B) = \frac{1}{2}(0 + 5) = 2.5, \quad v(C) = \frac{1}{2}(0 + 2) = 1, \quad \text{and}$$
$$v(A) = \frac{1}{2}(v(B) + v(C)) = 1.75. \tag{9.23}$$

These values are the average total future rewards that will be received during exploration of the maze when actions are chosen using the random policy. The temporal difference learning rule of equation 9.10 can be used to learn them. If the rat chooses action a at location u and ends up at

critic learning rule   location u', the temporal difference rule modifies the weight $w(u)$ by

$$w(u) \to w(u) + \epsilon\delta \quad \text{with} \quad \delta = r_a(u) + v(u') - v(u). \tag{9.24}$$

Figure 9.8 shows the result of applying the temporal difference rule to the maze task of figure 9.7. After a fairly short adjustment period, the weights $w(u)$ (and thus the predictions $v(u)$) fluctuate around the correct values for this policy, as given by equation 9.23. The size of the fluctuations could be reduced by making ϵ smaller, but at the expense of increasing the learning time.

In our earlier description of temporal difference learning, we included the possibility that the reward delivery might be stochastic. Here, that stochasticity is the result of a policy that makes use of the information provided by the critic. In the appendix, we discuss a Monte Carlo interpretation of the terms in the temporal difference learning rule that justifies its use.

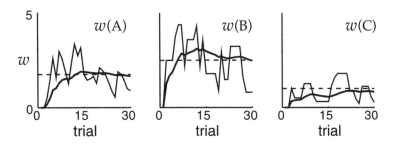

Figure 9.8 Policy evaluation. The thin lines show the course of learning of the weights $w(A)$, $w(B)$, and $w(C)$ over trials through the maze in figure 9.7, using a random unbiased policy ($\mathbf{m}(u) = 0$). Here $\epsilon = 0.5$, so learning is fast but noisy. The dashed lines show the correct weight values from equation 9.23. The thick lines are running averages of the weight values.

Policy Improvement

In policy improvement, the expected total future rewards at the different locations are used as surrogate immediate rewards. Suppose the rat is about to take action a at location u and move to location u'. The expected worth to the rat of that action is the sum of the actual reward received and the rewards that are expected to follow, which is $r_a(u) + v(u')$. For simplicity, we assume that the rat receives the reward for location u at the same time it decides to move on to location u'. The direct actor scheme of equation 9.22 uses the difference $r_a - \bar{r}$ between a sample of the worth of the action (r_a) and a reinforcement comparison term ($\bar{r}$), which might be the average value over all the actions that can be taken. Policy improvement uses $r_a(u) + v(u')$ as the equivalent of the sampled worth of the action, and $v(u)$ as the average value across all actions that can be taken at u. The difference between these is $\delta = r_a(u) + v(u') - v(u)$, which is exactly the same term as in policy evaluation (equation 9.24). The policy improvement or actor learning rule is then

actor learning rule

$$m_{a'}(u) \rightarrow m_{a'}(u) + \epsilon \left(\delta_{aa'} - P[a'; u] \right) \delta \qquad (9.25)$$

for all a', where $P[a'; u]$ is the probability of taking action a' at location u given by the softmax distribution of equation 9.12 with action value $m_{a'}(u)$.

To look at this more concretely, consider the temporal difference error starting from location $u = A$, using the true values of the locations given by equation 9.23 (i.e., assuming that policy evaluation is perfect). Depending on the action, δ takes the two values

$$\delta = 0 + v(B) - v(A) = \quad 0.75 \quad \text{for a left turn}$$
$$\delta = 0 + v(C) - v(A) = -0.75 \quad \text{for a right turn.}$$

The learning rule of equation 9.25 increases the probability that the action with $\delta > 0$ is taken and decreases the probability that the action with $\delta < 0$

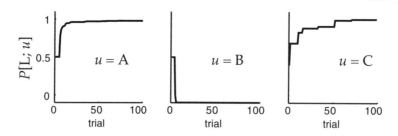

Figure 9.9 Actor-critic learning. The three curves show $P[L; u]$ for the three starting locations u = A, B, and C in the maze of figure 9.7. These rapidly converge to their optimal values, representing left turns at A and C and a right turn at B. Here, $\epsilon = 0.5$ and $\beta = 1$.

is taken. This increases the chance that the rat makes the correct turn (left) at A in the maze of figure 9.7.

As the policy changes, the values, and therefore the temporal difference terms, change as well. However, because the values of all locations can increase only if we choose better actions at those locations, this form of policy improvement inevitably leads to higher values and better actions. This monotonic improvement (or at least not worsening) of the expected future rewards at all locations is proved formally in the dynamic programming theory of policy iteration for a class of problems called Markov decision problems (which includes the maze task), as discussed in the appendix.

Markov decision problems

Strictly speaking, policy evaluation should be complete before a policy is improved. It is also most straightforward to improve the policy completely before it is re-evaluated. A convenient (though not provably correct) alternative is to interleave partial policy evaluation and policy improvement steps. This is called the actor-critic algorithm. Figure 9.9 shows the result of applying this algorithm to the maze task. The plots show the development over trials of the probability of choosing to go left, $P[L; u]$, for all three locations. The model rat quickly learns to go left at location A and right at B. Learning at location C is slow because the rat quickly learns that it is not worth going to C at all, so it rarely gets to try the actions there. Thus the algorithm makes an implicit choice of exploration strategy.

actor-critic algorithm

Generalizations of Actor-Critic Learning

The full actor-critic model for solving sequential action tasks includes three generalizations of the maze learner that we have presented. The first involves additional information that may be available at the different locations. If, for example, sensory information is available at a location u, we associate a state vector $\mathbf{u}(u)$ with that location. The vector $\mathbf{u}(u)$ parameterizes whatever information is available at location u that might help the animal decide which action to take. For example, the state vector might represent a faint scent of food that the rat might detect in the maze task. When a state vector is available, $v(u)$, which is the value at location

state vector **u**

u, can depend on it. The simplest dependence is provided by the linear form $v(u) = \mathbf{w} \cdot \mathbf{u}(u)$, similar to the input-output relationship used in linear feedforward network models. The learning rule for the critic (equation 9.24) is then generalized to include the information provided by the state vector,

$$\mathbf{w} \to \mathbf{w} + \epsilon \delta \mathbf{u}(u)\,, \tag{9.26}$$

with δ given as in equation 9.24. The maze task we discussed could be formulated in this way by using what is called a unary representation, $\mathbf{u}(\text{A}) = (1, 0, 0)$, $\mathbf{u}(\text{B}) = (0, 1, 0)$, and $\mathbf{u}(\text{C}) = (0, 0, 1)$.

unary representation

We must also modify the actor learning rule to make use of the information provided by the state vector. This is done by generalizing the action value vector $\mathbf{m}$ to a matrix $\mathbf{M}$, called an action matrix. $\mathbf{M}$ has as many columns as there are components of $\mathbf{u}$ and as many rows as there are actions. Given input $\mathbf{u}$, action a is chosen at location u with the softmax probability of equation 9.12, but using component a of the action value vector,

action matrix $\mathbf{M}$

$$\mathbf{m} = \mathbf{M} \cdot \mathbf{u}(u) \quad \text{or} \quad m_a = \sum_b M_{ab} u_b(u)\,. \tag{9.27}$$

In this case, the learning rule 9.25 must be generalized to specify how to change elements of the action matrix when action a is chosen at location u with state vector $\mathbf{u}(u)$, leading to location u'. A rule similar to equation 9.25 is appropriate, except that the change in $\mathbf{M}$ depends on the state vector $\mathbf{u}$,

three-term covariance rule

$$M_{a'b} \to M_{a'b} + \epsilon \left(\delta_{aa'} - P[a'; u] \right) \delta u_b(u) \tag{9.28}$$

for all a', with δ given again as in equation 9.24. This is called a three-term covariance learning rule.

We can speculate about the biophysical significance of the three-term covariance rule by interpreting $\delta_{aa'}$ as the output of cell a' when action a is chosen (which has mean value $P[a'; u]$) and interpreting $\mathbf{u}$ as the input to that cell. Compared with the Hebbian covariance rules studied in chapter 8, learning is gated by a third term, the reinforcement signal δ. It has been suggested that the dorsal striatum, which is part of the basal ganglia, is involved in the selection and sequencing of actions. Terminals of axons projecting from the substantia nigra pars compacta release dopamine onto synapses within the striatum, suggesting that they might play such a gating role. The activity of these dopamine neurons is similar to that of the VTA neurons discussed previously as a possible substrate for δ.

dorsal striatum
basal ganglia

The second generalization is to the case that rewards and punishments received soon after an action are more important than rewards and punishments received later. One natural way to accommodate this is a technique called exponential discounting. In computing the expected future reward, this amounts to multiplying a reward that will be received τ time steps after a given action by a factor γ^τ, where $0 \le \gamma \le 1$ is the discounting factor. The smaller γ, the stronger the effect (i.e., the less important temporally

discounting

distant rewards are). Discounting has a major influence on the optimal behavior in problems for which there are many steps to a goal. Exponential
discounting can be accommodated within the temporal difference framework by changing the prediction error δ to

$$\delta = r_a(u) + \gamma v(u') - v(u), \tag{9.29}$$

which is then used in the learning rules of equations 9.26 and 9.28.

In computing the amount to change a weight or action value, we defined
the worth of an action as the sum of the immediate reward delivered and
the estimate of the future reward arising from the next state. A final generalization of actor-critic learning comes from basing the learning rules on
the sum of the next two, three, or more immediate rewards delivered and
basing the estimate of the future reward on more temporally distant times
within a trial. As in discounting, we can use a factor λ to weight how
strongly the expected future rewards from temporally distant points affect
learning. Suppose that $\mathbf{u}(t) = \mathbf{u}(u(t))$ is the state vector used at time step
t of a trial. Such generalized temporal difference learning can be achieved
by computing new state vectors, defined by the recursive relation

$$\tilde{\mathbf{u}}(t) = \tilde{\mathbf{u}}(t-1) + (1 - \lambda)(\mathbf{u}(t) - \tilde{\mathbf{u}}(t-1)), \tag{9.30}$$

stimulus traces

TD(λ) rule

and using them instead of the original state vectors $\mathbf{u}$ in equations 9.26
and 9.28. These new state vectors $\tilde{\mathbf{u}}(t)$ are called stimulus traces, and the
resulting learning rule is called the TD(λ) rule. Use of this rule with an
appropriate value of λ can significantly speed up learning.

Learning the Water Maze

As an example of generalized reinforcement learning, we consider the water maze task. This is a navigation problem in which rats are placed in a
large pool of milky water and have to swim around until they find a small
platform that is submerged slightly below the surface of the water. The
opaqueness of the water prevents them from seeing the platform directly,
and their natural aversion to water (although they are competent swim-
mers) motivates them to find the platform. After several trials, the rats
learn the location of the platform and swim directly to it when placed in
the water.

Figure 9.10A shows the structure of the model, with the state vector $\mathbf{u}$
providing input to the critic and a collection of eight possible actions for
the actor, which are expressed as compass directions. The components of
$\mathbf{u}$ represent the activity of hippocampal place cells (which are discussed in
chapters 1 and 8). Figure 9.10B shows the activation of one of the input
units as a function of spatial position in the pool. The activity, like that of
a place cell, is spatially restricted.

During training, each trial consists of starting the model rat from a random location at the outside of the maze and letting it run until it finds

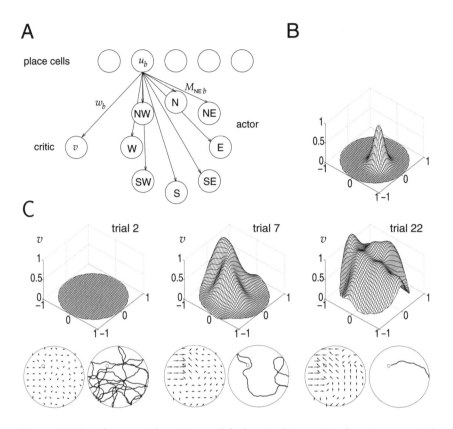

Figure 9.10 Reinforcement learning model of a rat solving a simple water maze task in a 2 m diameter circular pool. (A) There are 493 place cell inputs and 8 actions. The rat moves at 0.3 m/s and reflects off the walls of the maze if it hits them. (B) Gaussian place field for a single input cell with width 0.16 m. The centers of the place fields for different cells are uniformly distributed across the pool. (C) Upper: The development of the value function v as a function of the location in the pool over the first 22 trials, starting from $v = 0$ everywhere . Lower arrow plots: The action with the highest probability for each location in the maze. Lower path plots: Actual paths taken by the model rat from random starting points to the platform, indicated by a small circle. A slight modification of the actor learning rule was used to enforce generalization between spatially similar actions. (Adapted from Foster et al., 2000.)

the platform indicated by a small circle in the lower part of figure 9.10C. At that point a reward of 1 is provided. The reward is discounted with $\gamma = 0.9975$ to model the incentive for the rat to find the goal as quickly as possible. Figure 9.10C indicates the course of learning (trials 2, 7, and 22) of the expected future reward as a function of location (upper figures) and the policy (lower figures with arrows). The lower figures also show sample paths taken by the rat (lower figures with wiggly lines). The final value function (at trial 22) is rather inaccurate, but nevertheless the policy learned is broadly correct, and the paths to the platform are quite short and direct.

Judged by measures such as path length, initial learning proceeds in the

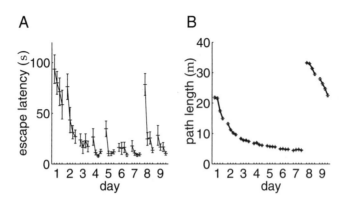

Figure 9.11 Comparison of rats and the model in the water maze task. (A) Average latencies of 12 rats in getting to a fixed platform in the water maze, using four trials per day. On the eighth day, the platform was moved to a new location, which is called reversal. (B) Average path length from 1000 simulations of the model performing the same task. Initial learning matches that of the rats, but performance is worse following reversal. (Adapted from Foster et al., 2000.)

model in a manner comparable to that of actual rats. Figure 9.11A shows the average performance of 12 real rats performing four trials per day in the water maze. The rats swam to a platform at a fixed location, starting from randomly chosen initial locations. The performance of the rats rapidly improves and levels off by about the sixth day. When the platform is moved on the eighth day, in what is called reversal training, the initial latency is long, because the rats search near the old platform position. However, they rapidly learn the new location. Figure 9.11B shows the performance of the model on the same task (though judged by path lengths rather than latencies). Initial learning is equally quick, with near perfect paths by the sixth day. However, performance during reversal training is poor, because the model has trouble forgetting the previous location of the platform. The rats are clearly better at handling this transition. Nevertheless the model shows something of the power of a primitive, but general, learning method.

9.5 Chapter Summary

We discussed reinforcement learning models for classical and instrumental conditioning, interpreting the former in terms of learning predictions about total future rewards and the latter in terms of optimization of those rewards. We introduced the Rescorla-Wagner or delta learning rule for classical conditioning, together with its temporal difference extension, and indirect and direct actor rules for instrumental conditioning given immediate rewards. Finally, we presented the actor-critic version of the dynamic programming technique of policy iteration, evaluating policies using temporal difference learning and improving them using the direct actor learning rule, based on surrogate immediate rewards from the evaluation step.

In the appendix, we show more precisely how temporal difference learning can be seen as a Monte Carlo technique for performing policy iteration.

9.6 Appendix

Markov Decision Problems

Markov decision problems offer a simple formalism for describing tasks such as the maze. A Markov decision problem is comprised of states, actions, transitions, and rewards. The states, labeled by u, are what we called locations in the maze task, and the actions, labeled by a, are the analogs of the choices of which direction to run. In the maze, each action taken at state u led uniquely and deterministically to a new state u'. Markov decision problems generalize this to include the possibility that the transitions from u due to action a may be stochastic, leading to state u' with a transition probability $P[u'|u;a]$. $\sum_{u'} P[u'|u;a] = 1$ for all u and a, because the animal has to end up somewhere. There can be absorbing states (like the shaded boxes in figure 9.7), which are u for which $P[u|u;a] = 1$ for all actions a (i.e., there is no escape for the animal from these locations). Finally, the rewards r can depend both on the state u and the action executed a, and they might be stochastic. We write $\langle r_a(u) \rangle$ for the mean reward in this case. The crucial Markov property is that, given the state at the current time step, the distribution over future states and rewards is independent of the past states. This defines the state sequence as the output of a controlled Markov chain. For convenience, we consider only Markov decision problems that are finite (finite numbers of actions and states) and absorbing (the animal always ends up in one of the absorbing states), and for which the rewards are bounded. We also require that $\langle r_a(u) \rangle = 0$ for all actions a at all absorbing states.

absorbing state

Markov property

The Bellman Equation

The task for a system or animal facing a Markov decision problem, starting in state u at time 0, is to choose a policy, denoted by $\mathbf{M}$, that maximizes the expected total future reward,

$$v^*(u) = \max_{\mathbf{M}} \left\langle \sum_{t=0}^{\infty} r_{a(t)}(u(t)) \right\rangle_{u,\mathbf{M}}, \tag{9.31}$$

where $u(0) = u$, actions $a(t)$ are determined (either deterministically or stochastically) on the basis of the state $u(t)$ according to policy $\mathbf{M}$, and the notation $\langle\rangle_{u,\mathbf{M}}$ implies taking an expectation over the actions and the states to which they lead, starting at state u and using policy $\mathbf{M}$.

The trouble with the sum in equation 9.31 is that the action $a(0)$ at time 0 affects not only $\langle r_{a(0)}(u(0)) \rangle$, but, by influencing the state of the system, also the subsequent rewards. It would seem that the animal would have to consider optimizing whole sequences of actions, the number of which grows exponentially with time. Bellman's (1957) insight was that the Markov property effectively solves this problem. He rewrote equation 9.31 to separate the first and subsequent terms, and used a recursive principle for the latter. The Bellman equation is

$$v^*(u) = \max_a \left\{ \langle r_a(u) \rangle + \sum_{u'} P[u'|u; a] v^*(u') \right\}.$$
(9.32)

This says that maximizing reward at u requires choosing the action a that maximizes the sum of the mean immediate reward $\langle r_a(u) \rangle$ and the average of the largest possible values of all the states u' to which a can lead the system, weighted by their probabilities.

Policy Iteration

The actor-critic algorithm is a form of a dynamic programming technique called policy iteration. Policy iteration involves interleaved steps of policy evaluation (updating the critic) and policy improvement (updating the actor). Evaluation of policy **M** requires working out the values for all states u. We call these values $v^M(u)$, to reflect explicitly their dependence on the policy. Each value is analogous to the quantity in 9.5. Using the same argument that led to the Bellman equation, we can derive the recursive formula

$$v^M(u) = \sum_a P_M[a; u] \left\{ \langle r_a(u) \rangle + \sum_{u'} P[u'|u; a] v^M(u') \right\}.$$
(9.33)

Equation 9.33, for all states u, is a set of linear equations for $v^M(u)$ that can be solved by matrix inversion. Reinforcement learning has been interpreted as a stochastic Monte Carlo method for performing this operation.

Monte Carlo method

Temporal difference learning uses an approximate Monte Carlo method to evaluate the right side of equation 9.33, and uses the difference between this approximation and the estimate of $v^M(u)$ as the prediction error. The first idea underlying the method is that $r_a(u) + v^M(u')$ is a sample whose mean is exactly the right side of equation 9.33. The second idea is bootstrapping, using the current estimate $v(u')$ in place of $v^M(u')$ in this sample. Thus $r_a(u) + v(u')$ is used as a sampled approximation to $v^M(u)$, and

$$\delta(t) = r_a(u) + v(u') - v(u)$$
(9.34)

is used as a sampled approximation to the discrepancy $v^M(u) - v(u)$, which is an appropriate error measure for training $v(u)$ to equal $v^M(u)$. Evaluating and improving policies from such samples without learning $P[u'|u; a]$ and $\langle r_a(u) \rangle$ directly is called an asynchronous, model-free approach to policy evaluation. It is possible to guarantee the convergence of the estimate v to its true value v^M under a set of conditions discussed in the texts mentioned in the annotated bibliography.

The other half of policy iteration is policy improvement. This normally works by finding an action a^* that maximizes the expression in the curly brackets in equation 9.33 and making the new $P_M[a^*; u]=1$. One can show that the new policy will be uniformly better than the old policy, making the expected long-term reward at every state no smaller than the old policy, or equally large, if it is already optimal. Further, because the number of different policies is finite, policy iteration is bound to converge.

Performing policy improvement like this requires knowledge of the transition probabilities and mean rewards. Reinforcement learning again uses an asynchronous, model-free approach to policy improvement, using Monte Carlo samples. First, note that any policy M' that improves the average value

$$\sum_a P_{M'}[u; a]\left\{\langle r_a(u)\rangle + \sum_{u'} P[u'|u; a]v^M(u')\right\} \qquad (9.35)$$

for every state u is guaranteed to be a better policy. The idea for a single state u is to treat equation 9.35 rather like equation 9.15, except replacing the average immediate reward $\langle r_a \rangle$ there by an effective average immediate reward $\langle r_a(u)\rangle + \sum_{u'} P[u'|u; a]v^M(u')$ to take long-term as well as current rewards into account. By the same reasoning as above, $r_a(u) + v(u')$ is used as an approximate Monte Carlo sample of the effective immediate reward, and $v(u)$ as the equivalent of the reinforcement comparison term $\bar{r}$. This leads directly to the actor learning rule of equation 9.25.

Note that there is an interaction between the stochasticity in the reinforcement learning versions of policy evaluation and policy improvement. This means that it is not known whether the two together are guaranteed to converge. One could perform temporal difference policy evaluation (which can be proven to converge) until convergence before attempting policy improvement, and this would be sure to work.

9.7 Annotated Bibliography

Dickinson (1980), **Mackintosh (1983)**, and **Shanks (1995)** review animal and human conditioning behavior, including alternatives to Rescorla & Wagner's (1972) rule. **Gallistel (1990)** and Gallistel & Gibbon (2000) discuss aspects of conditioning, in particular to do with timing, that we have omitted.

Our description of the temporal difference model of classical conditioning in this chapter is based on **Sutton (1988)** and **Sutton & Barto (1990)**. The treatment of static action choice comes from **Narendra & Thatachar (1989)** and **Williams (1992)**, and that of action choice with delayed rewards and the link to dynamic programming, from Barto, Sutton & Anderson (1983), Watkins (1989), Barto,et al. (1990), **Bertsekas & Tsitsiklis (1996)**, and **Sutton & Barto (1998)**. **Bertsekas & Tsitsiklis (1996)** and **Sutton & Barto (1998)** describe some of the substantial theory of temporal difference learning that has been developed. Dynamic programming as a computational tool of ethology is elucidated in Mangel & Clark (1988).

Schultz (1998) reviews the data on the activity of primate dopamine cells during appetitive conditioning tasks, together with the psychological and pharmacological rationale for studying these cells. The link with temporal difference learning was made in Montague, Dayan & Sejnowski (1996), Friston et al. (1994), and Houk et al. (1995). **Houk et al. (1995)** reviews the basal ganglia from a variety of perspectives. Wickens (1993) provides a theoretically motivated treatment. The model in Montague et al. (1995) for Real's (1991) experiments in bumblebee foraging was based on Hammer's (1993) description of an octopaminergic neuron in honeybees that appears to play, for olfactory conditioning, a somewhat similar role to the primate dopaminergic cells.

The kernel representation of the weight between a stimulus and reward can be seen as a form of a serial compound stimulus (Kehoe, 1977) or a spectral timing model (Grossberg & Schmajuk, 1989). Grossberg and colleagues (see Grossberg, 1982, 1987, 1988) have developed a sophisticated mathematical model of conditioning, including aspects of opponent processing (Konorski, 1967; Solomon & Corbit, 1974), which puts prediction of the absence of reward (or the presence of punishment) on a more equal footing with prediction of the presence of reward, and develops aspects of how animals pay differing amounts of attention to stimuli. There are many other biologically inspired models of conditioning, particularly of the cerebellum (e.g., Gluck et al., 1990; **Gabriel & Moore, 1990**; Raymond et al., 1996; Mauk & Donegan, 1997).

10 Representational Learning

10.1 Introduction

The response selectivities of individual neurons, and the way they are distributed across neuronal populations, define how sensory information is represented by neural activity. Sensory information is typically represented in multiple brain regions, the visual system being a prime example, with the nature of the representation shifting progressively along the sensory pathway. In previous chapters, we discussed how such representations can be generated by neural circuitry and developed by activity-dependent plasticity. In this chapter, we study neural representations from a computational perspective, asking what goals are served by particular representations, and how appropriate representations might be developed on the basis of input statistics.

Constructing new representations of, or re-representing, sensory input is important because sensory receptors often deliver information in a form that is unsuitable for higher-level cognitive tasks. For example, roughly 10^8 photoreceptors provide a pixelated description of the images that appear on our retinas. A list of the membrane potentials of each of these photoreceptors provides a bulky and awkward representation of the visual world, from which it would be difficult to identify directly the face of a friend or a familiar object. Instead, the information provided by photoreceptor outputs is processed in a series of stages involving increasingly sophisticated representations of the visual world. In this chapter, we consider how these more complex and useful representations can be constructed.

re-representation

The key to building useful representations for vision lies in determining the structure of visual images and capturing the constraints imposed on images by the natural world. The set of possible pixelated activities arising from natural scenes is richly structured and highly constrained, because images are not generated randomly, but arise from well-defined objects, such as rocks, trees, and people. We call these objects the "causes" of the images. In representational learning, we seek to identify causes by analyzing the statistical structure of visual images and building a model, called the generative model, that is able to reproduce this structure. Iden-

generative model

recognition model   tification of the causes of particular images (e.g., object recognition) is performed by a second model, called the recognition model, that is constructed on the basis of the generative model. This procedure is analogous to analyzing an experiment by building a model of the processes thought to underlie it, and using the model as a basis for extracting interesting features from the accumulated experimental data.

We follow the convention of previous chapters and use the variables **u** and v to represent the input and output of the models we consider. The input

input vector **u** vector **u** represents the data that we wish to analyze in terms of underlying
cause v causes. The output v, which is a variable that characterizes those causes, is
hidden or latent sometimes called the hidden or latent variable, but we call it the "cause".
variable In some models, causes may be associated with a vector **v**, rather than a single number v.

recognition   In terms of these variables, the ultimate goal of the models we consider is recognition, in which the model tells us something about the causes v

deterministic underlying a particular input **u**. Recognition can be either deterministic or
recognition probabilistic. In a model with deterministic recognition, the output $v(\mathbf{u})$ is the model's estimate of the cause underlying input **u**. In probabilistic

probabilistic recognition, the model estimates the probability that different values of
recognition v are associated with input **u**. In either case, the output is taken as the model's re-representation of the input.

We consider models that infer causes in an unsupervised manner. This means that the existence and identity of any underlying causes must be deduced solely from two sources of information. One is the set of general assumptions and guesses, collectively known as heuristics, concerning the

heuristics nature of the input data and the causes that might underly them. These heuristics determined the general form of the generative model. The other source of information is the statistical structure of the input data. In the absence of supervisory information or even reinforcement, causes are judged by their ability to explain and reproduce the statistical structure of the inputs they are designed to represent. The process is analogous to judging the validity of a model by comparing simulated data generated by it with the results of a real experiment. The basic idea is to use assumed causes to generate synthetic input data from a generative model. The statistical structure of the synthetic data is then compared with that of the real input data, and the parameters of the generative model are adjusted until the two are as similar as possible. If the final statistical match is good, the causes are judged trustworthy, and the model can be used as a basis for recognition.

Representational learning is a large and complex subject with a terminology and methodology that may be unfamiliar to many readers. Section 10.1 follows two illustrative examples to provide a general introduction. This should give the reader a basic idea of what representational learning attempts to achieve and how it works. Section 10.2 covers more technical aspects of the approach, and 10.3 surveys a number of examples.

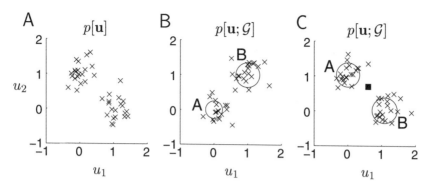

Figure 10.1 Clustering. (A) Input data points drawn from the distribution $p[\mathbf{u}]$ are indicated by the crosses. (B) Initialization for a generative model. The means and twice the standard deviations of the two Gaussians are indicated by the locations and radii of the circles. The crosses show synthetic data, which are samples from the distribution $p[\mathbf{u};\mathcal{G}]$ of the generative model. (C) Means, standard deviations, and synthetic data points generated by the optimal generative model. The square indicates a new input point that is assigned to either cluster A or cluster B with probabilities computed from the recognition model.

Causal Models

Figure 10.1A provides a simple example of structured data that suggests underlying causes. In this case, the input takes the form of a two-component vector, $\mathbf{u} = (u_1, u_2)$. A collection of sample inputs that we wish to represent in terms of underlying causes is indicated by the 40 crosses in figure 10.1A. These inputs are drawn from a probability density $p[\mathbf{u}]$ that we call the input distribution. Clearly, there are two clusters of points in figure 10.1A, one centered near $(0, 1)$ and the other near $(1, 0)$.

input distribution $p[\mathbf{u}]$

Many processes can generate clustered data. For example, u_1 and u_2 might be characterizations of the voltage recorded on an extracellular electrode in response to an action potential. Interpreted in this way, these data suggest that we are looking at spikes produced by two neurons (called A and B), which are the underlying causes of the two clusters seen in figure 10.1A. A more compact and causal description of the data can be provided by a single output variable v that takes the value A or B for each data point, representing which of the two neurons was responsible for a particular action potential. Directly reading the output of such a model would be an example of deterministic recognition, with $v(\mathbf{u}) = $ A or B providing the model's estimate of which neuron produced the spike associated with input $\mathbf{u}$. We consider, instead, a model with probabilistic recognition that estimates the probability that the spike with input data $\mathbf{u}$ was generated by either neuron A or neuron B.

In this example, we assume from the start that there are two possible, mutually exclusive causes for the data points, the two neurons A and B. By making this assumption, which is part of the heuristics underlying the generative model, we avoid the problem of identifying the number of possible causes (i.e., the number of clusters). Probabilistic methods can be used to make statistical inferences about the number of clusters in a data set, but they lie beyond the scope of this text.

Generative Models

To illustrate the concept of a generative model, we construct one (called a mixture of Gaussians model) for the data in figure 10.1A. The general form of the model is determined by the heuristics, assumptions about the nature of the causes and the way they generate inputs. However, the model has parameters that can be adjusted to fit the actual data that are observed. We begin by introducing parameters γ_A and γ_B that represent the proportions

mixing proportions (also known as mixing proportions) of action potentials generated by each of the neurons. These might account for the fact that one of the neurons has a higher firing rate than the other, for example. The parameter γ_v, with $v = A$ or B, specifies the probability $P[v; G]$ that a given spike was generated by neuron v in the absence of any knowledge about the input $\mathbf{u}$

prior $P[v; G]$ associated with that spike. $P[v; G] = \gamma_v$ is called the prior distribution over causes. The symbol G stands for all the parameters used to characterize the generative model. At this point, G consists of the two parameters γ_A

parameters G and γ_B, but more parameters will be added as we proceed. We start by assigning γ_A and γ_B random values consistent with the constraint that they must sum to 1.

To continue the construction of the generative model, we need to assume something about the distribution of $\mathbf{u}$ values arising from the action potentials generated by each neuron. An examination of figure 10.1A suggests that Gaussian distributions (with the same variance in both dimensions) might be appropriate. The probability density of $\mathbf{u}$ values given

generative that neuron v fired is $p[\mathbf{u}|v; G]$. This is set to a Gaussian distribution with
distribution a mean and variance that, initially, we guess. The parameter list G now
$p[\mathbf{u}|v; G]$ contains the prior probabilities, γ_A and γ_B, and the means and variances of the Gaussian distributions over $\mathbf{u}$ for $v = A$ and B, which we label $\mathbf{g}_v$ and Σ_v, respectively. Note that Σ_v is used to denote the variance of cluster v, not its standard deviation, and also that each cluster is characterized by a single variance because we consider only circularly symmetric Gaussian distributions.

Figure 10.1B shows synthetic data points (crosses) generated by this model. To create each point, we set $v = A$ with probability $P[v = A; G]$ (or otherwise set $v = B$) and then generated a point $\mathbf{u}$ randomly from the distribution $p[\mathbf{u}|v; G]$. This generative model clearly has the capacity to create a data distribution with two clusters, similar to the one in figure 10.1A. However, the values of the parameters G used in figure 10.1B are obviously inappropriate. They must be adjusted by a learning procedure that matches, as accurately as possible, the distribution of synthetic data points in figure 10.1B to the actual input distribution in figure 10.1A. We describe how this is done in a later section. After optimization, as seen in figure 10.1C, synthetic data points generated by the model (crosses) overlap well with the actual data points seen in figure 10.1A.

The distribution of synthetic data points in figures 10.1B and 10.1C is described by the probability density $p[\mathbf{u}; G]$ that the generative model synthesizes an input with the value $\mathbf{u}$. This can be computed from the conditional density $p[\mathbf{u}|v; G]$ and the prior distribution $P[v; G]$ that define the generative model,

$$p[\mathbf{u}; G] = \sum_v p[\mathbf{u}|v; G]P[v; G].\tag{10.1}$$

marginal distribution $p[\mathbf{u}; G]$

The process of summing over all causes is called marginalization, and $p[\mathbf{u}; G]$ is called the marginal distribution over $\mathbf{u}$. As in chapter 8, we use the additional argument G to distinguish the distribution of synthetic inputs produced by the generative model, $p[\mathbf{u}; G]$, from the distribution of actual inputs, $p[\mathbf{u}]$. The process of adjusting the parameters G to make the distributions of synthetic and real input data points match corresponds to making the marginal distribution $p[\mathbf{u}; G]$ approximate, as closely as possible, the distribution $p[\mathbf{u}]$ from which the input data points are drawn.

In a later section, we make use of an addition probability distribution associated with the generative model, the joint probability distribution over both causes and inputs, define by

$$p[v, \mathbf{u}; G] = p[\mathbf{u}|v; G]P[v; G].\tag{10.2}$$

joint distribution $p[v, \mathbf{u}; G]$

This describes the probability of cause v and input $\mathbf{u}$ both being produced by the generative model.

As mentioned previously, the choice of a particular structure for a generative model reflects our notions and prejudices (i.e., our heuristics) concerning the nature of the causes that underlie input data. Usually, the heuristics consist of biases toward certain types of representations, which are imposed through the choice of the prior distribution $p[v; G]$. For example, we may want the identified causes to be mutually independent (which leads to a factorial representation or code) or sparse, or of lower dimension than the input data. Many heuristics can be formalized using the information theoretic ideas we discuss in chapter 4.

factorial coding

sparse coding

dimensionality reduction

Recognition Models

Once the optimal generative model has been constructed, the culmination of representational learning is recognition, in which new input data are interpreted in terms of the causes established by the generative model. In probabilistic recognition models, this amounts to determining the probability that cause v is associated with input $\mathbf{u}$, $P[v|\mathbf{u}; G]$, which is called the posterior distribution over causes or the recognition distribution.

In the model of figure 10.1, and in many of the models discussed in this chapter, recognition falls directly out of the generative model. The probability of cause v, given input $\mathbf{u}$, $P[v|\mathbf{u}; G]$, is the statistical inverse of the

recognition distribution $P[v|\mathbf{u}; G]$

distribution $p[\mathbf{u}|v; \mathcal{G}]$ that defines the generative model. Using Bayes theorem, it can be expressed in terms of the distributions that define the generative model as

$$P[v|\mathbf{u}; \mathcal{G}] = \frac{p[\mathbf{u}|v; \mathcal{G}]P[v; \mathcal{G}]}{p[\mathbf{u}; \mathcal{G}]}. \tag{10.3}$$

Once the recognition distribution has been computed from this equation, the probability of various causes being associated with a given input can be determined. For instance, in the example of figure 10.1, equation 10.3 can be used to determine that the point indicated by the filled square in figure 10.1C has probability $P[v=\text{A}|\mathbf{u}; \mathcal{G}] = 0.8$ of being associated with neuron A and $P[v=\text{B}|\mathbf{u}; \mathcal{G}] = 0.2$ of being associated with neuron B.

Recall that constructing a generative model involves making a number of assumptions about the nature of the causes underlying a set of inputs. The recognition model provides a mechanism for checking the self-consistency of these assumptions. This is done by examining the distribution of causes produced by the recognition model in response to actual data. This distribution should match the prior distribution over causes, and thus share its desired properties, such as mutual independence. If the prior distribution of the generative model does not match the actual distribution of causes produced by the recognition model, this is an indication that the imposed heuristic does not apply accurately to the input data.

Expectation Maximization

EM

During our discussion of generative models, we skipped over the process by which the parameters $\mathcal{G}$ are refined to optimize the match between synthetic and real input data. There are various ways of doing this. In this chapter (except for one case), we use an approach called expectation maximization (EM). The general theory of EM is discussed in detail in the following section but, as an introduction to the method, we apply it here to the example of figure 10.1. Recall that the problem of optimizing the generative model in this case involves adjusting the mixing proportions, means, and variances of the two Gaussian distributions until the clusters of synthetic data points in figure 10.1B and 10.1C match the clusters of actual data points in figure 10.1A.

To optimize the match between synthetic and real input data, the parameters $\mathbf{g}_v$ and Σ_v, for $v=\text{A}$ and B, of the Gaussian distributions of the generative model should equal the means and variances of the data points associated with each cluster in figure 10.1A. If we knew which cluster each input point belonged to, it would be a simple matter to compute these means and variances and construct the optimal generative model. Similarly, we could set γ_v, the prior probability of a given spike being a member of cluster v, equal to the fraction of data points assigned to that cluster. Of course, we do not know the cluster assignments of the input points; that would amount to knowing the answer to the recognition problem. However, we can make an informed guess about which point belongs to which

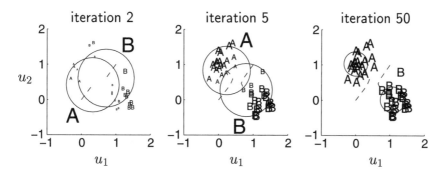

Figure 10.2 EM for clustering. Three stages during the course of EM learning of a generative model. The circles show the Gaussian distributions for clusters A and B (labeled with the largest A and B) as in figure 10.1B and 10.1C. The "trails" behind the centers of the circles plot the change in the mean since the last iteration. The data from figure 10.1A are plotted using the small labels. Label A is used if $P[v=A|\mathbf{u}; \mathcal{G}] > 0.5$ (and otherwise label B), with the font size proportional to $|P[v=A|\mathbf{u}; \mathcal{G}] - 0.5|$. This makes the fonts small in regions where the two distributions overlap, even inside one of the circles. The assignment of labels for the two Gaussians (i.e., which is A and which is B) depends on initial conditions.

cluster using the recognition distribution computed from equation 10.3. In other words, the recognition distribution $P[v|\mathbf{u}; \mathcal{G}]$ provides us with our best current guess about the cluster assignment, and this can be used in place of the actual knowledge about which neuron produced which spike. $P[v|\mathbf{u}; \mathcal{G}]$ is thus used to assign the data point $\mathbf{u}$ to cluster v in a probabilistic manner. In this context, $P[v|\mathbf{u}; \mathcal{G}]$ is also called the responsibility of v for $\mathbf{u}$. *responsibility*

Following this reasoning, the mean and variance of the Gaussian distribution corresponding to cause v are set equal to a weighted mean and variance of all the data points, with the weight for point $\mathbf{u}$ equal to the current estimate $P[v|\mathbf{u}; \mathcal{G}]$ of the probability that it belongs to cluster v. A similar argument is applied to the mixing proportions, resulting in the equations

$$\gamma_v = \langle P[v|\mathbf{u}; \mathcal{G}] \rangle, \quad \mathbf{g}_v = \frac{\langle P[v|\mathbf{u}; \mathcal{G}]\mathbf{u} \rangle}{\gamma_v}, \quad \Sigma_v = \frac{\langle P[v|\mathbf{u}; \mathcal{G}]|\mathbf{u} - \mathbf{g}_v|^2 \rangle}{2\gamma_v}.$$
(10.4)

The angle brackets indicate averages over all the input data points. The factors of γ_v dividing the last two expressions correct for the fact that the number of points in cluster v is expected to be γ_v times the total number of input data points, whereas the full averages denoted by the brackets involve dividing by the total number of data points.

The full EM algorithm consists of two phases that are applied in alternation. In the E (or expectation) phase, the responsibilities $P[v|\mathbf{u}; \mathcal{G}]$ are calculated from equation 10.3. In the M (or maximization) phase, the generative parameters $\mathcal{G}$ are modified according to equation 10.4. The process of determining the responsibilities and then averaging according to them repeats because the responsibilities change when $\mathcal{G}$ is modified. Figure 10.2 *E phase*

M phase

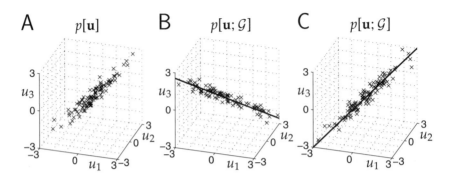

Figure 10.3 Factor analysis. (A) Input data points drawn from the distribution $p[\mathbf{u}]$ are indicated by the crosses. (B) The initial generative model. The solid line shows $\mathbf{g}$, and the crosses are synthetic data, which are samples from the generative distribution $p[\mathbf{u}; \mathcal{G}]$ (with $\Sigma_a = 0.0625$, for all a). (C) The line $\mathbf{g}$ and synthetic data points generated by the optimal generative model.

shows intermediate results at three different times during the running of the EM procedure, starting from the generative model in figure 10.1B and ending up with the fit shown in figure 10.1C.

Continuous Generative Models

The data of figure 10.1A consist of two separated clusters of points, so a cause v that takes only two different values is appropriate. Figure 10.3A shows data that suggest the need for a continuous variable v. We can think of these data as the outputs of three noisy sensors, each measuring the same quantity. In this case, the cause v represents the value of the quantity being measured, and recognition corresponds to extracting this value from the sensor outputs. Because v is a continuous variable, the prior and recognition distributions in this case are probability densities, $p[v; \mathcal{G}]$ and $p[v|\mathbf{u}; \mathcal{G}]$.

As in the case of clustering, the generative model is determined by the prior distribution $p[v; \mathcal{G}]$ and the generative distribution $p[\mathbf{u}|v; \mathcal{G}]$, where $\mathbf{u} = (u_1, u_2, u_3)$ represents the three sensor readings. A simple choice for the prior distribution over v is a Gaussian with mean 0 and variance 1. The generative distribution is designed to capture the fact that the data points in figure 10.3A lie along a line in the three-dimensional space. It is the product of three Gaussian functions, one for each of the sensors, with means $g_a v$ and variances Σ_a for $a = 1, 2, 3$. The vector $\mathbf{g} = (g_1, g_2, g_3)$ specifies the direction of the line along which synthetic data points produced by the generative model lie, and the variances determine how tightly the points hug this line in each input dimension. In the example of figure 10.3A, the sensors all measure the same quantity. Thus, from an arbitrary initial $\mathbf{g}$, the generative model must find the best fit, $\mathbf{g} \propto (1, 1, 1)$.

Figure 10.3B shows synthetic data points generated from the generative model, along with a solid line indicating the direction of **g**. As in figure 10.1B, although the generative model has the capacity to create a data distribution like that in figure 10.3A, the parameters underlying figure 10.3B are clearly inappropriate, and must be adjusted by a learning procedure. Figure 10.3C shows synthetic data after learning, indicating the close match between the marginal distribution $p[\mathbf{u}; G]$ from the model and the input distribution $p[\mathbf{u}]$.

This model is a simple case of factor analysis; the general case is discussed in section 10.3. The EM algorithm for factor analysis is similar in structure to that for clustering. As before, the basic idea is that if we knew the value of the cause that underlies each input point, we could find the parameters G easily. Here, the parameters would be determined by solving the linear regression problem that fits the observed inputs to the variable v. This mirrors the observation in our first example that if we knew the cluster assignment for each input point, we could easily find the optimal means and variances of the clusters. Of course, we do not know the values of the causes. Rather, as before, in the E phase of the EM algorithm, the distribution over causes $p[v|\mathbf{u}; G]$ is calculated from the continuous analog of equation 10.3 ($P[v; G]$ on the right side of equation 10.3 is replaced by $p[v; G]$), and this is used as our best current estimate of how likely cause v is associated with input **u**. Then the M phase consists of weighted linear regression, fitting the observations **u** to the variables v weighted by the current recognition probabilities. The result is analogous to equation 10.4; we set

$$g_a = \frac{\left\langle \int dv\, p[v|\mathbf{u}; G] v u_a \right\rangle}{\left\langle \int dv\, p[v|\mathbf{u}; G] v^2 \right\rangle} \quad \text{and} \quad \Sigma_a = \left\langle \int dv\, p[v|\mathbf{u}; G](u_a - v g_a)^2 \right\rangle .$$

$$(10.5)$$

Approximate Recognition

In the two examples we have considered, equation 10.3 was used to obtain the recognition distribution directly from the generative model. For some models, however, it is impractically difficult to evaluate the right side of equation 10.3 and obtain the recognition distribution in this way. We call models in which the recognition distribution can be computed tractably from equation 10.3 invertible, and those in which it cannot be computed tractably, noninvertible. In the latter case, because equation 10.3 cannot be used, recognition is based on an approximate recognition distribution. This is a function $Q[v; \mathbf{u}]$ that approximates the exact recognition distribution $P[v|\mathbf{u}; G]$. Often, as we discuss in the next section, the best approximation of the recognition distribution comes from adjusting parameters through an optimization procedure. Once this is done, $Q[v; \mathbf{u}]$ provides the model's estimate of the probability that input **u** is associated with cause v, and substitutes for the exact recognition distribution $P[v|\mathbf{u}; G]$.

invertible and noninvertible models

approximate recognition distribution $Q[v; \mathbf{u}]$

The E phase of the EM algorithm in a noninvertible model consists of making $Q[v; \mathbf{u}]$ approximate $P[v|\mathbf{u}; G]$ as accurately as possible, given the current parameters G. We can include invertible models within the same general formalism used to describe noninvertible models by noting that, in the E phase for an invertible model, we simple set $Q[v; \mathbf{u}] = P[v|\mathbf{u}; G]$ by solving equation 10.3.

Summary of Causal Models

In summary, causal models make use of the following probability distributions (for the case of continuous inputs and discrete causes).

- $p[\mathbf{u}]$, the input distribution

- $P[v; G]$, the prior distribution over causes

- $p[\mathbf{u}|v; G]$, the generative distribution

- $p[\mathbf{u}; G]$, the marginal distribution

- $P[v|\mathbf{u}; G]$, the recognition distribution

- $P[\mathbf{u}, v; G]$, the joint distribution over inputs and causes

- $Q[v; \mathbf{u}]$, the approximate recognition distribution.

The goal of generative modeling, which is implemented by successive M phases of the EM algorithm, is to make $p[\mathbf{u}; G] \approx p[\mathbf{u}]$ (as accurately as possible). This is done by using the marginal distribution obtained from prior E phases of EM and adjusting the parameters G to match it to the input distribution. The goal of each E phase is to make $Q[v; \mathbf{u}] \approx P[v|\mathbf{u}; G]$ (as accurately as possible) for the current values of the generative parameters G. Probabilistic recognition is carried out using the distribution $Q[v; \mathbf{u}]$ to determine the probability that cause v is responsible for input $\mathbf{u}$.

10.2 Density Estimation

density estimation

The process of matching the distribution $p[\mathbf{u}; G]$ produced by the generative model to the actual input distribution $p[\mathbf{u}]$ is a form of density estimation. This technique is discussed in chapter 8 in connection with the Boltzmann machine. As mentioned in the introduction, the parameters G of the generative model are fitted to the input data by minimizing the discrepancy between the probability density of the input data $p[\mathbf{u}]$ and the marginal probability density $p[\mathbf{u}; G]$ of equation 10.1. This discrepancy is measured using the Kullback-Leibler divergence (chapter 4),

$$D_{\mathrm{KL}}(p[\mathbf{u}], p[\mathbf{u}; G]) = \int d\mathbf{u} \, p[\mathbf{u}] \ln \frac{p[\mathbf{u}]}{p[\mathbf{u}; G]}$$

$$\approx -\langle \ln p[\mathbf{u}; G] \rangle + K, \tag{10.6}$$

where K is a term associated with the entropy of the distribution $p[\mathbf{u}]$, that is independent of G. In the second line, we have approximated the integral over all $\mathbf{u}$ values weighted by $p[\mathbf{u}]$ by the average over input data points generated from the distribution $p[\mathbf{u}]$. We assume there are sufficient input data to justify this approximation.

As in the case of the Boltzmann machine discussed in chapter 8, equation 10.6 implies that minimizing the discrepancy between $p[\mathbf{u}]$ and $p[\mathbf{u}; G]$ amounts to maximizing the log likelihood that the training data could have been created by the generative model,

log likelihood $L(G)$

$$L(G) = \langle \ln p[\mathbf{u}; G] \rangle . \tag{10.7}$$

$L(G)$ is the average log likelihood, and the method is known as maximum likelihood density estimation. A theorem due to Shannon describes circumstances under which the generative model that maximizes the likelihood over input data also provides the most efficient way of coding those data, so density estimation is closely related to optimal coding.

maximum likelihood density estimation

Theory of EM

Although stochastic gradient ascent can be used to adjust the parameters of the generative model to maximize the likelihood in equation 10.7 (as it was for the Boltzmann machine), the EM algorithm discussed in the introduction is an alternative procedure that is often more efficient. We applied this algorithm, on intuitive grounds, to the examples of figures 10.1 and 10.3, but we now present a more general and rigorous discussion. This is based on the connection of EM with maximization of the function

$\mathcal{F}(Q, G)$

$$\mathcal{F}(Q, G) = \left\langle \sum_v Q[v; \mathbf{u}] \ln \frac{p[v, \mathbf{u}; G]}{Q[v; \mathbf{u}]} \right\rangle , \tag{10.8}$$

where $Q[v; \mathbf{u}]$ is any nonnegative function of the discrete argument v and continuous input $\mathbf{u}$ that satisfies

$$\sum_v Q[v; \mathbf{u}] = 1 \tag{10.9}$$

for all $\mathbf{u}$. Although, in principle, $Q[v; \mathbf{u}]$ can be any function, we consider it to be an approximate recognition distribution, that is $Q[v; \mathbf{u}] \approx P[v|\mathbf{u}; G]$.

$\mathcal{F}$ is a useful quantity because, by a rearrangement of terms, it can be written as the difference of the average log likelihood and the average Kullback-Leibler divergence between $Q[v; \mathbf{u}]$ and $P[v|\mathbf{u}; G]$. This is done by noting that the joint distribution over inputs and causes satisfies $p[v, \mathbf{u}; G] = P[v|\mathbf{u}; G]p[\mathbf{u}; G]$, in addition to 10.2, and using 10.9 and the definition of the Kullback-Leibler divergence to obtain

$$\mathcal{F}(Q, \mathcal{G}) = \left\langle \sum_v Q[v; \mathbf{u}] \left(\ln p[\mathbf{u}; \mathcal{G}] + \ln \frac{P[v|\mathbf{u}; \mathcal{G}]}{Q[v; \mathbf{u}]} \right) \right\rangle$$

$$= \langle \ln p[\mathbf{u}; \mathcal{G}] \rangle - \left\langle \sum_v Q[v; \mathbf{u}] \left(\ln \frac{Q[v; \mathbf{u}]}{P[v|\mathbf{u}; \mathcal{G}]} \right) \right\rangle$$

$$= L(\mathcal{G}) - \langle D_{\mathrm{KL}}(Q[v; \mathbf{u}], P[v|\mathbf{u}; \mathcal{G}]) \rangle . \qquad (10.10)$$

Because the Kullback-Leibler divergence is never negative,

$$L(\mathcal{G}) \geq \mathcal{F}(Q, \mathcal{G}), \qquad (10.11)$$

and because $D_{\mathrm{KL}} = 0$ only if the two distributions being compared are identical, this inequality is saturated, becoming an equality, only if

$$Q[v; \mathbf{u}] = P[v|\mathbf{u}; \mathcal{G}]. \qquad (10.12)$$

free energy $-\mathcal{F}$ The negative of $\mathcal{F}$ is related to the free energy used in statistical physics.

Expressions 10.10, 10.11, and 10.12 are critical to the operation of EM. The two phases of EM are concerned with separately maximizing (or at least increasing) $\mathcal{F}$ with respect to one of its two arguments, keeping the other one fixed. When $\mathcal{F}$ increases, this increases a lower bound on the log likelihood of the input data (equation 10.11). In the M phase, $\mathcal{F}$ is increased with respect to $\mathcal{G}$, keeping Q constant. For the generative models considered as examples in the previous section, it is possible to maximize $\mathcal{F}$ with respect to $\mathcal{G}$ in a single step. For other generative models, this may require multiple steps that perform gradient ascent on $\mathcal{F}$. In the E phase, $\mathcal{F}$ is increased with respect to Q, keeping $\mathcal{G}$ constant. From equation 10.10, we see that increasing $\mathcal{F}$ by changing Q is equivalent to reducing the average Kullback-Leibler divergence between $Q[v; \mathbf{u}]$ and $P[v|\mathbf{u}; \mathcal{G}]$. This makes $Q[v; \mathbf{u}]$ a better approximation of $P[v|\mathbf{u}; \mathcal{G}]$. The E phase can proceed in at least three possible ways, depending on the nature of the generative model being considered. We discuss these separately.

One advantage of EM over likelihood maximization through gradient methods is that large steps toward the maximum can be taken during each M cycle of modification. Of course, the log likelihood may have multiple maxima, in which case neither gradient ascent nor EM is guaranteed to find the globally optimal solution.

Invertible Models

If the causal model being considered is invertible, the E step of EM simply consists of solving equation 10.3 for the recognition distribution, and setting Q equal to the resulting $P[v|\mathbf{u}; \mathcal{G}]$, as in equation 10.12. This maximizes $\mathcal{F}$ with respect to Q by setting the Kullback-Leibler term in equation 10.10 to 0, and it makes the function $\mathcal{F}$ equal to $L(\mathcal{G})$, the average log likelihood of the data points. However, the EM algorithm for maximizing $\mathcal{F}$ is not exactly the same as likelihood maximization by gradient ascent of $\mathcal{F}$. This is because the function Q is held constant during the M phase while the parameters of the generative model are modified. Although $\mathcal{F}$ is equal to L at the beginning of the M phase, exact equality ceases to be true as soon as the parameters are modified, making $P[v|\mathbf{u}; \mathcal{G}]$ different from Q. $\mathcal{F}$ is equal to $L(\mathcal{G})$ again only after the update of Q during the following E phase. At this point, $L(\mathcal{G})$ must have increased since the last E phase, because $\mathcal{F}$ has increased. This shows that the log likelihood increases monotonically during EM until the process converges.

For the example of figure 10.1, the joint probability over causes and inputs is

$$p[v, \mathbf{u}; \mathcal{G}] = \frac{\gamma_v}{2\pi\Sigma_v} \exp\left(-\frac{|\mathbf{u} - \mathbf{g}_v|^2}{2\Sigma_v}\right),\tag{10.13}$$

and thus

$$\mathcal{F} = \left\langle \sum_v Q[v; \mathbf{u}]\left(\ln\left(\frac{\gamma_v}{2\pi}\right) - \ln\Sigma_v - \frac{|\mathbf{u} - \mathbf{g}_v|^2}{2\Sigma_v} - \ln Q[v; \mathbf{u}]\right)\right\rangle.\tag{10.14}$$

The E phase amounts to computing $P[v|\mathbf{u}; \mathcal{G}]$ from equation 10.3 and setting Q equal to it, as in equation 10.12. The M phase involves maximizing $\mathcal{F}$ with respect to $\mathcal{G}$ for this Q. We leave it as an exercise for the reader to show that maximizing equation 10.14 with respect to the parameters γ_v (taking into account the constraint $\sum_v \gamma_v = 1$), $\mathbf{g}_v$, and Σ_v leads to the rules of equation 10.4. For the example of figure 10.3, the joint probability is

$$p[v, \mathbf{u}; \mathcal{G}] = \frac{\exp(-v^2/2)}{\sqrt{2\pi}} \frac{\exp(-\sum_a (u_a - g_a v)^2/2\Sigma_a)}{\sqrt{(2\pi)^3 \Sigma_1 \Sigma_2 \Sigma_3}},\tag{10.15}$$

from which it is straightforward to calculate the relevant $\mathcal{F}$ function and the associated learning rules of equation 10.5.

Noninvertible Deterministic Models

If the generative model is noninvertible, the E phase of the EM algorithm is more complex than simply setting Q equal to $P[v|\mathbf{u}; \mathcal{G}]$, because it is not practical to compute the recognition distribution exactly. The steps taken during the E phase depend on whether the approximation to the inverse of the model is deterministic or probabilistic, although the basic argument is the same in either case.

Deterministic recognition results in a prediction $v(\mathbf{u})$ of the cause underlying input $\mathbf{u}$. In terms of the function $\mathcal{F}$, this amounts to retaining only the single term $v = v(\mathbf{u})$ in the sum in equation 10.8, and for this single term $Q[v(\mathbf{u}); \mathbf{u}] = 1$. Thus, in this case $\mathcal{F}$ is a functional of the function $v(\mathbf{u})$ and a function of the parameters $\mathcal{G}$ given by

$$\mathcal{F}(Q, \mathcal{G}) = \mathcal{F}(v(\mathbf{u}), \mathcal{G}) = \langle \ln P[v(\mathbf{u}), \mathbf{u}; \mathcal{G}]\rangle.\tag{10.16}$$

The M phase of EM consists, as always, of maximizing this expression with respect to G. During the E phase we try to find the function $v(\mathbf{u})$ that maximizes $\mathcal{F}$. Because v is varied during the optimization procedure, the *variational method* approach is sometimes called a variational method. The E and M steps make intuitive sense; we are finding the input-output relationship that maximizes the probability that the generative model would have simultaneously produced the input $\mathbf{u}$ and cause $v(\mathbf{u})$.

Noninvertible Probabilistic Models

The alternative to using a deterministic approximate recognition model is to treat $Q[v; \mathbf{u}]$ as a full probability distribution over v for each input example $\mathbf{u}$. In this case, we choose a specific functional form for Q, expressed in terms of a set of parameters collectively labeled $\mathcal{W}$. Thus, we write the approximate recognition distribution as $Q[v; \mathbf{u}, \mathcal{W}]$. Like generative models, approximate recognition models can have different structures and parameters. $\mathcal{F}$ can now be treated as a function of $\mathcal{W}$, rather than of Q, so we write it as $\mathcal{F}(\mathcal{W}, G)$. As in all cases, the M phase of EM consists of maximizing $F(\mathcal{W}, G)$ with respect to G. The E phase now consists of maximizing $F(\mathcal{W}, G)$ with respect to $\mathcal{W}$. This has the effect of making $Q[v; \mathbf{u}, \mathcal{W}]$ as similar as possible to $P[v|\mathbf{u}; G]$, in the sense that the KL divergence between them, averaged over the input data, is minimized (see equation 10.10).

In some cases, $\mathcal{W}$ has separate parameters for each possible input $\mathbf{u}$. This means that each input has a separate approximate recognition distribution which is individually tailored, subject to the inherent simplifying assumptions, to its own causes. The mean-field approximation to the Boltzmann machine discussed in chapter 7 is an example of this type.

It is not necessary to maximize $F(\mathcal{W}, G)$ completely with respect to $\mathcal{W}$ and then G during successive E and M phases. Instead, gradient ascent steps that modify $\mathcal{W}$ and G by small amounts can be taken in alternation, in which case the E and M phases effectively overlap.

Because each E and M step separately increases the value of $\mathcal{F}$, the EM algorithm is guaranteed to converge to at least a local maximum of $\mathcal{F}$, except in rare cases when the process of maximizing a function one coordinate at a time (which is called coordinate ascent) finds local maxima that other optimization methods avoid (we encounter an example of this later in the chapter). In general, the maximum found does not correspond to a local maximum of the likelihood function because Q is not exactly equal to the actual recognition distribution (that is, $\mathcal{F}$ is guaranteed only to be a lower bound on $L(G)$). Nevertheless, a good generative model should be obtained if the lower bound is tight.

10.3   Causal Models for Density Estimation

In this section, we present a number of models in which representational learning is achieved through density estimation. The mixture of Gaussians and factor analysis models that we have already mentioned are examples of invertible generative models with probabilistic recognition. K-means is a limiting case of mixture of Gaussians with deterministic recognition, and principal components analysis is a limiting case of factor analysis with deterministic recognition. We consider two other models with deterministic recognition: independent components analysis, which is invertible; and sparse coding, which is noninvertible. Our final example, the Helmholtz machine, is noninvertible with probabilistic recognition. The Boltzmann machine, discussed in chapters 7 and 8, is an additional example that is closely related to the causal models discussed here. We summarize and interpret general properties of representations derived from causal models at the end of the chapter. The table in the appendix summarizes the generative and recognition distributions and the learning rules for all the models we discuss.

Mixture of Gaussians

The model applied in the introduction to the data in figure 10.1A is a mixture of Gaussians model. That example involved two causes and two Gaussian distributions, but we now generalize this to N_v causes, each associated with a separate Gaussian distribution. The model is defined by the probability distributions

$$P[v; \mathcal{G}] = \gamma_v \quad \text{and} \quad p[\mathbf{u}|v; \mathcal{G}] = \mathcal{N}(\mathbf{u}; \mathbf{g}_v, \Sigma_v), \qquad (10.17)$$

where v takes N_v values representing the different causes and, for an N_u component input vector,

$$\mathcal{N}(\mathbf{u}; \mathbf{g}, \Sigma) = \frac{1}{(2\pi\Sigma)^{N_u/2}} \exp\left(-\frac{|\mathbf{u} - \mathbf{g}|^2}{2\Sigma}\right) \qquad (10.18)$$

is a Gaussian distribution with mean $\mathbf{g}$ and variances for the individual components equal to Σ. The function $\mathcal{F}(Q, \mathcal{G})$ for this model is given by an expression similar to equation 10.14 (with slightly different factors if $N_u \neq 2$), leading to the M-phase learning rules given in the appendix. Once the generative model has been optimized, the recognition distribution is constructed from equation 10.3 as

$$P[v|\mathbf{u}; \mathcal{G}] = \frac{\gamma_v \mathcal{N}(\mathbf{u}; \mathbf{g}_v, \Sigma_v)}{\sum_{v'} \gamma_{v'} \mathcal{N}(\mathbf{u}; \mathbf{g}_{v'}, \Sigma_{v'})}. \qquad (10.19)$$

K-Means Algorithm

A special case of mixture of Gaussians can be derived in the limit that the variances of the Gaussians are equal and tend toward 0, $\Sigma_v = \Sigma \to 0$. We discuss this limit for two clusters, as in figure 10.1. When Σ is extremely small, the recognition distribution $P[v|\mathbf{u}; \mathcal{G}]$ of equation 10.19 degenerates because it takes essentially two values, 0 or 1, depending on whether $\mathbf{u}$ is closer to one cluster or the other. This provides a deterministic, rather than a probabilistic, classification of $\mathbf{u}$. In the degenerate case, EM consists of choosing two random values for the centers of the two cluster distributions, and then repeatedly finding all the inputs $\mathbf{u}$ that are closest to a given center $\mathbf{g}_v$, and then moving $\mathbf{g}_v$ to the average of these points. This is called the *K-means* algorithm (with $K = 2$ for two clusters). The mixing proportions γ_v do not play an important role for the *K*-means algorithm. New input points are recognized as belonging to the clusters to which they are closest.

Factor Analysis

The model used in figure 10.3 is an example of factor analysis. In general, factor analysis involves a continuous vector of causes, $\mathbf{v}$, drawn from a Gaussian prior distribution, and uses a Gaussian generative distribution with a mean that depends linearly on $\mathbf{v}$. We assume that the distribution $p[\mathbf{u}]$ has a mean of 0 (nonzero means can be accommodated simply by shifting the input data). The defining distributions for factor analysis are thus

$$p[\mathbf{v}; \mathcal{G}] = \mathcal{N}(\mathbf{v}; \mathbf{0}, 1) \quad \text{and} \quad p[\mathbf{u}|\mathbf{v}; \mathcal{G}] = \mathcal{N}(\mathbf{u}; \mathbf{G} \cdot \mathbf{v}, \Sigma), \qquad (10.20)$$

where the extension of equation 10.18, expressed in terms of the mean $\mathbf{g}$ and covariance matrix Σ, is

$$\mathcal{N}(\mathbf{u}; \mathbf{g}, \Sigma) = \frac{1}{((2\pi)^{N_u}|\det\Sigma|)^{1/2}} \exp\left(-\frac{1}{2}(\mathbf{u} - \mathbf{g}) \cdot \Sigma^{-1} \cdot (\mathbf{u} - \mathbf{g})\right). \tag{10.21}$$

The expression $|\det\Sigma|$ indicates the (absolute) value of the determinant of Σ. In factor analysis, Σ is taken to be diagonal, $\Sigma = \mathrm{diag}(\Sigma_1, \dots, \Sigma_{N_u})$ (see the Mathematical Appendix), with all the diagonal elements strictly positive, so its inverse is simply $\Sigma^{-1} = \mathrm{diag}(1/\Sigma_1, \dots, 1/\Sigma_{N_u})$ and $|\det\Sigma| = \Sigma_1\Sigma_2\dots\Sigma_{N_u}$.

Because Σ is diagonal, the individual components of $\mathbf{v}$ are mutually independent. Thus, any correlations between the components of $\mathbf{u}$ must arise from the mean values $\mathbf{G} \cdot \mathbf{v}$ of the generative distribution. To be well specified, the model requires $\mathbf{v}$ to have fewer dimensions than $\mathbf{u}$ ($N_v < N_u$). In terms of heuristics, factor analysis seeks a relatively small number of independent causes that account, in a linear manner, for collective Gaussian structure in the inputs.

The recognition distribution for factor analysis has the Gaussian form

$$p[\mathbf{v}|\mathbf{u}; \mathcal{G}] = \mathcal{N}(\mathbf{v}; \mathbf{W} \cdot \mathbf{u}, \Psi), \tag{10.22}$$

where expressions for $\mathbf{W}$ and $\boldsymbol{\Psi}$ are given in the appendix. These do not depend on the input $\mathbf{u}$, so factor analysis involves a linear relation between the input and the mean of the recognition distribution. EM, as applied to an invertible model, can be used to adjust $\mathcal{G} = (\mathbf{G}, \boldsymbol{\Sigma})$ on the basis of the input data. The resulting learning rules are given in the appendix. For the case of a single cause v, these reduce to equation 10.5.

In this case, we can understand the goal of density estimation in an additional way. By direct calculation, as in equation 10.1, the marginal distribution for $\mathbf{u}$ is

$$p[\mathbf{u}; \mathcal{G}] = \mathcal{N}(\mathbf{u}; 0, \mathbf{G} \cdot \mathbf{G}^{\mathrm{T}} + \boldsymbol{\Sigma}), \qquad (10.23)$$

where $[\mathbf{G}^{\mathrm{T}}]_{ab} = [\mathbf{G}]_{ba}$ and $[\mathbf{G} \cdot \mathbf{G}^{\mathrm{T}}]_{ab} = \sum_c G_{ac} G_{bc}$ (see the Mathematical Appendix). Maximum likelihood density estimation requires determining the $\mathcal{G}$ that makes $\mathbf{G} \cdot \mathbf{G}^{\mathrm{T}} + \boldsymbol{\Sigma}$ match, as closely as possible, the covariance matrix of the input distribution.

Principal Components Analysis

In the same way that setting the parameters Σ_v to 0 in the mixture of Gaussians model leads to the K-means algorithm, setting all the variances in factor analysis to 0 leads to another well-known method, principal components analysis (which we also discuss in chapter 8). To see this, consider the case of a single factor. This means that v is a single number, and that the mean of the distribution $p[\mathbf{u}|v; \mathcal{G}]$ is $v\mathbf{g}$, where the vector $\mathbf{g}$ replaces the matrix $\mathbf{G}$ of the general case. The elements of the diagonal matrix $\boldsymbol{\Sigma}$ are set to a single variance Σ, which we shrink to 0.

As $\Sigma \to 0$, the Gaussian distribution $p[\mathbf{u}|v; \mathcal{G}]$ in equation 10.20 approaches a δ function (see the Mathematical Appendix), and it can generate only the single vector $\mathbf{u}(v) = v\mathbf{g}$ from cause v. Similarly, the recognition distribution of equation 10.22 becomes a δ function, making the recognition process deterministic with $v(\mathbf{u}) = \mathbf{W} \cdot \mathbf{u}$ given by the mean of the recognition distribution of equation 10.22. Using the expression for $\mathbf{W}$ in the appendix in the limit $\Sigma \to 0$, we find

$$v(\mathbf{u}) = \frac{\mathbf{g} \cdot \mathbf{u}}{|\mathbf{g}|^2}. \qquad (10.24)$$

This is the result of the E phase of EM. In the M phase, we maximize

$$\mathcal{F}(v(\mathbf{u}), \mathcal{G}) = \langle \ln p[v(\mathbf{u}), \mathbf{u}; \mathcal{G}] \rangle = K - \frac{N_u \ln \Sigma}{2} - \left\langle \frac{v^2(\mathbf{u})}{2} + \frac{|\mathbf{u} - \mathbf{g}v(\mathbf{u})|^2}{2\Sigma} \right\rangle$$
$$(10.25)$$

with respect to $\mathbf{g}$, without changing the expression for $v(\mathbf{u})$. Here, K is a term independent of $\mathbf{g}$ and Σ. In this expression, the only term that depends on $\mathbf{g}$ is proportional to $|\mathbf{u} - \mathbf{g}v(\mathbf{u})|^2$. Minimizing this in the M phase produces a new value of **g** given by

$$\mathbf{g} = \frac{\langle v(\mathbf{u})\mathbf{u} \rangle}{\langle v^2(\mathbf{u}) \rangle}. \tag{10.26}$$

This depends only on the covariance matrix of the input distribution, as does the more general form given in the appendix. Under EM, equations 10.24 and 10.26 are alternated until convergence.

For principal components analysis, we can say more about the value of **g** at convergence. We consider the case $|\mathbf{g}|^2 = 1$ because we can always multiply **g** and divide $v(\mathbf{u})$ by the same factor to make this true without affecting the dominant term in $\mathcal{F}(v(\mathbf{u}), \mathcal{G})$ as $\Sigma \to 0$. Then, the **g** that maximizes this dominant term must minimize

$$\langle |\mathbf{u} - \mathbf{g}(\mathbf{g} \cdot \mathbf{u})|^2 \rangle = \langle |\mathbf{u}|^2 - (\mathbf{g} \cdot \mathbf{u})^2 \rangle. \tag{10.27}$$

Here, we have used expression 10.24 for $v(\mathbf{u})$. Minimizing 10.27 with respect to **g**, subject to the constraint $|\mathbf{g}|^2 = 1$, gives the result that **g** is the eigenvector of the covariance matrix $\langle \mathbf{u}\mathbf{u} \rangle$ with maximum eigenvalue. This is just the principal component vector and is equivalent to finding the vector of unit length with the largest possible average projection onto **u**. Note that there are ways other than EM for finding eigenvectors of this matrix.

The argument we have given shows that principal components analysis is a degenerate form of factor analysis. This is also true if more than one factor is considered, although maximizing $\mathcal{F}$ constrains the projections $\mathbf{G} \cdot \mathbf{u}$ and therefore is sufficient only to force **G** to represent the principal components subspace of the data. The same subspace emerges from full factor analysis provided that the variances of all the factors are equal, even when they are nonzero.

Figure 10.4 illustrates an important difference between factor analysis and principal components analysis. In this example, **u** is a three-component input vector, $\mathbf{u} = (u_1, u_2, u_3)$. Just as in figure 10.3, samples of input data were generated on the basis of a "true" cause, v_{true} according to

$$u_b = v_{\text{true}} + \epsilon_b, \tag{10.28}$$

where ϵ_b represents the noise added to component b of the input. Input data points were generated from this equation by choosing a value of v_{true} from a Gaussian distribution with mean 0 and variance 1, and values of ϵ_b from independent Gaussian distributions with 0 means. The variances of the distributions for ϵ_1, ϵ_2, and ϵ_3 were all are equal to 0.25 in figures 10.4A and 10.4B. However, in figures 10.4C and 10.4D, the variance for ϵ_3 is much larger (equal to 9), as if the third sensor was much noisier than the other two sensors. The graphs in figure 10.4 show the mean of the values of v extracted from sample inputs by factor analysis, or the value of v for principal components analysis, as a function of the true value used to generate the data. Perfect extraction of the underlying cause would have $v = v_{\text{true}}$, but this is impossible in this case because of the noise. The

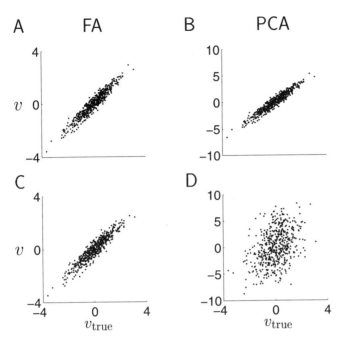

Figure 10.4 Factor analysis (FA) and principal components analysis (PCA) applied to 500 samples of noisy input reflecting a single underlying cause v_{true}. For A and B, $\langle u_i u_j \rangle = 1 + 0.25 \delta_{ij}$, whereas for C and D, one sensor is corrupted by independent noise with standard deviation 3 rather than 0.5. The plots compare the values of the true cause v_{true} and the cause v inferred by the model.

best we can expect is for the v values to be well correlated with the values of v_{true}. When the input components are equally variable (figure 10.4A and 10.4B), this is indeed what happens for both factor and principal components analysis. However, when u_3 is much more variable than the other components, principal components analysis (figure 10.4D) is fooled by the extra variance and finds a cause v that does not correlate very well with v_{true}. On the other hand, factor analysis (figure 10.4C) is affected only by the covariance between the input components and not by their individual variances (which are absorbed into $\boldsymbol{\Sigma}$), so the cause it finds is not significantly perturbed (merely slightly degraded) by the added sensor noise.

In chapter 8, we noted that principal components analysis maximizes the mutual information between the input and output under the assumption of a linear Gaussian model. This property, and the fact that principal components analysis minimizes the reconstruction error of equation 10.27, have themselves been suggested as goals for representational learning. We have now shown how they are also related to density estimation.

Both principal components analysis and factor analysis produce a marginal distribution $p[\mathbf{u}; \mathcal{G}]$ that is Gaussian. If the actual input distribution $p[\mathbf{u}]$ is non-Gaussian, the best that these models can do is to match the mean and covariance of $p[\mathbf{u}]$; they will fail to match higher-order moments. If the input is whitened to increase coding efficiency, as discussed

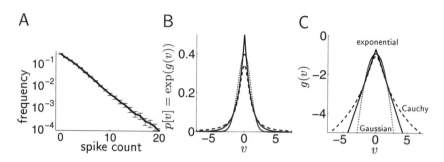

Figure 10.5 Sparse distributions. (A) Log frequency distribution for the activity of a macaque IT cell in response to video images. The fraction of times that various numbers of spikes appeared in a spike-counting window is plotted against the number of spikes. The size of the window was adjusted so that, on average, there were two spikes per window. (B) Three distributions $p[v] = \exp(g(v))$: double exponential ($g(v) = -|v|$, solid, kurtosis = 3); Cauchy ($g(v) = -\ln(1 + v^2)$, dashed, kurtosis = ∞); and Gaussian ($g(v) = -v^2/2$, dotted, kurtosis = 0). (C) The logarithms of the same three distributions. (A adapted from Baddeley et al., 1997.)

in chapter 4, so that the covariance matrix $\langle \mathbf{u}\mathbf{u} \rangle$ is equal to the identity matrix, neither method will extract any structure at all from the input data. By contrast, the generative models discussed in the following sections produce non-Gaussian marginal distributions and attempt to account for structure in the input data beyond merely covariance (and the mean).

Sparse Coding

The v values in response to input in factor and principal components analysis tend to be Gaussian distributed. If we attempt to relate such causal variables to the activities of cortical neurons, we find a discrepancy, because the activity distributions of cortical cells in response to natural inputs are not Gaussian. Figure 10.5A shows an example of the distribution of the numbers of spikes counted within a particular time window for a neuron in the inferotemporal (IT) area of the macaque brain recorded while a monkey freely viewed television shows. The distribution is close to being exponential. This means that the neurons are most likely to fire a small number of spikes in the counting interval, but that they can occasionally fire a large number of spikes.

sparse distributions   Distributions that generate values for the components of $\mathbf{v}$ close to 0 most of the time, but occasionally far from 0, are called sparse. Intuitively, sparse distributions are more likely than Gaussians of the same mean and variance to generate values near 0, and also more likely to generate values far from 0. These occasional high values can convey substantial information. Distributions with this character are also called heavy-tailed. Figures 10.5B and 10.5C compare two sparse distributions with a Gaussian distribution.

More formally, sparseness has been defined in a variety of ways. Sparseness of a distribution is sometimes linked to a high value of a measure called kurtosis. Kurtosis of a distribution $p[v]$ is defined as

$$k = \frac{\int dv\, p[v](v - \bar{v})^4}{\left(\int dv\, p[v](v - \bar{v})^2\right)^2} - 3 \quad \text{with} \quad \bar{v} = \int dv\, p[v]v, \qquad (10.29)$$

and it takes the value 0 for a Gaussian distribution. Positive values of k are taken to imply sparse distributions, which are also called super-Gaussian or leptokurtotic. Distributions with $k < 0$ are called sub-Gaussian or platykurtotic. This is a slightly different definition of sparseness from being heavy-tailed.

A sparse representation over a large population of neurons might more naturally be defined as one in which each input is encoded by a small number of the neurons in the population. Unfortunately, identifying this form of sparseness experimentally is difficult.

Sparse coding can arise in generative models that have sparse prior distributions over causes. Unlike factor analysis and principal components analysis, sparse coding does not stress minimizing the number of representing units (i.e., components of $\mathbf{v}$), and sparse representations may require large numbers of units. This is not a disadvantage for modeling the visual system because representations in visual areas are indeed greatly expanded at various steps along the pathway. For example, there are around 40 cells in primary visual cortex for each cell in the visual thalamus. Downstream processing can benefit greatly from sparse representations because, for one thing, they minimize interference between different patterns of input.

Because they employ Gaussian priors, factor analysis and principal components analysis do not generate sparse representations. The mixture of Gaussians model is extremely sparse because each input is represented by a single cause. This may be reasonable for relatively simple input patterns, but for complex stimuli such as images, we seek something between these extremes. Olshausen and Field (1996, 1997) suggested such a model by considering a nonlinear version of factor analysis. In this model, the distribution of $\mathbf{u}$ given $\mathbf{v}$ is Gaussian with a diagonal covariance matrix, as for factor analysis, but the prior distribution over causes is sparse. Defined in terms of a function $g(v)$ (as in figure 10.5), the model has

$$p[\mathbf{v}; \mathcal{G}] \propto \prod_{a=1}^{N_v} \exp(g(v_a)) \quad \text{and} \quad p[\mathbf{u}|\mathbf{v}; \mathcal{G}] = \mathcal{N}(\mathbf{u}; \mathbf{G} \cdot \mathbf{v}, \boldsymbol{\Sigma}). \qquad (10.30)$$

The prior $p[\mathbf{v}; \mathcal{G}]$ should be normalized so that its integral over $\mathbf{v}$ is 1, but we omit the normalization factor to simplify the equations.

The prior $p[\mathbf{v}; \mathcal{G}]$ in equation 10.30 makes the components of $\mathbf{v}$ mutually independent because it is a product. If we took $g(v) = -v^2$, $p[\mathbf{v}; \mathcal{G}]$ would be Gaussian (dotted lines in figures 10.5B and 10.5C), and the model would

double exponential distribution

perform factor analysis. An example of a function that provides a sparse prior is $g(v) = -\alpha|v|$. This generates a double exponential distribution (solid lines in figures 10.5B and 10.5C) similar to the activity distribution in figure 10.5A. Another commonly used form is

$$g(v) = -\ln(\beta^2 + v^2) \tag{10.31}$$

Cauchy distribution

with β a constant, which generates a Cauchy distribution (dashed lines in figures 10.5B and 10.5C).

For $g(v)$ such as equation 10.31, it is difficult to compute the recognition distribution $p[\mathbf{v}|\mathbf{u}; \mathcal{G}]$ exactly. This makes the sparse model noninvertible. Olshausen and Field chose a deterministic approximate recognition model. Thus, EM consists of finding $\mathbf{v}(\mathbf{u})$ during the E phase, and using it to adjust the parameters $\mathcal{G}$ during the M phase. To simplify the discussion, we make the covariance matrix of the generative model proportional to the identity matrix, $\boldsymbol{\Sigma} = \Sigma\mathbf{I}$. The function to be maximized is then

$$\mathcal{F}(\mathbf{v}(\mathbf{u}), \mathcal{G}) = \left\langle -\frac{1}{2\Sigma}|\mathbf{u} - \mathbf{G} \cdot \mathbf{v}(\mathbf{u})|^2 + \sum_{a=1}^{N_v} g(v_a(\mathbf{u})) \right\rangle + K, \tag{10.32}$$

where K is a term that is independent of $\mathbf{G}$ and $\mathbf{v}$. For convenience in discussing the EM procedure, we further take $\Sigma = 1$ and do not allow it to vary. Similarly, we assume that β in equation 10.31 is predetermined and held fixed. Then, $\mathcal{G}$ consists only of the matrix $\mathbf{G}$.

The E phase of EM involves maximizing $\mathcal{F}$ with respect to $\mathbf{v}(\mathbf{u})$ for every $\mathbf{u}$. This leads to the conditions (for all a)

$$\sum_{b=1}^{N_u} [\mathbf{u} - \mathbf{G} \cdot \mathbf{v}(\mathbf{u})]_b G_{ba} + g'(v_a) = 0. \tag{10.33}$$

The prime on $g(v_a)$ indicates a derivative. One way to solve this equation is to let $\mathbf{v}$ evolve over time according to the equation

$$\tau_\mathbf{v} \frac{dv_a}{dt} = \sum_{b=1}^{N_u} [\mathbf{u} - \mathbf{G} \cdot \mathbf{v}(\mathbf{u})]_b G_{ba} + g'(v_a), \tag{10.34}$$

where $\tau_\mathbf{v}$ is an appropriate time constant. This equation changes $\mathbf{v}$ so that it asymptotically approaches a value $\mathbf{v} = \mathbf{v}(\mathbf{u})$ that satisfies equation 10.33 and sets the right side of equation 10.34 to 0. We assume that the evolution of $\mathbf{v}$ according to equation 10.34 is carried out long enough during the E phase for this to happen. This process is guaranteed to find only a local, not a global, maximum of $\mathcal{F}$, and it is not guaranteed to find the same local maximum on each iteration.

Equation 10.34 resembles the equation used in chapter 7 for a firing-rate network model. The term $\sum_b u_b G_{ba}$, which can be written in vector form as $\mathbf{G}^\mathrm{T} \cdot \mathbf{u}$, acts as the total input arising from units with activities $\mathbf{u}$ fed through a feedforward coupling matrix $\mathbf{G}^\mathrm{T}$. The term $-\sum_b [\mathbf{G} \cdot \mathbf{v}]_b G_{ba}$ can

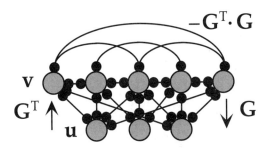

Figure 10.6 A network for sparse coding.   This network reproduces equation (10.34), using recurrent weights $-\mathbf{G}^{\mathrm{T}}\cdot\mathbf{G}$ in the $\mathbf{v}$ layer and weights connecting the input units to this layer that are given by the transpose of the matrix $\mathbf{G}$. The reverse connections from the $\mathbf{v}$ layer to the input layer indicate how the mean of the recognition distribution is computed.

be interpreted as a recurrent coupling of the $\mathbf{v}$ units through the matrix $-\mathbf{G}^{\mathrm{T}}\cdot\mathbf{G}$. Finally, the term $g'(v_a)$ plays the same role as the term $-v_a$ that would appear in the rate equations of chapter 7. If $g'(v) \neq -v$, this can be interpreted as a modified form of firing-rate dynamics. Figure 10.6 shows the resulting network. The feedback connections from the $\mathbf{v}$ units to the input units that determine the mean of the generative distribution, $\mathbf{G}\cdot\mathbf{v}$ (equation 10.30), are also shown in this figure.

After $\mathbf{v}(\mathbf{u})$ has been determined during the E phase of EM, a delta rule (chapter 8) is used during the M phase to modify $\mathbf{G}$ and improve the generative model. The full learning rule is given in the appendix. The delta rule follows from maximizing $\mathcal{F}(\mathbf{v}(\mathbf{u}), \mathcal{G})$ with respect to $\mathbf{G}$. A complication arises here because the matrix $\mathbf{G}$ always appears multiplied by $\mathbf{v}$. This means that the bias toward small values of v_a imposed by the prior can be effectively neutralized by scaling up $\mathbf{G}$. This complication results from the approximation of deterministic recognition. To prevent the weights from growing without bound, constraints are applied on the lengths of the generative weights for each cause, $\sum_b G_{ba}^2$, to encourage the variances of all the different v_a to be approximately equal (see the appendix). Further, it is conventional to precondition the inputs before learning by whitening them so that $\langle\mathbf{u}\rangle = 0$ and $\langle\mathbf{uu}\rangle = \mathbf{I}$. This typically makes learning faster, and it also ensures that the network is forced to find statistical structure beyond second order that would escape simpler methods such as factor analysis or principal components analysis. In the case that the input is created by sampling (e.g., pixelating an image), more sophisticated forms of preconditioning can be used to remove the resulting artifacts.

Applying the sparse coding model to inputs coming from the pixel intensities of small square patches of monochrome photographs of natural scenes leads to selectivities that resemble those of cortical simple cells. Before studying this result, we need to specify how the selectivities of generative models, such as the sparse coding model, are defined. The selectivities of sensory neurons are typically described by receptive fields, as in chapter 2. For a causal model, one definition of a receptive field for unit a is the set of inputs $\mathbf{u}$ for which v_a is likely to take large values. However, it may be *projective field*

impossible to construct receptive fields by averaging over these inputs in nonlinear models, such as sparse coding models. Furthermore, generative models are most naturally characterized by projective fields rather than receptive fields. The projective field associated with a particular cause v_a can be defined as the set of inputs that it frequently generates. This consists of all the **u** values for which $P[\mathbf{u}|v_a; \mathcal{G}]$ is sufficiently large when v_a is large. For the model of figure 10.1, the projective fields are simply the circles in figure 10.1C. It is important to remember that projective fields can be quite different from receptive fields.

Projective fields for the Olshausen and Field model trained on natural scenes are shown in figure 10.7A, with one picture for each component of **v**. In this case, the projective field for v_a is simply the matrix elements G_{ab} plotted for all b values. In figure 10.7A, the index b is plotted over a two-dimensional grid representing the location of the input u_b within the visual field. The projective fields form a Gabor-like representation for images, covering a variety of spatial scales and orientations. The resemblance of this representation to the receptive fields of simple cells in primary visual cortex is quite striking, although these are the projective fields of the model, not its receptive fields. Unfortunately, there is no simple form for the receptive fields of the **v** units. Figure 10.7B compares the projective field of one unit with receptive fields determined by presenting either dots or gratings as inputs and recording the responses. The responses to the dots directly determine the receptive field, while responses to the gratings directly determine the Fourier transform of the receptive field. Differences between the receptive fields calculated on the basis of these two types of input are evident in the figure. In particular, the receptive field computed from gratings shows more spatial structure than the one mapped by dots. Nevertheless, both show a resemblance to the projective field and to a typical simple-cell receptive field.

In a generative model, projective fields are associated with the causes underlying the visual images presented during training. The fact that the causes extracted by the sparse coding model resemble Gabor patches within the visual field is somewhat strange from this perspective. It is difficult to conceive of images as arising from such low-level causes, instead of causes couched in terms of objects within the images, for example. From the perspective of good representation, causes that are more like objects and less like Gabor patches would be more useful. To put this another way, although the prior distribution over causes biased them toward mutual independence, the causes produced by the recognition model in response to natural images are not actually independent. This is due to the structure in images arising from more complex objects than bars and gratings. It is unlikely that this high-order structure can be extracted by a model with only one set of causes. It is more natural to think of causes in a hierarchical manner, with causes at a higher level accounting for structure in the causes at a lower level. The multiple representations in areas along the visual pathway suggests such a hierarchical scheme, but the corresponding models are still in the rudimentary stages of development.

A

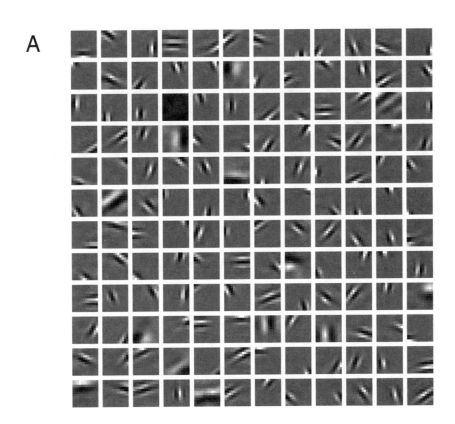

B

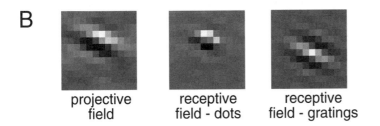

projective   receptive   receptive
   field   field - dots   field - gratings

Figure 10.7 Projective and receptive fields for a sparse coding network with $N_u = N_v = 144$. (A) Projective fields $\mathbf{G}_{ab}$ with a indexing representational units (the components of $\mathbf{v}$), and b indexing input units $\mathbf{u}$ on a 12×12 pixel grid. Each box represents a different a value, and the b values are represented within the box by the corresponding input location. Weights are represented by the gray-scale level, with gray indicating 0. (B) The relationship between projective and receptive fields. The left panel shows the projective field of one of the units in A. The middle and right panels show its receptive field mapped using inputs generated by dots and by gratings, respectively. (Adapted from Olshausen & Field, 1997.)

Independent Components Analysis

As for the case of the mixtures of Gaussians model and factor analysis, an interesting model emerges from sparse coding as $\Sigma \to 0$. In this limit, the generative distribution (equation 10.30) approaches a δ function and always generates $\mathbf{u}(\mathbf{v}) = \mathbf{G} \cdot \mathbf{v}$. Under the additional restriction that there are as many causes as inputs, the approximation we used for the sparse coding model of making the recognition distribution deterministic becomes exact, and the recognition distribution that maximizes $\mathcal{F}$ is

$$Q[\mathbf{v}; \mathbf{u}] = |\det \mathbf{W}|^{-1} \delta(\mathbf{u} - \mathbf{W}^{-1} \cdot \mathbf{v}), \qquad (10.35)$$

where $\mathbf{W} = \mathbf{G}^{-1}$ is the matrix inverse of the generative weight matrix. The factor $|\det \mathbf{W}|$ comes from the normalization condition on Q, $\int d\mathbf{v} \, Q(\mathbf{v}; \mathbf{u}) = 1$. At the maximum with respect to Q, the function $\mathcal{F}$ is

$$\mathcal{F}(Q, \mathcal{G}) = \left\langle -\frac{1}{2\Sigma} |\mathbf{u} - \mathbf{G} \cdot \mathbf{W} \cdot \mathbf{u}|^2 + \sum_a g\left([\mathbf{W} \cdot \mathbf{u}]_a\right) \right\rangle + \ln|\det \mathbf{W}| + K, \qquad (10.36)$$

where K is independent of $\mathbf{G}$. Under the conventional EM procedure, we would maximize this expression with respect to $\mathbf{G}$, keeping $\mathbf{W}$ fixed. However, the normal procedure fails in this case, because the minimum of the right side of equation 10.36 occurs at $\mathbf{G} = \mathbf{W}^{-1}$, and $\mathbf{W}$ is being held fixed, so $\mathbf{G}$ cannot change. This is an anomaly of coordinate ascent in this particular limit.

Fortunately, it is easy to fix this problem, because we know that $\mathbf{W} = \mathbf{G}^{-1}$ provides an exact inversion of the generative model. Therefore, instead of holding $\mathbf{W}$ fixed during the M phase of an EM procedure, we keep $\mathbf{W} = \mathbf{G}^{-1}$ at all times as we change $\mathbf{G}$. This sets $\mathcal{F}$ equal to the average log likelihood, and the process of optimizing with respect to $\mathbf{G}$ is equivalent to likelihood maximization. Because $\mathbf{W} = \mathbf{G}^{-1}$, maximizing with respect to $\mathbf{W}$ is equivalent to maximizing with respect to $\mathbf{G}$, and it turns out that this is easier to do. Therefore, we set $\mathbf{W} = \mathbf{G}^{-1}$ in equation 10.36, which causes the first term to vanish, and write the remaining terms as the log likelihood expressed as a function of $\mathbf{W}$ instead of $\mathbf{G}$,

$$L(\mathbf{W}) = \left\langle \sum_a g\left([\mathbf{W} \cdot \mathbf{u}]_a\right) \right\rangle + \ln|\det \mathbf{W}| + K. \qquad (10.37)$$

Direct stochastic gradient ascent on this log likelihood can be performed using the update rule

$$\mathbf{W}_{ab} \to \mathbf{W}_{ab} + \epsilon\left([\mathbf{W}^{-1}]_{ba} + g'(v_a)u_b\right), \qquad (10.38)$$

where ϵ is a small learning rate parameter, and we have used the fact that $\partial \ln|\det \mathbf{W}|/\partial \mathbf{W}_{ab} = [\mathbf{W}^{-1}]_{ba}$.

The update rule of equation 10.38 can be simplified by using a clever trick. Because $\mathbf{W}^T \mathbf{W}$ is a positive definite matrix (see the Mathematical

Appendix), the weight change can be multiplied by $\mathbf{W}^T\mathbf{W}$ without affecting the fixed points of the update rule. This means that the alternative learning rule

$$\mathbf{W}_{ab} \rightarrow \mathbf{W}_{ab} + \epsilon\left(\mathbf{W}_{ab} + g'(v_a)\left[\mathbf{v}\cdot\mathbf{W}\right]_b\right) \qquad (10.39)$$

has the same potential final weight matrices as equation 10.38. This is called a natural gradient rule, and it avoids the matrix inversion of $\mathbf{W}$ as well as providing faster convergence. Equation 10.39 can be interpreted as the sum of an anti-decay term that forces $\mathbf{W}$ away from 0, and a generalized type of anti-Hebbian term. The choice of prior $p[v] \propto 1/\cosh(v)$ makes $g'(v) = -\tanh(v)$ and produces the rule

$$\mathbf{W}_{ab} \rightarrow \mathbf{W}_{ab} + \epsilon\left(\mathbf{W}_{ab} - \tanh(v_a)\left[\mathbf{v}\cdot\mathbf{W}\right]_b\right). \qquad (10.40)$$

This algorithm is called independent components analysis. Just as the sparse coding network is a nonlinear generalization of factor analysis, independent components analysis is a nonlinear generalization of principal components analysis that attempts to account for non-Gaussian features of the input distribution. The generative model is based on the assumption that $\mathbf{u} = \mathbf{G}\cdot\mathbf{v}$. Some other technical conditions must be satisfied for independent components analysis to extract reasonable causes. Specifically, the prior distributions over causes $p[v] \propto \exp(g(v))$ must be non-Gaussian and, at least to the extent of being correctly super- or sub-Gaussian, must faithfully reflect the actual distribution over causes. The particular form $p[v] \propto 1/\cosh(v)$ is super-Gaussian, and thus generates a sparse prior. There are variants of independent components analysis in which the prior distributions are adaptive.

The independent components algorithm was suggested by Bell and Sejnowski (1995) from the different perspective of maximizing the mutual information between $\mathbf{u}$ and $\mathbf{v}$ when $v_a(\mathbf{u}) = f([\mathbf{W}\cdot\mathbf{u}]_a)$, with a particular, monotonically increasing nonlinear function f. Maximizing the mutual information in this context requires maximizing the entropy of the distribution over $\mathbf{v}$. This, in turn, requires the components of $\mathbf{v}$ to be as independent as possible because redundancy between them reduces the entropy. In the case that $f(v) = g'(v)$, the expression for the entropy is the same as that for the log likelihood in equation 10.37, up to constant factors. Thus, maximizing the entropy and performing maximum likelihood density estimation are identical.

An advantage of independent components analysis over other sparse coding algorithms is that, because the recognition model is an exact inverse of the generative model, receptive as well as projective fields can be constructed. Just as the projective field for v_a can be represented by the matrix elements G_{ab} for all b values, the receptive field is given by W_{ab} for all b.

To illustrate independent components analysis, figure 10.8 shows an (admittedly bizarre) example of its application to the sounds created by tapping a tooth while adjusting the shape of the mouth to reproduce a

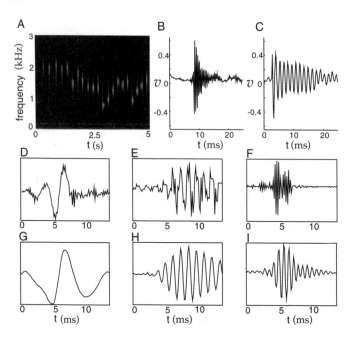

Figure 10.8 Independent components of tooth-tapping sounds. (A) Spectrogram of the input. (B-C) Waveforms for high- and low-frequency notes. The mouth acts as a damped resonant cavity in the generation of these tones. (D-F) Three independent components calculated on the basis of 1/80 s samples taken from the input at random times. The graphs show the receptive fields (from **W**) for three output units. D is reported to be sensitive to the sound of an air conditioner. E and F extract tooth taps of different frequencies. (G-I) The associated projective fields (from **G**), showing the input activity associated with the causes in D-F. (Adapted from Bell & Sejnowski, 1996.)

tune by Beethoven. The input, sampled at 8 kHz, has the spectrogram shown in figure 10.8A. In this example, we have some idea about likely causes. For example, the plots in figures 10.8B and 10.8C show high- and low-frequency tooth taps, although other causes arise from the imperfect recording conditions (e.g., the background sound of an air conditioner). A close variant of the independent components analysis method described above was used to extract $N_v = 100$ independent components. Figure 10.8D, 10.8E, and 10.8F show the receptive fields of three of these components. The last two extract particular frequencies in the input. Figure 10.8G, 10.8H, and 10.8I show projective fields. Note that the projective fields are much smoother than the receptive fields.

Bell and Sejnowski (1997) also used visual input data similar to those used in the example of figure 10.7, along with the prior $p[v] \propto 1/\cosh(v)$, and found that independent components analysis extracts Gabor-like receptive fields similar to the projective fields shown in figure 10.7A.

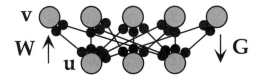

Figure 10.9 Network for the Helmholtz machine. In the bottom-up network, representational units **v** are driven by inputs **u** through feedforward weights **W**. In the top-down network, the inputs are driven by the **v** units through feedback weights **G**.

The Helmholtz Machine

The Helmholtz machine was designed to accommodate hierarchical architectures that construct complex multilayer representations. The model involves two interacting networks, one with parameters $\mathcal{G}$ that is driven in the top-down direction to implement the generative model, and the other, with parameters $\mathcal{W}$, driven bottom-up to implement the recognition model. The parameters are determined by a modified EM algorithm that results in roughly symmetric updates for the two networks.

We consider a simple, two-layer, nonlinear Helmholtz machine (figure 10.9) with binary units, so that u_b and v_a for all b and a take the values 0 or 1. For this model,

$$P[\mathbf{v}; \mathcal{G}] = \prod_a \left(f(g_a) \right)^{v_a} \left(1 - f(g_a) \right)^{1 - v_a} \tag{10.41}$$

$$P[\mathbf{u}|\mathbf{v}; \mathcal{G}] = \prod_b \left(f\left(h_b + [\mathbf{G} \cdot \mathbf{v}]_b \right) \right)^{u_b} \left(1 - f\left(h_b + [\mathbf{G} \cdot \mathbf{v}]_b \right) \right)^{1 - u_b}, \tag{10.42}$$

where g_a is a generative bias weight for output a that controls how frequently $v_a = 1$, h_b is the generative bias weight for u_b, and $f(g) = 1/(1 + \exp(-g))$ is the standard sigmoid function. The generative model is thus parameterized by $\mathcal{G} = (\mathbf{g}, \mathbf{h}, \mathbf{G})$. According to these distributions, the components of **v** are mutually independent, and the components of **u** are independent given a fixed value of **v**.

The generative model is noninvertible in this case, so an approximate recognition distribution must be constructed. This uses a form similar to equation 10.42, except with bottom-up weights **W** and biases **w**,

$$Q[\mathbf{v}; \mathbf{u}, \mathcal{W}] = \prod_a \left(f\left(w_a + [\mathbf{W} \cdot \mathbf{u}]_a \right) \right)^{v_a} \left(1 - f\left(w_a + [\mathbf{W} \cdot \mathbf{u}]_a \right) \right)^{1 - v_a}. \tag{10.43}$$

The parameter list for the recognition model is $\mathcal{W} = (\mathbf{w}, \mathbf{W})$. This distribution is only an approximate inverse of the generative model because it implies that the components of **v** are independent when, in fact, given a particular input **u**, they are conditionally dependent due to the way they interact in equation 10.42 to generate **u** (this is the same assumption as in the mean-field approximate distribution for the Boltzmann machine, except that the parameters of the distribution here are shared between all input cases).

The EM algorithm for this noninvertible model would consist of alternately maximizing the function $\mathcal{F}$ given by

$$\mathcal{F}(\mathcal{W}, \mathcal{G}) = \left\langle \sum_{\mathbf{v}} Q[\mathbf{v}; \mathbf{u}, \mathcal{W}] \ln \frac{P[\mathbf{v}, \mathbf{u}; \mathcal{G}]}{Q[\mathbf{v}; \mathbf{u}, \mathcal{W}]} \right\rangle \qquad (10.44)$$

with respect to the parameters $\mathcal{W}$ and $\mathcal{G}$. For the M phase of the Helmholtz machine, this is exactly what is done. However, during the E phase, maximizing with respect to $\mathcal{W}$ is problematic because the function $Q[\mathbf{v}; \mathbf{u}, \mathcal{W}]$ appears in two places in the expression for $\mathcal{F}$. This also makes the learning rule during the E phase take a different form from that during the M phase. Instead, the Helmholtz machine uses a simpler and more symmetric approximation to EM.

The approximation to EM used by the Helmholtz machine is constructed by re-expressing $\mathcal{F}$ from equation 10.10, explicitly writing out the average over input data and the expression for the Kullback-Leibler divergence,

$$\mathcal{F}(\mathcal{W}, \mathcal{G}) = L(\mathcal{G}) - \sum_{\mathbf{u}} P[\mathbf{u}] D_{\mathrm{KL}}(Q[\mathbf{v}; \mathbf{u}, \mathcal{W}], P[\mathbf{v}|\mathbf{u}; \mathcal{G}])$$

$$= L(\mathcal{G}) - \sum_{\mathbf{u}} P[\mathbf{u}] \sum_{\mathbf{v}} Q[\mathbf{v}; \mathbf{u}, \mathcal{W}] \ln \left(\frac{Q[\mathbf{v}; \mathbf{u}, \mathcal{W}]}{P[\mathbf{v}|\mathbf{u}; \mathcal{G}]} \right). \qquad (10.45)$$

This is the function that is maximized with respect to $\mathcal{G}$ during the M phase for the Helmholtz machine. However, the E phase is not based on maximizing equation 10.45 with respect to $\mathcal{W}$. Instead, an approximate $\mathcal{F}$ function that we call $\tilde{\mathcal{F}}$ is used. This is constructed by using $P[\mathbf{u}; \mathcal{G}]$ as an approximation for $P[\mathbf{u}]$ and $D_{\mathrm{KL}}(P[\mathbf{v}|\mathbf{u}; \mathcal{G}], Q[\mathbf{v}; \mathbf{u}, \mathcal{W}])$ as an approximation for $D_{\mathrm{KL}}(Q[\mathbf{v}; \mathbf{u}, \mathcal{W}], P[\mathbf{v}|\mathbf{u}; \mathcal{G}])$ in equation 10.45. These are likely to be good approximations if the generative and approximate recognition models are accurate. Thus, we write

$$\tilde{\mathcal{F}}(\mathcal{W}, \mathcal{G}) = L(\mathcal{G}) - \sum_{\mathbf{u}} P[\mathbf{u}; \mathcal{G}] D_{\mathrm{KL}}(P[\mathbf{v}|\mathbf{u}; \mathcal{G}], Q[\mathbf{v}; \mathbf{u}, \mathcal{W}])$$

$$= L(\mathcal{G}) - \sum_{\mathbf{u}} P[\mathbf{u}; \mathcal{G}] \sum_{\mathbf{v}} P[\mathbf{v}|\mathbf{u}; \mathcal{G}] \ln \left(\frac{P[\mathbf{v}|\mathbf{u}; \mathcal{G}]}{Q[\mathbf{v}; \mathbf{u}, \mathcal{W}]} \right). \qquad (10.46)$$

and maximize this, rather than $\mathcal{F}$, with respect to $\mathcal{W}$ during the E phase. This amounts to averaging the "flipped" Kullback-Leibler divergence over samples of $\mathbf{u}$ created by the generative model, rather than real data samples. The advantage of making these approximations is that the E and M phases become highly symmetric, as can be seen by examining the second equalities in equations 10.45 and 10.46.

Learning in the Helmholtz machine proceeds by using stochastic sampling to replace the weighted sums in equations 10.45 and 10.46. In the M phase, an input $\mathbf{u}$ from $P[\mathbf{u}]$ is presented, and a sample $\mathbf{v}$ is drawn from the current recognition distribution $Q[\mathbf{v}; \mathbf{u}, \mathcal{W}]$. Then the generative weights $\mathcal{G}$ are changed according to the discrepancy between $\mathbf{u}$ and the generative or top-down prediction $\mathbf{f}(\mathbf{h} + \mathbf{G} \cdot \mathbf{v})$ of $\mathbf{u}$ (see the appendix). Thus, the generative model is trained to make $\mathbf{u}$ more likely to be generated by the cause

v associated with it by the recognition model. In the E phase, samples of both **v** and **u** are drawn from the generative model distributions $P[\mathbf{v}; \mathcal{G}]$ and $P[\mathbf{u}|\mathbf{v}; \mathcal{G}]$, and the recognition parameters $\mathcal{W}$ are changed according to the discrepancy between the sampled cause **v** and the recognition or bottom-up prediction $\mathbf{f}(\mathbf{w} + \mathbf{W} \cdot \mathbf{u})$ of **v** (see the appendix). The rationale for this is that the **v** that was used by the generative model to create **u** is a good choice for its cause in the recognition model.

The two phases of learning are sometimes called wake and sleep because learning in the first phase is driven by real inputs **u** from the environment, while learning in the second phase is driven by values **v** and **u** "fantasized" by the generative model. This terminology is based on slightly different principles from the wake and sleep phases of the Boltzmann machine discussed in chapter 8. The sleep phase is only an approximation of the actual E phase, and general conditions under which learning converges appropriately are not known.

wake-sleep algorithm

10.4 Discussion

Because of the widespread significance of coding, transmitting, storing, and decoding visual images such as photographs and movies, substantial effort has been devoted to understanding the structure of this class of inputs. As a result, visual images provide an ideal testing ground for representational learning algorithms, allowing us to go beyond evaluating the representations they produce solely in terms of the log likelihood and qualitative similarities with cortical receptive fields.

Most modern image (and auditory) processing techniques are based on multi-resolution decompositions. In such decompositions, images are represented by the activity of a population of units with systematically varying spatial frequency and orientation preferences, centered at various locations on the image. The outputs of the representational units are generated by filters (typically linear) that act as receptive fields and are partially localized in both space and spatial frequency. The filters usually have similar underlying forms, but they are cast at different spatial scales and centered at different locations for the different units. Systematic versions of such representations, in forms such as wavelets, are important signal processing tools, and there is an extensive body of theory about their representational and coding qualities. Representation of sensory information in separated frequency bands at different spatial locations has significant psychophysical consequences as well.

The projective fields of the units in the sparse coding network shown in figure 10.7 suggest that they construct something like a multi-resolution decomposition of inputs, with multiple spatial scales, locations, and orientations. Thus, multi-resolution analysis gives us a way to put into sharper focus the issues arising from models such as sparse coding and independent components analysis. After a brief review of multi-resolution decom-

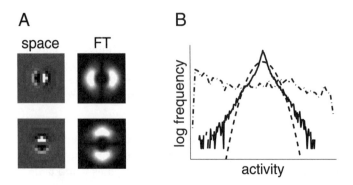

Figure 10.10 Multi-resolution filtering. (A) Vertical and horizontal filters (left) and their Fourier transforms (right) that are used at multiple positions and spatial scales to generate a multi-resolution representation. The rows of the matrix $\mathbf{W}$ are displayed here in gray scale on a two-dimensional grid representing the location of the corresponding input. (B) Log frequency distribution of the outputs of the highest spatial frequency filters (solid line) compared with a Gaussian distribution with the same mean and variance (dashed line) and the distribution of pixel values for the image shown in figure 10.11A (dot-dashed line). The pixel values of the image were rescaled to fit into the range. (Adapted from Simoncelli & Freeman, 1995; Karasaridis & Simoncelli, 1996.)

positions, we use them to consider various properties of representational learning from the perspective of information transmission and sparseness, overcompleteness, and residual dependencies between inferred causes.

Multi-resolution Decomposition

Many multi-resolution decompositions, with a variety of computational and representational properties, can be expressed as linear transformations $\mathbf{v} = \mathbf{W} \cdot \mathbf{u}$, where the rows of $\mathbf{W}$ describe filters, such as those illustrated in figure 10.10A. Figure 10.11 shows the result of applying multi-resolution filters, constructed by scaling and shifting the filters shown in figure 10.10A, to the photograph in figure 10.11A. Vertical and horizontal filters similar to those in figure 10.10A, but with different sizes, produce the decomposition shown in figures 10.11B-10.11D and 10.11F-10.11H when translated across the image. The level of gray indicates the output generated by placing the different filters over the corresponding point on the image. These outputs, plus the low-pass image in figure 10.11E and an extra high-pass image that is not shown, can be used to reconstruct the whole photograph almost perfectly through a generative process that is the inverse of the recognition process.

Coding

One reason for using multi-resolution decompositions is that they offer efficient ways of encoding visual images, whereas raw values of input pixels

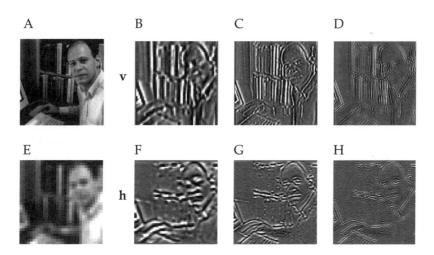

Figure 10.11 Multi-resolution image decomposition. A gray-scale image is decomposed, using the pair of vertical and horizontal filters shown in figure 10.10. (A) The original image. (B-D) The outputs of successively higher spatial frequency, vertically oriented filters translated across the image. (E) The image after passage through a low-pass filter. (F-H) The outputs of successively higher spatial frequency, horizontally oriented filters translated across the image.

provide an inefficient encoding. This is illustrated by the dot-dashed line in figure 10.10B, which shows that the distribution over the values of the input pixels of the image in figure 10.11A is approximately flat or uniform. Up to the usual additive constants related to the precision with which filter outputs are encoded, the contribution to the coding cost from a single unit is the entropy of the probability distribution of its output. The distribution over pixel intensities is flat, which is the maximum entropy distribution for a variable with a fixed range. Encoding the individual pixel values therefore incurs the maximum possible coding cost.

By contrast, the solid line in figure 10.10B shows the distribution of the outputs of the finest scale vertically and horizontally tuned filters (figures 10.11D and 10.11H) in response to figure 10.11A. The filter outputs have a sparse distribution similar to the double exponential distribution in figure 10.5B. This distribution has significantly lower entropy than the uniform distribution, so the filter outputs provide a more efficient encoding than pixel values.

In making these statements about the distributions of activities, we are equating the output distribution of a filter applied at many locations on a single image with the output distribution of a filter applied at a fixed location on many images. This assumes spatial translational invariance of the ensemble of visual images.

Images represented by multi-resolution filters can be further compressed by retaining only approximate values of the filter outputs. This is called lossy coding and may consist of reporting filter outputs as integer multiples of a basic unit. Making the multi-resolution code for an image lossy

lossy coding

by coarsely quantizing the outputs of the highest spatial frequency filters generally has quite minimal perceptual consequences, while saving substantial coding cost (because these outputs are most numerous). This fact illustrates the important point that trying to build generative models of all aspects of visual images may be unnecessarily difficult, because only certain aspects of images are actually relevant. Unfortunately, abstract principles are unlikely to tell us what information in the input can safely be discarded independent of details of how the representations are to be used.

Overcomplete Representations

Sparse representations often have more output units than input units. Such representations, called overcomplete, are the subject of substantial work in multi-resolution theory. Many reasons have been suggested for overcompleteness, although none obviously emerges from the requirement of fitting good probabilistic models to input data.

One interesting idea comes from the notion that the task of manipulating representations should be invariant to the groups of symmetry transformations of the input, which, for images, include rotation, translation, and scaling. Complete representations are minimal, and so do not densely sample orientations. This means that the operations required to manipulate images of objects presented at angles not directly represented by the filters are different from those required at the represented angles (such as horizontal and vertical for the example of figure 10.10). When a representation is overcomplete in such a way that different orientations are represented roughly equally, as in primary visual cortex, the computational operations required to manipulate images are more uniform as a function of image orientation. Similar ideas apply across scale, so that the operations required to manipulate large and small images of the same object (as if viewed from near and far) are likewise similar. However, it is impossible to generate representations that satisfy all these constraints perfectly.

In more realistic models that include noise, other rationales for overcompleteness come from considering population codes, in which many units redundantly report information about closely related quantities so that uncertainty can be reduced. Despite the ubiquity of overcomplete population codes in the brain, there are few representational learning models that produce them satisfactorily. The coordinated representations required to construct population codes are often incompatible with other heuristics such as factorial or sparse coding.

Interdependent Causes

One of the failings of multi-resolution decompositions for coding is that the outputs are not mutually independent. This makes encoding each of

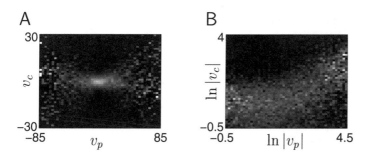

Figure 10.12 (A) Gray-scale plot of the conditional distribution of the output of a filter at the finest spatial scale (v_c) given the output of a coarser filter (v_p) with the same position and orientation (using the picture in figure 10.11A as input data). Each column is separately normalized. The plot has a characteristic bow-tie shape. (B) The same data plotted as the conditional distribution of $\ln|v_c|$ given $\ln|v_p|$. (Adapted from Simoncelli & Adelson, 1990; Simoncelli & Schwartz, 1999.)

the redundant filter outputs wasteful. Figure 10.12 illustrates such an interdependence by showing the conditional distribution for the output v_c of a horizontally tuned filter at a fine scale, given the output v_p of a horizontally tuned unit at the next coarser scale. The plots show gray-scale values of the conditional probability density $p[v_c|v_p]$. The mean of this distribution is roughly 0, but there is a clear correlation between the magnitude of $|v_p|$ and the variance of v_c. This means that structure in the image is coordinated across different spatial scales, so that high outputs from a coarse scale filter are typically accompanied by substantial output (of one sign or the other) at a finer scale. Following Simoncelli (1997), we plot the conditional distribution of $\ln|v_c|$ given $\ln|v_p|$ in figure 10.12B. For small values of $\ln|v_p|$, the distribution of $\ln|v_c|$ is flat, but for larger values of $\ln|v_p|$ the growth in the value of $|v_c|$ is clear.

The interdependence shown in figure 10.12 suggests a failing of sparse coding to which we have alluded before. Although the prior distribution for sparse coding stipulates independent causes, the causes identified as underlying real images are not independent. The dependence apparent in figure 10.12 can be removed by a nonlinear transformation in which the outputs of the units normalize each other (similar to the model introduced to explain contrast saturation in chapter 2). This transformation can lead to more compact codes for images. However, the general problem suggests that something is amiss with the heuristic of seeking independent causes for representations early in the visual pathway.

The most important dependencies as far as causal models are concerned are those induced by the presence in images of objects with large-scale coordinated structure. Finding and building models of these dependencies is the goal for more sophisticated, hierarchical representational learning schemes aimed ultimately at object recognition within complex visual scenes.

10.5   Chapter Summary

We have presented a systematic treatment of exact and approximate maximum likelihood density estimation as a way of fitting probabilistic generative models and thereby performing representational learning. Recognition models, which are the statistical inverses of generative models, specify the causes underlying an input and play a crucial role in learning. We discussed the expectation maximization (EM) algorithm applied to invertible and noninvertible models, including the use of deterministic and probabilistic approximate recognition models and a lower bound on the log likelihood.

We presented a variety of models for continuous inputs with discrete, continuous, or vector-valued causes. These include mixture of Gaussians, K-means, factor analysis, principal components analysis, sparse coding, and independent components analysis. We also described the Helmholtz machine and discussed general issues of multi-resolution representation and coding.

10.6   Appendix

Summary of Causal Models

Model	Generative Model	Recognition Model	Learning Rules
mixture of Gaussians	$P[v;\mathcal{G}] = \gamma_v$ $P[u\|v;\mathcal{G}] = \mathcal{N}(u; g_v, \Sigma_v)$	$P[v\|u;\mathcal{G}] \propto \gamma_v \mathcal{N}(u; g_v, \Sigma_v)$	$\mu_v = \langle P[v\|u;\mathcal{G}]\rangle$ $g_v = \langle P[v\|u;\mathcal{G}]u\rangle/\gamma_v$ $\Sigma_v = \langle P[v\|u;\mathcal{G}]\|u - g_v\|^2\rangle/(N_u\gamma_v)$
factor analysis	$P[v;\mathcal{G}] = \mathcal{N}(v; 0, 1)$ $P[u\|v;\mathcal{G}] = \mathcal{N}(u; G\cdot v, \Sigma)$ $\Sigma = \text{diag}(\Sigma_1, \Sigma_2, \ldots, \Sigma_{N_u})$	$P[v\|u;\mathcal{G}] = \mathcal{N}(v; W\cdot u, \Psi)$ $\Psi = (I + G^T\cdot\Sigma^{-1}\cdot G)^{-1}$ $W = \Psi\cdot G^T\cdot\Sigma^{-1}$	$G = C\cdot W^T\cdot(W\cdot C\cdot W^T + \Psi)^{-1}$ $\Sigma = \text{diag}(G\cdot\Psi\cdot G^T + (I - G\cdot W)\cdot C\cdot(I - G\cdot W)^T)$ $C = \langle uu\rangle$
principal components analysis	$P[v;\mathcal{G}] = \mathcal{N}(v; 0, 1)$ $u = G\cdot v$	$v = W\cdot u$ $W = (G^T\cdot G)^{-1}\cdot G^T$	$G = C\cdot W^T\cdot(W\cdot C\cdot W^T)^{-1}$ $C = \langle uu\rangle$
sparse coding	$P[v;\mathcal{G}] \propto \prod_a \exp(g(v_a))$ $P[u\|v;\mathcal{G}] = \mathcal{N}(u; G\cdot v, \Sigma)$	$G^T\cdot(u - G\cdot v) + g'(v) = 0$	$G \to G + \epsilon(u - G\cdot v)v$ $(\sum_b G_{ba}^2) \to (\sum_b G_{ba}^2)((v_a^2 - \langle v_a\rangle^2)/\sigma^2)^{0.01}$
independent components analysis	$P[v;\mathcal{G}] \propto \prod_a \exp(g(v_a))$ $u = G\cdot v$	$v = W\cdot u$ $W = G^{-1}$	$W_{ab} \to W_{ab} + \epsilon(W_{ab} + g'(v_a)[v\cdot W]_b)$ $g'(v) = -\tanh(v)$ if $g(v) = -\ln\cosh(v)$
binary Helmholtz machine	$P[v;\mathcal{G}] = \prod_a (f(g_a))^{v_a}(1 - f(g_a))^{1-v_a}$ $P[u\|v;\mathcal{G}] = \prod_b (f_b(h + G\cdot v))^{u_b} \times$ $\qquad (1 - f_b(h + G\cdot v))^{1-u_b}$ $f_b(h + G\cdot v) = f(h_b + [G\cdot v]_b)$	$Q[v;u,\mathcal{W}] = \prod_a (f_a(w + W\cdot u))^{v_a} \times$ $\qquad (1 - f_a(w + W\cdot u))^{1-v_a}$ $f_a(w + W\cdot u) = f(w_a + [W\cdot u]_a)$	wake: $u \sim P[u]$, $\quad v \sim Q[v; u, \mathcal{W}]$ $g \to g + \epsilon(v - f(g))$ $h \to h + \epsilon(u - f(h + G\cdot v))$ $G \to G + \epsilon(u - f(h + G\cdot v))v$ sleep: $v \sim P[v; \mathcal{G}]$, $\quad u \sim P[u\|v; \mathcal{G}]$ $w \to w + \epsilon(v - f(w + W\cdot u))$ $W \to W + \epsilon(v - f(w + W\cdot u))u$

Table 1: All models are discussed in detail in the text, and the forms quoted are just for the simplest cases. $\mathcal{N}(u; g, \Sigma)$ is a multivariate Gaussian distribution with mean g and covariance matrix Σ (for $\mathcal{N}(u; g, \Sigma)$, the variance of each component is Σ). For the sparse coding network, σ^2 is a target for the variances of each output unit. For the Helmholtz machine, $f(c) = 1/(1 + \exp(-c))$, and the symbol $\sim$ indicates that the indicated variable is drawn from the indicated distribution. Other symbols and distributions are defined in the text.

10.7   Annotated Bibliography

The literature on unsupervised representational learning models is extensive. Recent reviews, from which we have borrowed include **Hinton (1989)**, **Bishop (1995)**, **Hinton & Ghahramani (1997)**, and **Becker & Plumbley (1996)**. These references also describe unsupervised learning methods such as IMAX (Becker & Hinton, 1992) that find statistical structure in the inputs directly rather than through causal models (see also projection pursuit, Huber, 1985). The field of belief networks or graphical statistical models (**Pearl, 1988**; Lauritzen, 1996; **Jordan, 1998**) provides an even more general framework for probabilistic generative models. Apart from **Barlow** (1961, **1989**), early inspiration for unsupervised learning models came from Uttley (1979) and Marr (1970), and from the adaptive resonance theory (ART) of Carpenter & Grossberg (1991).

Analysis by synthesis (e.g., Neisser, 1967), to which generative and recognition models are closely related, was developed in a statistical context by Grenander (1995), and was suggested by Mumford (1994) as a way of understanding hierarchical neural processing. Suggestions made in MacKay (1956), Pece (1992), Kawato et al. (1993), and Rao & Ballard (1997) can be seen in a similar light.

Nowlan (1991) introduced the mixtures of Gaussians architecture into neural networks. Mixture models are commonplace in statistics and are described by Titterington et al. (1985).

Factor analysis is described by Everitt (1984). Some of the differences and similarities between factor analysis and principal components analysis are brought out in Jolliffe (1986), Tipping & Bishop (1999), and Roweis & Ghahramani (1999). Rubin & Thayer (1982) discusses the use of EM for factor analysis. Roweis (1998) presents EM for principal components analysis.

Neal & Hinton (1998) describes $\mathcal{F}$ and its role in the EM algorithm (Baum et al., 1970; Dempster et al., 1977). EM is closely related to mean field methods in physics, as discussed in Jordan et al. (1998) and Saul & Jordan (2000). Hinton & Zemel (1994) and Zemel (1994) use $\mathcal{F}$ for unsupervised learning in a backpropagation network called the autoencoder, and these results are related to minimum description length coding (Risannen, 1989). Hinton et al. (1995) and Dayan et al. (1995) use $\mathcal{F}$ in the Helmholtz machine and the associated wake-sleep algorithm.

Olshausen & Field (1996) presents the sparse coding network based on Field's (1994) general analysis of sparse representations, and Olshausen (1996) develops some of the links to density estimation. Independent components analysis (ICA) was introduced as a problem by Herrault & Jutten (1986). The version of the ICA algorithm that we described is due to Bell & Sejnowski (1995) and Roth & Baram (1996), using the natural gradient trick of Amari (1999). The derivation we used is from MacKay (1996). Pearlmutter & Parra (1996) and Olshausen (1996) also derive maximum likelihood interpretations of ICA. Multi-resolution decompositions were introduced into computer vision in Witkin (1983) and Burt & Adelson (1983). Wavelet analysis is reviewed in Daubechies (1992), **Simoncelli et al. (1992)**, and **Mallat (1998)**.

Mathematical Appendix

This book assumes a familiarity with basic methods of linear algebra, differential equations, and probability theory, as covered in standard texts. Here, we describe the notation we use and briefly sketch highlights of various techniques. The references in the bibliography at the end of this appendix provides further information.

A.1 Linear Algebra

An operation O on a quantity z is called linear if, applied to any two instances z_1 and z_2, $O(\alpha z_1 + \beta z_2) = \alpha O(z_1) + \beta O(z_2)$ for any constants α and β. In this section, we consider linear operations on vectors and functions. We define a vector $\mathbf{v}$ as an array of N numbers $(v_1, v_2, \ldots, v_N)$. The numbers v_a for $a = 1, 2, \ldots, N$ are called the components of the vector. These are sometimes listed in a single N-row column

linear operator
vector $\mathbf{v}$

$$\mathbf{v} = \begin{pmatrix} v_1 \\ v_2 \\ \vdots \\ v_N \end{pmatrix}. \tag{A.1}$$

When necessary, we write component a of $\mathbf{v}$ as $[\mathbf{v}]_a = v_a$. We use $\mathbf{0}$ to denote the vector with all its components equal to 0. Spatial vectors, which are related to displacements in space, are a special case, and we denote them by $\vec{v}$ with components v_x and v_y in two-dimensional space or v_x, v_y, and v_z in three-dimensional space.

zero vector $\mathbf{0}$

spatial vector $\vec{v}$

The length or norm of $\mathbf{v}$, $|\mathbf{v}|$, when squared, can be written as a dot product,

norm

$$|\mathbf{v}|^2 = \mathbf{v} \cdot \mathbf{v} = \sum_{a=1}^{N} v_a^2 = v_1^2 + v_2^2 + \ldots + v_N^2. \tag{A.2}$$

The dot product of two different N-component vectors, $\mathbf{v}$ and $\mathbf{u}$, is

dot product

$$\mathbf{v} \cdot \mathbf{u} = \sum_{a=1}^{N} v_a u_a \,. \tag{A.3}$$

matrix $\mathbf{W}$

Matrix multiplication is a basic linear operation on vectors. An N_r by N_c matrix $\mathbf{W}$ is an array of N_r rows and N_c columns

$$\mathbf{W} = \begin{pmatrix} W_{11} & W_{12} & \dots & W_{1N_c} \\ W_{21} & W_{22} & \dots & W_{2N_c} \\ & & \vdots & \\ W_{N_r 1} & W_{N_r 2} & \dots & W_{N_r N_c} \end{pmatrix} \tag{A.4}$$

with elements W_{ab} for $a = 1, \dots, N_r$ and $b = 1, \dots, N_c$. In this text, the product of a matrix and a vector is written as $\mathbf{W} \cdot \mathbf{v}$. The dot implies multiplication and summation over a shared index, as it does for the dot product. If $\mathbf{W}$ is an N_r by N_c matrix and $\mathbf{v}$ is a N_c-component vector, $\mathbf{W} \cdot \mathbf{v}$ is an *matrix-vector* N_r-component vector with components
product

$$[\mathbf{W} \cdot \mathbf{v}]_a = \sum_{b=1}^{N_c} W_{ab} v_b \,. \tag{A.5}$$

In conventional matrix notation, the product of a matrix and a vector is written as $\mathbf{W}\mathbf{v}$, but we prefer to use the dot notation to avoid frequent occurrences of matrix transposes (see below). We similarly denote a matrix *matrix product* product as $\mathbf{W} \cdot \mathbf{M}$. Matrices can be multiplied in this way only if the number of columns of $\mathbf{W}$, N_c, is equal to the number of rows of $\mathbf{M}$. Then, $\mathbf{W} \cdot \mathbf{M}$ is a matrix with the same number of rows as $\mathbf{W}$ and the same number of columns as $\mathbf{M}$, and with elements

$$[\mathbf{W} \cdot \mathbf{M}]_{ab} = \sum_{c=1}^{N_c} W_{ac} M_{cb} \,. \tag{A.6}$$

A vector, written as in equation A.1, is equivalent to a one-column, N-row matrix, and the rules for various matrix operations can thus be applied to vectors as well.

square matrix

identity matrix

Square matrices are those for which $N_r = N_c = N$. An important square matrix is the identity matrix $\mathbf{I}$ with elements

$$[\mathbf{I}]_{ab} = \delta_{ab} \,, \tag{A.7}$$

Kronecker delta

where the Kronecker delta is defined as

$$\delta_{ab} = \begin{cases} 1 & \text{if } a = b \\ 0 & \text{otherwise} \,. \end{cases} \tag{A.8}$$

diagonal matrix

Another important type of square matrix is the diagonal matrix, defined by

$$\mathbf{W} = \text{diag}(h_1, h_2, \dots, h_N) = \begin{pmatrix} h_1 & 0 & \dots & 0 \\ 0 & h_2 & \dots & 0 \\ & & \vdots & \\ 0 & 0 & \dots & h_N \end{pmatrix}, \tag{A.9}$$

which has components $W_{ab} = h_a \delta_{ab}$ for some set of h_a, $a = 1, 2, \dots, N$.

The transpose of an N_r by N_c matrix $\mathbf{W}$ is an N_c by N_r matrix $\mathbf{W}^T$ with elements $[\mathbf{W}^T]_{ab} = W_{ba}$. The transpose of a column vector is a row vector, $\mathbf{v}^T = (v_1 v_2 \dots v_N)$. This is distinguished by the absence of commas from $(v_1, v_2, \dots, v_N)$ which, for us, is a listing of the components of a column vector. In the following table, we define a number of operations involving vectors and matrices. In the definitions, we provide our notation and the corresponding expressions in terms of vector components and matrix elements. We also provide the conventional matrix notation for these quantities as well as the notation used by MATLAB®, a computer software package commonly used to perform these operations numerically. For the MATLAB® notation (which does not use bold or italic symbols), we denote two column vectors by u and v, assuming they are defined within MATLAB® by instructions such as v =[v(1) v(2) ... v(N)]′.

transpose

Quantity	Definition	Matrix	MATLAB®
norm	$\|\mathbf{v}\|^2 = \mathbf{v} \cdot \mathbf{v} = \sum_a v_a^2$	$\mathbf{v}^T \mathbf{v}$	v′*v
dot product	$\mathbf{v} \cdot \mathbf{u} = \sum_a v_a u_a$	$\mathbf{v}^T \mathbf{u}$	v′*u
outer product	$[\mathbf{vu}]_{ab} = v_a u_b$	$\mathbf{vu}^T$	v*u′
matrix-vector product	$[\mathbf{W} \cdot \mathbf{v}]_a = \sum_b W_{ab} v_b$	$\mathbf{Wv}$	W*v
vector-matrix product	$[\mathbf{v} \cdot \mathbf{W}]_a = \sum_b v_b W_{ba}$	$\mathbf{v}^T \mathbf{W}$	v′*W
quadratic form	$\mathbf{v} \cdot \mathbf{W} \cdot \mathbf{u} = \sum_{ab} v_a W_{ab} u_b$	$\mathbf{v}^T \mathbf{W} \mathbf{u}$	v′*W*u
matrix-matrix product	$[\mathbf{W} \cdot \mathbf{M}]_{ab} = \sum_c W_{ac} M_{cb}$	$\mathbf{WM}$	W*M
transpose	$[\mathbf{W}^T]_{ab} = W_{ba}$	$\mathbf{W}^T$	W′

Several important definitions for square matrices are given below.

Operation	Notation	Definition	MATLAB®
inverse	$\mathbf{W}^{-1}$	$\mathbf{W} \cdot \mathbf{W}^{-1} = \mathbf{I}$	inv(W)
trace	$\text{tr}\mathbf{W}$	$\sum_a W_{aa}$	trace(W)
determinant	$\det \mathbf{W}$	see references	det(W)

A square matrix has an inverse only if its determinant is nonzero. Square matrices with certain properties are given special names (table below).

Property	Definition
symmetric	$\mathbf{W}^T = \mathbf{W}$ or $W_{ba} = W_{ab}$
orthogonal	$\mathbf{W}^T = \mathbf{W}^{-1}$ or $\mathbf{W}^T \cdot \mathbf{W} = \mathbf{I}$
positive-definite	$\mathbf{v} \cdot \mathbf{W} \cdot \mathbf{v} > 0$ for all $\mathbf{v} \neq \mathbf{0}$
Töplitz	$W_{ab} = f(a - b)$

where $f(a-b)$ is any function of the single variable $a-b$.

del operator ∇

For any real-valued function $E(\mathbf{v})$ of a vector $\mathbf{v}$, we can define the vector derivative (which is sometimes called del) of $E(\mathbf{v})$ as the vector $\nabla E(\mathbf{v})$ with components

$$[\nabla E(\mathbf{v})]_a = \frac{\partial E(\mathbf{v})}{\partial v_a}.\qquad (A.10)$$

directional derivative

The derivative of $E(\mathbf{v})$ in the direction $\mathbf{u}$ is then

$$\lim_{\epsilon \to 0}\left(\frac{E(\mathbf{v}+\epsilon\mathbf{u})-E(\mathbf{v})}{\epsilon}\right) = \mathbf{u}\cdot\nabla E(\mathbf{v}).\qquad (A.11)$$

Eigenvectors and Eigenvalues

eigenvector

An eigenvector of a square matrix $\mathbf{W}$ is a nonzero vector $\mathbf{e}$ that satisfies

$$\mathbf{W}\cdot\mathbf{e} = \lambda\mathbf{e}\qquad (A.12)$$

eigenvalue

for some number λ called the eigenvalue. Possible values of λ are determined by solving the polynomial equation

$$\det(\mathbf{W}-\lambda\mathbf{I}) = 0.\qquad (A.13)$$

Typically, but not always, this has N solutions if $\mathbf{W}$ is an N by N matrix, and these can be either real or complex. Complex eigenvalues come in complex-conjugate pairs if $\mathbf{W}$ has real-valued elements. We use the index μ to label the different eigenvalues and eigenvectors, λ_μ and $\mathbf{e}_\mu$. Note that μ identifies the eigenvector (and eigenvalue) to which we are referring; it does not signify a component of the eigenvector $\mathbf{e}_\mu$.

degeneracy

If $\mathbf{e}$ is an eigenvector, $\alpha\mathbf{e}$ is also an eigenvector for any nonzero value of α. We can use this freedom to normalize eigenvectors so that $|\mathbf{e}| = 1$. If two eigenvectors, say $\mathbf{e}_1$ and $\mathbf{e}_2$, have the same eigenvalues $\lambda_1 = \lambda_2$, they are termed degenerate. Then, $\alpha\mathbf{e}_1 + \beta\mathbf{e}_2$ is also an eigenvector with the same eigenvalue, for any α and β that are not both 0. Apart from such degeneracies, an N by N matrix can have at most N eigenvectors, although some matrices have fewer. If $\mathbf{W}$ has N nondegenerate eigenvalues, the

linear independence eigenvectors $\mathbf{e}_1, \ldots, \mathbf{e}_N$ are linearly independent, meaning that $\sum_\mu c_\mu \mathbf{e}_\mu = 0$ only if the coefficients $c_\mu = 0$ for all μ. These eigenvectors can be used to represent any N component vector $\mathbf{v}$ through the relation

$$\mathbf{v} = \sum_{\mu=1}^{N} c_\mu \mathbf{e}_\mu,\qquad (A.14)$$

basis

with a unique set of coefficients c_μ. They are thus said to form a basis for the set of vectors $\mathbf{v}$.

symmetric matrix

The eigenvalues and eigenvectors of symmetric matrices (for which $\mathbf{W}^T = \mathbf{W}$) have special properties, and for the remainder of this section, we consider this case. The eigenvalues of a symmetric matrix are real, and the eigenvectors are real and orthogonal (or can be made orthogonal in the case of degeneracy). This means that, if they are normalized to unit length, the eigenvectors satisfy

orthonormal eigenvectors

$$\mathbf{e}_\mu \cdot \mathbf{e}_\nu = \delta_{\mu\nu}. \tag{A.15}$$

To derive this result we note that, if $\mathbf{W}$ is a symmetric matrix, we can write $\mathbf{e}_\mu \cdot \mathbf{W} = \mathbf{W} \cdot \mathbf{e}_\mu = \lambda_\mu \mathbf{e}_\mu$. Therefore, allowing the matrix to act in both directions, we find $\mathbf{e}_\nu \cdot \mathbf{W} \cdot \mathbf{e}_\mu = \lambda_\mu \mathbf{e}_\nu \cdot \mathbf{e}_\mu = \lambda_\nu \mathbf{e}_\nu \cdot \mathbf{e}_\mu$. If $\lambda_\mu \neq \lambda_\nu$, this requires $\mathbf{e}_\nu \cdot \mathbf{e}_\mu = 0$. For orthogonal and normalized (orthonormal) eigenvectors, the coefficients in equation A.14 take the values

$$c_\mu = \mathbf{v} \cdot \mathbf{e}_\mu. \tag{A.16}$$

Let $\mathbf{E} = (\mathbf{e}_1 \, \mathbf{e}_2 \, \ldots \, \mathbf{e}_N)$ be an N by N matrix with columns formed from the orthonormal eigenvectors of a symmetric matrix. From equation A.15, this satisfies $[\mathbf{E}^{\mathrm{T}} \cdot \mathbf{E}]_{\mu\nu} = \mathbf{e}_\mu \cdot \mathbf{e}_\nu = \delta_{\mu\nu}$. Thus, $\mathbf{E}^{\mathrm{T}} = \mathbf{E}^{-1}$, making $\mathbf{E}$ an orthogonal matrix. $\mathbf{E}$ generates a transformation from the original matrix $\mathbf{W}$ to a diagonal form, which is called matrix diagonalization,

matrix diagonalization

$$\mathbf{E}^{-1} \cdot \mathbf{W} \cdot \mathbf{E} = \mathrm{diag}(\lambda_1, \ldots, \lambda_N). \tag{A.17}$$

Conversely,

$$\mathbf{W} = \mathbf{E} \cdot \mathrm{diag}(\lambda_1, \ldots, \lambda_N) \cdot \mathbf{E}^{-1}. \tag{A.18}$$

The transformation to and back from a diagonal form is extremely useful because computations with diagonal matrices are easy. Defining $\mathbf{L} = \mathrm{diag}(\lambda_1, \ldots, \lambda_N)$, we find, for example, that

$$\begin{aligned} \mathbf{W}^n &= (\mathbf{E} \cdot \mathbf{L} \cdot \mathbf{E}^{-1}) \cdot (\mathbf{E} \cdot \mathbf{L} \cdot \mathbf{E}^{-1}) \cdots (\mathbf{E} \cdot \mathbf{L} \cdot \mathbf{E}^{-1}) \\ &= \mathbf{E} \cdot \mathbf{L}^n \cdot \mathbf{E}^{-1} = \mathbf{E} \cdot \mathrm{diag}(\lambda_1^n, \ldots, \lambda_N^n) \cdot \mathbf{E}^{-1}. \end{aligned} \tag{A.19}$$

This result serves as a basis for defining functions of matrices. For any function f that can be written as a power or expanded in a power series (including, for example, exponentials and logarithms),

$$f(\mathbf{W}) = \mathbf{E} \cdot \mathrm{diag}(f(\lambda_1), \ldots, f(\lambda_N)) \cdot \mathbf{E}^{-1}. \tag{A.20}$$

Functional Analogs

A function $v(t)$ can be treated as if it were a vector with a continuous label. *functions as vectors* In other words, the function value $v(t)$ parameterized by the continuously varying argument t takes the place of the component v_a labeled by the integer-valued index a. In applying this analogy, sums over a for vectors are replaced by integrals over t for functions, $\sum_a \to \int dt$. For example, the functional analog of the squared norm and dot product are

$$\int_{-\infty}^{\infty} dt \, v^2(t) \quad \text{and} \quad \int_{-\infty}^{\infty} dt \, v(t)u(t). \tag{A.21}$$

linear integral operator

The analog of matrix multiplication for a function is the linear integral operator

$$\int_{-\infty}^{\infty} dt'\, W(t, t')v(t') \tag{A.22}$$

with the function values $W(t, t')$ playing the role of the matrix elements W_{ab}. The analog of the identity matrix is the Dirac δ function $\delta(t - t')$ discussed at the end of this section. The analog of a diagonal matrix is a function of two variables that is proportional to a δ function, $W(t, t') = h(t)\delta(t - t')$, for some function h.

functional inverse

All of the vector and matrix operations and properties defined above have functional analogs. Of particular importance are the functional inverse (which is not equivalent to an inverse function) that satisfies

$$\int_{-\infty}^{\infty} dt''\, W^{-1}(t, t'')W(t'', t') = \delta(t - t'), \tag{A.23}$$

translation invariance

and the analog of the Töplitz matrix, which is a linear integral operator that is translational-invariant, and thus can be written as

$$W(t, t') = K(t - t'). \tag{A.24}$$

linear filter

The linear integral operator then takes the form of a linear filter,

$$\int_{-\infty}^{\infty} dt'\, K(t - t')v(t') = \int_{-\infty}^{\infty} d\tau\, K(\tau)v(t - \tau), \tag{A.25}$$

where we have made the replacement $t' \to t - \tau$.

The δ Function

Despite its name, the Dirac δ function is not a properly defined function, but rather the limit of a sequence of functions. In this limit, the δ function approaches 0 everywhere except where its argument is 0, and there it grows without bound. The infinite height and infinitesimal width of this function are matched so that its integral is 1. Thus,

$$\int dt\, \delta(t) = 1, \tag{A.26}$$

provided only that the limits of integration surround the point $t = 0$ (otherwise the integral is 0). The integral of the product of a δ function with any continuous function f is

$$\int dt'\, \delta(t - t')f(t') = f(t) \tag{A.27}$$

for any value of t contained within the integration interval (if t is not within this interval, the integral is 0). These two identities normally provide enough information to use the δ function in calculations despite its unwieldy definition.

The sequence of functions used to construct the δ function as a limit is not unique. In essence, any function that integrates to 1 and has a single peak that gets continually narrower and taller as the limit is taken can be used. For example, the δ function can be expressed as the limit of a square pulse

$$\delta(t) = \lim_{\Delta t \to 0} \begin{cases} 1/\Delta t & \text{if } -\Delta t/2 < t < \Delta t/2 \\ 0 & \text{otherwise} \end{cases} \qquad (A.28)$$

or a Gaussian function

$$\delta(t) = \lim_{\Delta t \to 0} \frac{1}{\sqrt{2\pi}\Delta t} \exp\left[-\frac{1}{2}\left(\frac{t}{\Delta t}\right)^2\right]. \qquad (A.29)$$

It is most often expressed as

$$\delta(t) = \frac{1}{2\pi} \int_{-\infty}^{\infty} d\omega \, \exp(i\omega t). \qquad (A.30)$$

δ function definition

This underlies the inverse Fourier transform, as discussed below.

Eigenfunctions

The functional analog of the eigenvector (equation A.12) is the eigenfunction $e(t)$ that satisfies

$$\int dt' \, W(t, t')e(t') = \lambda e(t). \qquad (A.31)$$

For translationally invariant integral operators, $W(t, t') = K(t - t')$, the eigenfunctions are complex exponentials,

$$\int dt' \, K(t - t') \exp(i\omega t') = \left(\int d\tau \, K(\tau) \exp(-i\omega\tau)\right) \exp(i\omega t), \qquad (A.32)$$

as can be seen by making the change of variables $\tau = t - t'$. Here $i = \sqrt{-1}$, and the complex exponential is defined by the identity

$$\exp(i\omega t) = \cos(\omega t) + i\sin(\omega t). \qquad (A.33)$$

complex exponential

Comparing equations A.31 and A.32, we see that the eigenvalue for this eigenfunction is

$$\lambda(\omega) = \int d\tau \, K(\tau) \exp(-i\omega\tau). \qquad (A.34)$$

In this case, the continuous label ω takes the place of the discrete label μ used to identify the different eigenvalues of a matrix.

A functional analog of expanding a vector using eigenvectors as a basis (equation A.14) is the inverse Fourier transform, which expresses a function in an expansion using complex exponential eigenfunctions as a basis. The analog of equation A.16 for determining the coefficient functions of this expansion is the Fourier transform.

Fourier Transforms

As outlined in the previous section, Fourier transforms provide a useful representation for functions when they are acted upon by translation-invariant linear operators.

Fourier transform

The Fourier transform of a function $f(t)$ is a complex function of a real argument ω given by

$$\tilde{f}(\omega) = \int_{-\infty}^{\infty} dt\, f(t) \exp(i\omega t)\,. \tag{A.35}$$

inverse Fourier transform

The Fourier transform $\tilde{f}(\omega)$ provides an alternative representation of the original function $f(t)$ because it can be inverted through

$$f(t) = \frac{1}{2\pi} \int_{-\infty}^{\infty} d\omega\, \tilde{f}(\omega) \exp(-i\omega t)\,. \tag{A.36}$$

This provides an inverse because

$$\frac{1}{2\pi} \int_{-\infty}^{\infty} d\omega \exp(-i\omega t) \int_{-\infty}^{\infty} dt'\, f(t') \exp(i\omega t') \tag{A.37}$$

$$= \int_{-\infty}^{\infty} dt'\, f(t') \frac{1}{2\pi} \int_{-\infty}^{\infty} d\omega \exp(i\omega(t' - t)) = \int_{-\infty}^{\infty} dt'\, f(t') \delta(t' - t) = f(t)$$

by the definition of the δ function in equation A.30. The function $f(t)$ has to satisfy a set of criteria called the Dirichlet conditions for the inversion of the Fourier transform to be exact.

convolution

The convolution of two functions f and g is the integral

$$h(t) = \int_{-\infty}^{\infty} d\tau\, f(\tau) g(t - \tau)\,. \tag{A.38}$$

This is sometimes denoted by $h = f * g$. Note that the operation of multiplying a function by a linear filter and integrating, as in equation A.25, is a convolution. Fourier transforms are useful for dealing with convolutions because the Fourier transform of a convolution is the product of the Fourier transforms of the two functions being convolved,

$$\tilde{h}(\omega) = \tilde{f}(\omega)\tilde{g}(\omega)\,. \tag{A.39}$$

To show this, we note that

$$\tilde{h}(\omega) = \int_{-\infty}^{\infty} dt \exp(i\omega t) \int_{-\infty}^{\infty} d\tau\, f(\tau) g(t - \tau) \tag{A.40}$$

$$= \int_{-\infty}^{\infty} d\tau\, f(\tau) \exp(i\omega \tau) \int_{-\infty}^{\infty} dt\, g(t - \tau) \exp(i\omega(t - \tau))$$

$$= \int_{-\infty}^{\infty} d\tau\, f(\tau) \exp(i\omega \tau) \int_{-\infty}^{\infty} dt'\, g(t') \exp(i\omega t') \quad \text{where } t' = t - \tau,$$

which is equivalent to equation A.39. A related result is Parseval's theorem,   *Parseval's theorem*

$$\int_{-\infty}^{\infty} dt \, |f(t)|^2 = \frac{1}{2\pi} \int_{-\infty}^{\infty} d\omega \, |\tilde{f}(\omega)|^2 \, . \tag{A.41}$$

If $f(t)$ is periodic with period T (so that $f(t+T)=f(t)$ for all t), it can be represented by a Fourier series rather than a Fourier integral. That is,   *periodic function* / *Fourier series*

$$f(t) = \sum_{k=-\infty}^{\infty} \tilde{f}_k \exp(-i2\pi kt/T) \, , \tag{A.42}$$

where $\tilde{f}_k$ is given by

$$\tilde{f}_k = \frac{1}{T} \int_0^T dt \, f(t) \exp(i2\pi kt/T) \, . \tag{A.43}$$

As in the case of Fourier transforms, certain conditions have to hold for the series to converge and to be exactly invertible. The Fourier series has properties similar to Fourier transforms, including a convolution theorem and a version of Parseval's theorem. The real and imaginary parts of a Fourier series are often separated, giving the alternative form

$$f(t) = \tilde{f}_0 + \sum_{k=1}^{\infty} \left(\tilde{f}_k^c \cos(2\pi kt/T) + \tilde{f}_k^s \sin(2\pi kt/T) \right) \tag{A.44}$$

with

$$\tilde{f}_0 = \frac{1}{T} \int_0^T dt \, f(t) \, , \quad \tilde{f}_k^c = \frac{2}{T} \int_0^T dt \, f(t) \cos(2\pi kt/T) \, ,$$

$$\tilde{f}_k^s = \frac{2}{T} \int_0^T dt \, f(t) \sin(2\pi kt/T) \, . \tag{A.45}$$

When computed numerically, a Fourier transform is typically based on a certain number, N_t, of samples of the function, $f_n = f(n\delta)$ for $n = 0, 1, \ldots N_t - 1$. The discrete Fourier transform of these samples is then used as an approximation of the continuous Fourier transform. The discrete Fourier transform is defined as   *discrete Fourier transform*

$$\tilde{f}_m = \sum_{n=0}^{N_t-1} f_n \exp\left(i2\pi nm/N_t\right) \, . \tag{A.46}$$

Note that $\tilde{f}_{N_t+m} = \tilde{f}_m$. An approximation of the continuous Fourier transform is provided by the relation $\tilde{f}(2\pi m/(N_t\delta)) \approx \delta \tilde{f}_m$. The inverse discrete Fourier transform is

$$f_n = \frac{1}{N_t} \sum_{m=0}^{N_t-1} \tilde{f}_m \exp\left(-i2\pi mn/N_t\right) \, . \tag{A.47}$$

This equation implies a periodic continuation of f_n outside the range $0 \le n < N_t$, so that $f_{n+N_t} = f_n$ for all n. Consult the references in the bibliography for an analysis of the properties of the discrete Fourier transform and the quality of its approximation to the continuous Fourier transform. Note in particular that there is a difference between the discrete-time Fourier transform, which is the Fourier transform of a signal that is inherently discrete (i.e., is defined only at discrete points), and the discrete Fourier transform, given above, which is based on a finite number of samples of an underlying continuous function. If $f(t)$ is band-limited, *sampling theorem*   meaning that $\tilde{f}(\omega) = 0$ for $|\omega| > \pi/\delta$, the sampling theorem states that $f(t)$ is completely determined by regular samples spaced at intervals $1/\delta$.

Fourier transforms of functions of more than one variable involve a direct extension of the equations given above to multi-dimensional integrals. For example,

$$\tilde{f}(\omega_x, \omega_y) = \int dx \int dy\, f(x, y) \exp(i(\omega_x x + \omega_y y)). \qquad \text{(A.48)}$$

The properties of multi-dimensional transforms are similar to those of one-dimensional transforms.

A.2   Finding Extrema and Lagrange Multipliers

An operation frequently encountered in the text is minimizing a quadratic form. In terms of vectors, this typically amounts to finding the matrix $\mathbf{W}$ that makes the product $\mathbf{W} \cdot \mathbf{v}$ closest to another vector $\mathbf{u}$ when averaged over a number of presentations of $\mathbf{v}$ and $\mathbf{u}$. The function to be minimized is the average squared error $\langle |\mathbf{W} \cdot \mathbf{v} - \mathbf{u}|^2 \rangle$, where the brackets denote averaging over all the different samples $\mathbf{v}$ and $\mathbf{u}$. Setting the derivative of this expression with respect to $\mathbf{W}$ (or equivalently its elements W_{ab}) to 0 gives *minimization of*   the equation
quadratic form

$$\mathbf{W} \cdot \langle \mathbf{vv} \rangle = \langle \mathbf{uv} \rangle \quad \text{or} \quad \sum_{c=1}^{N} W_{ac} \langle v_c v_b \rangle = \langle u_a v_b \rangle . \qquad \text{(A.49)}$$

Many variants of this equation, solved by a number of techniques, appear in the text.

Often, when a function $f(\mathbf{v})$ has to be minimized or maximized with respect to a vector $\mathbf{v}$, there is an additional constraint on $\mathbf{v}$ that requires another function $g(\mathbf{v})$ to be held constant. The standard way of dealing with this situation is to find the extrema of the function $f(\mathbf{v}) + \lambda g(\mathbf{v})$ where λ is *Lagrange*   a free parameter called a Lagrange multiplier. Once this is done, the value *multiplier*   of λ is determined by requiring $g(\mathbf{v})$ to take the specified value. This procedure can appear a bit mysterious when first encountered, so we provide a rather extended discussion.

The condition that characterizes an extreme value of the function $f(\mathbf{v})$ is that small changes $\Delta\mathbf{v}$ (with components Δv_a) in the vector $\mathbf{v}$ should not change the value of the function to first order in $\Delta\mathbf{v}$. This results in the condition

$$\sum_{a=1}^{N}[\nabla f]_a \Delta v_a = 0\,,\qquad\qquad(A.50)$$

where we use the notation

$$[\nabla f]_a = \frac{\partial f}{\partial v_a}\qquad\qquad(A.51)$$

to make the equations more compact. Without a constraint, equation A.50 must be satisfied for all $\Delta\mathbf{v}$, which can occur only if each term in the sum vanishes separately. Thus, we find the usual condition for an extremum

$$[\nabla f]_a = 0\qquad\qquad(A.52)$$

for all a. However, with a constraint such as $g(\mathbf{v})=$constant, equation A.50 does not have to hold for all possible $\Delta\mathbf{v}$, only for those that satisfy the constraint. The condition on $\Delta\mathbf{v}$ imposed by the constraint is that g cannot change to first order in $\Delta\mathbf{v}$. Therefore,

$$\sum_{a=1}^{N}[\nabla g]_a \Delta v_a = 0\qquad\qquad(A.53)$$

with the same notation for the derivative used for g as for f.

The most obvious way to deal with the constraint equation A.53 is to solve for one of the components of $\Delta\mathbf{v}$, say Δv_c, writing

$$\Delta v_c = -\frac{1}{[\nabla g]_c}\sum_{a\neq c}[\nabla g]_a \Delta v_a\,.\qquad\qquad(A.54)$$

Then we substitute this expression into equation A.50 to obtain

$$\sum_{a\neq c}[\nabla f]_a \Delta v_a - \frac{[\nabla f]_c}{[\nabla g]_c}\sum_{a\neq c}[\nabla g]_a \Delta v_a = 0\,.\qquad\qquad(A.55)$$

Because we have eliminated the constraint, this equation must be satisfied for all values of the remaining components of $\Delta\mathbf{v}$, those with $a\neq c$, and thus we find

$$[\nabla f]_a - \frac{[\nabla f]_c}{[\nabla g]_c}[\nabla g]_a = 0\qquad\qquad(A.56)$$

for all $a\neq c$. The derivatives of f and g are functions of $\mathbf{v}$, so these equations can be solved to determine where the extremum point is located.

In the above derivation, we have singled out component c for special treatment. We have no way of knowing until we get to the end of the calculation whether the particular c we chose leads to a simple or a complex set of final equations. The clever idea of the Lagrange multiplier is to notice that the whole problem is symmetric with respect to the different components of $\Delta \mathbf{v}$. Choosing one c value, as we did above, breaks this symmetry and often complicates the algebra. To introduce the Lagrange multiplier, we simply define it as

$$\lambda = -\frac{[\nabla f]_c}{[\nabla g]_c} \, . \tag{A.57}$$

With this notation, the final set of equations (A.56) can be written as

$$[\nabla f]_a + \lambda [\nabla g]_a = 0 \, . \tag{A.58}$$

Before, we had to say that these equations held only for $a \neq c$ because c was treated differently. Now, however, notice that the above equation when a is set to c is algebraically equivalent to the definition in equation A.57. Thus, we can say that equation A.58 applies for all a, and this provides a symmetric formulation of the problem of finding an extremum that often results in simpler algebra.

The final realization is that equation A.58 for all a is precisely what we would have derived if we had set out in the first place to find an extremum of the function $f(\mathbf{v}) + \lambda g(\mathbf{v})$ and forgotten about the constraint entirely. Of course this lunch is not completely free. From equation A.58, we derive a set of extremum points parameterized by the undetermined variable λ. To fix λ, we must substitute this family of solutions back into $g(\mathbf{v})$ and find the value of λ that satisfies the constraint that $g(\mathbf{v})$ equals the specified value. This provides the solution to the constrained problem.

A.3   Differential Equations

The most general differential equation we consider takes the form

$$\frac{d\mathbf{v}}{dt} = \mathbf{f}(\mathbf{v}) \, , \tag{A.59}$$

fixed point

limit cycle

chaos

strange attractor

equilibrium point

where $\mathbf{v}$ is an N-component vector of time-dependent variables, and $\mathbf{f}$ is a vector of functions of $\mathbf{v}$. Unless it is unstable, allowing the absolute value of one or more of the components of $\mathbf{v}$ to grow without bound, this type of equation has three classes of solutions. For one class, called stable fixed points or point attractors, $\mathbf{v}(t)$ approaches a time-independent vector $\mathbf{v}_\infty$ ($\mathbf{v}(t) \to \mathbf{v}_\infty$) as $t \to \infty$. In a second class of solutions, called limit cycles, $\mathbf{v}(t)$ becomes periodic at large times and repeats itself indefinitely. For the third class of solutions, the chaotic ones, $\mathbf{v}(t)$ never repeats itself but the trajectory of the system lies in a limited subspace of the total space of allowed configurations called a strange attractor. Chaotic solutions are extremely sensitive to initial conditions.

We focus most of our analysis on fixed point solutions, which are also called equilibrium points. For $\mathbf{v}_\infty$ to be a time-independent solution of equation A.59, we must have $\mathbf{f}(\mathbf{v}_\infty) = 0$. General solutions of equation A.59 when $\mathbf{f}$ is nonlinear cannot be constructed, but we can use linear techniques to study the behavior of $\mathbf{v}$ near a fixed point $\mathbf{v}_\infty$. If $\mathbf{f}$ is linear, the techniques we use and the solutions we obtain as approximations in the nonlinear case are exact. Near the fixed point $\mathbf{v}_\infty$, we write

$$\mathbf{v}(t) = \mathbf{v}_\infty + \boldsymbol{\epsilon}(t) \tag{A.60}$$

and consider the case when all the components of the vector $\boldsymbol{\epsilon}$ are small. Then, we can expand $\mathbf{f}$ in a Taylor series,

Taylor series

$$\mathbf{f}(\mathbf{v}(t)) \approx \mathbf{f}(\mathbf{v}_\infty) + \mathbf{J} \cdot \boldsymbol{\epsilon}(t) = \mathbf{J} \cdot \boldsymbol{\epsilon}(t) \,, \tag{A.61}$$

where $\mathbf{J}$ is the called the Jacobian matrix and has elements

Jacobian matrix

$$J_{ab} = \left. \frac{\partial f_a(\mathbf{v})}{\partial v_b} \right|_{\mathbf{v}=\mathbf{v}_\infty} . \tag{A.62}$$

In the second equality of equation A.61, we have used the fact that $\mathbf{f}(\mathbf{v}_\infty) = 0$.

Using the approximation of equation A.61, equation A.59 becomes

$$\frac{d\boldsymbol{\epsilon}}{dt} = \mathbf{J} \cdot \boldsymbol{\epsilon} \,. \tag{A.63}$$

The temporal evolution of $\mathbf{v}(t)$ is best understood by expanding $\boldsymbol{\epsilon}$ in the basis provided by the eigenvectors of $\mathbf{J}$. Assuming that $\mathbf{J}$ is real and has N linearly independent eigenvectors $\mathbf{e}_1, \dots, \mathbf{e}_N$ with different eigenvalues $\lambda_1, \dots, \lambda_N$, we write

$$\boldsymbol{\epsilon}(t) = \sum_{\mu=1}^{N} c_\mu(t)\mathbf{e}_\mu \,. \tag{A.64}$$

Substituting this into equation A.63, we find that the coefficients must satisfy

$$\frac{dc_\mu}{dt} = \lambda_\mu c_\mu \,. \tag{A.65}$$

This produces the solution

$$\boldsymbol{\epsilon}(t) = \sum_{\mu=1}^{N} c_\mu(0) \exp(\lambda_\mu t)\mathbf{e}_\mu \,, \tag{A.66}$$

where $\boldsymbol{\epsilon}(0) = \sum_\mu c_\mu(0)\mathbf{e}_\mu$. The individual terms in the sum on the right side of equation A.66 are called modes. This solution is exact for equation A.63, but is only a valid approximation when applied to equation A.59 if $\boldsymbol{\epsilon}$ is small. Note that the different coefficients c_μ evolve over time, independently of each other. This does not require the eigenvectors to be orthogonal. If the eigenvalues and eigenvectors are complex, $\mathbf{v}(t)$ will nonetheless remain real if $\mathbf{v}(0)$ is real, because the complex modes come

modes

in conjugate pairs that combine to form a real function. Expression A.66 is not the correct solution if some of the eigenvalues are equal. The reader should consult the references for the solution in this case.

Equation A.66 determines how the evolution of $\mathbf{v}(t)$ in the neighborhood of $\mathbf{v}_\infty$ depends on the eigenvalues of $\mathbf{J}$. If we write $\lambda_\mu = \alpha_\mu + i\omega_\mu$,

$$\exp(\lambda_\mu t) = \exp(\alpha_\mu t)\left(\cos(\omega_\mu t) + i\sin(\omega_\mu t)\right). \tag{A.67}$$

This implies that modes with real eigenvalues ($\omega_\mu = 0$) evolve exponentially over time, and modes with complex eigenvalues ($\omega_\mu \neq 0$) oscillate with a frequency ω_μ. Recall that the eigenvalues are always real if $\mathbf{J}$ is a symmetric matrix. Modes with negative real eigenvalues ($\alpha_\mu < 0$ and $\omega_\mu = 0$) decay exponentially to 0, while those with positive real eigenvalues ($\alpha_\mu > 0$ and $\omega_\mu = 0$) grow exponentially. Similarly, the oscillations for modes with complex eigenvalues are damped exponentially to 0 if the real part of the eigenvalue is negative ($\alpha_\mu < 0$ and $\omega_\mu \neq 0$), and grow exponentially if the real part is positive ($\alpha_\mu > 0$ and $\omega_\mu \neq 0$).

attractor

unstable fixed point

marginal stability

Stability of the fixed point $\mathbf{v}_\infty$ requires the real parts of all the eigenvalues to be negative ($\alpha_\mu < 0$ for all μ). In this case, the point $\mathbf{v}_\infty$ is a stable fixed-point attractor of the system, meaning that $\mathbf{v}(t)$ will approach $\mathbf{v}_\infty$ if it starts from any point in the neighborhood of $\mathbf{v}_\infty$. If any real part is positive ($\alpha_\mu > 0$ for any μ), the fixed point is unstable. Almost any $\mathbf{v}(t)$ initially in the neighborhood of $\mathbf{v}_\infty$ will move away from that neighborhood. If $\mathbf{f}$ is linear, the exponential growth of $|\mathbf{v}(t) - \mathbf{v}_\infty|$ never stops in this case. For a nonlinear f, equation A.66 determines what happens only in the neighborhood of $\mathbf{v}_\infty$, and the system may ultimately find a stable attractor away from $\mathbf{v}_\infty$, either a fixed point, a limit cycle, or a chaotic attractor. In all these cases, the mode for which the real part of λ_μ takes the largest value dominates the dynamics as $t \to \infty$. If this real part is equal to 0, the fixed point is called marginally stable.

As mentioned previously, the analysis presented above as an approximation for nonlinear differential equations near a fixed point is exact if the original equation is linear. In the text, we frequently encounter linear equations of the form

$$\tau\frac{dv}{dt} = v_\infty - v. \tag{A.68}$$

This can be solved by setting $z = v - v_\infty$, rewriting the equation as $dz/z = -dt/\tau$, and integrating both sides:

$$\int_{z(0)}^{z(t)} dz'\, \frac{1}{z'} = \ln\left(\frac{z(t)}{z(0)}\right) = -\frac{t}{\tau}. \tag{A.69}$$

This gives $z(t) = z(0)\exp(-t/\tau)$ or

$$v(t) = v_\infty + (v(0) - v_\infty)\exp(-t/\tau). \tag{A.70}$$

In some cases, we consider discrete rather than continuous dynamics defined over discrete steps $n = 1, 2, \ldots$ through a difference rather than a differential equation. Linearization about equilibrium points can be used to analyze nonlinear difference equations as well as differential equations, and this reveals similar classes of behavior. We illustrate difference equations by analyzing a linear case,

difference equation

$$\mathbf{v}(n+1) = \mathbf{v}(n) + \mathbf{W} \cdot \mathbf{v}(n). \tag{A.71}$$

The strategy for solving this equation is similar to that for solving differential equations. Assuming $\mathbf{W}$ has a complete set of linearly independent eigenvectors $\mathbf{e}_1, \ldots, \mathbf{e}_N$ with different eigenvalues $\lambda_1, \ldots, \lambda_N$, the modes separate, and the general solution is

$$\mathbf{v}(n) = \sum_{\mu=1}^{N} c_\mu (1 + \lambda_\mu)^n \mathbf{e}_\mu, \tag{A.72}$$

where $\mathbf{v}(0) = \sum_\mu c_\mu \mathbf{e}_\mu$. This has characteristics similar to equation A.66. Writing $\lambda_\mu = \alpha_\mu + i\omega_\mu$, mode μ is oscillatory if $\omega_\mu \neq 0$. In the discrete case, stability of the system is controlled by the magnitude

$$|1 + \lambda_\mu|^2 = (1 + \alpha_\mu)^2 + (\omega_\mu)^2. \tag{A.73}$$

If this is greater than 1 for any value of μ, $|\mathbf{v}(n)| \to \infty$ as $n \to \infty$. If it is less than 1 for all μ, $\mathbf{v}(n) \to \mathbf{0}$ in this limit.

A.4 Electrical Circuits

Biophysical models of single cells involve equivalent circuits composed of resistors, capacitors, and voltage and current sources. We review here basic results for such circuits. Figures A.1A and A.1B show the standard symbols for resistors and capacitors, and define the relevant voltages and currents. A resistor (figure A.1A) satisfies Ohm's law, which states that the voltage $V_R = V_1 - V_2$ across a resistance R carrying a current I_R is

Ohm's law

$$V_R = I_R R. \tag{A.74}$$

Resistance is measured in ohms (Ω); 1 ohm is the resistance through which 1 ampere of current causes a voltage drop of 1 volt (1 V = 1 A × 1 Ω).

A capacitor (figure A.1B) stores charge across an insulating medium, and the voltage across it $V_C = V_1 - V_2$ is related to the charge it stores, Q_C, by

$$CV_C = Q_C, \tag{A.75}$$

where C is the capacitance. Electrical current cannot cross the insulating medium, but charges can be redistributed on each side of the capacitor, which leads to the flow of current. We can take a time derivative of both sides of equation A.75 and use the fact that current is equal to the rate of

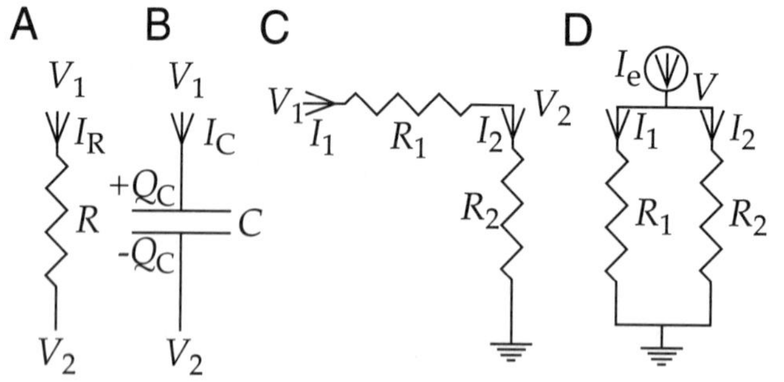

Figure A.1 Electrical circuit elements and resistor circuits. (A) Current I_R flows through a resistance R, producing a voltage drop $V_1 - V_2 = V_R$. (B) Charge $\pm Q_C$ is stored across a capacitance C, leading to a voltage $V_C = V_1 - V_2$ and a current I_C. (C) Series resistor circuit called a voltage divider. (D) Parallel resistor circuit. I_e represents an external current source. The lined triangle symbol at the bottom of the circuits in C and D represents an electrical ground, which is defined to be at 0 voltage.

change of charge, $I_C = dQ_C/dt$, to obtain the basic voltage-current relationship for a capacitor,

$$C\frac{dV_C}{dt} = I_C.\tag{A.76}$$

V-I relation for capacitor

Capacitance is measured in units of farads (F), defined as the capacitance for which 1 ampere of current causes a voltage change of 1 volt per second ($1\,\mathrm{F} \times 1\,\mathrm{V/s} = 1\,\mathrm{A}$).

Kirchhoff's laws

The voltages at different points in a circuit and the currents flowing through various circuit elements can be computed using equations A.74 and A.76 and rules called Kirchhoff's laws. These state that voltage differences around any closed loop in a circuit must sum to 0, and that the sum of all the currents entering any point in a circuit must be 0. Applying the second of these rules to the circuit in figure A.1C, we find that $I_1 = I_2$. Ohm's law tells us that $V_1 - V_2 = I_1 R_1$ and $V_2 = I_2 R_2$. Solving these gives $V_1 = I_1(R_1 + R_2)$, which tells us that resistors arranged in series add, and $V_2 = V_1 R_2/(R_1 + R_2)$, which is why this circuit is called a voltage divider.

In the circuit of figure A.1D, we have added an external source passing the current I_e. For this circuit, Kirchhoff's and Ohm's laws tells us that $I_e = I_1 + I_2 = V/R_1 + V/R_2$. This indicates how resistors add in parallel, $V = I_e R_1 R_2/(R_1 + R_2)$.

Next, we consider the electrical circuit in figure A.2A, in which a resistor and capacitor are connected together. Kirchhoff's laws require that $I_C + I_R = 0$. Putting this together with equations A.74 and A.76, we find

$$C\frac{dV}{dt} = I_C = -I_R = -\frac{V}{R}.\tag{A.77}$$

Solving this gives

$$V(t) = V(0)\exp(-t/RC),\tag{A.78}$$

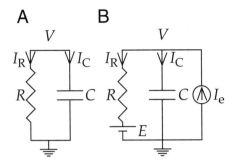

Figure A.2 RC circuits. (A) Current $I_C = -I_R$ flows in the resistor-capacitor circuit as the stored charge is released. (B) Simple passive membrane model including a potential E and current source I_e. As in figure A.1, the lined triangles represent a ground or point of 0 voltage.

showing the exponential decay (with time constant $\tau = RC$) of the initial voltage $V(0)$ as the charge on the capacitor leaks out through the resistor.

Figure A.2B includes two extra components needed to build a simple model neuron, the voltage source E and the current source I_e. Using Kirchhoff's laws, $I_e - I_C - I_R = 0$, and the equation for the voltage V is

$$C\frac{dV}{dt} = \frac{E - V}{R} + I_e.$$ (A.79)

If I_e is constant, the solution of this equation is

$$V(t) = V_\infty + (V(0) - V_\infty)\exp(-t/\tau),$$ (A.80)

where $V_\infty = E + RI_e$ and $\tau = RC$. This shows exponential relaxation from the initial potential $V(0)$ to the equilibrium potential V_∞ at a rate governed by the time constant τ of the circuit.

For the case $I_e = I\cos(\omega t)$, once an initial transient has decayed to 0, we find

$$V(t) = E + \frac{RI\cos(\omega t - \phi)}{\sqrt{1 + \omega^2\tau^2}},$$ (A.81)

where $\tan(\phi) = \omega\tau$. Equation A.81 shows that the cell membrane acts as a low-pass filter, because the higher the frequency ω of the input current, the greater the attenuation of the oscillations of the potential due to the factor $1/(1 + \omega^2\tau^2)^{1/2}$. The phase shift ϕ is an increasing function of frequency that approaches $\pi/2$ as $\omega \to \infty$.

A.5   Probability Theory

Probability distributions and densities are discussed extensively in the text. Here, we present a slightly more formal treatment. At the heart of *sample space*

probability measure

probability theory lie two objects: a sample space, Ω, and a measure. We begin by considering the simplest case of a finite sample space. Here, each element ω of the full sample space Ω can be thought of as one of the possible outcomes of a random process, for example, one of the 6^5 possible results of rolling five dice. The measure assigns a number γ_ω to each outcome ω, and these must satisfy $0 \leq \gamma_\omega \leq 1$ and $\sum_\omega \gamma_\omega = 1$.

random variable

We are primarily interested in random variables (which are infamously neither random nor variable). A random variable is a mapping from a random outcome ω to a space such as the space of integers. An example is the number of ones that appear when five dice are rolled. Typically, a capital letter, such as S, is used for the random variable, and the corresponding lowercase letter, s in this case, is used for a particular value it might take. The probability that S takes the value s is then written as $P[S = s]$. In the text, we typically shorten this to $P[s]$, but here we keep the full notation (except in the following table). $P[S = s]$ is determined by the measures of the events for which $S = s$ and takes the value

$$P[S = s] = \sum_{\substack{\omega \text{ with} \\ S(\omega)=s}} \gamma_\omega \,. \tag{A.82}$$

The notation $S(\omega)$ refers to the value of S generated by the random event labeled by ω, and the sum is over all events for which $S(\omega) = s$.

Some key statistics for discrete random variables include the following.

Quantity	Definition	Alias
mean	$\langle s \rangle = \sum_s P[s]s$	$\bar{s}$, $\mathcal{E}[S]$
variance	$\mathrm{var}(S) = \langle s^2 \rangle - \langle s \rangle^2 = \sum_s P[s]s^2 - \langle s \rangle^2$	σ_s^2, $\mathcal{V}[S]$
covariance	$\langle s_1 s_2 \rangle - \langle s_1 \rangle \langle s_2 \rangle = \sum_{s_1 s_2} P[s_1, s_2]s_1 s_2 - \langle s_1 \rangle \langle s_2 \rangle$	$\mathrm{cov}(S_1, S_2)$

where S_1 and S_2 are two random variables defined over the same sample space. This links the two random variables, in that

$$P[S_1 = s_1, S_2 = s_2] = \sum_{\substack{\omega \text{ with} \\ S_1(\omega)=s_1, \\ S_2(\omega)=s_2}} \gamma_\omega \,, \tag{A.83}$$

and provides a basis for them to be correlated. Means are additive,

$$\langle s_1 + s_2 \rangle = \langle s_1 \rangle + \langle s_2 \rangle \,, \tag{A.84}$$

but other quantities typically are not, for example,

$$\mathrm{var}(S_1 + S_2) = \mathrm{var}(S_1) + \mathrm{var}(S_2) + 2\mathrm{cov}(S_1, S_2) \,. \tag{A.85}$$

independence

Two random variables are independent if $P[S_1 = s_1, S_2 = s_2] = P[S_1 = s_1]P[S_2 = s_2]$ for all s_1 and s_2. If S_1 and S_2 are independent, $\mathrm{cov}(S_1, S_2) = 0$, but the converse is not true in general.

In addition to finite, sample spaces can be either countably or uncountably infinite. In the latter case, there are various technical complications that are discussed in the references. Under suitable conditions, a continuous random variable S, which is a mapping from a sample space to a continuous space such as the real numbers, has a probability density function $p[s]$ defined by

continuous random variable

probability density

$$p[s] = \lim_{\Delta s \to 0} \left(\frac{P[s \le S \le s + \Delta s]}{\Delta s} \right). \tag{A.86}$$

Quantities such as the mean and variance of a continuous random variable are defined as for a discrete random variable, but involve integrals over probability densities rather than sums over probabilities.

Some commonly used discrete and continuous distributions are listed in the table below.

Name	Range of s	Probability	Mean	Variance
Bernoulli	$s = 0$ or 1	$p^s(1-p)^{1-s}$	p	$p(1-p)$
Poisson	$s = 0, 1, 2, \ldots$	$\alpha^s \exp(-\alpha)/s!$	α	α
Exponential	$s > 0$	$\alpha \exp(-\alpha s)$	$1/\alpha$	$1/\alpha^2$
Gaussian	$-\infty < s < \infty$	$\mathcal{N}[s; g, \Sigma]$	g	Σ
Cauchy	$-\infty < s < \infty$	$\beta/(\pi((s-\alpha)^2 + \beta^2))$	$* \alpha *$	$* 1/\beta^2 *$

where

$$\mathcal{N}(s; g, \Sigma) = \frac{1}{\sqrt{2\pi\Sigma}} \exp\left(-\frac{(s-g)^2}{2\Sigma} \right). \tag{A.87}$$

Here, we use Σ to denote the variance of the Gaussian distribution, which is more often written as σ^2. The asterisks in the entries for the Cauchy distribution reflect the fact that it has such heavy tails that the integrals defining its mean and variance do not converge. Nevertheless, α and $1/\beta^2$ play similar roles, and are called location and scale parameters respectively.

The Gaussian distribution is particularly important because of the central limit theorem. Consider m continuous random variables $S_1, S_2, S_3, \ldots S_m$ that are independent and have identical distributions with finite mean g and variance σ^2. Defining

central limit theorem

$$Z_m = \frac{1}{m} \sum_{k=1}^{m} S_k, \tag{A.88}$$

the central limit theorem states that, under rather general conditions,

$$\lim_{m \to \infty} P\left[\frac{\sqrt{m}(Z_m - g)}{\sigma} \le s \right] = \frac{1}{\sqrt{2\pi}} \int_{-\infty}^{s} dz \, \exp(-z^2/2) \tag{A.89}$$

for every s. This implies that for large m, Z_m should be approximately Gaussian distributed with mean g and variance σ^2/m.

A.6   Annotated Bibliography

Most of the material in this appendix is covered in standard texts on mathematical methods such as **Mathews & Walker (1970)** and **Boas (1996)**. Discussion of relevant computational techniques, and code for implementing them, is available in **Press et al. (1992)**. Linear algebra is covered by **Strang (1976)**; linear and nonlinear differential equations, by **Jordan & Smith (1977)**; probability theory, by **Feller (1968)**; and Fourier transforms and the analysis of linear systems and electrical circuits, by **Siebert (1986)** and **Oppenheim & Willsky (1997)**. Mathematical approaches to biological problems are described in **Edelstein-Keshet (1988)** and **Murray (1993)**. Modern techniques of mathematical modeling are described by **Gershenfeld (1999)**.

General references for the other bodies of techniques used in the book include, for statistics, **Lindgren (1993)** and **Cox & Hinckley (1974)**, and for information theory, **Cover & Thomas (1991)**.

References

Abbott, LF (1992) Simple diagrammatic rules for solving dendritic cable problems. *Physica* **A185**:343–356.

Abbott, LF (1994) Decoding neuronal firing and modeling neural networks. *Quarterly Review of Biophysics* **27**:291–331.

Abbott, LF, & Blum, KI (1996) Functional significance of long-term potentiation for sequence learning and prediction. *Cerebral Cortex* **6**:406–416.

Abbott, LF, Farhi, E, & Gutmann, S (1991) The path integral for dendritic trees. *Biological Cybernetics* **66**:49–60.

Abbott, LF, Varela, JA, Sen, K, & Nelson, SB (1997) Synaptic depression and cortical gain control. *Science* **275**:220–224.

Adelson, EH, & Bergen, JR (1985) Spatiotemporal energy models for the perception of motion. *Journal of the Optical Society of America* **A2**:284–299.

Ahmed, B, Anderson, JC, Douglas, RJ, Martin, KAC, & Whitterage, D (1998) Estimates of the net excitatory currents evoked by visual stimulation of identified neurons in cat visual cortex. *Cerebral Cortex* **8**:462–476.

Amari, S (1999) Natural gradient learning for over- and under-complete bases in ICA. *Neural Computation* **11**:1875–1883.

Amit, DJ (1989) *Modeling Brain Function*. New York: Cambridge University Press.

Amit, DJ, & Tsodyks, MV (1991a) Quantitative study of attractor neural networks retrieving at low spike rates. I. Substrate-spikes, rates and neuronal gain. *Network: Computation in Neural Systems* **2**:259–273.

Amit, DJ, & Tsodyks, MV (1991b) Quantitative study of attractor neural networks retrieving at low spike rates. II. Low-rate retrieval in symmetric networks. *Network: Computation in Neural Systems* **2**:275–294.

Andersen, RA (1989) Visual and eye movement functions of posterior parietal cortex. *Annual Review of Neuroscience* **12**:377–403.

Atick, JJ (1992) Could information theory provide an ecological theory of sensory processing? *Network: Computation in Neural Systems* **3**:213–251.

Atick, JJ, Li, Z, & Redlich, AN (1992) Understanding retinal color coding from first principles. *Neural Computation* **4**:559–572.

Atick, JJ, & Redlich, AN (1990) Towards a theory of early visual processing. *Neural Computation* **2**:308–320.

Atick, JJ, & Redlich, AN (1993) Convergent algorithm for sensory receptive field development. *Neural Computation* **5**:45–60.

Baddeley, R, Abbott, LF, Booth, MJA, Sengpiel, F, Freeman, T, Wakeman, EA, & Rolls, ET (1997) Responses of neurons in primary and interior temporal visual cortices to natural scenes. *Proceedings of the Royal Society of London*. **B264**:1775–1783.

Bair, W, & Koch, C (1996) Temporal precision of spike trains in extrastriate cortex of the behaving macaque monkey. *Neural Computation* **8**:1185–1202.

Bair, W, Koch, C, Newsome, WT, & Britten, KH (1994) Power spectrum analysis of bursting cells in area MT in the behaving monkey. *Journal of Neuroscience* **14**:2870–2892.

Baldi, P, & Heiligenberg, W (1988) How sensory maps could enhance resolution through ordered arrangements of broadly tuned receivers. *Biological Cybernetics* **59**:313–318.

Barlow, HB (1961) Possible principles underlying the transformation of sensory messages. In WA Rosenblith, ed., *Sensory Communication*. Cambridge, MA: MIT Press.

Barlow, HB (1989) Unsupervised learning. *Neural Computation* **1**:295–311.

Barlow, HB, & Levick, WR (1965) The mechanism of directionally selective units in the rabbit's retina. *Journal of Physiology* **193**:327–342.

Barto, AG, & Duff, M (1994) Monte Carlo matrix inversion and reinforcement learning. In G Tesauro, JD Cowan, & J Alspector, eds., *Advances in Neural Information Processing Systems, 6*, 598–605. San Mateo, CA: Morgan Kaufmann.

Barto, AG, Sutton, RS, & Anderson, CW (1983) Neuronlike elements that can solve difficult learning problems. *IEEE Transactions on Systems, Man, and Cybernetics* **13**:834–846.

Barto, AG, Sutton, RS, & Watkins, CJCH (1990) Learning and sequential decision making. In M Gabriel, & J Moore, eds., *Learning and Computational Neuroscience: Foundations of Adaptive Networks*, 539–602. Cambridge, MA: MIT Press.

Battaglia, FP, & Treves, A (1998) Attractor neural networks storing multiple space representations: A model for hippocampal place fields. *Physical Review* **E58**:7738–7753.

Baum, LE, Petrie, E, Soules, G, & Weiss, N (1970) A maximization technique occurring in the statistical analysis of probabilistic functions of Markov chains. *Annals of Mathematical Statistics* **41**:164–171.

Bear, MF, Connors, BW, & Paradiso, MA (1996) *Neuroscience: Exploring the Brain*. Baltimore: Williams & Wilkins.

Becker, S, & Hinton, GE (1992) A self-organizing neural network that discovers surfaces in random-dot stereograms. *Nature* **355**:161–163.

Becker, S, & Plumbley, M (1996) Unsupervised neural network learning procedures for feature extraction and classification. *International Journal of Applied Intelligence* **6**:185–203.

Bell, AJ, & Sejnowski, TJ (1995) An information maximisation approach to blind separation and blind deconvolution. *Neural Computation* **7**:1129–1159.

Bell, AJ, & Sejnowski, TJ (1996) Learning the higher-order structure of a natural sound. *Network: Computation in Neural Systems* **7**:261–267.

Bell, AJ, & Sejnowski, TJ (1997) The "independent components" of natural scenes are edge filters. *Vision Research* **37**:3327–3338.

Bellman, RE (1957) *Dynamic Programming*. Princeton, NJ: Princeton University Press.

Ben-Yishai, R, Bar-Or, RL, & Sompolinsky, H (1995) Theory of orientation tuning in visual cortex. *Proceedings of the National Academy of Sciences of the United States of America* **92**:3844–3848.

Berry, MJ, & Meister, M (1998) Refractoriness and neural precision. *Journal of Neuroscience* **18**: 2200–2211.

Bertsekas, DP, & Tsitsiklis, JN (1996) *Neuro-Dynamic Programming.* Belmont, MA: Athena Scientific.

Bialek, W, DeWeese, M, Rieke, F, & Warland, D (1993) Bits and brains: Information flow in the nervous system. *Physica* **A200**:581–593.

Bialek W, Rieke F, de Ruyter van Steveninck RR, & Warland D (1991) Reading a neural code. *Science* **252**:1854–1857.

Bienenstock, EL, Cooper, LN, & Munro, PW (1982) Theory for the development of neuron selectivity: Orientation specificity and binocular interaction in visual cortex. *Journal of Neuroscience* **2**:32–48.

Bishop, CM (1995) *Neural Networks for Pattern Recognition.* Oxford: Oxford University Press.

Blum, KI, & Abbott, LF (1996) A model of spatial map formation in the hippocampus of the rat. *Neural Computation* **8**:85–93.

Boas, ML (1966) *Mathematical Methods in the Physical Sciences.* New York: Wiley.

de Boer, E, & Kuyper, P (1968) Triggered correlation. *IEEE Biomedical Engineering* **15**:169–179.

Bower, JM, & Beeman, D (1998) *The Book of GENESIS: Exploring Realistic Neural Models with the GEneral NEural SImulation System.* Santa Clara, CA: Telos.

Braitenberg, V, & Schuz, A (1991) *Anatomy of the Cortex.* Berlin: Springer-Verlag.

Bressloff, PC, & Coombes, S (2000) Dynamics of strongly coupled spiking neurons. *Neural Computation* **12**:91–129.

Britten, KH, Shadlen, MN, Newsome, WT, & Movshon, JA (1992) The analysis of visual motion: A comparison of neuronal and psychophysical performance. *Journal of Neuroscience* **12**:4745–4765.

Brotchie PR, Andersen RA, Snyder LH, & Goodman SJ (1995) Head position signals used by parietal neurons to encode locations of visual stimuli. *Nature* **375**:232–235.

Burt, PJ, & Adelson, EH (1983) The Laplacian pyramid as a compact image code. *IEEE Transactions on Communications* **31**:532–540.

Bussgang, JJ (1952) Cross-correlation functions of amplitude-distorted Gaussian signals. *MIT Research Laboratory for Electronic Technology Report* **216**:1–14.

Bussgang, JJ (1975) Cross-correlation functions of amplitude-distorted Gaussian inputs. In AH Haddad, ed., *Nonlinear Systems.* Stroudsburg, PA: Dowden, Hutchinson & Ross.

Cajal, S Ramón y (1911) *Histologie du Système Nerveux de l'Homme et des Vertébrés.* (Translated by L Azoulay). Paris: Maloine. English translation by N Swanson, & LW Swanson (1995) *Histology of the Nervous Systems of Man and Vertebrates.* New York: Oxford University Press.

Campbell, FW, & Gubisch, RW (1966) Optical quality of the human eye. *Journal of Physiology* **186**:558–578.

Carandini M, Heeger DJ, & Movshon JA (1996) Linearity and gain control in V1 simple cells. In EG Jones, & PS Ulinski, eds. *Cerebral Cortex. Vol. 10, Cortical Models.* New York: Plenum Press.

Carandini, M, & Ringach, DL (1997). Predictions of a recurrent model of orientation selectivity. *Vision Research* **37**:3061–3071.

Carpenter, GA, & Grossberg, S, eds. (1991) *Pattern Recognition by Self-Organizing Neural Network*. Cambridge, MA: MIT Press.

Chance, FS (2000) *Modeling Cortical Dynamics and the Responses of Neurons in the Primary Visual Cortex*. Ph.D. dissertation, Brandeis University.

Chance, FS, Nelson, SB, & Abbott, LF (1998) Synaptic depression and the temporal response characteristics of V1 simple cells. *Journal of Neuroscience* **18**:4785–4799.

Chance, FS, Nelson, SB, & Abbott, LF (1999) Complex cells as cortically amplified simple cells. *Nature Neuroscience* **2**:277–282.

Chauvin, Y, & Rumelhart, DE, eds. (1995) *Back Propagation: Theory, Architectures, and Applications*. Hillsdale, NJ: Erlbaum.

Churchland, PS, & Sejnowski, TJ (1992) *The Computational Brain*. Cambridge, MA: MIT Press.

Cohen, MA, & Grossberg, S (1983). Absolute stability of global pattern formation and parallel memory storage by competitive neural networks. *IEEE Transactions on Systems, Man, and Cybernetics* **13**:815–826.

Compte, A, Brunel, N, Goldman-Rakic, PS, & Wang, XJ (2000) Synaptic mechanisms and network dynamics underlying spatial working memory in a cortical network model. *Cerebral Cortex* **10**:910–923.

Connor, JA, & Stevens, CF (1971) Prediction of repetitive firing behaviour from voltage clamp data on an isolated neurone soma. *Journal of Physiology* **213**:31–53.

Connor, JA, Walter, D, & McKown, R (1977) Neural repetitive firing: modifications of the Hodgkin-Huxley axon suggested by experimental results from crustacean axons. *Biophysical Journal* **18**:81–102.

Cover, TM (1965) Geometrical and statistical properties of systems of linear inequalities with application in pattern recognition. *IEEE Transactions on Electronic Computers* **EC14**:326–334.

Cover, TM, & Thomas, JA (1991) *Elements of Information Theory*. New York: Wiley.

Cox, DR (1962) *Renewal Theory*. London: Methuen; New York: Wiley.

Cox, DR, & Hinckley, DV (1974) *Theoretical Statistics*. London: Chapman & Hall.

Cox, DR, & Isham, V (1980) *Point Processes*. New York: Chapman & Hall.

Crair, MC, Gillespie, DC, & Stryker, MP (1998) The role of visual experience in the development of columns in cat visual cortex. *Science* **279**:566–570.

Crowley, JC, & Katz, LC (1999) Development of ocular dominance columns in the absence of retinal input. *Nature Neuroscience* **2**:1125–1130.

Dan, Y, Atick, JJ, & Reid, RC (1996) Efficient coding of natural scenes in the lateral geniculate nucleus: Experimental test of a computational theory. *Journal of Neuroscience* **16**:3351–3362.

Daubechies, I (1992) *Ten Lectures on Wavelets*. Philadelphia: Society for Industrial and Applied Mathematics.

Daubechies, I, Grossmann, A, & Meyer, Y (1986) Painless nonorthogonal expansions. *Journal of Mathematical Physics* **27**:1271–1283.

Daugman, JG (1985) Uncertainty relation for resolution in space, spatial frequency, and orientation optimization by two-dimensional visual cortical filters. *Journal of the Optical Society of America* **2**:1160–1169.

Dayan, P, Hinton, GE, Neal, RM, & Zemel, RS (1995) The Helmholtz machine. *Neural Computation* **7**:889–904.

DeAngelis, GC, Ohzawa, I, & Freeman, RD (1995) Receptive field dynamics in the central visual pathways. *Trends in Neuroscience* **18**:451–458.

Dempster, AP, Laird, NM, & Rubin, DB (1977) Maximum likelihood from incomplete data via the EM algorithm. *Journal of the Royal Statistical Society* **B39**:1–38.

Destexhe, A, Mainen, Z, & Sejnowski, T (1994) Synthesis of models for excitable membranes, synaptic transmission and neuromodulation using a common kinetic formalism. *Journal of Computational Neuroscience* **1**:195–230.

De Valois, RL, & De Valois, KK (1990) *Spatial Vision*. New York: Oxford University Press.

Dickinson, A (1980) *Contemporary Animal Learning Theory*. Cambridge: Cambridge University Press.

Dong, DW, & Atick, JJ (1995) Temporal decorrelation: A theory of lagged and nonlagged responses in the lateral geniculate nucleus. *Network: Computation in Neural Systems* **6**:159–178.

Douglas, RJ, Koch, C, Mahowald, M, Martin, KAC, & Suarez, HH (1995) Recurrent excitation in neocortical circuits. *Science* **269**:981–985.

Douglas, RJ, & Martin, KAC (1998). Neocortex. In GM Shepherd, ed., *The Synaptic Organisation of the Brain. 4th ed.*, 459–509. Oxford: Oxford University Press.

Dowling, JE (1987) *The Retina: An Approachable Part of the Brain*. Cambridge, MA: Bellknap Press.

Dowling, JE (1992) *An Introduction to Neuroscience*. Cambridge, MA: Bellknap Press.

Duda, RO, & Hart, PE (1973) *Pattern Classification and Scene Analysis*. New York: Wiley.

Duda, RO, Hart, PE, & Stork, DG (2000) *Pattern Classification*. New York: Wiley.

Durbin, R, & Mitchison, G (1990) A dimension reduction framework for cortical maps. *Nature* **343**:644–647.

Durbin, R, & Willshaw, DJ (1987) An analogue approach to the travelling salesman problem using an elastic net method. *Nature* **326**:689–691.

Edelstein-Keshet, L (1988) *Mathematical Models in Biology*. New York: Random House.

Engel, AK, Konig, P, & Singer, W (1991) Direct physiological evidence for scene segmentation by temporal coding. *Proceedings of the National Academy of Sciences of the United States of America* **88**:9136–9140.

Enroth-Cugell, C, & Robson, JG (1966) The contrast sensitivity of retinal ganglion cells of the cat. *Journal of Physiology* **187**:517–522.

Ermentrout, GB (1998) Neural networks as spatio-temporal pattern-forming systems. *Reports on Progress in Physics* **64**:353–430.

Ermentrout, GB, & Cowan, J (1979) A mathematical theory of visual hallucination patterns. *Biological Cybernetics* **34**:137–150.

Erwin, E, Obermayer, K, & Schulten, K (1995) Models of orientation and ocular dominance columns in the visual cortex: A critical comparison. *Neural Computation* **7**:425–468.

Everitt, BS (1984) *An Introduction to Latent Variable Models*. London: Chapman & Hall.

Feller, W (1968) *An Introduction to Probability Theory and Its Application*. New York: Wiley.

Ferster, D (1994) Linearity of synaptic interactions in the assembly of receptive fields in cat visual cortex. *Current Opinion in Neurobiology* **4**:563–568.

Field, DJ (1987) Relations between the statistics of natural images and the response properties of cortical cells. *Journal of the Optical Society of America* **A4**:2379–2394.

Field, DJ (1994) What is the goal of sensory coding? *Neural Computation* **6**:559–601.

Földiák, P (1989) Adaptive network for optimal linear feature extraction. In *Proceedings of the IEEE/INNS International Joint Conference on Neural Networks*, 401–405. New York: IEEE Press.

Földiák, P (1991) Learning invariance from transformed sequences. *Neural Computation* **3**:194–200.

Foster, DJ, Morris, RGM, & Dayan, P (2000) Models of hippocampally dependent navigation using the temporal difference learning rule. *Hippocampus* **10**:1–16.

Freeman, WJ, & Schneider, W (1982) Changes in spatial patterns of rabbit olfactory EEG with conditioning to odors. *Psychophysiology* **19**:44–56.

Friston, KJ, Tononi, G, Reeke, GN Jr, Sporns, O, & Edelman, GM (1994) Value-dependent selection in the brain: Simulation in a synthetic neural model. *Neuroscience* **59**:229–243.

Fuster, JM (1995) *Memory in the Cerebral Cortex*. Cambridge, MA: MIT Press.

Gabbiani, F, & Koch, C (1998) Principles of spike train analysis. In C Koch, & I Segev, eds., *Methods of Neuronal Modeling*, 313–360. Cambridge, MA: MIT Press.

Gabbiani, F, Metzner, W, Wessel, R, & Koch, C (1996) From stimulus encoding to feature extraction in weakly electric fish. *Nature* **384**:564–567.

Gabor D (1946) Theory of communication. *Journal of the Institution of Electrical Engineers* **93**:429–457.

Gabriel, M, & Moore, JW, eds. (1990) *Learning and Computational Neuroscience*. Cambridge, MA: MIT Press.

Gallistel, CR (1990) *The Organization of Learning*. Cambridge, MA: MIT Press.

Gallistel, CR, & Gibbon, J (2000) Time, rate and conditioning. *Psychological Review* **107**:289–344.

Georgopoulos, AP, Kalaska, JF, Caminiti, R, & Massey, JT (1982) On the relations between the directions of two-dimensional arm movements and cell discharge in primate motor cortex. *Journal of Neuroscience* **2**:1527–1537.

Georgopoulos, AP, Kettner, RE, & Schwartz, AB (1988) Primate motor cortex and free arm movements to visual targets in three-dimensional space. II. Coding of the direction of movement by a neuronal population. *Neuroscience* **8**:2928–2937.

Georgopoulos, AP, Schwartz, AB, & Kettner, RE (1986) Neuronal population coding of movement direction. *Science* **243**:1416–1419.

Gershenfeld, NA (1999) *The Nature of Mathematical Modeling*. Cambridge: Cambridge University Press.

Gerstner, W (1998) Spiking neurons. In W Maass, & CM Bishop, eds., *Pulsed Neural Networks*. Cambridge, MA: MIT Press, 3–54.

van Gisbergen, JAM, Van Opstal, AJ, & Tax, AMM (1987) Collicular ensemble coding of saccades based on vector summation. *Neuroscience* **21**:541–555.

Gluck, MA, Reifsnider, ES, & Thompson, RF (1990) Adaptive signal processing and the cerebellum: Models of classical conditioning and VOR adaptation. In MA Gluck, & DE Rumelhart, eds., *Neuroscience and Connectionist Theory. Developments in Connectionist Theory*, 131–185. Hillsdale, NJ: Erlbaum.

Gluck, MA, & Rumelhart, DE, eds. (1990) *Neuroscience and Connectionist Theory*. Hillsdale, NJ: Erlbaum.

Goldman-Rakic, PS (1994) Cellular basis of working memory. *Neuron* **14**:477–485.

Goodall, MC (1960) Performance of a stochastic net. *Nature* **185**:557–558.

Goodhill, GJ (1993) Topography and ocular dominance: A model exploring positive correlations. *Biological Cybernetics* **69**:109–118.

Goodhill, GJ, & Richards, LJ (1999) Retinotectal maps: Molecules, models and misplaced data. *Trends in Neuroscience* **22**:529–534.

Goodhill, GJ, & Willshaw, DJ (1990) Application of the elastic net algorithm to the formation of ocular dominance stripes. *Network: Computation in Neural Systems* **1**:41–61.

Graham, NVS (1989) *Visual Pattern Analyzers*. New York: Oxford University Press.

Graziano, MSA, Hu, XT, & Gross, CG (1997) Visuospatial properties of ventral premotor cortex. *Journal of Neurophysiology* **77**:2268–2292.

Green, DM, & Swets, JA (1966) *Signal Detection Theory and Psychophysics*. Los Altos, CA: Peninsula Publishing.

Grenander, U (1995) *Elements of Pattern Theory*. Baltimore: Johns Hopkins University Press.

Grossberg S (1982) Processing of expected and unexpected events during conditioning and attention: A psychophysiological theory. *Psychological Review* **89**:529–572.

Grossberg, S, ed. (1987) *The Adaptive Brain. Vols. 1 and 2*. Amsterdam: Elsevier.

Grossberg, S, ed. (1988) *Neural Networks and Natural Intelligence*. Cambridge, MA: MIT Press.

Grossberg, S, & Schmajuk, NA (1989) Neural dynamics of adaptive timing and temporal discrimination during associative learning. *Neural Networks* **2**:79–102.

Haberly, LB (1990) Olfactory cortex. In GM Shepherd, ed., *The Synaptic Organization of the Brain*. New York: Oxford University Press.

Hahnloser, RH, Sarpeshkar, R, Mahowald, MA, Douglas, RJ, & Seung, HS (2000) Digital selection and analogue amplification coexist in a cortex-inspired silicon circuit. *Nature* **405**:947–951.

Hammer, M (1993) An identified neuron mediates the unconditioned stimulus in associative olfactory learning in honeybees. *Nature* **336**:59–63.

van Hateren, JH (1992) A theory of maximizing sensory information. *Biological Cybernetics* **68**:23–29.

van Hateren, JH (1993) Three modes of spatiotemporal preprocessing by eyes. *Journal of Comparative Physiology* **A172**:583–591.

Hebb, DO (1949) *The Organization of Behavior: A Neuropsychological Theory*. New York: Wiley.

Heeger, DJ (1992) Normalization of cell responses in cat striate cortex. *Visual Neuroscience* **9**:181–198.

Heeger, DJ (1993) Modeling simple-cell direction selectivity with normalized, half-squared, linear operators. *Journal of Neurophysiology* **70**:1885–1898.

Henry, GH, Dreher, B, & Bishop, PO (1974) Orientation specificity of cells in cat striate cortex. *Journal of Neurophysiology* **37**:1394–1409.

Herrault, J, & Jutten, C (1986) Space or time adaptive signal processing by neural networks. In JS Denker, ed., *Neural Networks for Computing*. New York: American Institute for Physics.

Hertz, J, Krogh, A, & Palmer, RG (1991) *Introduction to the Theory of Neural Computation*. Redwood City, CA: Addison-Wesley.

Hille, B (1992) *Ionic Channels of Excitable Membranes*. Sunderland, MA: Sinauer Associates.

Hines, ML (1984) Efficient computation of branched nerve equations. *International Journal of Biomedical Computation* **15**:69–76.

Hines, ML, & Carnevale, NT (1997) The NEURON simulation environment. *Neural Computation* **9**:1179–1209.

Hinton, GE (1981) Shape representation in parallel systems. In *Proceedings of the Seventh International Joint Conference on Artificial Intelligence*, 1088–1096. Vancouver, BC.

Hinton, GE (1984) *Distributed Representations*. Technical report CMU-CS-84-157, Computer Science Department, Carnegie-Mellon University.

Hinton, GE (1989) Connectionist learning procedures. *Artificial Intelligence* **40**:185–234.

Hinton, GE (2000) *Training Products of Experts by Minimizing Contrastive Divergence*. Gatsby Computational Neuroscience Unit, University College London, technical report 2000-004.

Hinton, GE, Dayan, P, Frey, BJ, & Neal, RM (1995) The wake-sleep algorithm for unsupervised neural networks. *Science* **268**:1158–1160.

Hinton, GE, & Ghahramani, Z (1997) Generative models for discovering sparse distributed representations. *Philosophical Transactions of the Royal Society of London*. **B352**:1177–1190.

Hinton, GE, & Sejnowski, TJ (1986) Learning and relearning in Boltzmann machines. In DE Rumelhart, & JL McClelland, eds., *Parallel Distributed Processing: Explorations in the Microstructure of Cognition. Vol. 1, Foundations*. Cambridge, MA: MIT Press, 282-317.

Hinton, GE, & Zemel, RS (1994) Autoencoders, minimum description length and Helmholtz free energy. In JD Cowan, G Tesauro, & J Alspector, eds., *Advances in Neural Information Processing Systems, 6*, 3–10. San Mateo, CA: Morgan Kaufmann.

Hodgkin, AL, & Huxley, AF (1952) A quantitative description of membrane current and its application to conduction and excitation in nerve. *Journal of Physiology* **117**:500–544.

Holt, GR, Softky, GW, Koch, C, & Douglas, RJ (1996) Comparison of discharge variability in vitro and in vivo in cat visual cortex neurons. *Journal of Neurophysiology* **75**:1806–1814.

Hopfield, JJ (1982) Neural networks and systems with emergent selective computational abilities. *Proceedings of the National Academy of Sciences of the United States of America* **79**:2554–2558.

Hopfield, JJ (1984) Neurons with graded response have collective computational properties like those of two-state neurons. *Proceedings of the National Academy of Sciences of the United States of America* **81**:3088–3092.

Houk, JC, Adams, JL, & Barto, AG (1995) A model of how the basal ganglia generate and use neural signals that predict reinforcement. In JC Houk, JL Davis, & DG Beiser, eds., *Models of Information Processing in the Basal Ganglia*, 249–270. Cambridge, MA: MIT Press.

Houk, JC, Davis, JL, & Beiser, DG, eds. (1995) *Models of Information Processing in the Basal Ganglia*. Cambridge, MA: MIT Press.

Hubel, DH (1988) *Eye, Brain, and Vision*. New York: WH Freeman.

Hubel, DH, & Wiesel, TN (1962) Receptive fields, binocular interaction and functional architecture in the cat's visual cortex. *Journal of Physiology* **160**:106–154.

Hubel, DH, & Wiesel, TN (1968) Receptive fields and functional architecture of the monkey striate cortex. *Journal of Physiology* **195**:215–243.

Hubel, DH, & Wiesel, TN (1977) Functional architecture of macaque monkey visual cortex. *Proceedings of the Royal Society of London* **B198**:1–59.

Hubener, M, Shoham, D, Grinvald, A, & Bonhoeffer, T (1997) Spatial relationships among three columnar systems in cat area 17. *Journal of Neuroscience* **17**:9270–9284.

Huber, PJ (1985) Projection pursuit. *Annals of Statistics* **13**:435–475.

Huguenard, JR, & McCormick, DA (1992) Simulation of the currents involved in rhythmic oscillations in thalamic relay neurons. *Journal of Neurophysiology* **68**:1373–1383.

Humphrey, DR, Schmidt, EM, & Thompson, WD (1970) Predicting measures of motor performance from multiple cortical spike trains. *Science* **170**:758–761.

Intrator, N, & Cooper, LN (1992) Objective function formulation of the BCM theory of visual cortical plasticity: Statistical connections, stability conditions. *Neural Networks* **5**:3–17.

Jack, JJB, Noble, D, & Tsien, RW (1975) *Electrical Current Flow in Excitable Cells*. Oxford: Oxford University Press.

Jahr, CE, & Stevens, CF (1990) A quantitative description of NMDA receptor channel kinetic behavior. *Journal of Neuroscience* **10**:1830–1837.

Johnston, D, & Wu, SM (1995) *Foundations of Cellular Neurophysiology*. Cambridge, MA: MIT Press.

Jolliffe, IT (1986) *Principal Component Analysis*. New York: Springer-Verlag.

Jones, J, & Palmer, L (1987a) The two-dimensional spatial structure of simple receptive fields in cat striate cortex. *Journal of Neurophysiology* **58**:1187–1211.

Jones, J, & Palmer, L (1987b) An evaluation of the two-dimensional Gabor filter model of simple receptive fields in cat striate cortex. *Journal of Neurophysiology* **58**:1233–1258.

Jordan, DW, & Smith, P (1977) *Nonlinear Ordinary Differential Equations*. Oxford: Clarendon Press.

Jordan, MI, ed. (1998) *Learning in Graphical Models*. Dordrecht: Kluwer Academic Publishers.

Jordan, MI, Ghahramani, Z, Jaakkola, TS, & Saul, LK (1998). An introduction to variational methods for graphical models. In MI Jordan, ed., *Learning in Graphical Models*, 105–162. Dordrecht: Kluwer Academic Publishers.

Kalaska, JF, Caminiti, R, & Georgopoulos, AP (1983) Cortical mechanisms related to the direction of two-dimensional arm movements: Relations in parietal area 5 and comparison with motor cortex. *Experimental Brain Research* **51**:247–260.

Kandel, ER, & Schwartz, JH, eds. (1985) *Principles of Neural Science*. 2nd ed. New York: McGraw-Hill.

Kandel, ER, Schwartz, JH, & Jessel, TM, eds. (1991) *Principles of Neural Science*. 3rd ed. New York: McGraw-Hill.

Kandel, ER, Schwartz, JH, & Jessel, TM, eds. (2000) *Principles of Neural Science*. 4th ed. New York: McGraw-Hill.

Karasaridis, A, & Simoncelli, EP (1996) A filter design technique for steerable pyramid image transforms. *Proceedings of the International Conference on Acoustics, Speech and Signal Processing*, 2387–2390. New York: IEEE.

Kawato, M, Hayakama, H, & Inui, T (1993) A forward-inverse optics model of reciprocal connections between visual cortical areas. *Network: Computation in Neural Systems* **4**:415–422.

Kearns, MJ, & Vazirani, UV (1994) *An Introduction to Computational Learning Theory*. Cambridge, MA: MIT Press.

Kehoe, EJ (1977) *Effects of Serial Compound Stimuli on Stimulus Selection in Classical Conditioning of the Rabbit Nictitating Membrane Response*. Ph.D. dissertation, University of Iowa.

Kempter R, Gerstner W, & van Hemmen JL (1999) Hebbian learning and spiking neurons. *Physical Review* **E59**:4498–4514.

Koch, C (1998) *Biophysics of Computation: Information Processing in Single Neurons*. New York: Oxford University Press.

Koch, C, & Segev, I, eds. (1998) *Methods in Neuronal Modeling: From Synapses to Networks*. Cambridge, MA: MIT Press.

Konorski, J (1967) *Integrative Activity of the Brain*. Chicago: University of Chicago Press.

Lapicque, L (1907) Recherches quantitatives sur l'excitation electrique des nerfs traitee comme une polarization. *Journal de Physiologie et Pathologie Général* **9**:620–635.

Laughlin, S (1981) A simple coding procedure enhances a neuron's information capacity. *Zeitschrift für Naturforschung* **36**:910–912.

Lauritzen, SL (1996) *Graphical Models*. Oxford: Clarendon Press.

Lee, C, Rohrer, WH, & Sparks, DL (1988) Population coding of saccadic eye movements by neurons in the superior colliculus. *Nature* **332**:357–360.

Leen, TK (1991) Dynamics of learning in recurrent feature-discovery networks. In RP Lippmann, JE Moody, & DS Touretzky, eds., *Advances in Neural Information Processing Systems, 3*. San Mateo, CA: Morgan Kaufmann, 70–76.

LeMasson, G, Marder, E, & Abbott, LF (1993) Activity-dependent regulation of conductances in model neurons. *Science* **259**:1915–1917.

Lewis, JE, & Kristan, WB (1998) A neuronal network for computing population vectors in the leech. *Nature* **391**:76–79.

Li, Z (1995) Modeling the sensory computations of the olfactory bulb. In E Domany, JL van Hemmen, & K Schulten, eds., *Models of Neural Networks, Volume 2*. New York: Springer-Verlag, 221–251.

Li, Z (1996) A theory of the visual motion coding in the primary visual cortex. *Neural Computation* **8**:705–730.

Li, Z (1998) A neural model of contour integration in the primary visual cortex. *Neural Computation* **10**:903–940.

Li, Z (1999) Visual segmentation by contextual influences via intra-cortical interactions in the primary visual cortex. *Network: Computation in Neural Systems* **10**:187–212.

Li, Z, & Atick, JJ (1994a) Efficient stereo coding in the multiscale representation. *Network: Computation in Neural Systems* **5**:157–174.

Li, Z, & Atick, JJ (1994b) Toward a theory of the striate cortex. *Neural Computation* **6**:127–146.

Li, Z, & Dayan, P (1999) Computational differences between asymmetrical and symmetrical networks. *Network: Computation in Neural Systems* **10**:59–78.

Li, Z, & Hopfield, JJ (1989) Modeling the olfactory bulb and its neural oscillatory processings. *Biological Cybernetics* **61**:379–392.

Lindgren, BW (1993) *Statistical Theory*. 4th ed. New York: Chapman & Hall.

Linsker, R (1986) From basic network principles to neural architecture. *Proceedings of the National Academy of Sciences of the United States of America* **83**:7508–7512, 8390–8394, 8779–8783.

Linsker, R (1988) Self-organization in a perceptual network. *Computer* **21**:105–117.

Lisman, JE (1997) Bursts as a unit of neural information: making unreliable synapses reliable. *Trends in Neuroscience* **20**:38–43.

Liu, Z, Golowasch, J, Marder, E, & Abbott, LF (1998) A model neuron with activity-dependent conductances regulated by multiple calcium sensors. *Journal of Neuroscience* **18**:2309–2320.

MacKay, DJC (1996) *Maximum Likelihood and Covariant Algorithms for Independent Components Analysis*. Unpublished manuscript.

MacKay, DJC, & Miller, KD (1990) Analysis of Linsker's application of Hebbian rules to linear networks. *Network: Computation in Neural Systems* **1**:257–299.

MacKay, DM (1956) The epistemological problem for automata. In CE Shannon, & J McCarthy, eds., *Automata Studies*, 235–251. Princeton, NJ: Princeton University Press.

Mackintosh, NJ (1983) *Conditioning and Associative Learning*. Oxford: Oxford University Press.

Magleby, KL (1987) Short-term changes in synaptic efficacy. In G Edelman, W Gall, & W Cowan, eds., *Synaptic Function*, 21–56. New York: Wiley.

Mangel, M, & Clark, CW (1988) *Dynamic Modeling in Behavioral Ecology*. Princeton, NJ: Princeton University Press.

Mallat, SG (1998) *A Wavelet Tour of Signal Processing*. San Diego, CA: Academic Press.

Marder, E, & Calabrese, RL (1996) Principles of rhythmic motor pattern generation. *Physiological Review* **76**:687–717.

Markram, H, Lubke, J, Frotscher, M, & Sakmann, B (1997) Regulation of synaptic efficacy by coincidence of postsynaptic APs and EPSPs. *Science* **275**:213–215.

Markram, H, & Tsodyks, M (1996) Redistribution of synaptic efficacy between neocortical pyramidal neurons. *Nature* **382**:807–810.

Markram, H, Wang, Y, & Tsodyks, MV (1998) Differential signalling via the same axon of neocortical pyramidal neurons. *Proceedings of the National Academy of Sciences of the United States of America* **95**:5323–5328.

Marmarelis, PZ, & Marmarelis, VZ (1978) *Analysis of Physiological Systems: The White-Noise Approach.* New York: Plenum Press.

Marom, S, & Abbott, LF (1994) Modeling state-dependent inactivation of membrane currents. *Biophysical Journal* **67**:515–520.

Marr, D (1970) A theory for cerebral neocortex. *Proceedings of the Royal Society of London.* **B176**:161–234.

Mascagni, M, & Sherman, A (1998) Numerical methods for neuronal modeling. In C Koch, & I Segev, eds., *Methods in Neuronal Modeling: From Synapses to Networks*, 569–606. Cambridge, MA: MIT Press.

Mathews, J, & Walker, RL (1970) *Mathematical Methods of Physics.* New York: Benjamin.

Mauk, MD, & Donegan, NH (1997) A model of Pavlovian conditioning based on the synaptic organization of the cerebellum. *Learning and Memory* **4**:130–158.

McCormick, DA (1990) Membrane properties and neurotransmitter actions. In GM Shepherd, ed., *The Synaptic Organization of the Brain.* New York: Oxford University Press.

Mehta, MR, Barnes, CA, & McNaughton, BL (1997) Experience-dependent, asymmetric expansion of hippocampal place fields. *Proceedings of the National Academy of Sciences of the United States of America* **94**:8918–8921.

Mehta, MR, Quirk, MC, & Wilson, M (2000) Experience dependent asymmetric shape of hippocampal receptive fields. *Neuron* **25**:707–715.

Miller, KD (1994) A model for the development of simple cell receptive fields and the ordered arrangement of orientation columns through activity-dependent competition between on- and off-center inputs. *Journal of Neuroscience* **14**:409–441.

Miller, KD (1996a) Receptive fields and maps in the visual cortex: Models of ocular dominance and orientation columns. In E Domany, JL van Hemmen, & K Schulten, eds., *Models of Neural Networks, Volume 3*, 55–78. New York: Springer-Verlag.

Miller, KD (1996b) Synaptic economics: competition and cooperation in synaptic plasticity. *Neuron* **17**:371–374.

Miller, KD, Keller, JB, & Stryker, MP (1989) Ocular dominance column development: Analysis and simulation. *Science* **245**:605–615.

Miller, KD, & MacKay, DJC (1994) The role of constraints in Hebbian learning. *Neural Computation* **6**:100–126.

Minai, AA, & Levy, WB (1993) Sequence learning in a single trial. *International Neural Network Society World Congress of Neural Networks II.* Portland, OR: International Neural Network Society, 505–508.

Minsky, M, & Papert, S (1969) *Perceptrons.* Cambridge, MA: MIT Press.

Mirenowicz, J, & Schultz, W (1994) Importance of unpredictability for reward responses in primate dopamine neurons. *Journal of Neurophysiology* **72**:1024–1027.

Montague, PR, Dayan, P, Person, C, & Sejnowski TJ (1995) Bee foraging in uncertain environments using predictive Hebbian learning. *Nature* **377**:725–728.

Montague, PR, Dayan, P, & Sejnowski, TJ (1996) A framework for mesencephalic dopamine systems based on predictive Hebbian learning. *Journal of Neuroscience* **16**:1936–1947.

Movshon JA, Thompson ID, & Tolhurst DJ (1978a) Spatial summation in the receptive fields of simple cells in the cat's striate cortex. *Journal of Neurophysiology* **283**:53–77.

Movshon JA, Thompson ID, & Tolhurst DJ (1978b) Spatial and temporal contrast sensitivity of neurones in areas 17 and 18 of the cat's visual cortex. *Journal of Neurophysiology* **283**:101–120.

Mumford, D (1994) Neuronal architectures for pattern-theoretic problems. In C Koch, & J Davis, eds., *Large-Scale Theories of the Cortex*, 125–152. Cambridge, MA: MIT Press.

Murray, JD (1993) *Mathematical Biology*. New York: Springer-Verlag.

Narendra, KS, & Thatachar, MAL (1989) *Learning Automata: An Introduction*. Englewood Cliffs, NJ: Prentice-Hall.

Neal, RM (1993) *Probabilistic Inference Using Markov Chain Monte Carlo Methods*. Department of Computer Science, University of Toronto, technical teport CRG-TR-93-1.

Neal, RM, & Hinton, GE (1998) A view of the EM algorithm that justifies incremental, sparse, and other variants. In MI Jordan, ed., *Learning in Graphical Models*, 355–368. Dordrecht: Kluwer Academic Publishers.

Neisser, U (1967) *Cognitive Psychology*. New York: Appleton-Century-Crofts.

Newsome, WT, Britten, KH, & Movshon, JA (1989) Neural correlates of a perceptual decision. *Nature 341*:52–54.

Nicholls, JG, Martin, R, & Wallace, BG (1992) *From Neuron to Brain: A Cellular and Molecular Approach to the Function of the Nervous System*. Sunderland, MA: Sinauer Associates.

Nowlan, SJ (1991) *Soft Competitive Adaptation: Neural Network Learning Algorithms Based on Fitting Statistical Mixtures*. Ph.D. dissertation, Carnegie-Mellon University.

Obermayer, K, & Blasdel, GG (1993) Geometry of orientation and ocular dominance columns in monkey striate cortex. *Journal of Neuroscience 13*:4114–4129.

Obermayer, K, Blasdel, GG, & Schulten, K (1992) Statistical-mechanical analysis of self-organization and pattern formation during the development of visual maps. *Physical Review A 45*:7568–7589.

Oja, E (1982) A simplified neuron model as a principal component analyzer. *Journal of Mathematical Biology 16*:267–273.

O'Keefe, J, & Recce, ML (1993) Phase relationship between hippocampal place units and the EEG theta rhythm. *Hippocampus 3*:317–330.

O'Keefe, LP, Bair, W, & Movshon, JA (1997) Response variability of MT neurons in macaque monkey. *Society for Neuroscience Abstracts* **23**:1125.

Olshausen, B (1996) *Learning Linear, Sparse, Factorial Codes*. MIT AI Lab, MIT, AI-memo 1580.

Olshausen, BA, & Field, DJ (1996) Emergence of simple-cell receptive field properties by learning a sparse code for natural images. *Nature 381*:607–609.

Olshausen, BA, & Field, DJ (1997) Sparse coding with an overcomplete basis set: A strategy employed by V1? *Vision Research 37*:3311–3325.

Oppenheim, AV, & Willsky, AS, with Nawab, H (1997) *Signals and Systems*. 2nd ed. Upper Saddle River, NJ: Prentice-Hall.

Oram, MW, Földiák, P, Perrett, DI, & Sengpiel, F (1998) The "ideal homunculus": Decoding neural population signals. *Trends in Neuroscience* **21**:259–265.

Orban, GA (1984) *Neuronal Operations in the Visual Cortex*. Berlin: Springer-Verlag.

O'Reilly, RC (1996) Biologically plausible error-driven learning using local activation differences: The generalised recirculation algorithm. *Neural Computation* **8**:895–938.

Paradiso, MA (1988) A theory for the use of visual orientation information which exploits the columnar structure of striate cortex. *Biological Cybernetics* **58**:35–49.

Parker, AJ, & Newsome, WT (1998) Sense and the single neuron: Probing the physiology of perception. *Annual Review of Neuroscience* **21**:227–277.

Patlak, J (1991) Molecular kinetics of voltage-dependent Na^+ channels. *Physiological Review* **71**:1047–1080.

Pearl, J (1988) *Probabilistic Reasoning in Intelligent Systems: Networks of Plausible Inference*. San Mateo, CA: Morgan Kaufmann.

Pearlmutter, BA, & Parra, LC (1996) A context-sensitive generalization of ICA. In S-I Amari, L Xu, L-W Chan, & I King, eds., *Proceedings of the International Conference on Neural Information Processing 1996*, 151–157. Singapore: Springer-Verlag.

Pece, AEC (1992) Redundancy reduction of a Gabor representation: A possible computational role for feedback from primary visual cortex to lateral geniculate nucleus. In I Aleksander, & J Taylor, eds., *Artificial Neural Networks, 2*, 865–868. Amsterdam: Elsevier.

Percival, DB, & Waldron, AT (1993) *Spectral Analysis for Physical Applications*. Cambridge: Cambridge University Press.

Piepenbrock, C, & Obermayer, K (1999) The role of lateral cortical competition in ocular dominance development. In MS Kearns, SA Solla, & DA Cohn, eds., *Advances in Neural Information Processing Systems, 11*. Cambridge, MA: MIT Press.

Plumbley, MD (1991) *On Information Theory and Unsupervised Neural Networks*. Cambridge University Engineering Department, Cambridge, technical report CUED/F-INFENG/TR.78.

Poggio, GF, & Talbot WH (1981) Mechanisms of static and dynamic stereopsis in foveal cortex of the rhesus monkey. *Journal of Physiology* **315**:469–492.

Poggio, T (1990) A theory of how the brain might work. *Cold Spring Harbor Symposium on Quantitative Biology* **55**:899–910.

Pollen, D, & Ronner, S (1982) Spatial computations performed by simple and complex cells in the visual cortex of the cat. *Vision Research* **22**:101–118.

Poor, HV (1994) *An Introduction to Signal Detection and Estimation, Second Edition*. New York: Springer-Verlag.

Pouget A, & Sejnowski TJ (1995) Spatial representations in the parietal cortex may use basis functions. In G Tesauro, DS Touretzky & TK Leen, eds., *Advances in Neural Information Processing Systems, 7*, 157–164. San Mateo, CA: Morgan Kaufmann.

Pouget, A, & Sejnowski, TJ (1997) Spatial transformations in the parietal cortex using basis functions. *Journal of Cognitive Neuroscience* **9**:222–237.

Pouget, A, Zhang, KC, Deneve, S, & Latham, PE (1998) Statistically efficient estimation using population coding. *Neural Computation* **10**:373–401.

Press, WH, Teukolsky, SA, Vetterling, WT, & Flannery, BP (1992) *Numerical Recipes in C.* Cambridge: Cambridge University Press.

Price, DJ, & Willshaw, DJ (2000) *Mechanisms of Cortical Development.* Oxford: Oxford University Press.

Purves, D, Augustine, GJ, Fitzpatrick, D, Katz, LC, LaMantia, A-S, McNamara, JO, & Williams, SM, eds. (2000) *Neuroscience.* Sunderland MA: Sinauer Associates.

Rall, W (1959) Branching dendritic trees and motoneuron membrane resistivity. *Experimental Neurology* **2**:503–532.

Rall, W (1977) Core conductor theory and cable properties of neurons. In Kandel, ER, ed., *Handbook of Physiology*, vol. 1, 39–97. Bethesda, MD: American Physiology Society.

Rao, RPN, & Ballard, DH (1997) Dynamic model of visual recognition predicts neural response properties in the visual cortex. *Neural Computation* **9**:721–763.

Raymond, JL, Lisberger, SG, & Mauk, MD (1996) The cerebellum: A neuronal learning machine? *Science* **272**:1126–1131.

Real, LA (1991) Animal choice behavior and the evolution of cognitive architecture. *Science* **253**:980–986.

Reichardt, W (1961) Autocorrelation: A principle for the evaluation of sensory information by the central nervous system. In WA Rosenblith, ed., *Sensory Communication.* New York: Wiley.

Rescorla, RA, & Wagner, AR (1972) A theory of Pavlovian conditioning: The effectiveness of reinforcement and non-reinforcement. In AH Black & WF Prokasy, eds., *Classical Conditioning II: Current Research and Theory*, 64–69. New York: Appleton-Century-Crofts.

Rieke F, Bodnar, DA, & Bialek, W (1995) Naturalistic stimuli increase the rate and efficiency of information transmission by primary auditory afferents. *Proceedings of the Royal Society of London.* **B262**:259–265.

Rieke, FM, Warland, D, de Ruyter van Steveninck, R, & Bialek, W (1997) *Spikes: Exploring the Neural Code.* Cambridge, MA: MIT Press.

Rinzel, J, & Ermentrout, B (1998) Analysis of neural excitability and oscillations. In C Koch & I Segev, eds., *Methods in Neuronal Modeling: From Synapses to Networks*, 251–292. Cambridge, MA: MIT Press.

Rissanen, J (1989) *Stochastic Complexity in Statistical Inquiry.* Singapore: World Scientific Press.

Robinson, DA (1989) Integrating with neurons. *Annual Review of Neuroscience* **12**:33–45.

Rodieck, R (1965) Quantitative analysis of cat retinal ganglion cell responses to visual stimuli. *Vision Research* **5**:583–601.

Rolls, ET, & Treves, A (1998) *Neural Networks and Brain Function.* New York: Oxford University Press.

Rosenblatt, F (1958) The perceptron: A probabilistic model for information storage and organization in the brain. *Psychological Review* **65**:386–408.

Roth, Z, & Baram, Y (1996) Multidimensional density shaping by sigmoids. *IEEE Transactions on Neural Networks* **7**:1291–1298.

Rovamo, J, & Virsu, V (1984) Isotropy of cortical magnification and topography of striate cortex. *Vision Research* **24**:283–286.

Roweis, S (1998) EM Algorithms for PCA and SPCA. In MI Jordan, M Kearns, & SA Solla, eds., *Advances in Neural Information Processing Systems, 10*, 626–632. Cambridge, MA: MIT Press.

Roweis, S, & Ghahramani, Z (1999) A unifying review of linear Gaussian models. *Neural Computation* **11**:305–345

Rubin, DB, & Thayer, DT (1982) EM algorithms for ML factor analysis. *Psychometrika* **47**:69–76.

de Ruyter van Steveninck, R, & Bialek, W (1988) Real-time performance of a movement-sensitive neuron in the blowfly visual system: Coding and information transfer in short spike sequences. *Proceedings of the Royal Society of London.* **B234**:379–414.

Sakmann, B, & Neher, E (1983) *Single Channel Recording.* New York: Plenum.

Salinas, E, & Abbott, LF (1994) Vector reconstruction from firing rates. *Journal of Computational Neuroscience* **1**:89–107.

Salinas, E, & Abbott, LF (1995) Transfer of coded information from sensory to motor networks. *Journal of Neuroscience* **15**:6461–6474.

Salinas, E, & Abbott, LF (1996). A model of multiplicative neural responses in parietal cortex. *Proceedings of the National Academy of Sciences of the United States of America* **93**:11956–11961.

Salinas, E, & Abbott, LF (2000) Do simple cells in primary visual cortex form a tight frame? *Neural Computation* **12**:313–336.

Salzman, CA, Shadlen, MN, & Newsome, WT (1992) Microstimulation in visual area MT: Effects on directional discrimination performance. *Journal of Neuroscience* **12**:2331–2356.

Samsonovich, A, & McNaughton, BL (1997) Path integration and cognitive mapping in a continuous attractor neural network model. *Journal of Neuroscience* **17**:5900–5920.

Sanger, TD (1994) Theoretical considerations for the analysis of population coding in motor cortex. *Neural Computation* **6**:29–37.

Sanger, TD (1996) Probability density estimation for the interpretation of neural population codes. *Journal of Neurophysiology* **76**:2790–2793.

Saul, AB, & Humphrey, AL (1990) Spatial and temporal properties of lagged and nonlagged cells in the cat lateral geniculate nucleus. *Journal of Neurophysiology* **68**:1190–1208.

Saul, LK, & Jordan, ML (2000) Attractor dynamics in feedforward neural networks. *Neural Computation* **12**:1313–1335.

Schultz, W (1998) Predictive reward signal of dopamine neurons. *Journal of Neurophysiology* **80**:1–27.

Schultz, W, Romo, R, Ljungberg, T, Mirenowicz, J, Hollerman, JR, & Dickinson, A (1995) Reward-related signals carried by dopamine neurons. In JC Houk, JL Davis, & DG Beiser, eds., *Models of Information Processing in the Basal Ganglia*, 233–248. Cambridge, MA: MIT Press.

Schwartz, EL (1977) Spatial mapping in the primate sensory projection: Analytic structure and relevance to perception. *Biological Cybernetics* **25**:181–194.

Sclar, G, & Freeman, R (1982) Orientation selectivity in cat's striate cortex is invariant with stimulus contrast. *Experimental Brain Research* **46**:457–461.

Scott, DW (1992) *Multivariate Density Estimation: Theory, Practice, and Visualization.* New York: Wiley.

Sejnowski, TJ (1977) Storing covariance with nonlinearly interacting neurons. *Journal of Mathematical Biology* **4**:303–321.

Sejnowski, TJ (1999) The book of Hebb. *Neuron* **24**:773–776.

Seung, HS (1996) How the brain keeps the eyes still. *Proceedings of the National Academy of Sciences of the United States of America* **93**:13339–13344.

Seung, HS, Lee, DD, Reis, BY, & Tank DW (2000) Stability of the memory for eye position in a recurrent network of conductance-based model neurons. *Neuron* **26**:259–271.

Seung, HS, & Sompolinsky, H (1993) Simple models for reading neuronal population codes. *Proceedings of the National Academy of Sciences of the United States of America* **90**:10749–10753.

Shadlen, MN, Britten, KH, Newsome, WT, & Movshon, JA (1996) A computational analysis of the relationship between neuronal and behavioral responses to visual motion. *Journal of Neuroscience* **16**:1486–1510.

Shadlen, MN, & Newsome WT (1998) The variable discharge of cortical neurons: Implications for connectivity, computation, and information coding. *Journal of Neuroscience* **18**:3870–3896.

Shanks, DR (1995) *The Psychology of Associative Learning*. Cambridge: Cambridge University Press.

Shannon, CE, & Weaver, W (1949) *The Mathematical Theory of Communications*. Urbana, IL: University of Illinois Press.

Shepherd, GM (1997) *Neurobiology*. Oxford: Oxford University Press.

Siebert, WMcC (1986) *Circuits, Signals, and Systems*. Cambridge, MA: MIT Press; New York: McGraw-Hill.

Simoncelli, EP (1997) Statistical models for images: Compression, restoration and synthesis. *Proceedings of the 31st Asilomar Conference on Signals, Systems and Computers*. Pacific Grove, CA: IEEE Computer Society, 673–678.

Simoncelli, EP, & Adelson, EH (1990) Subband image coding. In JW Woods, ed., *Subband Transforms.*, 143–192 Norwell, MA: Kluwer Academic Publishers.

Simoncelli, EP, & Freeman, WT (1995) The steerable pyramid: A flexible architecture for derivative computation. *Proceedings of the International Conference on Image Processing*, 444–447. Los Alamitos, CA: IEEE Computer Society Press.

Simoncelli, EP, Freeman, WT, Adelson, EH, & Heeger, DJ (1992) Shiftable multiscale transforms. *IEEE Transactions on Information Theory* **38**:587–607.

Simoncelli, EP, & Schwartz, O (1999) Modeling non-specific suppression in V1 neurons with a statistically-derived normalization model. In MS Kearns, SA Solla, & DA Cohn, eds. *Advances in Neural Information Processing Systems, 11*, 153–159. Cambridge, MA: MIT Press.

Snippe, HP (1996) Parameter extraction from population codes: a critical assessment. *Neural Computation* **8**:511–530.

Snippe, HP & Koenderink, JJ (1992a) Discrimination thresholds for channel-coded systems. *Biological Cybernetics* **66**:543-551.

Snippe, HP, & Koenderink, JJ (1992b) Information in channel-coded systems: Correlated receivers. *Biological Cybernetics* **67**:183–190.

Softky, WR, & Koch, C (1992) Cortical cells should spike regularly but do not. *Neural Computation* **4**:643–646.

Solomon, RL, & Corbit, JD (1974) An opponent-process theory of motivation. I. Temporal dynamics of affect. *Psychological Review* **81**:119–145.

Somers, DC, Nelson, SB, & Sur, M (1995) An emergent model of orientation selectivity in cat visual cortical simple cells. *Journal of Neuroscience* **15**:5448–5465.

Sompolinsky, H, & Shapley, R (1997) New perspectives on the mechanisms for orientation selectivity. *Current Opinion in Neurobiology* **7**:514–522.

Song, S, Miller, KD, & Abbott, LF (2000) Competitive Hebbian learning through spike-timing dependent synaptic plasticity. *Nature Neuroscience* **3**:919–926.

Stemmler, M, & Koch, C (1999) How voltage-dependent conductances can adapt to maximize the information encoded by neuronal firing rate. *Nature Neuroscience* **2**:521–527.

Stevens, CM, & Zador, AM (1998) Novel integrate-and-fire-like model of repetitive firing in cortical neurons. In *Proceedings of the 5th Joint Symposium on Neural Computation*. La Jolla, CA: University of California at San Diego.

Strang, G (1976) *Linear Algebra and Its Applications*. New York: Academic Press.

Strong, SP, Koberle, R, de Ruyter van Steveninck, RR, & Bialek, W (1998) Entropy and information in neural spike trains. *Physical Review Letters* **80**:197–200.

Stuart, GJ, & Sakmann, B (1994) Active propagation of somatic action potentials into neocortical pyramidal cell dendrites. *Nature* **367**:69–72.

Stuart, GJ, & Spruston, N (1998) Determinants of voltage attenuation in neocortical pyramidal neuron dendrites. *Journal of Neuroscience* **18**:3501–3510.

Sutton, RS (1988) Learning to predict by the methods of temporal difference. *Machine Learning* **3**:9–44.

Sutton, RS, & Barto, AG (1990) Time-derivative models of Pavlovian conditioning. In M Gabriel, & JW Moore, eds., *Learning and Computational Neuroscience*, 497–537. Cambridge, MA: MIT Press.

Sutton, RS, & Barto, AG (1998) *Reinforcement Learning*. Cambridge, MA: MIT Press.

Swindale, NV (1996) The development of topography in the visual cortex: A review of models. *Network: Computation in Neural Systems* **7**:161–247.

Theunissen, FE, & Miller, JP (1991) Representation of sensory information in the cricket cercal sensory system. II. Information theoretic calculation of system accuracy and optimal tuning-curve widths of four primary interneurons. *Journal of Neurophysiology* **66**:1690–1703.

Tipping, ME, & Bishop, CM (1999) Mixtures of probabilistic principal component analyzers. *Neural Computation* **11**:443–482.

Titterington, DM, Smith, AFM, & Makov, UE (1985) *Statistical Analysis of Finite Mixture Distributions*. New York: Wiley.

Tootell, RB, Silverman, MS, Switkes, E, & De Valois, RL (1982) Deoxyglucose analysis of retinotopic organization in primate striate cortex. *Science* **218**:902–904.

Touretzky, DS, Redish, AD, & Wan, HS (1993) Neural representation of space using sinusoidal arrays. *Neural Computation* **5**:869–884.

Troyer, TW, & Miller, KD (1997) Physiological gain leads to high ISI variability in a simple model of a cortical regular spiking cell. *Neural Computation* **9**:971–983.

Tsai, KY, Carnevale, NT, Claiborne, BJ, & Brown, TH (1994) Efficient mapping from neuroanatomical to electrotonic space. *Network: Computation in Neural Systems* **5**:21–46.

Tsodyks, MV, & Markram, H (1997) The neural code between neocortical pyramidal neurons depends on neurotransmitter release probability. *Proceedings of the National Academy of Sciences of the United States of America* **94**:719–723.

Tuckwell, HC (1988) *Introduction to Theoretical Neurobiology.* Cambridge, UK: Cambridge University Press.

Turing, AM (1952) The chemical basis of morphogenesis. *Philosophical Transactions of the Royal Society of London* **B237**:37–72.

Turrigiano, G, LeMasson, G, & Marder, E (1995) Selective regulation of current densities underlies spontaneous changes in the activity of cultured neurons. *Journal of Neuroscience* **15**:3640–3652.

Uttley, AM (1979) *Information Transmission in the Nervous System.* London: Academic Press.

Van Essen, DC, Newsome, WT, & Maunsell, JHR (1984) The visual field representation in striate cortex of the macaque monkey: Asymmetries, anisotropies, and individual variability. *Vision Research* **24**:429–448.

Van Santen, JP, & Sperling, G (1984) Temporal covariance model of human motion perception. *Journal of the Optical Society of America* **A1**:451–473.

Varela, J, Sen, K, Gibson, J, Fost, J, Abbott, LF, Nelson, SB (1997) A quantitative description of short-term plasticity at excitatory synapses in layer 2/3 of rat primary visual cortex. *Journal of Neuroscience* **17**:7926–7940.

Venkatesh, SS (1986) Epsilon capacity of a neural network. In J Denker, ed., *Proceedings of Neural Networks for Computing.* AIP Conference Proceedings Volume 151, New York: American Institute of Physics, 440–445.

Vogels, R (1990) Population coding of stimulus orientation by cortical cells. *Journal of Neuroscience* **10**:3543–3558.

van Vreeswijk, C, Abbott, LF, & Ermentrout, GB (1994) When inhibition not excitation synchronizes neuronal firing. *Journal of Computational Neuroscience* **1**:313–321.

Wallis, G, & Baddeley, R (1997) Optimal, unsupervised learning in invariant object recognition. *Neural Computation* **9**:883–894.

Wandell, BA (1995) *Foundations of Vision.* Sunderland, MA: Sinauer Associates.

Wang, X-J (1994) Multiple dynamical modes of thalamic relay neurons: Rhythmic bursting and intermittent phase-locking. *Neuroscience* **59**:21–31.

Wang, X-J (1998) Calcium coding and adaptive temporal computation in cortical pyramidal neurons. *Journal of Neurophysiology* **79**:1549–1566.

Wang, X-J, & Rinzel, J (1992) Alternating and synchronous rhythms in reciprocally inhibitory model neurons. *Neural Computation* **4**:84–97.

Watkins, CJCH (1989) *Learning from Delayed Rewards.* Ph.D. dissertation, University of Cambridge.

Watson, AB, & Ahumada, AJ (1985) Model of human visual-motion sensing. *Journal of the Optical Society of America* **A2**:322–342.

Weliky, M (2000) Correlated neuronal activity and visual cortical development. *Neuron* **27**:427–430.

Werblin, FS, & Dowling, JE (1969) Organization of the retina of the mudpuppy, *Necturus maculosus*. II. Intracellular recording. *Journal of Neurophysiology* **32**:339–355.

Wickens, J (1993) *A Theory of the Striatum*. New York: Pergamon Press.

Widrow, B, & Hoff, ME (1960) Adaptive switching circuits. *WESCON Convention Report* **4**:96–104.

Widrow, B, & Stearns, SD (1985) *Adaptive Signal Processing*. Englewood Cliffs, NJ: Prentice-Hall.

Wiener, N (1958) *Nonlinear Problems in Random Theory*. New York: Wiley.

Williams, RJ (1992) Simple statistical gradient-following algorithms for connectionist reinforcement learning. *Machine Learning* **8**:229–256.

Wilson, HR, & Cowan, JD (1972) Excitatory and inhibitory interactions in localized populations of model neurons. *Biophysical Journal* **12**:1–24.

Wilson, HR, & Cowan, JD (1973) A mathematical theory of the functional dynamics of cortical and thalamic nervous tissue. *Kybernetik* **13**:55–80.

Witkin, A (1983) Scale space filtering. In *Proceedings of the International Joint Conference on Artificial Intelligence*, vol. 2, 1019–1022. San Mateo, CA: Morgan Kaufmann.

Wörgötter, F, & Koch, C (1991) A detailed model of the primary visual pathway in the cat: Comparison of afferent excitatory and intracortical inhibitory connection schemes for orientation selectivity. *Journal of Neuroscience* **11**:1959–1979.

Yuste, R, & Sur, M (1999) Development and plasticity of the cerebral cortex: From molecules to maps. *Journal of Neurobiology* **41**:1–6.

Zador, A, Agmon-Snir, H, & Segev, I (1995) The morphoelectrotonic transform: A graphical approach to dendritic function. *Journal of Neuroscience* **15**:1669–1682.

Zemel, RS (1994) *A Minimum Description Length Framework for Unsupervised Learning*. Ph.D. dissertation, University of Toronto.

Zhang, K (1996). Representation of spatial orientation by the intrinsic dynamics of the head-direction cell ensemble: A theory. *Journal of Neuroscience* **16**:2112–2126.

Zhang, K, & Sejnowski, T (1999) Neural tuning: To sharpen or broaden? *Neural Computation* **11**:75–84.

Zhang, LI, Tao, HW, Holt, CE, Harris, WA, & Poo M-m (1998) A critical window for cooperation and competition among developing retinotectal synapses. *Nature* **395**:37–44.

Zigmond, MJ, Bloom, FE, Landis, SC, & Squire, LR, eds. (1998) *Fundamental Neuroscience*. San Diego: Academic Press.

Zipser, D, & Andersen, RA (1988) A back-propagation programmed network that simulates response properties of a subset of posterior parietal neurons. *Nature* **331**:679–684.

Zohary, E (1992) Population coding of visual stimuli by cortical neurons tuned to more than one dimension. *Biological Cybernetics* **66**:265–272.

Zucker, RS (1989). Short-term synaptic plasticity. *Annual Review of Neuroscience* **12**:13–31.

Index

bold fonts mark sections or subsections in the text
italic fonts mark margin labels